Lecture Notes in Computer Science   14605

Founding Editors

Gerhard Goos
Juris Hartmanis

The series Lecture Notes in Computer Science (LNCS), including its subseries Lecture Notes in Artificial Intelligence (LNAI) and Lecture Notes in Bioinformatics (LNBI), has established itself as a medium for the publication of new developments in computer science and information technology research, teaching, and education.

LNCS enjoys close cooperation with the computer science R & D community, the series counts many renowned academics among its volume editors and paper authors, and collaborates with prestigious societies. Its mission is to serve this international community by providing an invaluable service, mainly focused on the publication of conference and workshop proceedings and postproceedings. LNCS commenced publication in 1973.

Sara Brunetti · Andrea Frosini · Simone Rinaldi
Editors

Discrete Geometry and Mathematical Morphology

Third International Joint Conference, DGMM 2024
Florence, Italy, April 15–18, 2024
Proceedings

Springer

Editors
Sara Brunetti 🆔
University of Siena
Siena, Italy

Andrea Frosini 🆔
University of Firenze
Firenze, Italy

Simone Rinaldi 🆔
University of Siena
Siena, Italy

ISSN 0302-9743   ISSN 1611-3349 (electronic)
Lecture Notes in Computer Science
ISBN 978-3-031-57792-5   ISBN 978-3-031-57793-2 (eBook)
https://doi.org/10.1007/978-3-031-57793-2

This Springer imprint is published by the registered company Springer Nature Switzerland AG
The registered company address is: Gewerbestrasse 11, 6330 Cham, Switzerland

Paper in this product is recyclable.

Preface

This volume contains the papers presented at DGMM 2024: the IAPR International Conference on Discrete Geometry and Mathematical Morphology held during April 15–18, 2024, in Florence. DGMM is sponsored by the International Association for Pattern Recognition (IAPR), and is associated with the IAPR Technical Committee on Discrete Geometry and Mathematical Morphology (TC18). This is the third joint event between the two main conference series of IAPR TC18, the International Conference on Discrete Geometry for Computer Imagery (DGCI), with 21 previous editions, and the International Symposium on Mathematical Morphology (ISMM), with 14 previous editions. This third DGMM edition attracted 51 submissions by authors from 14 countries: Austria, Brazil, Canada, China, Philippines, France, Germany, Hungary, Italy, Lebanon, Netherlands, Serbia, United Arab Emirates, and USA. Out of these 51 submissions, 34 were selected for presentation at the conference after a review and rebuttal process where each submission received 3 reports, and a meta-review. The DGMM 2024 papers highlight the current trends and advances in discrete geometry and mathematical morphology, encompassing purely theoretical contributions, algorithmic developments, or novel applications in image processing, computer vision, and pattern recognition.

In addition, three internationally well-known researchers were invited for keynote lectures:

- Dominique Attali, Gipsa – lab, Grenoble (FR), on "Shape Reconstruction with Guarantees";
- Massimo Caccia, Università dell'Insubria (IT), on "Random Power: in-silco quantum generation of random bit streams";
- Laurent Najman, Université Gustave Eiffel Paris (FR), on "Power-Watershed: a graph-based optimization framework for image and data processing".

These invited speakers proposed an abstract of their communications that can be found in this volume. Following the tradition of both DGCI and ISMM, the DGMM 2024 proceedings appear in Springer's LNCS series and a special issue of the Journal of Mathematical Imaging and Vision, with extended versions of selected outstanding contributions, is planned. We wish to thank the members of the Program Committee and all the volunteer reviewers for their efforts in reviewing all submissions on time and giving extensive feedback. We would like to thank all the other people involved in this conference: first, the Steering Committee for giving us the chance to organize DGMM 2024; second, the three invited speakers, Dominique Attali, Massimo Caccia, and Laurent Najman, for accepting to share their recognized expertise; and finally, the most important component of any scientific conference, the authors for producing the high-quality and original contributions.

We are thankful to the IAPR for its sponsorship and we acknowledge the EasyChair conference management system that was invaluable in handling the paper submission, the review process, and putting this volume together. We also acknowledge Springer for making possible the publication of these proceedings in the LNCS series.

February 2024

Sara Brunetti
Andrea Frosini
Simone Rinaldi

Organization

Organizing Committee

Sara Brunetti (Program Chair)	University of Siena, Italy
Andrea Frosini (General Chair)	University of Florence, Italy
Simone Rinaldi (Program Chair)	University of Siena, Italy
Michela Ascolese (Advisor)	University of Florence, Italy
Leonardo Bindi (Webmaster)	University of Siena, Italy
Niccolò Di Marco (Advisor)	University of Roma, Italy
Veronica Guerrini (Advisor)	University of Pisa, Italy
Giulia Palma (Advisor)	University of Siena, Italy
Elisa Pergola (Local Chair)	University of Florence, Italy

Steering Committee

Jesús Angulo-Lopez	Ecole des Mines de Paris, France
Étienne Baudrier	Université de Strasbourg, France
Isabelle Bloch	Université - LIP6, France
Gunilla Borgefors	Uppsala University, Sweden
Srécko Brlek	Université du Québec à Montréal, Canada
Sara Brunetti	University of Siena, Italy
Bernhard Burgeth	Università des Saarlandes, Germany
David Coeurjolly	CNRS - LIRIS, France
Isabelle Debled-Rennesson	Université de Lorraine - LORIA, France
Andrea Frosini	University of Florence, Italy
Maria Jose Jiménez	University of Seville, Spain
Yukiko Kenmochi (Chair)	CNRS - GREYC, France
Walter Kropatsch	TU Wien, Austria
Jacques-Olivier Lachaud	Université Savoie Mont Blanc - LAMA, France
Cris Luengo Hendriks	Flagship Biosciences, USA
Petros Maragos	National Technical University of Athens, Greece
Laurent Najman	ESIEE Paris - LIGM, France
Dan Schonfeld	University of Illinois, USA
Hugues Talbot	CentralSupelec - Université Paris-Saclay, France
Michael H. F. Wilkinson (Vice Chair)	University of Groningen, Netherlands

Program Committee

Jesús Angulo-Lopez	Ecole des Mines de Paris, France
Samy Blusseau	CMM, Mines Paris, PSL Research University, France
Nicolas Boutry	LRDE, EPITA, France
Michael Breuß	Brandenburgische Technische Universität, Germany
Srécko Brlek	Université du Québec à Montréal, Canada
Lidija Comic	University of Novi Sad, Serbia
Jean Cousty	LIGM, Université Gustave Eiffel, ESIEE, France
Paolo Dulio	Politecnico di Milano, Italy
Yan Gerard	Université Clermont Auvergne, LIMOS, France
Rocio Gonzalez-Diaz	University of Seville, Spain
Bertrand Kerautret	LIRIS, France
Walter G. Kropatsch	TU Wien, Austria
Jacques-Olivier Lachaud	LAMA, University of Savoie Mont Blanc, France
Laurent Najman	LIGM, Université Gustave Eiffel, ESIEE, France
Phuc Ngo	Université Nancy, LORIA, France
Akihiro Sugimoto	National Institute of Informatics, Japan
Michael H. F. Wilkinson	University of Groningen, Netherlands

Additional Reviewers

Eric Andres	Thierry Géraud
Andreas Alpers	Silvio Guimaraes
Juan Humberto Sossa Azuela	Carolin Hannusch
Mohammad Babai	VladimirIlić AtsushiImiya
Péter Balázs	Damien Jamet
Martin Bauer	Marvin Kahra
Michael Biehl	Christer Oscar Kiselman
Stéphane Breuils	Alexander Köhler
Luc Brun	Adrien Krähenbühl
Edwin Carlinet	Sebastien Labbe
Mathieu Desbrun	Bastien Laboureix
Niccolò Di Marco	Nabil Laiche
Eric Domenjoud	Mélodie Lapointe
Mohammad Hashem Faezi	Etienne Le Quentrec
Alexandre Falcão	Olivier Lezoray
Thomas Fernique	David Mosquera
Fabien Feschet	Lois Tibor Lukic
Simon Gazagnes	Paola Magillo

Filip Malmberg
Beatriz Marcotegui
Alexandre Blondin Massé
Benoît Naegel
Benedek Nagy
Thanh Phuong Nguyen
Guillaume Noyel
Adam Onus
Kalman Palagyi
Eduardo Paluzo-Hidalgo
Nicolas Passat
Valentin Penaud-Polge
Benjamin Perret
Attila Pethö
Kacper Pluta
Elodie Puybareau
Christophe Reutenauer
Vanessa Robins

Christian Ronse
Tristan Roussillon
Philippe Salembier
Mateus Sangalli
Shima Shabani
Isabelle Sivignon
Ryan Slechta
John Stell
Lama Tarsissi
Guillaume Tochon
ÁlvaroTorras-Casas
Antoine Vacavant
Marcos Eduardo Valle
Santiago Velasco-Forero
José Antonio Vilches
Martin Welk
Petra Wiederhold
Yongchao Xu

Contents

Discrete and Combinatorial Topology

Mathematical Morphology and Digital Geometry for Applications

Digital Geometry - Models, Transforms, and Visualization

Bijectivity Analysis of Finite Rotations on $\mathbb{Z}^2$: A Hierarchical Approach

Nicolas Passat[1]($\boxtimes$) , Phuc Ngo[2] , and Yukiko Kenmochi[3]

[1] Université de Reims Champagne-Ardenne, CReSTIC, Reims, France
`nicolas.passat@univ-reims.fr`
[2] Université de Lorraine, CNRS, LORIA, 54000 Nancy, France
[3] Normandie Univ., UNICAEN, ENSICAEN, CNRS, GREYC, Caen, France

Abstract. In this article, we investigate the rotations on $\mathbb{Z}^2$ (a.k.a discrete rotations). In particular, we focus on the finite rotations that act on finite subsets of $\mathbb{Z}^2$, especially Euclidean balls. We shed light on the hierarchical structure of these rotations, induced by their discontinuity (characterized by hinge angles) and the size of the considered ball. We propose efficient algorithmic schemes leading to the construction of combinatorial models (trees) of the bijective finite rotations. These algorithms and structures open the way to a better understanding of the notion of bijectivity with respect to finite vs. infinite discrete rotations.

Keywords: Finite rotations · Cartesian grid · Bijectivity · Hinge angles · Hierarchical models

1 Introduction

Rotation is a fundamental operation in geometry. In the Euclidean spaces $\mathbb{R}^d$ ($d \geqslant 2$), rotations preserve distances and angles; they are also bijective. By contrast, their discrete analogues in $\mathbb{Z}^d$, the so-called discrete rotations, exhibit challenges. Not only they do no longer preserve distances and angles, but they are also non-bijective in most cases.

Early in the study of discrete rotations, efforts were geared towards understanding under which hypotheses discrete rotations could be bijective. In [11], a sufficent condition was proposed for bijectivity of discrete rotations on $\mathbb{Z}^2$. In [13,18], a characterization of bijective discrete rotations on $\mathbb{Z}^2$ was further proposed (whereas it was proved in [10] that there is no bijective discrete rotation on $\mathbb{Z}^2$ under the specific case where the floor digitization paradigm is considered, see Sect. 2.3). Extensions of these results were investigated in the hexagonal grid [17] and in $\mathbb{Z}^3$ [6,16]. In [3] a family of rotations on $\mathbb{Z}^2$ based on quasi-shears was proposed, also fulfilling bijectivity properties. This paradigm was extended to hexagonal grids in [5]. In [4], bijective rotations were handled via the composition of bijective reflections. A similar paradigm was investigated via the framework of geometric algebra on $\mathbb{Z}^2$ [7] and $\mathbb{Z}^3$ [6].

This work was supported by the French Agence Nationale de la Recherche (Grants ANR-22-CE45-0034 and ANR-23-CE45-0015).

In this article, we focus on the discrete rotations applied on finite sets (Euclidean balls) of $\mathbb{Z}^2$, and we study their bijectivity. Indeed, in practical cases, rotations act on images with a finite support, and in this context, the notion of bijectivity may be considered with regards to this support instead of the whole space $\mathbb{Z}^2$. We propose an algorithmic framework that allows to build a combinatorial, hierarchical structure which models the space of all the (bijective) discrete rotations on finite Euclidean balls of $\mathbb{Z}^2$.

This article is organized as follows. Section 2 provides basic notions on continuous and discrete rotations required to make this article self-contained. In Sect. 3, we describe the hierarchical structure of the finite discrete rotations, and we provide an efficient algorithm for building it. In Sect. 4, we describe the hierarchical structures of two families of bijective finite discrete rotations and we provide efficient algorithms for building them. Section 5 illustrates results obtained by applying these algorithms. Section 6 concludes this article. A version of this article with additional proofs of the properties is also available [15].

2 Basics on Rotations

A point of $\mathbb{R}^2$ is noted in bold, e.g. $\mathbf{p} \in \mathbb{R}^2$. Its coordinates are noted in italic and subscripted by x and y, respectively, e.g. $\mathbf{p} = {}^t(p_x, p_y)$ (or (p_x, p_y) by abuse of notation). If it is defined by its coordinates, it is noted as the associated couple, e.g. $(a, b) \in \mathbb{R}^2$.

2.1 Continuous Rotations

We note $\mathbb{U} = [0, 2\pi)$ and we consider $\mathbb{U}$ as cyclic, i.e. we identify 0 and 2π. Let $\theta \in \mathbb{U}$. The rotation (with center $\mathbf{0} = (0, 0)$) of angle θ is the function $\mathcal{R}_\theta : \mathbb{R}^2 \to \mathbb{R}^2$ defined, for all $\mathbf{p} \in \mathbb{R}^2$, by

$$\mathcal{R}_\theta(\mathbf{p}) = \begin{pmatrix} \cos\theta & -\sin\theta \\ \sin\theta & \cos\theta \end{pmatrix} \cdot \begin{pmatrix} p_x \\ p_y \end{pmatrix} = \begin{pmatrix} p_x \cos\theta - p_y \sin\theta \\ p_x \sin\theta + p_y \cos\theta \end{pmatrix} \tag{1}$$

Such rotation is called a *continuous rotation*. We note $\mathfrak{R}_{\mathbb{R}^2}$ the set of all the continuous rotations.

2.2 Hinge Angles

Let $\theta \in \mathbb{U}$ and $\mathcal{R}_\theta$ the induced continuous rotation. Let $\mathbf{p} \in \mathbb{Z}^2$. Let us suppose that there exists $k \in \mathbb{Z}$ such that one of the following two equations is satisfied

$$p_x \cos\theta - p_y \sin\theta = k + 1/2 \tag{2}$$
$$p_x \sin\theta + p_y \cos\theta = k + 1/2 \tag{3}$$

Then we say that θ is a *hinge angle* [14] (induced by $\mathbf{p}$). We note $\mathbb{H} \subset \mathbb{U}$ the set of all the hinge angles ($\mathbb{H}$ is dense in $\mathbb{U}$ [12]). We note $\mathring{\mathbb{U}} = \mathbb{U} \backslash \mathbb{H}$ the set of non-hinge angles.

A hinge angle $\theta \in \mathbb{H}$ is determined by a triplet $(p_x, p_y, k) \in \mathbb{Z}^3$. However any triplet in $\mathbb{Z}^3$ does not necessarily define a hinge angle. In particular, $(p_x, p_y, k) \in \mathbb{Z}^3$ defines a hinge angle iff $|k + 1/2| \leqslant \|(p_x, p_y)\|_2$. We note $\mathbb{T} \subset \mathbb{Z}^3$ the set of all the triplets that define hinge angles.

We define the function $\eta : \mathbb{T} \rightarrow \mathbb{H}$ such that for all $t = (p_x, p_y, k) \in \mathbb{T}$, $\eta(t)$ is the hinge angle induced by t. The function η is non-injective: many triplets of $\mathbb{T}$ define the same hinge angle of $\mathbb{H}$. The following proposition clarifies this many-to-one relation.

Proposition 1 ([12]). *Let $h \in \mathbb{H}$. There exists a prime generator triplet $(p_x, p_y, k) \in \mathbb{T}$ such that $\eta^{-1}(\{h\}) = \{((2n + 1)p_x, (2n + 1)p_y, (2n + 1)k + n) \mid n \in \mathbb{Z}\}$.*

2.3   Discrete Rotations

We now consider rotations (with center $\mathbf{0}$) from $\mathbb{Z}^2$ to $\mathbb{Z}^2$. Let $\theta \in \mathbb{U}$. Let $\mathcal{R}_\theta \in \mathfrak{R}_{\mathbb{R}^2}$ be a continuous rotation (see Eq. (1)). Except in very few cases (i.e. when θ is a multiple of $\pi/2$), we have $\mathcal{R}_\theta(\mathbb{Z}^2) \not\subseteq \mathbb{Z}^2$. To tackle this issue, i.e. to ensure that the result of a rotation applied on $\mathbb{Z}^2$ lies in $\mathbb{Z}^2$, it is common to compose the result of the continuous rotation with a discretization operator, which is generally set as

$$\left| \begin{array}{rcl} D : \mathbb{R}^2 & \longrightarrow & \mathbb{Z}^2 \\ (p_x, p_y) & \longmapsto & ([p_x], [p_y]) \end{array} \right. \tag{4}$$

where $[\cdot]$ is the rounding function on $\mathbb{R}$. (As observed in [10], the floor function, and by symmetry, the ceil one are less interesting from the bijectivity point of view that we adopt in this study.)

Then, we can define a rotation of angle θ from $\mathbb{Z}^2$ to $\mathbb{Z}^2$ as a function $R_\theta : \mathbb{Z}^2 \rightarrow \mathbb{Z}^2$ such that, for all $\mathbf{p} \in \mathbb{Z}^2$, we have

$$R_\theta(\mathbf{p}) = (D \circ \mathcal{R}_\theta)(\mathbf{p}) = \begin{pmatrix} [p_x \cos \theta - p_y \sin \theta] \\ [p_x \sin \theta + p_y \cos \theta] \end{pmatrix} \tag{5}$$

This rotation is called a *discrete rotation*. Such discrete rotation R_θ is well defined iff $\theta \in \mathring{\mathbb{U}} = \mathbb{U} \backslash \mathbb{H}$ (due to the ambiguous definition of $[\cdot]$ on $\mathbb{Z} + 1/2$). We note $\mathfrak{R}_{\mathbb{Z}^2} = \{R_\theta \mid \theta \in \mathring{\mathbb{U}}\}$ the set of all the discrete rotations. We set $R : \mathring{\mathbb{U}} \rightarrow \mathfrak{R}_{\mathbb{Z}^2}$ the (surjective) function defined for each $\theta \in \mathring{\mathbb{U}}$ by $R(\theta) = R_\theta$.

By contrast with a continuous rotation $\mathcal{R}_\theta : \mathbb{R}^2 \rightarrow \mathbb{R}^2$ which is bijective, a discrete rotation $R_\theta : \mathbb{Z}^2 \rightarrow \mathbb{Z}^2$ may be bijective or not, depending on the value of θ. This fact is clarified by the following proposition.

Proposition 2 ([13]). *Let $\theta \in \mathring{\mathbb{U}}$. The discrete rotation $R_\theta : \mathbb{Z}^2 \rightarrow \mathbb{Z}^2$ is bijective if and only if there exists $p \in \mathbb{N}$ and $\varepsilon \in \{-1, 1\}$ such that*

$$\sin \theta \in \left\{ \varepsilon \cdot \frac{2p(p + 1)}{2p^2 + 2p + 1}, \varepsilon \cdot \frac{2p + 1}{2p^2 + 2p + 1} \right\} \tag{6}$$

The set $\mathbb{B}$ of the angles characterized by Eq. (6) is composed of specific Pythagorean angles (i.e. angles with rational sine and cosine determined by twin Pythagorean triplets). Pythagorean angles do not intersect $\mathbb{H}$ [14], and we then have $\mathbb{B} \cap \mathbb{H} = \emptyset$.

2.4   Discrete Finite Rotations

Let $\theta \in \overset{\circ}{\mathbb{U}}$. Let $R_\theta \in \mathfrak{R}_{\mathbb{Z}^2}$. Let $\rho \in \mathbb{R}_+$. Let $B^\rho = \{\mathbf{q} \in \mathbb{Z}^2 \mid \|\mathbf{q}\|_2 \leqslant \rho\}$. We note $R_\theta^\rho : B^\rho \to \mathbb{Z}^2$ the restriction of the discrete rotation R_θ to the discrete Euclidean ball B^ρ. Such rotation R_θ^ρ is called a discrete ρ-*rotation* (or simply a ρ-*rotation*). We note $\mathfrak{R}_{\mathbb{Z}^2}^\rho = \{R_\theta^\rho \mid R_\theta \in \mathfrak{R}_{\mathbb{Z}^2}\}$ the set of all the ρ-rotations. We set $R^\rho : \overset{\circ}{\mathbb{U}} \to \mathfrak{R}_{\mathbb{Z}^2}^\rho$ the (surjective) function defined for each $\theta \in \overset{\circ}{\mathbb{U}}$ by $R^\rho(\theta) = R_\theta^\rho$.

We define the subset

$$\mathbb{T}^\rho = \left\{(p_x, p_y, k) \in \mathbb{Z}^3 \mid |k + 1/2| \leqslant \|\mathbf{p}\|_2 \leqslant \rho\right\} \subset \mathbb{T} \tag{7}$$

which gathers the triplets of $\mathbb{T}$ induced by the points inside B^ρ. We have $|\mathbb{T}^\rho| = O(\rho^3)$. We define the subset $\mathbb{H}^\rho = \eta(\mathbb{T}^\rho) \subset \mathbb{H}$ of the hinge angles induced the triplets of $\mathbb{T}^\rho$. We have $|\mathbb{H}^\rho| = O(\rho^3)$ [2]. Note that we have $\rho < 1$ iff $\mathbb{T}^\rho = \mathbb{H}^\rho = \emptyset$.

We consider on $\mathbb{U}$ (viewed here as non-cyclic) the restriction $\leqslant_\mathbb{U}$ (simply noted $\leqslant$) of the total order $\leqslant$ on $\mathbb{R}$. By assuming that $\mathbb{H}^\rho$ is ordered by $\leqslant$, we note $\mathbb{H}^\rho = \{h_j^\rho\}_{j=0}^{\sigma^\rho - 1}$ with $\sigma^\rho \in \mathbb{N}$.

We consider on $\mathbb{T}$ the preorder $\leqslant_\mathbb{T}$ (simply noted $\leqslant$) defined, for any $t_1, t_2 \in \mathbb{T}$ by

$$t_1 \leqslant_\mathbb{T} t_2 \Longleftrightarrow \eta(t_1) \leqslant_\mathbb{U} \eta(t_2) \tag{8}$$

By assuming that $\mathbb{T}^\rho$ is sorted with respect to $\leqslant$, we set $\mathbb{T}^\rho = \{(p_{x,i}, p_{y,i}, k_i)\}_{i=0}^{s^\rho - 1}$ with $s^\rho \in \mathbb{N}$. Note that we have $\sigma^\rho \leqslant s^\rho$ and $\sigma^\rho \geqslant 1$ iff $\rho \geqslant 1$.

If $\rho < 1$, we set $\mathbb{H}_\bullet^\rho = \emptyset$ and $\mathbb{H}_{\circ\!-\!\circ}^\rho = \mathbb{H}_{\bullet\!-\!\circ}^\rho = \{\mathbb{U}\}$. If $\rho \geqslant 1$, we set

$$\mathbb{H}_\bullet^\rho = \left\{\left\{h_j^\rho\right\}\right\}_{j=0}^{\sigma^\rho - 1} \tag{9}$$

$$\mathbb{H}_{\circ\!-\!\circ}^\rho = \left\{\left(h_j^\rho, h_{j+1[\sigma^\rho]}^\rho\right)\right\}_{j=0}^{\sigma^\rho - 1} \tag{10}$$

$$\mathbb{H}_{\bullet\!-\!\circ}^\rho = \left\{\left[h_j^\rho, h_{j+1[\sigma^\rho]}^\rho\right)\right\}_{j=0}^{\sigma^\rho - 1} \tag{11}$$

where $n[k]$ is n modulo k. These three sets can be seen as faces in a complex model of $\mathbb{U}$, namely 0-dimensional elements ($\mathbb{H}_\bullet^\rho$), 1-dimensional open elements ($\mathbb{H}_{\circ\!-\!\circ}^\rho$) and 1-dimensional semi-open elements ($\mathbb{H}_{\bullet\!-\!\circ}^\rho$).

Property 3. *There is a trivial bijection between $\mathbb{H}^\rho$ and $\mathbb{H}_\bullet^\rho$, and another between $\mathbb{H}_{\circ\!-\!\circ}^\rho$ and $\mathbb{H}_{\bullet\!-\!\circ}^\rho$. When $\rho \geqslant 1$, there are trivial bijections between $\mathbb{H}^\rho$, $\mathbb{H}_\bullet^\rho$, $\mathbb{H}_{\circ\!-\!\circ}^\rho$ and $\mathbb{H}_{\bullet\!-\!\circ}^\rho$.*

We set $\overset{\circ}{\mathbb{U}}^\rho = \mathbb{U} \backslash \mathbb{H}^\rho$. The set $\mathbb{H}_{\circ\!-\!\circ}^\rho$ is a partition of $\overset{\circ}{\mathbb{U}}^\rho$. The set $\mathbb{H}_{\bullet\!-\!\circ}^\rho$ is a partition of $\mathbb{U}$. The union of sets $\mathbb{H}_\bullet^\rho \cup \mathbb{H}_{\circ\!-\!\circ}^\rho$ is also a partition of $\mathbb{U}$ (that refines the partition $\mathbb{H}_{\bullet\!-\!\circ}^\rho$).

Let $H \in \mathbb{H}_{\circ\!-\!\circ}^\rho$. Let $\theta_1, \theta_2 \in \overset{\circ}{\mathbb{U}}$ be two distinct angles such that $\theta_1 \in H$. Let $R_{\theta_1}, R_{\theta_2} \in \mathfrak{R}_{\mathbb{Z}^2}$ be the two (distinct) discrete rotations associated to θ_1, θ_2, respectively. Let $R_{\theta_1}^\rho, R_{\theta_2}^\rho \in \mathfrak{R}_{\mathbb{Z}^2}^\rho$ be the two ρ-rotations associated to θ_1, θ_2, respectively. We have $\theta_2 \in H$ if and only if $R_{\theta_1}^\rho = R_{\theta_2}^\rho$. This justifies the following property.

Property 4. *There exists a bijection between $\mathbb{H}_{\circ\!-\!\circ}^\rho$ and $\mathfrak{R}_{\mathbb{Z}^2}^\rho$.*

It follows that we can symbolically model $\mathfrak{R}_{\mathbb{Z}^2}^\rho$ by $\mathbb{H}_{\circ\!-\!\circ}^\rho$ (or any other set in bijection with $\mathbb{H}_{\circ\!-\!\circ}^\rho$, see Property 3).

Remark 5. *Although the function R^ρ is initially defined on $\mathring{\mathbb{U}}$ (i.e. only for non-hinge angles), it is possible to extend it, without loss of correctness, so that it is defined on $\mathbb{U}$ (i.e. for all angles). Let $H \in \mathbb{H}^\rho_{\circ-\circ}$. Let $h \in H \cap \mathbb{H}$. Since h is a hinge angle, the discrete rotation R_h is undefined, and so is the ρ-rotation R^ρ_h. Nonetheless, the hinge angle $h \in \mathbb{H}$ does not belong to $\mathbb{H}^\rho$. We can extend R^ρ_h by continuity, by setting $R^\rho_h = R^\rho_\theta$ with $\theta \in H \cap \mathring{\mathbb{U}}$. Now, let $h = h^\rho_j \in \mathbb{H}^\rho$ ($0 \leqslant j \leqslant \sigma^\rho - 1$). The ρ-rotation R^ρ_h is undefined. However, R^ρ_θ is defined (and constant) for all $\theta \in (h^\rho_j, h^\rho_{j+1[\sigma^\rho]})$. We can then extend R^ρ_h by continuity, by setting $R^\rho_h = R^\rho_\theta$. Doing so, we can extend the function $R^\rho : \mathring{\mathbb{U}} \to \mathfrak{R}^\rho_{\mathbb{Z}^2}$ as $R^\rho : \mathbb{U} \to \mathfrak{R}^\rho_{\mathbb{Z}^2}$.*

3 Hierarchical Structure of Finite Discrete Rotations

In this section, we describe a combinatorial (tree) structure for modeling the ρ-rotations (Sect. 3.1) and we provide an algorithmic scheme for building it (Sect. 3.2).

3.1 Description

For any $\rho \in \mathbb{R}_+$, we set $\mathbb{I}^\rho = \mathbb{H}^\rho_\bullet \cup \mathbb{H}^\rho_{\circ-\circ}$. We set $\mathbb{I} = \{\mathbb{I}^\rho\}_{\rho \in \mathbb{R}_+}$. Let $\rho_1, \rho_2 \in \mathbb{R}_+$ and $r \in \mathbb{N}$. If $\sqrt{r} \leqslant \rho_1, \rho_2 < \sqrt{r+1}$ then we have $\mathbb{I}^{\rho_1} = \mathbb{I}^{\rho_2}$. It follows that $\mathbb{I} = \{\mathbb{I}^{\sqrt{r}}\}_{r \in \mathbb{N}} = \{\mathbb{I}^\rho\}_{\rho \in \sqrt{\mathbb{N}}}$.

Let $\rho_1, \rho_2 \in \sqrt{\mathbb{N}}$. If $\rho_1 \leqslant \rho_2$, then the partition $\mathbb{I}^{\rho_2}$ refines the partition $\mathbb{I}^{\rho_1}$. We note $\sqsubseteq_\mathbb{I}$ (or simply $\sqsubseteq$) this refinement relation on $\mathbb{I}$. In particular, for all $\rho_1, \rho_2 \in \sqrt{\mathbb{N}}$ we have $\rho_1 \leqslant \rho_2 \Rightarrow \mathbb{I}^{\rho_1} \sqsubseteq \mathbb{I}^{\rho_2}$. It is plain that $(\mathbb{I}, \sqsubseteq)$ is a totally ordered set which admits an infimum and a supremum. The infimum (actually, the minimum) is the trivial partition $\mathbb{I}^0 = \{\mathbb{U}\}$. The (unique) discrete 0-rotation associated to $\mathbb{I}^0$ is the trivial rotation $R^0_0 : \{\mathbf{0}\} \to \{\mathbf{0}\}$ that maps $\mathbf{0}$ onto itself. The supremum is the partition of $\mathbb{U}$ noted $\mathbb{I}^\infty = \{\{\theta\} \mid \theta \in \mathbb{U}\}$. For each $\theta \in \mathring{\mathbb{U}}$, the associated discrete rotation is R_θ. (The other values $\theta \in \mathbb{H}$, i.e. the hinge angles, are not associated to discrete rotations, by definition.)

For any $\rho \in \sqrt{\mathbb{N}}$, each set $\mathbb{I}^\rho$ models both the ρ-rotations (via $\mathbb{H}^\rho_{\circ-\circ}$) and the hinge angles between these ρ-rotations (via $\mathbb{H}^\rho_\bullet$). Except for $\rho = 0$, there is a trivial bijection between $\mathbb{H}^\rho_\bullet$, $\mathbb{H}^\rho_{\circ-\circ}$ and $\mathbb{H}^\rho_{\bullet-\circ}$ (Property 3). There is then a trivial two-to-one function from $\mathbb{I}^\rho$ to $\mathbb{H}^\rho_{\bullet-\circ}$ for $\rho > 0$ (plus a one-to-one function from $\mathbb{I}^\rho$ to $\mathbb{H}^\rho_{\bullet-\circ}$, for $\rho = 0$). One can then equivalently model the structure of the ρ-rotations either by $\mathbb{I}^\rho$ or $\mathbb{H}^\rho_{\bullet-\circ}$. From now on, we consider $\mathbb{H}^\rho_{\bullet-\circ}$ (noted $\mathbb{S}^\rho$) instead of $\mathbb{I}^\rho$, without loss of generality.

We set $\mathbb{K} = \bigcup_{\rho \in \sqrt{\mathbb{N}}} \mathbb{S}^\rho$. Each element $K \in \mathbb{K}$ is a semi-open interval of $\mathbb{U}$ that defines a specific discrete ρ-rotation. By definition, there exists $r \in \mathbb{N}$ such that $K \in \mathbb{S}^{\sqrt{r}}$. This value r may be non-unique. More precisely, there exists $\alpha(K), \omega(K) \in \mathbb{N}$, with $\alpha(K) \leqslant \omega(K)$, such that for all $r \in [\![\alpha(K), \omega(K)]\!]$, we have $K \in \mathbb{S}^{\sqrt{r}}$.

We endow $\mathbb{K}$ with the inclusion order relation $\subseteq$, and we note $\mathfrak{T} = (\mathbb{K}, \lhd)$ the Hasse diagram of the ordered set $(\mathbb{K}, \subseteq)$. The maximum of $(\mathbb{K}, \subseteq)$ is $\mathbb{U}$.

Property 6. *Let $K \in \mathbb{K} \backslash \{\mathbb{U}\}$. There is a unique $K' \in \mathbb{K}$ such that $K \lhd K'$.*

It follows from this property that $\mathfrak{T} = (\mathbb{K}, \lhd)$ has a tree structure with $\mathbb{U}$ as root.

The set $\mathbb{K}$ is discrete but infinite, and so is the tree $\mathfrak{T}$. Its whole construction is then impossible. However, one may build a finite part of it from its root $\mathbb{U}$ until a finite, but arbitrary large depth μ.

Let $\mu = \sqrt{m} \in \sqrt{\mathbb{N}}$. We set $\mathbb{K}^\mu = \bigcup_{r=0}^{m} \mathbb{S}^{\sqrt{r}} \subset \mathbb{K}$. We endow $\mathbb{K}^\mu$ with the inclusion order relation $\subseteq$, and we note $\mathfrak{T}^\mu = (\mathbb{K}^\mu, \lhd)$ the Hasse diagram of $(\mathbb{K}^\mu, \subseteq)$. The order set $(\mathbb{K}^\mu, \subseteq)$ still has $\mathbb{U}$ as maximum. By contrast with $(\mathbb{K}, \subseteq)$ it also has minimal elements gathered in $\mathbb{S}^\mu$. It is plain that $\mathfrak{T}^\mu$ is still a tree. By contrast with $\mathfrak{T}$, the tree $\mathfrak{T}^\mu$ is finite, with $\mathbb{U}$ as root and the elements of $\mathbb{S}^\mu$ as set of leaves. More precisely, we have the following property.

Property 7. *The tree $\mathfrak{T}^\mu$ is a partition tree (a.k.a. a dendrogram).*

This property means that each total cut of this tree (i.e. each maximal set of non-overlapping elements) is a partition of $\mathbb{U}$. A popular example of partition tree is the binary partition tree [19] usually considered in mathematical morphology. Here, the tree $\mathfrak{T}^\mu$ is not a binary tree but a general partition tree. Indeed, for each non-leaf element $K \in \mathbb{K}$, there are $k \geqslant 2$ elements $K' \in \mathbb{K}$ such that $K' \lhd K$ (by contrast with $k = 2$ for the binary partition tree).

The number of vertices of a partition tree is lower than $2\lambda - 1$ where λ is the number of leaves (this bound is reached if the partition tree is binary). The size of a general partition tree is then $O(\lambda)$. In the case of the tree $\mathfrak{T}^\mu$, the set of leaves is $\mathbb{S}^\mu$, of size $O(\mu^3)$. This is also the size of the tree $\mathfrak{T}^\mu$, that represents all the ρ-rotations for $0 \leqslant \rho \leqslant \mu$.

Algorithm 1: Construction of the structure of the ρ-rotations ($0 \leqslant \rho \leqslant \mu$).

Input : $\mu = \sqrt{m} \in \sqrt{\mathbb{N}}$
Output: $\mathfrak{T}^\mu$

1 **begin**
2   Build B^μ (of size $O(\mu^2)$)
3   Build $\mathbb{T}^\mu$ (of size $O(\mu^3)$) from B^μ
4   Sort $\mathbb{T}^\mu$
5   Build $\mathbb{H}^\mu$ (sorted, of size $O(\mu^3)$) from $\mathbb{T}^\mu$
6   Build $\mathcal{G}$ (of size $O(\mu^3)$) from $\mathbb{H}^\mu$
7   Build $\varDelta$ (of size $O(\mu^3)$) from $\mathbb{H}^\mu$
8   Build a binary partition tree $\widehat{\mathfrak{T}^\mu}$ (of size $O(\mu^3)$) from $(\mathcal{G}, \varDelta)$
9   Build $\mathfrak{T}^\mu$ (of size $O(\mu^3)$) from $\widehat{\mathfrak{T}^\mu}$

3.2 Construction

Let $\mu = \sqrt{m} \in \sqrt{\mathbb{N}}$, with $\mu \neq 0$. Let us focus on the partition $\mathbb{S}^\mu$ that gathers the leaves of the tree $\mathfrak{T}^\mu$. Any element of $\mathbb{S}^\mu = \mathbb{H}^\mu_{\bullet\multimap\circ}$ is a semi-open interval $S_j^\mu = [h_j^\mu, h_{j+1[\sigma^\mu]}^\mu)$ ($0 \leqslant j \leqslant \sigma^\mu - 1$). This semi-open interval is adjacent to exactly two other elements of $\mathbb{S}^\mu$, namely $S_{j+1[\sigma^\mu]}^\mu$ and $S_{j-1[\sigma^\mu]}^\mu$ (this adjacency is defined by the fact that the closures of the sets overlap). In other words, the topological structure of $\mathbb{S}^\mu$ is defined by a graph $\mathcal{G}^\mu = (\mathbb{V}^\mu, \mathbb{E}^\mu)$ which is isomorphic to $(\mathbb{H}^\mu_{\circ\multimap\circ}, \mathbb{H}^\mu_\bullet)$. In particular, it is a σ^μ-cycle graph.

The edges of $\mathcal{G}^\mu$ can be endowed with a valuation $\varDelta : \mathbb{H}^\mu \to \mathbb{N}$ that associates each hinge angle $h \in \mathbb{H}^\mu$, to the value $\varDelta(h)$ at which it appears in the combinatorial structure of the discrete rotations, i.e.

$$\varDelta(h) = \min\left\{r \in \mathbb{N} \mid \{h\} \in \mathbb{H}_\bullet^{\sqrt{r}}\right\} \tag{12}$$

Note that for any $K = [h, h') \in \mathbb{K}^\mu$, we have $\alpha(K) = \max\{\varDelta(h), \varDelta(h')\}$. We also have the following property, that derives from Proposition 1.

Property 8. *Let $h \in \mathbb{H}^\mu$. Let $(p_x, p_y, k) \in \mathbb{T}$ be the prime generator of h, associated to the point $\mathbf{p} = (p_x, p_y) \in \mathbb{Z}^2$. We have $\varDelta(h) = \|\mathbf{p}\|_2^2$.*

The edge-valued graph $(\mathcal{G}^\mu, \varDelta)$ can be seen as a saliency map [9], where the saliency measure is defined by $\varDelta$ on the ordered set $(\mathbb{N}, \geqslant)$ (i.e. higher saliencies correspond to lower values). We then have the following property.

Property 9. *The tree $\mathfrak{T}^\mu$ is the watershed tree [8] of $(\mathcal{G}^\mu, \varDelta)$.*

It follows that the tree $\mathfrak{T}^\mu$ can be built from $(\mathcal{G}^\mu, \varDelta)$ as a watershed tree. This can be done by using the standard binary partition tree construction proposed in [19] (by defining the priority of vertex merging from the saliency measure) followed by a straightforward collapsing procedure dedicated to turn the obtained binary tree into a general tree.

Based on these considerations, a procedure for building $\mathfrak{T}^\mu$, namely the combinatorial structure of all the ρ-rotations ($0 \leqslant \rho \leqslant \mu$) is given in Algorithm 1.

The construction of the Euclidean ball B^μ (line 2) has a time cost $O(\mu^2)$. The construction of the set $\mathbb{T}^\mu$ of hinge angle triplets from B^μ (line 3) has a time cost $O(\mu^3)$. Two hinge angles of $\mathbb{H}$ can be compared in $(\mathbb{R}, \leqslant)$ with a time cost $O(1)$ [20], based on their generating triplets in $\mathbb{T}$. It follows that sorting $\mathbb{T}^\mu$ (line 4) with respect to $\leqslant$ (see Eq. (8)) has a time cost $O(\mu^3 \log \mu)$. The construction of $\mathbb{H}^\mu$ (line 5) is done by choosing in $\mathbb{T}^\mu$ the triplets which are prime generators of hinge angles. Since $\mathbb{T}^\mu$ is sorted with respect to $\leqslant$, this procedure has a time cost $O(\mu^3)$. Due to the bijective links between $\mathbb{H}^\mu$, $\mathbb{H}_\bullet^\mu$, $\mathbb{H}_{\circ-\circ}^\mu$ and $\mathbb{H}_{\bullet-\circ}^\mu$, the construction of $\mathcal{G}$ from $\mathbb{H}^\mu$ (line 6) has a time cost $O(\mu^3)$. Based on Property 8, the construction of $\varDelta$ (line 7) from $\mathbb{H}^\mu$ (modeled by the prime generators of $\mathbb{T}^\mu$) has a time cost $O(\mu^3)$. The construction of a binary partition tree $\widetilde{\mathfrak{T}}^\mu$ (line 8) from the edge-valued graph $(\mathcal{G}, \varDelta)$ presents a time cost $O(\mu^3 \log \mu)$ [19]. The final conversion of this binary partition tree $\widetilde{\mathfrak{T}}^\mu$ into the general partition tree $\mathfrak{T}^\mu$ (line 9) is done by collapsing the redundant vertices within $\widetilde{\mathfrak{T}}^\mu$. (These vertices are characterized by the fact that both their creation and their merging are carried out for a same saliency value of $\varDelta$.) This conversion has a time cost $O(\mu^3)$. This justifies the following property.

Property 10. *The construction of $\mathfrak{T}^\mu$ (of size $O(\mu^3)$) has a time cost $O(\mu^3 \log \mu)$.*

Based on Algorithm 1, the hierarchical structure of all the ρ-rotations (with $0 \leqslant \rho \leqslant \mu \in \sqrt{\mathbb{N}}$) can then be built with a quasi-linear time cost with respect to its size.

4   Hierarchical Structure of Bijective Discrete Rotations

Based on the tree $\mathfrak{T}^\mu$, we now investigate the structure of the discrete rotations with regard to the bijectivity property. In particular, we adopt two points of view. First, we focus on the ρ-rotations which are the restrictions of bijective discrete rotations (Sect. 4.1). Second, we focus on the injective ρ-rotations, which are then bijective from their support to their finite image set (Sect. 4.2).

4.1 Structure of the Restrictions of Bijective Discrete Rotations

In a first time, we consider the ρ-rotations which are restrictions of bijective discrete rotations. We note $\mathcal{B} \subset \mathfrak{R}_{\mathbb{Z}^2}$ the set of the bijective discrete rotations, characterized by Proposition 2.

Let $\mu = \sqrt{m} \in \sqrt{\mathbb{N}}$. Our purpose is to define, for all $0 \leqslant \rho \leqslant \mu$, the subset $\mathcal{B}^\rho = \{R_\theta^\rho \in \mathfrak{R}_{\mathbb{Z}^2}^\rho \mid R_\theta \in \mathcal{B}\}$, i.e. the subset of the ρ-rotations which are restrictions of bijective discrete rotations. (We bijectively associate the set $\mathcal{B}^\rho \subseteq \mathfrak{R}_{\mathbb{Z}^2}^\rho$ to the subset $\mathfrak{S}^\rho \subseteq \mathbb{S}^\rho$.)

In a first time, we investigate the specific subset $\mathcal{B}^\mu \subset \mathfrak{R}_{\mathbb{Z}^2}^\mu$. This subset $\mathcal{B}^\mu$ can be defined by determining which rotations in $\mathfrak{R}_{\mathbb{Z}^2}^\mu$ (corresponding to leaves of the tree $\mathfrak{T}^\mu$, i.e. to intervals of $\mathbb{H}_{\bullet\circ}^\mu$) are bijective.

From the definition of $\mathcal{B}^\mu$, a rotation corresponding to the interval $[h_j^\mu, h_{j+1[\sigma^\mu]}^\mu)$ is bijective iff there exists an angle $\theta \in \mathbb{B}$ (see Sect. 2.3) such that $h_j^\mu < \theta < h_{j+1[\sigma^\mu]}^\mu$. It is then sufficient to sort the set $\mathbb{H}^\mu \cup \mathbb{B}$ to determine the bijective μ-rotations, i.e. to build $\mathfrak{S}^\mu$. (We recall that $\mathbb{H}^\mu \cap \mathbb{B} = \emptyset$.) Since $\mathbb{B}$ is infinite, such sorting is not tractable. Nonetheless, we can consider only a finite subset of $\mathbb{B}$, as shown by the following discussion.

From Proposition 2, we know that the sine of angles that define bijective discrete rotations are characterized by specific rational values, namely twin Pythagorean triplets. Due to symmetry considerations we restrict, without loss of generality, our discussion to the subset of angles $(0, \pi/4]$, where the angles θ leading to bijective discrete rotations are characterized by

$$\sin \theta \in \left\{ s_p = \frac{2p + 1}{2p^2 + 2p + 1} \right\}_{p \in \mathbb{N}^\star} \tag{13}$$

Let $\theta \in (0, \pi/4]$ such that $\sin \theta = s_p$ for a given $p \in \mathbb{N}^\star$. Let us suppose that $p^2 > 4m$. Then, we have $\sin \theta < 1/p < 1/2\mu$. In such conditions, for any point $\mathbf{p} \in B^\mu$, we have $\|\mathbf{p} - \mathcal{R}_\theta(\mathbf{p})\|_2 < 1/2$, and thus $R_\theta^\mu(\mathbf{p}) = \mathbf{p}$. In other words, we have $R_\theta^\mu = R_0^\mu$. It follows that for all $p \in \mathbb{N}^\star$ such that $p > 2\mu$, the angles θ_p generating bijective discrete rotations belong to exactly one interval of $\mathbb{S}^\mu$, namely $[h_{\sigma^\mu-1}^\mu, h_0^\mu)$ (which trivially corresponds to the bijective discrete rotation of angle 0).

As a consequence, we can restrict the bijectivity analysis to a finite subset noted $\mathbb{B}^\mu$ generated by the values of $p \leqslant 2\sqrt{m} = 2\mu$. This set $\mathbb{B}^\mu$ has a size $O(\mu)$, and it is thus sufficient to sort the (finite) set $\mathbb{H}^\mu \cup \mathbb{B}^\mu$ to define $\mathcal{B}^\mu$. We recall that from Algorithm 1, $\mathbb{H}^\mu$ is already sorted. Sorting (by dichotomy) $\mathbb{H}^\mu \cup \mathbb{B}^\mu$ then has a time cost $O(\mu \log \mu)$.

Let $r \in [\![0, m-1]\!]$. Let $S \in \mathbb{S}^{\sqrt{r}}$. From the definition of $\mathcal{B}^{\sqrt{r}}$, we have

$$S \in \mathfrak{S}^{\sqrt{r}} \iff \exists S' \in \mathfrak{S}^{\sqrt{r+1}}, S' \lhd S \tag{14}$$

where $\lhd$ is the Hasse relation of the tree $\mathfrak{T}^\mu$. Once the bottom structure $\mathfrak{S}^\mu$ (that defines the bijective leaves of the tree $\mathfrak{T}^\mu$) is obtained, we can define the successive sets $\mathfrak{S}^{\sqrt{r}}$ for r from $m - 1$ to 0 by a propagation from the leaves of the tree $\mathfrak{T}^\mu$ to all the other vertices in a bottom-up fashion, with a time cost $O(\mu)$. We then derive the whole set $\mathfrak{S}_\mu = \bigcup_{r=0}^m \mathfrak{S}^{\sqrt{r}}$ (and equivalently $\mathfrak{B}_\mu = \bigcup_{r=0}^m \mathfrak{B}^{\sqrt{r}}$), the structure of which is given by the restriction of the relation $\lhd$ to $\mathfrak{S}_\mu$. In particular, $(\mathfrak{S}_\mu, \lhd_{\mathfrak{S}_\mu})$ is a partial tree of $\mathfrak{T}^\mu$. This analysis justifies the following property.

Property 11. *The tree* $(\mathfrak{S}_\mu, \lhd_{\mathfrak{S}_\mu})$ *that models the structure of all the ρ-rotations ($0 \leqslant \rho \leqslant \mu$), which are restrictions of bijective discrete rotations, has a size $O(\mu)$ and can be built with a time cost $O(\mu \log \mu)$.*

4.2 Structure of the Injective/Bijective ρ-Rotations

In a second time, we consider the ρ-rotations which are injective, and thus bijective from their support set to their image set. For any $\rho \in \sqrt{\mathbb{N}}$, we note $\mathfrak{I}^\rho \subseteq \mathfrak{R}_{\mathbb{Z}^2}^\rho$ the set of these injective ρ-rotations.

Let $\rho \in \sqrt{\mathbb{N}}$. Let $\theta \in \mathring{\mathbb{U}}$. Let $R_\theta^\rho \in \mathfrak{R}_{\mathbb{Z}^2}^\rho$ be the ρ-rotation induced by θ. We have

$$R_\theta^\rho \in \mathfrak{B}^\rho \implies R_\theta^\rho \in \mathfrak{I}^\rho \tag{15}$$

However, the reciprocal is not always true. We have $\mathfrak{I}^\rho \supseteq \mathfrak{B}^\rho$ but we may have $\mathfrak{I}^\rho \not\subseteq \mathfrak{B}^\rho$.

Let $\mu = \sqrt{m} \in \sqrt{\mathbb{N}}$. We set $\mathfrak{I}_\mu = \bigcup_{r=0}^m \mathfrak{I}^{\sqrt{r}}$. By contrast with $\mathfrak{B}_\mu$ which was built in a bottom-up fashion from $\mathfrak{B}^\mu$ to $\mathfrak{B}^0$ (i.e. from the leaves of $\mathfrak{T}^\mu$ to its root), here $\mathfrak{I}_\mu$ will be built in a top-down fashion from $\mathfrak{I}^0$ to $\mathfrak{I}^\mu$ (i.e. from the root of $\mathfrak{T}^\mu$ to its leaves).

Let $\beta^\mu : \mathbb{K}^\mu \to \{\bot, \top\}$ be the Boolean function that characterizes the (non-)injectivity of any $K \in \mathbb{K}^\mu$. Building $\mathfrak{I}_\mu$ is then equivalent to building β^μ. In the sequel, we then build β^ρ for ρ from 0 to μ. The definition of β^0 is trivial, since $\mathbb{K}^0 = \{\mathbb{U}\}$ contains a unique element that models the rotation R_0^0. Since $R_0^0 \in \mathfrak{B}^0 \subseteq \mathfrak{I}^0$, we have $\beta^0(\mathbb{U}) = \top$.

Now, let $\rho = \sqrt{r} \in \sqrt{\mathbb{N}}$ with $r \geqslant 1$, and let us suppose that $\beta^{\sqrt{r-1}}$ is already known. The current iteration of the algorithm consists of building $\beta^{\sqrt{r}}$ from $\beta^{\sqrt{r-1}}$. Given a Euclidean ball B^ρ with radius ρ, a ρ-rotation is defined as a function $R_\theta^\rho : B^\rho \to \mathbb{Z}^2$. In particular, its support $B^{\sqrt{r}}$ is larger than the support $B^{\sqrt{r-1}}$ of the discrete $\sqrt{r-1}$-rotations. We note $C^\rho = B^{\sqrt{r}} \backslash B^{\sqrt{r-1}} = \{\mathbf{p} \in \mathbb{Z}^2 \mid \|\mathbf{p}\|_2 = \rho\}$.

Let $\mathbf{q} \in C^\rho$. This point $\mathbf{q}$ generates triplets $(q_x, q_y, k^\mathbf{q}) \in \mathbb{T}^\rho$ that induce hinge angles. We note $\mathbb{H}^\mathbf{q} = \{h_i^\mathbf{q}\}_{i=1}^{\sigma^\mathbf{q}}$ ($\sigma^\mathbf{q} \geqslant 0$) the set of all these (sorted) hinge angles associated with $\mathbf{q}$. The set $\mathbb{H}^\mathbf{q}$ allows to define a partition $\mathbb{H}_{\bullet\circ}^\mathbf{q} = \{H_j^\mathbf{q} = [h_j^\mathbf{q}, h_{j+1[\sigma^\mathbf{q}]}^\mathbf{q})\}_{j=0}^{\sigma^\mathbf{q}-1}$ the same way as in Eq. (11). The set $\mathbb{H}_{\bullet\circ}^\mathbf{q}$ can be endowed with a valuation function $v^\mathbf{q} : \mathbb{H}_{\bullet\circ}^\mathbf{q} \to \mathbb{Z}^2$ defined, for all $j \in [\![0, \sigma^\mathbf{q} - 1]\!]$, by $v^\mathbf{q}(H_j^\mathbf{q}) = R_\theta(\mathbf{q})$ for $\theta \in H_j^\mathbf{q}$.

Let $\mathbf{b} \in \mathbb{Z}^2$ be a point such that $\|\mathbf{q} - \mathbf{b}\|_2 = 1$ and $\mathbf{b} \in B^{\sqrt{r-1}}$. There exists one or two such points. Hereafter, the procedure is described by assuming that $\mathbf{b}$ is unique; it is simply repeated if there exist two points $\mathbf{b}$. The point $\mathbf{b}$ is (one of) the only (two) point(s) for which we may have $\mathbf{q}, \mathbf{b} \in B^{\sqrt{r}}$ and $R_\theta(\mathbf{q}) = R_\theta(\mathbf{b})$, leading to a non-injective configuration with respect to $\mathbf{q}$. We define $\mathbb{T}^\mathbf{b}, \mathbb{H}^\mathbf{b}, \mathbb{H}_{\bullet\circ}^\mathbf{b}$ and $v^\mathbf{b}$ the same way as for $\mathbf{p}$. From the union set $\mathbb{H}^\mathbf{b} \cup \mathbb{H}^\mathbf{q}$, we can build the partition $\overline{\mathbb{H}}_{\bullet\circ}^\mathbf{q}$ that refines $\mathbb{H}_{\bullet\circ}^\mathbf{q}$ and $\mathbb{H}_{\bullet\circ}^\mathbf{b}$ (as their supremum for $\sqsubseteq$). We extend the valuation functions $v^\mathbf{q}$ and $v^\mathbf{b}$ from their initial support to $\overline{\mathbb{H}}_{\bullet\circ}^\mathbf{q}$ so that for all $j \in [\![0, \overline{\sigma}^\mathbf{q} - 1]\!]$, we have $v^\mathbf{q}(\overline{H}_j^\mathbf{q}) = v^\mathbf{q}(H_{j'}^\mathbf{q})$ with $\overline{H}_j^\mathbf{q} \subseteq H_{j'}^\mathbf{q} \in \mathbb{H}_{\bullet\circ}^\mathbf{q}$ and $v^\mathbf{b}(\overline{H}_j^\mathbf{q}) = v^\mathbf{b}(H_{j'}^\mathbf{b})$ with $\overline{H}_j^\mathbf{q} \subseteq H_{j'}^\mathbf{b} \in \mathbb{H}_{\bullet\circ}^\mathbf{b}$. We can endow the new partition $\overline{\mathbb{H}}_{\bullet\circ}^\mathbf{q}$

with the local injectivity characterization function $\iota^q : \overline{\mathbb{H}}_{\bullet\circ}^q \to \{\bot, \top\}$, derived from ν^q and ν^b and defined, for all $j \in [\![0, \overline{\sigma}^q - 1]\!]$, by $\iota^q(\overline{H}_j^q) = (\nu^q(\overline{H}_j^q) \neq (\nu^b(\overline{H}_j^q))$.

For all the points $\mathbf{q} \in C^\rho$, we then have access to a partition $\overline{\mathbb{H}}_{\bullet\circ}^q$ of $\mathbb{U}$ and a local injectivity characterization function $\iota^q : \overline{\mathbb{H}}_{\bullet\circ}^q \to \{\bot, \top\}$. Following the same approach as described above, one can define the partition $\mathbb{S}^{\sqrt{r}} = \mathbb{H}_{\bullet\circ}^{\sqrt{r}}$ as the supremum of the partition $\mathbb{S}^{\sqrt{r-1}} = \mathbb{H}_{\bullet\circ}^{\sqrt{r-1}}$ and all the partitions $\overline{\mathbb{H}}_{\bullet\circ}^q$ induced by all the points $\mathbf{q} \in C^\rho$. The function $\beta^{\sqrt{r-1}}$ and all the functions ι^q can be extended from their initial support to $\mathbb{H}_{\bullet\circ}^{\sqrt{r}}$. The function $\beta^{\sqrt{r}} : \mathbb{H}_{\bullet\circ}^{\sqrt{r}} \to \{\bot, \top\}$ is finally defined, for all $j \in [\![0, \sigma^{\sqrt{r}} - 1]\!]$ by

$$\beta^{\sqrt{r}}(H_j^{\sqrt{r}}) = \beta^{\sqrt{r-1}}(H_j^{\sqrt{r-1}}) \wedge \bigwedge_{\mathbf{q} \in C^{\sqrt{r}}} \iota^q(H_j^{\sqrt{r-1}}) \tag{16}$$

For each $r \in [\![0, m]\!]$, the set $\mathfrak{I}^\rho$ is given by $(\beta^{\sqrt{r}})^{-1}(\{\top\})$.

The overall procedure is summarized in Algorithm 2. For the sake of efficiency, it is relevant to design partial partitions $\mathbb{H}_{\bullet\circ}^{\sqrt{r}}$ that only contain the intervals for which $\beta^{\sqrt{r}}$ has the $\top$ value, instead of total partitions of $\mathbb{U}$ (storing the intervals for which $\beta^{\sqrt{r}}$ has the $\bot$ value is useless, since a function is non-injective whenever one of its restricted functions is non-injective). We assume that we handle partial partitions in the above complexity analysis. Under this hypothesis, the space cost of the partial partitions associated to $\mathbb{H}^{\sqrt{r}}, \mathbb{H}_{\bullet\circ}^{\sqrt{r}}$ is assumed to be $O(\sqrt{r})$. (At this stage, this assumption relies on our experimental analysis of the evolution of the population of injective rotations; this remains to be theoretically confirmed.) The construction of $\mathbb{H}^0, \mathbb{K}^0$ and β^0 (line 2) has a time cost $O(1)$. The external loop (line 3) iterates $O(m)$ times. The construction of $C^{\sqrt{r}}$ (of size $O(1)$) (line 4) has a time cost $O(1)$. The medial loop (line 5) iterates $O(1)$ times. At iteration r, the construction of $\mathbb{H}^q, \mathbb{H}_{\bullet\circ}^q$ and ν^q, of size $O(\sqrt{r})$, (line 6) has a time cost $O(\sqrt{r})$. The internal loop (line 7) iterates $O(1)$ times. At iteration r of the medial loop, the construction of $\mathbb{H}^b, \mathbb{H}_{\bullet\circ}^b$ and ν^b (of size $O(\sqrt{r})$) (line 8) while the updating of $\overline{\mathbb{H}}^q, \overline{\mathbb{H}}_{\bullet\circ}^q$ and ι^q (line 9), that consists of merging pairs of partitions of size $O(\sqrt{r})$ also

---

Algorithm 2: Construction of the injective / bijective ρ-rotations.

Input : $\mu = \sqrt{m} \in \sqrt{\mathbb{N}}$
Output: $\mathfrak{I}_\mu$ (given by β^ρ for $0 \leqslant \rho \leqslant \mu$)

1 **begin**
2 | Build $\mathbb{H}^0, \mathbb{H}_{\bullet\circ}^0$ and β^0
3 | **for** $r \in [\![1, m]\!]$ **do**
4 | | Build $C^{\sqrt{r}}$
5 | | **foreach** $\mathbf{q} \in C^{\sqrt{r}}$ **do**
6 | | | Build $\mathbb{H}^q, \mathbb{H}_{\bullet\circ}^q$ and ν^q
7 | | | **foreach** $\mathbf{b} \in \mathbb{Z}^2$ s.t. $\|\mathbf{b}\|_2 < \sqrt{r}$ *and* $\|\mathbf{p} - \mathbf{b}\|_2 = 1$ **do**
8 | | | | Build $\mathbb{H}^b, \mathbb{H}_{\bullet\circ}^b$ and ν^b
9 | | | | Build / update $\overline{\mathbb{H}}^q, \overline{\mathbb{H}}_{\bullet\circ}^q$ and ι^q
10 | | Build / update $\mathbb{H}^{\sqrt{r}}, \mathbb{H}_{\bullet\circ}^{\sqrt{r}}$ and $\beta^{\sqrt{r}}$

---

has a time cost $O(\sqrt{r})$. At iteration r of the medial loop, the update of $\mathbb{H}^{\sqrt{r}}$, $\mathbb{H}^{\sqrt{r}}_{\bullet\circ}$ and $\beta^{\sqrt{r}}$ that consists of merging pairs partitions of size $O(\sqrt{r-1})$ and $O(\sqrt{r})$, has a time cost $O(\sqrt{r})$. This analysis motivates the following conjecture.

Conjecture 12. *The tree $(\mathfrak{I}_\mu, \lhd_{\mathfrak{I}_\mu})$ that models the structure of all the injective ρ-rotations ($0 \leqslant \rho \leqslant \mu$) has a size $O(\mu)$ and can be built with a time cost $O(\mu^3)$.*

5   Experimental Results

We now illustrate some results obtained by application of the proposed algorithms.

In Fig. 1, we show the first stages of the hierarchical structure of the bijective and injective ρ-rotations. The non-bijective ρ-rotations were also computed, but they are not visualized for the sake of readability.

Figure 2 provides the growth of the size of the various families of ρ-rotations with respect to the radius ρ of the Euclidean balls. Theoretically, the numbers of the ρ-rotations and the non-bijective ρ-rotations grow in $O(\rho^3)$, while the number of the bijective ρ-rotations (i.e. the restrictions of bijective discrete rotations) grows in $O(\rho)$. These behaviours are experimentally confirmed. The number of the injective ρ-rotations is experimentally assessed as growing in $O(\rho)$. This result, although intuitive, remains to be proved. This is why the results stated at the end of Sect. 4 are given as a conjecture.

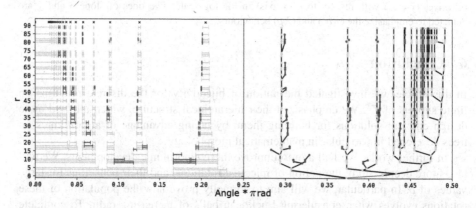

Fig. 1. Hierarchical structure of the bijective ρ-rotations. Here, the values of $\rho = \sqrt{r}$ are given for $0 \leqslant r \leqslant 90$ (y axis). The visualized angles of $\mathbb{U}$ are restricted to $[0, \pi/2]$ (x axis), due to symmetry considerations. For the sake of visualisation, only the bijective (Sect. 4.1) and injective (Sect. 4.2) ρ-rotations are visualized. On the left side (angles in $[0, \pi/4]$), the intervals of $\mathbb{S}^\rho$ corresponding to the ρ-rotations are depicted. On the (symmetric) right side (angles in $[\pi/4, \pi/2]$), the hierarchical structure between these intervals/ρ-rotations is depicted. The pink elements correspond to ρ-rotations which are the restrictions of discrete rotations (in particular, the bijective angles associated to these rotations are depicted by black $\times$ in the upper part of the right side). The blue elements correspond to ρ-rotations which are injective on B^ρ and thus bijective from their finite support to their image. (Color figure online)

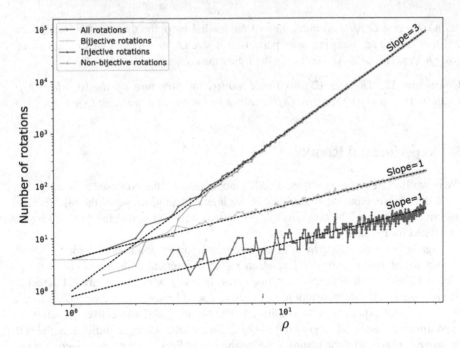

Fig. 2. Evolution of the size of the different families of ρ-rotations (all, bijective, injective, non-bijective) (y-axis) with respect to ρ (x-axis) in log-log scale. The lines of slope 3 and 1 are depicted to emphasize the $O(\rho^3)$ and $O(\rho)$ behaviours.

6 Conclusion

In this article, we investigated the notion of bijectivity for the discrete rotations on finite subsets of $\mathbb{Z}^2$. We emphasized their hierarchical structure, which allowed us to design efficient solutions for building them, by taking advantage of algorithmics on trees developed in particular in mathematical morphology.

In further works, we will aim to improve these algorithmic developments. We will also more deeply study the notion of injective rotations, and their behaviour for large values of ρ. In particular, we will aim to formally prove how the population of these rotations evolves when considering Euclidean balls of increasing radii. To complete this framework, we will also investigate alternative definitions of bijectivity. Finally, we will build upon this framework to deal with the problem of approximation of non-bijective rotations [7]. We may also further investigate other properties of finite discrete rotations, that were also pioneered e.g. in [1] from the point of view of dynamic systems.

References

1. Akiyama, S., Pethö, A.: Discretized rotation has infinitely many periodic orbits. Nonlinearity **26**, 871 (2013)
2. Amir, A., Kapah, O., Tsur, D.: Faster two-dimensional pattern matching with rotations. Theor. Comput. Sci. **368**, 196–204 (2006)

3. Andres, É.: The Quasi-Shear rotation. In: Miguet, S., Montanvert, A., Ubéda, S. (eds.) DGCI 1996. LNCS, vol. 1176, pp. 307–314. Springer, Heidelberg (1996). https://doi.org/10.1007/3-540-62005-2_26

4. Andres, É., Dutt, M., Biswas, A., Largeteau-Skapin, G., Zrour, R.: Digital two-dimensional bijective reflection and associated rotation. In: Couprie, M., Cousty, J., Kenmochi, Y., Mustafa, N. (eds.) DGCI 2019. LNCS, vol. 11414, pp. 3–14. Springer, Cham (2019). https://doi.org/10.1007/978-3-030-14085-4_1

5. Andres, É., Largeteau-Skapin, G., Zrour, R.: Shear based bijective digital rotation in hexagonal grids. In: Lindblad, J., Malmberg, F., Sladoje, N. (eds.) DGMM 2021. LNCS, vol. 12708, pp. 217–228. Springer, Cham (2021). https://doi.org/10.1007/978-3-030-76657-3_15

6. Breuils, S., Kenmochi, Y., Andres, É., Sugimoto, A.: Conjecture on characterisation of bijective 3D digitized reflections and rotations. In: Hitzer, E., Papagiannakis, G., Vasik, P. (eds.) ENGAGE 2022. LNCS, vol. 13862, pp. 41–53. Springer, Cham (2022). https://doi.org/10.1007/978-3-031-30923-6_4

7. Breuils, S., Kenmochi, Y., Sugimoto, A.: Visiting bijective digitized reflections and rotations using geometric algebra. In: Lindblad, J., Malmberg, F., Sladoje, N. (eds.) DGMM 2021. LNCS, vol. 12708, pp. 242–254. Springer, Cham (2021). https://doi.org/10.1007/978-3-030-76657-3_17

8. Cousty, J., Bertrand, G., Najman, L., Couprie, M.: Watershed cuts: minimum spanning forests and the drop of water principle. IEEE Trans. Pattern Anal. Mach. Intell. **31**, 1362–1374 (2009)

9. Cousty, J., Najman, L., Kenmochi, Y., Ferzoli Guimarães, S.J.: Hierarchical segmentations with graphs: quasi-flat zones, minimum spanning trees, and saliency maps. J. Math. Imaging Vis. **60**, 479–502 (2018)

10. Hannusch, C., Pethö, A.: Rotation on the digital plane. Period. Math. Hung. **86**, 564–577 (2023)

11. Jacob, M.A., Andres, É.: On discrete rotations. In: DGCI, pp. 161–174 (1995)

12. Nouvel, B.: Rotation discrètes et automates cellulaires. Ph.D. thesis, École Normale Supérieure de Lyon (2010)

13. Nouvel, B., Rémila, É.: Characterization of bijective discretized rotations. In: Klette, R., Zunić, J. (eds.) IWCIA 2004. LNCS, vol. 3322, pp. 248–259. Springer, Cham (2004). https://doi.org/10.1007/978-3-540-30503-3_19

14. Nouvel, B., Rémila, É.: Incremental and transitive discrete rotations. In: Reulke, R., Eckardt, U., Flach, B., Knauer, U., Polthier, K. (eds.) IWCIA 2006. LNCS, vol. 4040, pp. 199–213. Springer, Heidelberg (2006). https://doi.org/10.1007/11774938_16

15. Passat, N., Ngo, P., Kenmochi, Y.: Bijectivity analysis of finite rotations on $\mathbb{Z}^2$: a hierarchical approach. Technical report, hal-04371613 (2024)

16. Pluta, K., Romon, P., Kenmochi, Y., Passat, N.: Bijectivity certification of 3D digitized rotations. In: Bac, A., Mari, J.L. (eds.) CTIC 2016. LNCS, vol. 9667, pp. 30–41. Springer, Cham (2016). https://doi.org/10.1007/978-3-319-39441-1_4

17. Pluta, K., Roussillon, T., Coeurjolly, D., Romon, P., Kenmochi, Y., Ostromoukhov, V.: Characterization of bijective digitized rotations on the hexagonal grid. J. Math. Imaging Vis. **60**, 707–716 (2018)

18. Roussillon, T., Coeurjolly, D.: Characterization of bijective discretized rotations by Gaussian integers. Technical report (2016). https://hal.archives-ouvertes.fr/hal-01259826

19. Salembier, P., Garrido, L.: Binary partition tree as an efficient representation for image processing, segmentation, and information retrieval. IEEE Trans. Image Process. **9**, 561–576 (2000)

20. Thibault, Y.: Rotations in 2D and 3D discrete spaces. Ph.D. thesis, Université Paris-Est (2010)

Recognition of Arithmetic Line Segments and Hyperplanes Using the Stern-Brocot Tree

Bastien Laboureix[✉] and Isabelle Debled-Rennesson

Université de Lorraine, CNRS, LORIA, 54000 Nancy, France
{bastien.laboureix,isabelle.debled-rennesson}@loria.fr

Abstract. The classic problem of discrete structure recognition is revisited in this article. We focus on naive digital straight segments (DSS) and, more generally, naive arithmetic hyperplanes, and we present a new approach to recognise these discrete structures based on the Stern-Brocot tree. The algorithm for DSS recognition proposes an alternative method to the state of the art, keeping the linear complexity and incremental character. While most of the concepts can be generalised to planes in dimension 3 and hyperplanes in higher dimensions, certain points in the process of descending in the Stern-Brocot tree need to be explored further. The proposed algorithm calculates separating chords characterising the membership of planes to cones generated by the branch of the Stern-Brocot tree. This generalisation shows the close link between arithmetic hyperplanes and the generalised Stern-Brocot tree and opens up interesting perspectives for the recognition of pieces of arithmetic hyperplanes.

Keywords: digital straight segments · arithmetic hyperplanes · recognition algorithm · Stern-Brocot tree

1 Introduction

Discrete geometry is concerned with various structures of space $\mathbb{Z}^d$. This article focuses on arithmetic discrete lines, introduced in [17] and their generalisation to any dimension: arithmetic hyperplanes [1]. Our problem is to decide, given a finite subset S of $\mathbb{Z}^d$, whether S is a piece of arithmetic hyperplane and, if so, to compute its minimal parameters.

In dimension 2, the problem of naive digital straight segment (DSS) recognition was first addressed from the point of view of symbolic dynamics and the study of Sturmian words: see [16] for the initial article, [3,10] for a review of the links between discrete geometry and symbolic dynamics. A history of the DSS recognition problem can be found in [12]. The arithmetic definition of discrete lines and their geometric structures were used to obtain the incremental and linear algorithm for DSS recognition presented by I. Debled-Rennesson and J.-P. Réveillès in [8].

© The Author(s), under exclusive license to Springer Nature Switzerland AG 2024
S. Brunetti et al. (Eds.): DGMM 2024, LNCS 14605, pp. 16–28, 2024.
https://doi.org/10.1007/978-3-031-57793-2_2

While the hyperplane recognition problem is largely solved in dimension 2, the problem remains difficult in dimensions 3 and higher. In dimension 3, numerous studies of discrete planes have been carried out using different approaches (see the survey [4]). A generalisation of I. Debled-Rennesson's 2D algorithm was presented in 1994 in [9,15], but the algorithm, restricted to pieces of rectangular planes, loses in simplicity and the number of cases to be processed rapidly explodes. In 2005, Y.Gerard et al.'s algorithm [11] solves the recognition problem for any finite set of points and in any dimension, using the properties of the convex hull of the chord space. However, the algorithm announces a high complexity for dimension 3 and does not guarantee to obtain the minimum parameters. In 2008, the Charrier et al.'s algorithm [6] proposed a linear optimisation approach to the problem. This method provides a quasi-linear algorithm in terms of the number of points, but does not allow the minimal parameters of the hyperplane to be obtained (the complexity would then revert to that of the simplex algorithm: exponential). Finally, a serie of articles has been written on plane probing ([13] for the first version), which makes it possible to obtain the characteristics of the plane from successive oracles.

In this paper, we propose another DSS recognition algorithm based on the Stern-Brocot tree (introduced in [18] and [5]). The successive points of the segment are then taken into account, allowing us to go down the tree until we find the slope corresponding to the minimal parameters of the segment. While I. Debled-Rennesson's algorithm can be interpreted as a descent down the tree (see [7] and [19]), the method proposed in this article differs. We also obtain linear complexity while maintaining the incremental character.

The approach of recognition by descent down the Stern-Brocot tree can also be extended to higher dimensions. In [14], H. Lennerstad generalises the Stern-Brocot tree to any finite dimension d. We introduce a new concept: the notion of separating chord, used to determine the branch of descent down the tree corresponding to a location of the normal vector of the plane to be recognised. Each step of the algorithm now requires $\binom{d}{2}$ tests to determine in which of the $d!$ branches to continue the search. The proposed method recognises discrete hyperplanes and opens up interesting prospects for recognising pieces of discrete planes.

A formal proof and a detailed algorithm are in an appendix to the article, available in full version on HAL.

2 Discrete Lines

We are working in $\mathbb{R}^d$ with its canonical scalar product. In particular, in this section $d = 2$. In 1991, J.-P. Réveillès defined the concept of a discrete line in [17]. We are interested here in a version of the naive discrete line where the thickness parameter is fixed to guarantee good connectedness properties:

Definition 1 (Naive discrete line). Let $a, b, \mu \in \mathbb{R}$ with $(a, b) \neq (0, 0)$.
The naive discrete line with slope $\frac{a}{b}$ (we agree that a slope $\frac{1}{0}$ is a vertical slope) and shift μ is the set $\mathcal{D}(a, b, \mu) \stackrel{def}{=} \{(x, y) \in \mathbb{Z}^2 \mid 0 \leqslant ax - by + \mu < \|(a, b)\|_\infty\}$

Definition 2 (8-neighbourhood and 8-connectedness). *We say that 2 points $p, q \in \mathbb{Z}^2$ are 8-neighbours iff $\|p - q\|_\infty = 1$. A set $A \subset \mathbb{Z}^2$ is then 8-connected iff every pair of points of A is connected by a path for the 8-neighbourhood. A naive digital straight segment (DSS) is an 8-connected part of a naive line.*

Naive lines are 8-connected and minimise thickness for this property [17], so they are widely used in discrete geometry. The terms "naive" and "8-connected" will be omitted for convenience. The DSS recognition problem is then to decide, given a set $S \subset \mathbb{Z}^2$, whether or not S is a DSS. Note that there is no uniqueness of the parameters (a, b, μ) of a segment: we therefore ask, if necessary, to calculate the minimal integer parameters of a naive DSS, i.e. those minimising $\|(a, b)\|_\infty$.

I. Debled-Rennesson and J.-P. Réveillès's recognition algorithm presented in [8] proposes a linear algorithm in $|S|$ and incremental, in the sense that adding a point updates the parameters of the segment in $O(1)$. The algorithm is based on the notion of leaning point:

Definition 3 (Leaning points). *Let D be a naive discrete line with parameters (a, b, μ) and $p = (x, y)$ a point of S. We say that p is a lower (resp. upper) leaning point of D iff $ax - by + \mu = \|(a, b)\|_\infty - 1$ (resp. $ax - by + \mu = 0$). In a DSS, It is said to be extremal iff it is the lower (or upper) leaning point with the minimum or maximum abscissa.*

3   Recognition Using the Stern-Brocot Tree

Our segment recognition algorithm is based on a descent into the Stern-Brocot tree, presented in [18] and [5]. This tree lists all positive irreducible fractions and presents them in tree form (see Fig. 1). We define the Stern-Brocot tree B_n truncated at level n by recurrence on n:

- B_0 consists of 2 nodes labelled $\frac{0}{1}$ and $\frac{1}{0}$ called inputs.
- given a truncated tree B_n, we list all the fractions it contains using the prefix depth traversal. Between 2 fractions $\frac{a}{b}$ and $\frac{c}{d}$, we insert into the tree the fraction $\frac{a}{b} \oplus \frac{c}{d} \stackrel{\text{def}}{=} \frac{a+c}{b+d}$.

Remark 4. *This addition of fractions is actually the addition of the pairs (a, b) and (c, d). The irreducible fraction $\frac{a}{b}$ and the pair (a, b) will later be happily confused.*

Theorem 5 (Stern-Brocot [5,18]). *The Stern-Brocot tree obtained by the union of $(B_n)_{n \in \mathbb{N}}$ contains exactly once each irreducible fraction of $\mathbb{Q}_+$.*

The principle of the algorithm is then to find the slope a/b of the segment S in the Stern-Brocot tree. By symmetry and translation, we can always assume that S lies in the first octant (i.e. verifies $0 \leqslant a \leqslant b$) and has point of minimum

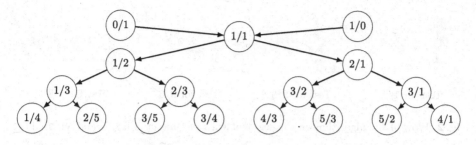

Fig. 1. Stern-Brocot tree

abscissa $(0, 0)$. Note that, from then on, the abscissas of the points of S are indexed on $[\![0, n-1]\!]$ where $n \overset{\text{def}}{=} |S|$.

We first propose a naive version of the algorithm, before optimising it. If the segment is neither horizontal nor diagonal (extremal cases), we start with the $1/2$ slope. At each stage of the algorithm, we look at points $p_{\min}$ and $p_{\max}$ of minimum and maximum scalar products for the slope a/b under consideration, and $\delta = (x, y) \overset{\text{def}}{=} p_{\max} - p_{\min}$. If $r_{\frac{a}{b}}(x, y) \overset{\text{def}}{=} ax - by < b$, the difference between the maximum and minimum scalar products is small enough: the slope is therefore suitable and the shift is calculated using $p_{\min}$. Otherwise, to determine whether the slope is too small or too large, we look at the sign of the δ coordinates. The 2 coordinates have the same sign because S is in the first octant. Note also that, despite the non-uniqueness of $p_{\min}$ and $p_{\max}$, the sign of δ does not depend on the choice of $p_{\min}$ and $p_{\max}$, according to the algorithm's proof of correctness. If this is positive, $p_{\max}$ is to the right of $p_{\min}$: decreasing the difference in the scalar product between $p_{\min}$ and $p_{\max}$ means decreasing the slope, which is done by a descent to the left in Stern-Brocot. Similarly, if the sign of the coordinates of δ is negative, $p_{\max}$ is to the left of $p_{\min}$, so we increase the slope by going to the right in the Stern-Brocot tree. The detailed algorithm can be found in the full version.

For example, let's look at the flow of the naive algorithm (see Fig. 2) on the segment with parameters $(5, 8, 3)$ and length 11. The algorithm starts with the slope $\frac{1}{2}$ in the Stern-Brocot tree. The minimum $r_{\frac{1}{2}}$ value of -3 is obtained in $(9, 6)$ and the maximum $r_{\frac{1}{2}}$ value of 0 is obtained in $(0, 0)$. So $\delta = (-9, -6)$. Then $r_{\frac{1}{2}}(\delta) = 3$, which is greater than 2, so the slope is unsuitable. As δ has negative coordinates, we increase the slope by going down the right-hand side of the tree to arrive at a slope of $\frac{2}{3}$. For the slope $\frac{2}{3}$, $\delta = (3, 1)$. The slope is still unsuitable ($r_{\frac{2}{3}}(\delta) = 3 \geqslant 3$) and δ is positive, so the slope is too steep: we go down the left-hand side of the tree to $\frac{3}{5}$. For this new slope, $\delta = (-5, -4)$, $r_{\frac{3}{5}}(\delta) = 5 \geqslant 5$. As δ is negative, we increase the slope by going down to the right towards $\frac{5}{8}$. Finally, for $\frac{5}{8}$, $r_{\frac{5}{8}}(\delta) = 7$, which is strictly smaller than 8, so the slope is appropriate. The shift is then the opposite of the minimum $r_{\frac{5}{8}}$ value, so the DSS parameters are $(5, 8, 3)$.

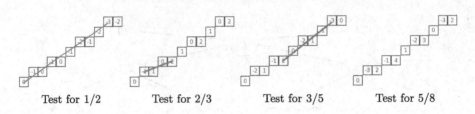

Test for 1/2   Test for 2/3   Test for 3/5   Test for 5/8

Fig. 2. Flow of the algorithm on the segment with parameters $(5, 8, 3)$ and length 11

Theorem 6. *The naive recognition algorithm decides whether a finite set S of the first octant is a DSS, returns its minimum parameters if it is, and then terminates in $O(nh)$ where $n = |S|$ and h is the height of the parameters of S in the Stern-Brocot tree.*

Sketch of Proof: The formal proof is in the full version. An invariant ensures that the slope of the piece studied is always between the lower and upper slopes given by the Stern-Brocot tree. The sign of the observed chord can then be used to choose the right descent branch in the Stern-Brocot tree.

This naive version is not particularly interesting in terms of complexity and is not an incremental method. However, it is easily optimised (see Algorithm 1 for the full detailled incremental algorithm) based on the following observations:

- you can add the points of the segment one by one and calculate the parameters of the new segment by starting again from the old parameters in the Stern-Brocot tree: the method then becomes incremental;
- there is no point in calculating the remainders (i.e. values of $r_{\frac{a}{b}}(M)$) for all the points M in the segment. The remainders associated with the at most 4 extremal upper and lower leaning points (which can be calculated in $O(1)$) are sufficient to guarantee that an equation is satisfied for all points on the segment (by convexity). The most that can be done is to calculate a constant number of remainders in the Extremal-rests function: the call is therefore made in $O(1)$. The complexity therefore drops to $O(n + h) = O(n)$.

The incremental Algorithm 1 thus provides an efficient version of segment recognition using the Stern-Brocot tree. However, the advantage of the naive algorithm is that it can be generalised to finite subsets of $\mathbb{Z}^2$ that are not necessarily connected, using the same method of descent in the tree.

4   Stern-Brocot Tree in Higher Dimensions

The notion of a discrete line naturally extends to higher dimensions: The notion of a discrete line extends naturally into higher dimensions, as does the notion of a multidimensional Sturmian word presented in [2]:

Definition 7 (Naive arithmetic hyperplane [1]). *Let $v \in \mathbb{R}^d$ be non-zero and $\mu \in \mathbb{R}$. The naive arithmetic (or discrete) hyperplane with normal vector v and shift μ is the set $\mathbb{P}(v, \mu) \stackrel{def}{=} \{x \in \mathbb{Z}^d \mid 0 \leqslant \langle x, v \rangle + \mu < \|v\|_\infty\}$.*

The general problem of plane or hyperplane recognition then consists in deciding whether a finite subset S of $\mathbb{Z}^d$ is part of a naive discrete plane or hyperplane, and computing, if so, its minimal integer parameters i.e. minimising $\|v\|_\infty$. We then propose an algorithm based on an extension of the Stern-Brocot tree in d dimension.

The Stern-Brocot tree naturally extends into 3D and higher dimensions, as presented in [14]. This tree lists the d-uplets of natural integers that are prime to each other in their set. Instead of using 2 end points as before, we use d. These points are represented in $(d-1)$-simplex form as follows (illustrated in Fig. 3(a) and (b); with normalisation, the sum of 2 vectors is represented by their barycentric sum):

- Initially, the d endpoints are the points e_i of the canonical basis of $\mathbb{Z}^d$.
- Given d endpoints $p_1, ..., p_d$ (affinely independent in $\mathbb{R}^{d-1}$ by induction on the tree), we construct the $(d-1)$-simplex Γ with endpoints $p_1, ..., p_d$.
- Given a permutation σ of $\mathfrak{S}_d$, consider the simplex $\Gamma[\sigma]$ formed by the points
$$q_1, ..., q_d \text{ where } q_j \stackrel{\text{def}}{=} \sum_{i=1}^{j} p_{\sigma(i)}.$$
- The children of the simplex Γ in the tree are then the simplexes $\Gamma[\sigma]$ for $\sigma \in \mathfrak{S}_d$.

By symmetry, we can assume that the normal vector v of the piece of hyperplane under study has positive coordinates. The d initial ends of the Stern-Brocot tree are then the e_i vectors of the canonical basis. The question is then, given d ends of the tree, to know into which of the $d!$ subtrees to descend. For example, initially, the permutation to choose is the coordinate sorting permutation. For the rest of this paper, we will consider dimension 3, as the results obtained can be generalised without any problem to any finite dimension.

In order to choose the right sub-simplex, we define the notions of chord and separation (by chords) as follows:

Definition 8 (Chord). *Let $v \in \mathbb{Z}^d$ and $\mu \in \mathbb{Z}$. We say that the chord $\delta \in \mathbb{Z}^d$ appears (see Fig. 4) in the plane $\mathbb{P}(v, \mu)$ iff there exist $x, y \in \mathbb{P}(v, \mu)$ such that $\delta = y - x$. Note that, by Bézout's theorem (in the rational v case) or by density (in the irrational v case), this definition does not depend on the μ shift considered.*

Remark 9. *Let us note that a chord δ appears in a plane of normal vector v iff there exist $x, y \in \mathbb{P}(v, \mu)$ such that $\delta = y - x$ and $\begin{cases} 0 \leqslant \langle x, v \rangle + \mu < \|v\|_\infty \\ 0 \leqslant \langle y, v \rangle + \mu < \|v\|_\infty \end{cases}$. So, by subtracting, δ appears in a plane with normal vector v iff $|\langle \delta, v \rangle| < \|v\|_\infty$.*

Definition 10 (Separation (by chord)). *Let $p_1, p_2, p_3 \in \mathbb{N}^3$ be Stern-Brocot endpoints. We say that a pair $(\delta_-^{(i)}, \delta_+^{(i)})$ is a separation (see Fig. 3(c)) for (p_1, p_2, p_3) according to p_i iff for all $v \in \mathcal{C}(p_1, p_2, p_3)$ (convex cone generated by p_1, p_2, p_3):*

- *the $\delta_-^{(i)}$ chord appears in $\mathbb{P}(v, \mu)$ iff $v \in \mathcal{C}(p_i, m_i, p_{i-1}) \backslash \mathcal{C}(p_i, m_i)$ (triangle p_i, m_i, p_{i-1} without side p_i, m_i).*

Algorithm 1: Incremental DSS recognition algorithm

Input: $S \subset \mathbb{Z}^2$, set of points indexed from 0 to $n - 1$ in the first octant

Output: Decide whether S is a segment and, if so, return its minimum parameters (a, b, μ).

Incremental recognition(S):

$i \leftarrow 0$ (maximum points index considered) ;

while *the piece of segment considered is horizontal or diagonal and $i < n$.* **do**
  | $i \leftarrow i + 1$;

if $i = n$ **then**
  | Return $(0, 1, 0)$ or $(1, 1, 0)$ depending on whether the piece in question is horizontal or diagonal. ;

else
  | $test \leftarrow \{(0, 0), (i - 1, 0)\}$ or $\{(0, 0), (i - 1, i - 1)\}$ either horizontal or diagonal ;

pente$_{\text{inf}} \leftarrow \frac{0}{1}$ (lower limit) ;

pente$_{\text{sup}} \leftarrow \frac{1}{1}$ (upper limit) ;

$\frac{a}{b} \leftarrow$ pente$_{\text{inf}} \oplus$ pente$_{\text{sup}}$ (current slope initially 1/2) ;

$\mu \leftarrow 0$;

while $i < n$ *(we have not yet considered the entire segment)* **do**
  | correct $\leftarrow$ FALSE (indicates whether the current slope is suitable) ;
  | **while** $b < n$ *and not correct* **do**
  | | $p_{\min}, p_{\max} \leftarrow$ Extremal-rests($a, b, test \cup \{S[i]\}$) (minimum and maximum residual points for the a/b slope) ;
  | | $\delta \leftarrow p_{\max} - p_{\min}$;
  | | **if** $\langle \delta, (a, -b) \rangle < b$ *(the slope a/b is suitable)* **then**
  | | | $\mu \leftarrow - \langle p_{\min}, (a, -b) \rangle$;
  | | | $i \leftarrow i + 1$ (if the slope is suitable, move on to the next point);
  | | | $test \leftarrow$ all the extremal leaning points using the parameters (a, b, μ) of the equation for the current segment piece ;
  | | | correct $\leftarrow$ TRUE ;
  | | **if** *not correct* **then**
  | | | **if** δ *has an abscissa > 0 (the current slope is too steep)* **then**
  | | | | pente$_{\text{sup}} \leftarrow \frac{a}{b}$;
  | | | **if** δ *has an abscissa < 0 (the current slope is too shallow)* **then**
  | | | | pente$_{\text{inf}} \leftarrow \frac{a}{b}$;
  | | | $\frac{a}{b} \leftarrow$ pente$_{\text{inf}} \oplus$ pente$_{\text{sup}}$ (new current slope) ;
  | **if** *not correct and $b \geqslant n$* **then**
  | | Return "Not a segment"

Return(a, b, μ)

Algorithm 2: Extremal rests

Input: a, b parameters of the slope tested, *test* set of points to be tested

Output: $p_{\min}, p_{\max}$ points of *test* having the minimum and maximum residues with the slope $\frac{a}{b}$.

Extremal-rests($test, a, b$):

$p_{\min}, p_{\max} \leftarrow$ first element of *test* ;

for p *in test* **do**
  | $p \leftarrow S[i]$;
  | **if** $\langle p, (a, -b) \rangle < \langle p_{\min}, (a, -b) \rangle$ **then**
  | | $p_{\min} \leftarrow p$;
  | **if** $\langle p, (a, -b) \rangle > \langle p_{\max}, (a, -b) \rangle$ **then**
  | | $p_{\max} \leftarrow p$;

Return($p_{\min}, p_{\max}$) ;

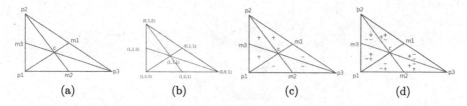

Fig. 3. (a) a step in the construction of the tree in dimension 3. The points p_1, p_2, p_3 are the extremities of the simplex. The points m_i represent the points $p_{i+1} + p_{i+2}$ (where the indices are taken modulo 3). Point c represents $p_1 + p_2 + p_3$. The permutation σ such that $\sigma(i) = (i+1) \mod 3$ then gives the simplex with extremities $p_{\sigma(1)} = p_2$, $p_{\sigma(1)} + p_{\sigma(2)} = p_2 + p_3 = m_1$ and $p_{\sigma(1)} + p_{\sigma(2)} + p_{\sigma(3)} = c$, i.e. the triangle p_2, m_1, c at the top right of the figure. (b) an example with initialisation at $p_i = e_i$. (c) illustration of a separation along p_1. In the sub-triangles marked $-$, the plane contains the chord $\delta_-^{(1)}$ but not the chord $\delta_+^{(1)}$. In the triangles marked $+$, the plane contains the chord $\delta_+^{(1)}$ but not the chord $\delta_-^{(1)}$. On the dividing line in green, neither chord appears. (d) partition of the triangle according to the chords appearing in the planes. In green, blue and red, the boundary lines. The signs then correspond to the chords appearing in the planes of each sub-zone. (Color figure online)

– the chord $\delta_+^{(i)}$ appears in $\mathbb{P}(v, \mu)$ iff $v \in \mathcal{C}(p_i, m_i, p_{i+1}) \backslash \mathcal{C}(p_i, m_i)$.

The $\delta_-^{(i)}$ and $\delta_+^{(i)}$ chords are then said to be separating.

By symmetry, we place ourselves in the case where the vector v has increasing positive coordinates. Thus, the chord δ appears in the plane of normal vector v iff $0 \leqslant |\langle v, \delta \rangle| < |\langle v, e_3 \rangle|$. Given extremities p_1, p_2, p_3, in order to find a separation for p_i, it suffices to place the following constraints on $(\delta_-^{(i)}, \delta_+^{(i)})$:

$$\begin{cases} \langle p_i, \delta_-^{(i)} \rangle = \langle p_i, e_3 \rangle \\ \langle m_i, \delta_-^{(i)} \rangle = \langle m_i, e_3 \rangle \\ \langle p_{i-1}, \delta_-^{(i)} \rangle = \langle p_{i-1}, e_3 \rangle - 1 \end{cases} \text{ and } \begin{cases} \langle p_i, \delta_+^{(i)} \rangle = \langle p_i, e_3 \rangle \\ \langle m_i, \delta_+^{(i)} \rangle = \langle m_i, e_3 \rangle \\ \langle p_{i+1}, \delta_+^{(i)} \rangle = \langle p_{i+1}, e_3 \rangle - 1 \end{cases}$$

The first two conditions for each system ensure that the limit of appearance of the chords $\delta_-^{(i)}$ and $\delta_+^{(i)}$ is the green line passing through p_i and m_i (see Fig. 3(c)). The last condition ensures that $\delta_-^{(i)}$ appears in the plane with normal vector p_{i-1} (resp. $\delta_+^{(i)}$ in the plane with normal vector p_{i+1}). The characterisation by equivalence of the separation $(\delta_-^{(i)}, \delta_+^{(i)})$ then follows immediately from these observations by convexity. Finally, let us add that the both systems each admit a unique solution because they have determinant ± 1 (by induction on the tree). By separating the zones according to each p_i, we obtain a partition of the triangle p_1, p_2, p_3 into 6 sub-triangles (see Fig. 3(d)).

Remark 11. *The condition* $\langle p_{i+1}, \delta_+^{(i)} \rangle = \langle p_{i+1}, e_3 \rangle - 1$ *can be replaced by* $\langle p_{i+1}, \delta_+^{(i)} \rangle = \xi$ *for any ξ in the interval* $[\![-(\langle p_{i+1}, e_3 \rangle - 1), \langle p_{i+1}, e_3 \rangle - 1]\!]$. *How-*

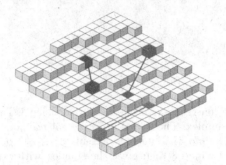

Fig. 4. Chords in the plane with normal vector $(7, 17, 57)$. The plane contains, among others, the chords $(7, 0, 0)$ (green), $(2, 3, -1)$ (blue) and $(6, 4, -2)$ (red). However, the plane does not contain the chords $(0, 0, 1)$ or $(0, 4, 0)$. (Color figure online)

ever, choosing $\xi = \langle p_{i+1}, e_3 \rangle - 1$ *results a priori in smaller separating chords and is therefore generally more relevant.*

The separations can therefore be used to characterise the membership of planes to certain cones. For example, we can look at the separation for the Stern-Brocot tree in dimension 2, with extremities 0 and $1/2$ and center $1/3$. After solving the system, we obtain $\delta_- = (3, 0)$ and $\delta_+ = (3, 2)$. The straight lines whose slope is between 0 and $1/2$ are therefore:

- in the interval $[0, 1/3[$ iff they contain the chord $(3, 0)$, i.e. a level of size at least 4.
- in the interval $]1/3, 1/2]$ iff they contain the chord $(3, 2)$, i.e. a level of size exactly 2
- the $1/3$ slope is the only one not to contain any of the 2 chords: the levels are all exactly 3 in size.

We can then return to the Stern-Brocot tree in dimension 2 with the various separations obtained for the interval $[0, 1]$, as in Fig. 5.

5   Recognition Algorithm in Arbitrary Dimension

Using the separations, we can choose the appropriate sub-triangle for descending the Stern-Brocot tree. We then obtain a recognition algorithm (see Algorithm 3).

Before looking at the tricky points of the algorithm (in bold), let's look at 2 runs of the algorithm on square pieces of planes with parameters $v = (4, 7, 11), \mu = 2$, size $= 12$ and $v = (7, 8, 10), \mu = 8$, size $= 10$ in Fig. 6 and 7. Note that the algorithm can also be applied to any piece of plan, not necessarily rectangular. The chords detected are indicated in the plane. The sub-triangle to descend to is indicated by a point. In the second example, the chords $(7, -10, 4)$

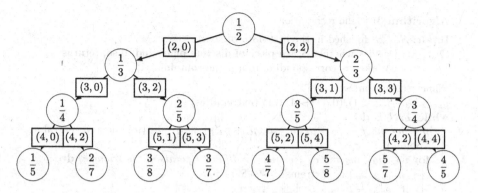

Fig. 5. Stern-Brocot tree labelled by separations

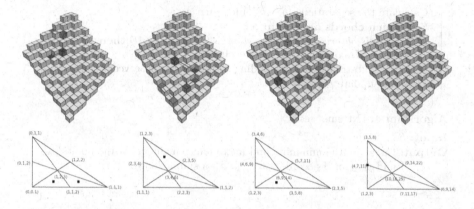

Fig. 6. Flow of the algorithm on the piece of square plane with normal vector $(4, 7, 11)$, shift 2 and size 12.

and $(9, -9, 0)$ belong to the whole $\mathbb{P}(v, \mu)$ plane but not to the piece shown, which is too small for this. The algorithm therefore fails and returns "Inconsistency in the chords", even though the piece is indeed that of a plane. Finally, the algorithm may only detect 2 chords and no third: this means that there is a dividing line and 2 possible sub-triangles. The algorithm presented here then chooses one of the 2 triangles, but it would also be possible to write a version that performs a Stern-Brocot in dimension 2 on the line in question.

The above algorithm has a few tricky points that need to be explained in detail:

In its current version, the algorithm necessarily stops, either because it reaches the parameters of the plane under consideration, or because the separating chords it has to deal with lead to inconsistency. In fact, their length increases at each stage, so they end up exceeding the size of the considered piece of plane. The "While" stopping condition deserves a simple bound on the coordinates of v. The leaning points algorithm presented in [15] allows us to obtain

Algorithm 3: Plane recognition

Input: $S \subset \mathbb{Z}^3$ finished in the first 48th of space

Output: Decides whether S is a piece of discrete plane and, if so, returns parameters corresponding to a plane containing it

Plane-recognition(S):

$p_1, p_2, p_3 \leftarrow (0,0,1), (0,1,1), (1,1,1)$ (extremities) ;

while *TRUE* do

    $m_1, m_2, m_3, c \leftarrow p_2 + p_3, p_1 + p_3, p_1 + p_2, p_1 + p_2 + p_3$ (middle and central points) ;

    for $v \in \{m_1, m_2, m_3, c\}$ *(testing the different points of the triangle)* **do**

        $(p_{\min}, p_{\max}) \leftarrow$ Extremal-rests(S, v) ;

        if $\langle p_{\max} - p_{\min}, v \rangle < \|v\|_\infty$ **then**

            $\mu \leftarrow - \langle p_{\min}, v \rangle$;

            Return(v, μ)

    Calculate the separations $(\delta_-^{(i)}, \delta_+^{(i)})$ for each p_i. ;

    See which chords appear in S ;

    if *the chords observed are inconsistent (see Figure 3 (d))* **then**

        Return(**"Inconsistency in the chords"**)

    Select the sub-triangle corresponding to the chords observed (see Figure 3 (d)) and update p_1, p_2, p_3 ;

Algorithm 4: Extremal rests

Input: $S \subset \mathbb{Z}^3$ finite, $v \in \mathbb{Z}^3$

Output: Calculate the minimum and maximum remainders of the points $p_{\min}$ and $p_{\max}$ of S according to v (with a simple loop).

a bound in dimension 3 for rectangular pieces of plane. The simplex algorithm used in [6] gives a bound in the general case, but much larger. This part of our algorithm is therefore easy to modify.

Detecting chords in a finite set is much more problematic. First of all, the naive method is not optimal (all pairs of points are tested). However, this stage can be improved by taking into account only the leaning points of each level (zone of constant z height), thus bringing us back to a constant number of points per level. Nevertheless, after a certain stage, the chords force us to leave the S subset under consideration. We therefore observe the same problem as with plane probing algorithms (see [13] for the first version). Generally speaking, the deeper the algorithm dives into the Stern-Brocot tree, the larger the separating chords become and the greater the risk of them leaving the study space. When the algorithm returns "Incoherence in the chords", there are two possible outcomes:

- the piece under consideration is not a piece of plane. Such an entry inevitably leads to the set of separating chords which appear in the plane not having an associated sub-triangle (see Fig. 3(d)), hence the inconsistency.
- the piece under consideration is indeed a piece of plane but is too small for some separating chords to appear on the piece. The incoherence of chords

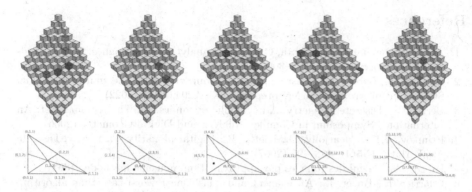

Fig. 7. Flow of the algorithm on the piece of square plane with normal vector $(7, 8, 10)$, shift 2 and size 10.

detected then simply means that the algorithm cannot conclude without considering more points.

Note that, on a sufficiently large piece of plane (a fortiori on an infinite plane), the second case mentioned cannot occur. Therefore, the reference "Incoherence in chords" implies that the input is not a plane.

The set of solution parameters of the problem forms a convex. In dimension 2, this property demonstrates that the algorithm will necessarily stop at the minimal parameters, and that there is uniqueness of the descent branch in the algorithm. In dimension 3, a convex may not intersect any remarkable point of the triangle and may straddle several zones. There are therefore configurations where the zone to be chosen for descent is not unique.

6   Conclusion

This paper presents a new approach for detecting pieces of arithmetic hyperplanes from the Stern-Brocot tree. In dimension 2, for DSS, the proposed algorithm is incremental and has linear complexity. The Stern-Brocot tree can also be used to characterise the various slope intervals of discrete straight lines using separating chords. The extension of the algorithm to dimensions 3 and higher retains the idea of separations, with greater combinatorial complexity. Unfortunately, although they appear in the infinite planes studied, the chords considered may be too large to be recognisable in a piece of plane. Even if this algorithm deserves a more accomplished version, it still creates a deep link between arithmetic hyperplanes and the generalised Stern-Brocot tree, via the separating chords.

In our opinion, the notion of separating chord deserves to be explored in greater depth, in particular to make more explicit the fact that hyperplanes belong to Stern-Brocot cones. In addition, as the enumeration of the tree vertices is governed by a simple induction relation, we would like in a future work to take an interest in the induction relations induced on the set of separating chords.

References

1. Andres, E., Acharya, R., Sibata, C.: Discrete analytical hyperplanes. Graph. Models Image Process. **59**(5), 302–309 (1997)
2. Barbieri, S., Labbé, S.: Indistinguishable asymptotic pairs and multidimensional Sturmian configurations. arXiv preprint arXiv:2204.06413 (2022)
3. Berthé, V.: Discrete geometry and symbolic dynamics. In: The Kiselmanfest: An International Symposium in Complex Analysis and Digital Geometry (2006)
4. Brimkov, V.E., Coeurjolly, D., Klette, R.: Digital planarity - a review. Discrete Appl. Math. **155**(4), 468–495 (2007)
5. Brocot, A.: Calcul des rouages par approximation: nouvelle méthode (1862)
6. Charrier, E., Buzer, L.: An efficient and quasi linear worst-case time algorithm for digital plane recognition. In: Coeurjolly, D., Sivignon, I., Tougne, L., Dupont, F. (eds.) DGCI 2008. LNCS, vol. 4992, pp. 346–357. Springer, Heidelberg (2008). https://doi.org/10.1007/978-3-540-79126-3_31
7. Debled-Rennesson, I.: Etude et reconnaissance des droites et plans discrets. Ph.D. thesis, Université Louis Pasteur (Strasbourg) (1971–2008) (1995)
8. Debled-Rennesson, I., Reveillès, J.-P.: A linear algorithm for segmentation of digital curves. Int. J. Pattern Recogn. Artif. Intell. **9**(4), 635–662 (1995)
9. Debled, I., Reveillès, J.P.: An incremental algorithm for digital plane recognition. In: 4th International Conference DGCI 1994 (1994)
10. Fernique, T.: Pavages, fractions continues et géométrie discrete. Ph.D. thesis, Université Montpellier II-Sciences et Techniques du Languedoc (2007)
11. Gérard, Y., Debled-Rennesson, I., Zimmermann, P.: An elementary digital plane recognition algorithm. Discrete Appl. Math. **151**(1–3), 169–183 (2005)
12. Klette, R., Rosenfeld, A.: Digital straightness - a review. Discrete Appl. Math. **139**(1–3), 197–230 (2004)
13. Lachaud, J.-O., Provençal, X., Roussillon, T.: An output-sensitive algorithm to compute the normal vector of a digital plane. Theor. Comput. Sci. **624**, 73–88 (2016)
14. Lennerstad, H.: The n-dimensional Stern-Brocot tree (2012)
15. Mesmoudi, M.M.: A simplified recognition algorithm of digital planes pieces. In: Braquelaire, A., Lachaud, J.O., Vialard, A. (eds.) DGCI 2002. LNCS, vol. 2301, pp. 404–416. Springer, Cham (2002). https://doi.org/10.1007/3-540-45986-3_36
16. Morse, M., Hedlund, G.A.: Symbolic dynamics II. Sturmian trajectories. Am. J. Math. **62**(1), 1–42 (1940)
17. Reveillès, J.-P.: Géométrie discrete, calcul en nombres entiers et algorithmique. Ph.D. thesis, Université Louis Pasteur (1991)
18. Stern, M.: Über eine zahlentheoretische funktion (1858)
19. De Vieilleville, F., Lachaud, J.-O.: Revisiting digital straight segment recognition. In: Kuba, A., Nyúl, L.G., Palágyi, K. (eds.) DGCI 2006. LNCS, vol. 4245, pp. 355–366. Springer, Cham (2006). https://doi.org/10.1007/11907350_30

Bijective Digitized 3D Rotation Based on Beam Shears

Lidija Čomić[1]([✉]) [iD], Rita Zrour[2] [iD], Eric Andres[2] [iD],
and Gaëlle Largeteau-Skapin[2] [iD]

[1] University of Novi Sad, Faculty of Technical Sciences, Novi Sad, Serbia
`comic@uns.ac.rs`
[2] University of Poitiers, Laboratory XLIM, ASALI, UMR CNRS, 7252, Futuroscope,
BP 30179, 86962 Chasseneuil, France
`{rita.zrour,eric.andres,gaelle.largeteau.skapin}@univ-poitiers.fr`

Abstract. Each 3D rotation can be decomposed into three 2D rotations parametrized by three Euler angles. Each of the three 2D rotations can be expressed as a sequence of three 2D shears along coordinate axes, leading to a decomposition of a 3D rotation into nine (beam) shears in total. We define a 3D digitized rotation using nine digitized beam shears, i.e., we round the result of each shear before applying the next one. As digitized shears are bijective, our 3D digitized rotation inherits the same property. Experiments show that the average error of our digitized rotation compared to the continuous one is kept under 1 (around 0.8).

Keywords: Digital geometry · Bijective digitized rotation · Shear · Beam shear

1 Introduction

The ability to apply rigid motions (rotation composed with a translation) to discrete objects is essential in discrete geometry and digital image analysis and processing. This is especially true since most geometric transformations in the discrete world can be decomposed into a sequence of rigid motions. This supposes however that we have digitized rigid motions that are bijective or otherwise we would loose points at each step of the decomposition and bijectivity is an important property to seek anyway. The obvious way to define a digitized rigid motion is to apply the continuous Euclidean transformation on each integer point and digitize it by rounding the point coordinates. This method produces the least error, but in general it is non-bijective, non-invertible, and uses floating point arithmetic.

In 2D, there are angles for which the (continuous) rotated image of a digital point is again a digital point [3,9,14]. These angles do not cover all possible angles however. In 3D, we have reasons to believe that except 2D rotations lifted to 3D, there are no other digitized rotations that are bijective [4].

© The Author(s), under exclusive license to Springer Nature Switzerland AG 2024
S. Brunetti et al. (Eds.): DGMM 2024, LNCS 14605, pp. 29–40, 2024.
https://doi.org/10.1007/978-3-031-57793-2_3

In 2D, a feasible way to achieve bijectivity is to decompose a digitized rotation into a sequence of three digitized shears [1]. We consider a 3D rotation as a sequence of three 2D rotations in the (planes parallel to the) coordinate planes about an arbitrary point, not necessarily the origin. We use the decomposition of each of the three 2D rotations into a sequence of three shears and we define the 3D digitized rotation as a composition of nine digitized shears: we apply the continuous shears and round the obtained points after each shear.

We compute the average and maximal distance between the continuous and the digitized rotation. Our experiments show that the average distance between the continuously rotated point and its discrete counterpart is always less than 1 and in general close to 0.8. The maximal distance between a continuously rotated point and its discrete counterpart is around 2.

At the end of the paper, we briefly discuss another shear-based approach proposed in the continuous case by Chen and Kaufman [6] and explain why the digitization of this solution poses difficulties that we hope to overcome in the future.

2 Preliminaries

We briefly review various types of shears defined in the literature, ways to express rotations through shears and related work on bijective digitized rotations.

2.1 Shears

In 2D, a (2×2) shear matrix has 1's on the mail diagonal and one other non-zero (off-diagonal) element.

In 3D, various types of shears have been defined, mainly with the aim to minimize the number of factors in the decomposition of the rotation and more general affine transformations. For rotations, the shear matrices have 1's on the main diagonal and

- one other non-zero coefficient for beam shears [6];
- two other non-zero coefficients in one column for plane or slice shears [6];
- three other non-zero coefficients above or below the main diagonal for general (unrestricted) shears [17].

A 3D rotation can then be expressed as a product of nine beam shears, and a minimum number of four plane shears [6] or three unrestricted shears [17]. Alternatively, a 3D rotation can be decomposed into only two pseudoshears, which include scaling as their matrices are upper or lower triangular [7]. Thus, with pseudoshears, bijectivity is sacrificed for minimality. For 3D affine transformations (which include also reflection, scaling and translation), a decomposition into three matrices (in homogenous coordinates) which differ from the identity matrix in one row, has also been proposed [8].

2.2 Continuous Rotations Through Beam Shears

A rotation $R_{xy}(\psi)$ in the xy plane about the origin by an angle $\psi \neq (2k+1)\pi$, $k \in \mathbb{Z}$, can be decomposed into three (beam) shears [11,15] along e.g. x, y, x axes respectively as

$$R_{xy}(\psi) = \begin{bmatrix} \cos\psi & -\sin\psi \\ \sin\psi & \cos\psi \end{bmatrix} = S_x(a)S_y(b)S_x(a) = \begin{bmatrix} 1 & a \\ 0 & 1 \end{bmatrix}\begin{bmatrix} 1 & 0 \\ b & 1 \end{bmatrix}\begin{bmatrix} 1 & a \\ 0 & 1 \end{bmatrix}.$$

Multiplying the shear matrices and equating gives $a = -\tan\frac{\psi}{2}$, $b = \sin\psi$, i.e.,

$$R_{xy}(\psi) = \begin{bmatrix} 1 & -\tan\frac{\psi}{2} \\ 0 & 1 \end{bmatrix}\begin{bmatrix} 1 & 0 \\ \sin\psi & 1 \end{bmatrix}\begin{bmatrix} 1 & -\tan\frac{\psi}{2} \\ 0 & 1 \end{bmatrix}.$$

In 3D, each rotation can be parametrized in terms of the Euler angles α, θ, ϕ about the axes e.g. z, y, z, respectively [17] as $R(\alpha, \theta, \phi) = R_z(\alpha)R_y(\theta)R_z(\phi)$. We use the right coordinate system, so the rotation about the (positive) y-axis by angle θ is equivalent to the rotation by the angle $-\theta$ in the xz plane. Thus, a 3D rotation about an axis through the origin can be written as:

$$R(\alpha, \theta, \phi) = \begin{bmatrix} \cos\alpha & -\sin\alpha & 0 \\ \sin\alpha & \cos\alpha & 0 \\ 0 & 0 & 1 \end{bmatrix}\begin{bmatrix} \cos\theta & 0 & \sin\theta \\ 0 & 1 & 0 \\ -\sin\theta & 0 & \cos\theta \end{bmatrix}\begin{bmatrix} \cos\phi & -\sin\phi & 0 \\ \sin\phi & \cos\phi & 0 \\ 0 & 0 & 1 \end{bmatrix},$$

and (for $\alpha, \theta, \phi \neq (2k+1)\pi$, $k \in \mathbb{Z}$) each of the three 2D rotations about coordinate axes can be expressed through three beam shears as

$$R_z(\alpha) = \begin{bmatrix} 1 & -\tan\frac{\alpha}{2} & 0 \\ 0 & 1 & 0 \\ 0 & 0 & 1 \end{bmatrix}\begin{bmatrix} 1 & 0 & 0 \\ \sin\alpha & 1 & 0 \\ 0 & 0 & 1 \end{bmatrix}\begin{bmatrix} 1 & -\tan\frac{\alpha}{2} & 0 \\ 0 & 1 & 0 \\ 0 & 0 & 1 \end{bmatrix}$$

$$R_y(\theta) = \begin{bmatrix} 1 & 0 & \tan\frac{\theta}{2} \\ 0 & 1 & 0 \\ 0 & 0 & 1 \end{bmatrix}\begin{bmatrix} 1 & 0 & 0 \\ 0 & 1 & 0 \\ -\sin\theta & 0 & 1 \end{bmatrix}\begin{bmatrix} 1 & 0 & \tan\frac{\theta}{2} \\ 0 & 1 & 0 \\ 0 & 0 & 1 \end{bmatrix}$$

$$R_z(\phi) = \begin{bmatrix} 1 & -\tan\frac{\phi}{2} & 0 \\ 0 & 1 & 0 \\ 0 & 0 & 1 \end{bmatrix}\begin{bmatrix} 1 & 0 & 0 \\ \sin\phi & 1 & 0 \\ 0 & 0 & 1 \end{bmatrix}\begin{bmatrix} 1 & -\tan\frac{\phi}{2} & 0 \\ 0 & 1 & 0 \\ 0 & 0 & 1 \end{bmatrix}$$

giving a final decomposition of a 3D rotation into a sequence of nine beam shears in total.

2.3 Related Work on Digitized Bijective Rotations

Digitized rigid transformations, and especially digitized rotations, received a considerable amount of attention in the literature, including expressing them through digitized shears in 2D (first used in antialiasing rotations [11] and later adapted to bijective rotations [1]) or 3D [6,17], various types of decompositions [2,13], certifying the bijectivity of 2D [10,14] or 3D [12] digitized rational rotations, decomposing the interval $(-\pi, \pi]$ into subintervals (by hinge angles) such that the rotations by any two angles in the same subinterval produce the same result for the given rotation line and the given image space [16].

In a recent work, S. Breuils et al. made the hypothesis that the only digitized bijective 3D rotations are 2D rotations lifted up to 3D [4]. If the conjecture is confirmed, it would mean that a direct approach to bijective digitized rotations will not work.

3   Digitized Shear-Based Rotation in 2D

As opposed to (linear) shear which leaves fixed one of the coordinate axes, (affine) horizontal and vertical quasi-shears [1] leave fixed a horizontal and vertical line $y = y_0$ and $x = x_0$, respectively. In the continuous case, they are defined by

$$CQH(a, y_0) : (x, y) \rightarrow (x + a(y - y_0), y)$$
$$CQV(b, x_0) : (x, y) \rightarrow (x, y + b(x - x_0))$$

Choosing $a = -\tan \frac{\psi}{2}$, $b = \sin \psi$ gives a decomposition of the 2D rotation $R_{xy}(\psi, (x_0, y_0))$ about the center $(x_0, y_0) \in \mathbb{R}^2$ by angle $\psi \neq (2k + 1)\pi$, $k \in \mathbb{Z}$, into three quasi-shears, i.e.,

$$R_{xy}(\psi, (x_0, y_0)) = CQH(a, y_0)CQV(b, x_0)CQH(a, y_0).$$

In [1], a more complicated method has been proposed to handle the non-integer center, which involves only integer parameters. Here, we consider a more direct approach, which leads to the same error measures. We define the digitized quasi-shears by rounding the coordinates at each step:

$$HQS(a, y_0) : (x, y) \rightarrow (x + Round(a(y - y_0)), y)$$
$$VQS(b, x_0) : (x, y) \rightarrow (x, y + Round(b(x - x_0)))$$

where $Round(t) = -sgn(t)\lfloor -|t| + 0.5 \rfloor$ denotes (symmetric) rounding half towards zero, and we define the digitized 2D rotation about an arbitrary point $(x_0, y_0) \in \mathbb{R}^2$ through three digitized quasi-shears

$$QSR(\psi, x_0, y_0) = HQS(-\tan \frac{\psi}{2}, y_0)VQS(\sin \psi, x_0)HQS(-\tan \frac{\psi}{2}, y_0)$$

as illustrated in Algorithm 3 (shearRot2D).

The three shears define a rotation, however one of the coefficients in the shear matrices is equal to $-\tan \frac{\psi}{2}$ which means that it is not defined for $\psi = (2k+1)\pi$, $k \in \mathbb{Z}$, and moreover, the value of this coefficient grows the closer we get to $(2k + 1)\pi$. There lies a problem when considering a sequence of several digitized shears. To explain the problem, let us consider the shear $x' = x + ay$ with for instance $a = 100.0$, $x = 0.0$ and $y = 0.45$. Then we have $x' = 45.0$ for the continuous shear. Now, in the digitized version, we'd have still $a = 100.0$, $x = 0$ but, if y is the result obtained by the previous shear, after rounding we get $y = 0$ and consequently $x' = 0$. The (small) difference between the continuous value

---

Algorithm 1: shearRot2D($pt, \psi, (x_0, y_0)$)

Input : initial point $pt(x, y)$, rotation angle ψ, rotation center (x_0, y_0)
Output: rotated point (x_r, y_r)

1 $x_1 = x - Round((y - y_0) \tan \frac{\psi}{2})$;
2 $y_r = y + Round((x_1 - x_0) \sin \psi)$;
3 $x_r = x_1 - Round((y_r - y_0) \tan \frac{\psi}{2})$;
4 **return** (x_r, y_r)

Algorithm 2: normalize(ψ)

Input : initial angle ψ
Output: normalized angle ψ_n, quadrant q

1 $\psi_n = \psi - 2\pi \left\lfloor \frac{\psi + \frac{\pi}{4}}{2\pi} \right\rfloor$;
2 $q = \left\lfloor \frac{1}{2} + \frac{2\psi_n}{\pi} \right\rfloor$ // q is the quadrant of ψ;
3 $\psi_n = \psi_n - q\frac{\pi}{2}$ // $\psi_n \in [-\pi/4, \pi/4]$ differs from ψ by $k\pi/2$, $k \in \mathbb{Z}$;
4 **return** ψ_n, q

---

and the digitized counterpart of the input value of y leads to important error of the output x'. To control the error, the absolute values of the coefficients in the shear matrices must be small, and not exceed 1.0. This is one of the problems that makes the $3D$ plane-shear solution problematic (see Sect. 6).

The absolute value of the coefficient $-\tan \frac{\psi}{2}$ is smaller than 1.0 for angles $\psi \in [-\pi/4, \pi/4]$. To overcome this problem, the shear-based rotation algorithm [1] normalizes the angle ψ by expressing it as $\psi = \psi_n + k\pi/2$, $\psi_n \in [-\pi/4, \pi/4]$ (Algorithm 2 (normalize)), rotates by the normalized angle ψ_n, and then by angle $k\pi/2$ (or vice versa, as rotations in 2D with the same center are commutative). This last rotation amounts to simple permutation and/or sign change of coordinate values, as illustrated in Algorithm 3 (symmetry). For angles in $[-\pi/4, \pi/4]$, the maximum distance between a point p and the point p' obtained by rotating p (using three digitized shears) and then rotating it back (using inverse digitized shears in reverse order) is at most $\sqrt{5}/2$ [5].

4   Digitized Beam-Shear-Based Rotation in 3D

We define and test the digitized 3D rotation based on applying nine 2D quasi-shears and rounding at each step. To control the error, we use the normalization of the three Euler angles α, θ, ϕ. Algorithm 4 (shearRot3D) rotates a single 3D point. After normalizing the angles, the 2D quasi-shear-based rotation algorithm (Algorithm 1 (shearRot2D)) is called three times, to rotate the point about z, y, z axes, respectively. In the right-handed coordinate system that we use, for rotation about the y-axis by angle θ, Algorithm 2 (shearRot2D) is called with

---

Algorithm 3: symmetry($pt, q, (x_0, y_0)$)

    Input   : input point $pt(x, y)$, the quadrant q, rotation center (x_0, y_0)
    Output: rotated point by $q\pi/2$, (x_{new}, y_{new})

1 $xtemp = x - x_0$; $ytemp = y - y_0$;
2 case q:
3 1: $temp = xtemp$; $xtemp = -ytemp$; $ytemp = temp$;
4 2: $xtemp = -xtemp$; $ytemp = -ytemp$;
5 3: $temp = xtemp$; $xtemp = ytemp$; $ytemp = -temp$;
6 $x_{new} = xtemp + x_0$;
7 $y_{new} = ytemp + y_0$;
8 **return** (x_{new}, y_{new})

Algorithm 4: shearRot3D($pt, (\alpha_n, q_\alpha), (\theta_n, q_\theta), (\phi_n, q_\phi), (x_0, y_0, z_0)$)

    Input   : initial point $pt(x, y, z)$, normalized rotation angles and their quadrants
           $(\alpha_n, q_\alpha), (\theta_n, q_\theta), (\phi_n, q_\phi)$ and the rotation center (x_0, y_0, z_0)
    Output: rotated point (x_r, y_r, z_r)

1 (x_1, y_1)=shearRot2D($(x, y), \phi_n, (x_0, y_0)$) ;
2 (xf_1, yf_1)=symmetry($(x_1, y_1), q_\phi, (x_0, y_0)$) ;
3 (x_2, z_2)=shearRot2D($(xf_1, z), -\theta_n, (x_0, z_0)$);
4 (xf_2, z_r)=symmetry($(x_2, z_2), -q_\theta \pmod 4, (x_0, y_0)$) ;
5 (x_3, y_3)=shearRot2D($(xf_2, yf_1), \alpha_n, (x_0, y_0)$) ;
6 (x_r, y_r)=symmetry($(x_3, y_3), q_\alpha, (x_0, y_0)$) ;
7 **return** (x_r, y_r, z_r)

Algorithm 5: Final3DRotation(Initial_image, (α, θ, ϕ), (x_0, y_0, z_0)): rotation of a 3D image by angle (α, θ, ϕ) with rotation center (x_0, y_0, z_0)

    Input   : Initial_image, α, θ, ϕ, rotation center (x_0, y_0, z_0)
    Output: Final_image

1 (α_n, q_α)=normalize(α);
2 (θ_n, q_θ)=normalize(θ);
3 (ϕ_n, q_ϕ)=normalize(ϕ);
4 **foreach** *point pt(x,y,z) of Initial_image* **do**
5   (x_r, y_r, z_r)=shearRot3D($pt, (\alpha_n, q_\alpha), (\theta_n, q_\theta), (\phi_n, q_\phi), (x_0, y_0, z_0)$);
6   write (x_r, y_r, z_r) to Final_image;
7 **return** *Final_image*

---

the second (angle) argument equal to $-\theta$, and Algorithm 3 (symmetry) is called with the second (quadrant) argument equal to $-q \pmod 4$. To rotate an object, the final algorithm (Algorithm 5 (Final3DRotation)) is applied on each of its points.

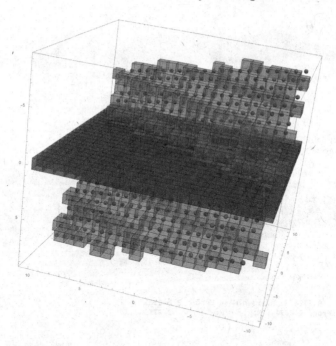

Fig. 1. Rotation of the xy plane (in blue) by angles $\alpha = \frac{5\pi}{3}$, $\theta = \frac{3\pi}{4}$ and $\phi = \frac{-2\pi}{5}$. The output of Algorithm 5 is in green, and the continuously rotated rounded plane is in red. (Color figure online)

Figure 1 presents an example of a rotated plane. The original plane is the xy plane in blue voxels, the result of our shear-based digitized rotation is in green voxels and the continuous rounded result corresponds to the red spheres.

5   Results and Error Measures

To evaluate the quality of our bijective digitized rotation we compute the distance between the digitized rotated point obtained by Algorithm 4 (shearRot3D) and the continuous rotated one.

5.1   Error Measure Depending on Angle

In order to measure the error depending on angle, the data are generated as follows: 1000 discrete points are chosen randomly in $[-10000, 10000]^3$, a center of rotation is chosen randomly in $[-0.5, 0.5]^3$. The average of the average errors and the average of the maximum errors are computed on ten values of ϕ chosen randomly (between 0 and 2π), while varying α and θ between 0 and 2π by a step of 0.2. At each step, a random center of rotation is chosen in $[-0.5, 0.5]^3$. Figure 2 (first line) shows the results of the error measure depending on angle.

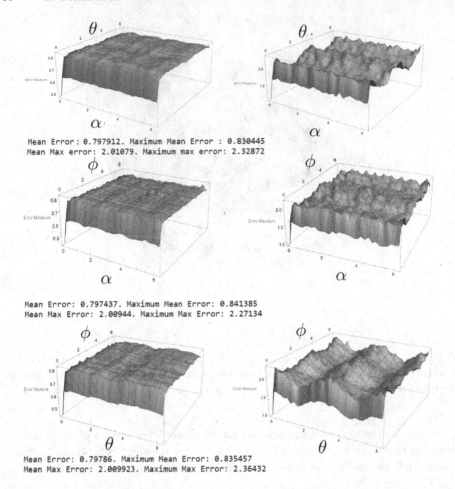

Mean Error: 0.797912. Maximum Mean Error : 0.830445
Mean Max error: 2.01079. Maximum max error: 2.32872

Mean Error: 0.797437. Maximum Mean Error: 0.841385
Mean Max Error: 2.00944. Maximum Max Error: 2.27134

Mean Error: 0.79786. Maximum Mean Error: 0.835457
Mean Max Error: 2.009923. Maximum Max Error: 2.36432

Fig. 2. Error measure depending on angle. The average of the average errors and the average of the maximum errors are computed on ten values of ϕ (first line), θ (second line) or α (third line), chosen randomly (between 0 and 2π), while varying the other two angles between 0 and 2π by a step of 0.2.

The same tests have been done by computing the average of the average errors and the average of the maximum errors on ten values of θ chosen randomly (between 0 and 2π) while varying ϕ and α between 0 and 2π by a step of 0.2 (Fig. 2 second line) and then on ten values of α chosen randomly (between 0 and 2π) while varying ϕ and θ between 0 and 2π by a step of 0.2 (Fig. 2 third line).

5.2   Error Measure Depending on Rotation Axis

In order to measure the error depending on rotation axis, the data are generated as follows: 1000 points are chosen randomly in $[-10000, 10000]^3$, then 3000 directions are chosen randomly around a surface of a sphere. For each direction (rotation axis), we consider a range of angles between 0 and 2π by a step of 0.1. For each couple (axis, angle), we choose a random center in $[-0.5, 0.5]^3$. From the axis and angle, we compute the three corresponding Euler angles. We then apply both our shear based digitized rotation and the continuous rotation on the initial set of points. The error is measured as the mean and maximum Euclidean distances between both results on the 1000 points.

The results are displayed by drawing for each axis the average of the average errors for all angles (Fig. 3 (a)) and the average of the maximum errors for all angles (Fig. 3 (b)) and encoding them as the distance from a point in the axis direction to the origin. The obtained interpolated surface looks like a sphere, showing that the error is uniform along different directions of rotation axes and different rotation angles, with no specific values with error peaks. Figure 4 shows both the average of the average errors in red and the average of the maximum errors in green for all angles. One can also take the maximum of the average errors over all directions and the maximum of the maximum errors over all directions that gives for the described data 0.82 for the average error and 2.19 for the maximum error.

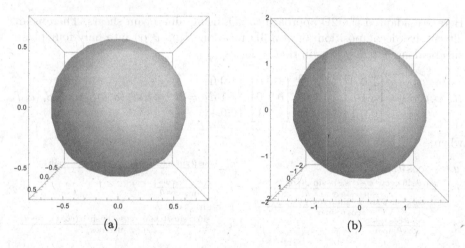

(a)   (b)

Fig. 3. Error measure depending on rotation axis. (a) The average of the average errors and (b) the average of the maximum errors, encoded as the distance (in the corresponding direction) from the origin. The shading is only for visualization purposes.

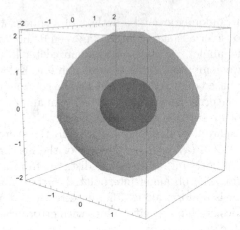

Fig. 4. The same errors as in Fig. 3 with merged (a) (in red) and (b) (in green). (Color figure online)

6 Discussion

Here, we adapted the 2D approach to 3D, using nine beam shears. There is an alternative decomposition [6] of a 3D rotation $R_3(\varphi, \theta, \alpha)$ into only four plane shears along y, z, x, y axes in that order as

$$R_3(\varphi, \theta, \alpha) = \begin{bmatrix} 1 & 0 & 0 \\ g & 1 & h \\ 0 & 0 & 1 \end{bmatrix} \begin{bmatrix} 1 & a & b \\ 0 & 1 & 0 \\ 0 & 0 & 1 \end{bmatrix} \begin{bmatrix} 1 & 0 & 0 \\ 0 & 1 & 0 \\ e & f & 1 \end{bmatrix} \begin{bmatrix} 1 & 0 & 0 \\ c & 1 & d \\ 0 & 0 & 1 \end{bmatrix} = S_y(g, h) S_x(a, b) S_z(e, f) S_y(c, d)$$

where

$$a = -\cos\theta \sin\alpha \qquad b = \tfrac{\cos\theta \sin\alpha + \sin\varphi \sin\theta \cos\alpha - \cos\varphi \sin\alpha}{\sin\varphi \cos\theta}$$

$$c = \tfrac{\sin\theta \sin\alpha(\cos\varphi - \cos\theta) + \sin\varphi(\cos\theta - \cos\alpha)}{\sin\varphi \sin\alpha \cos^2\theta} \qquad d = \tfrac{\cos\varphi \cos\theta - 1}{\sin\varphi \cos\theta}$$

$$e = -\tfrac{\cos\varphi \sin\theta \sin\alpha - \sin\varphi \cos\alpha + \sin\varphi \cos\theta}{\cos\theta \sin\alpha} \qquad f = \sin\varphi \cos\theta$$

$$g = -\tfrac{\cos\theta \cos\alpha - 1}{\cos\theta \sin\alpha} \qquad h = \tfrac{\sin\varphi \sin\theta(\cos\theta - \cos\alpha) + \sin\alpha(\cos\varphi - \cos\theta)}{\sin\varphi \sin\alpha \cos^2\theta}$$

For 3D digitized rotation, one can, after each shear, perform a rounding before applying the next shear, i.e.,

$$Round(S_y(g, h)) Round(S_x(a, b)) Round(S_z(e, f)) Round(S_y(c, d)).$$

The problem with this approach is that some coefficients may be arbitrarily large. As explained at the end of Sect. 3, this may induce large errors due to the successive roundings. We established sectors where all the coefficients of the shear matrices are within acceptable bounds, however as 3D rotations are not commutative, normalizing all three rotation angles ϕ, θ, α at once to intervals with small error is not possible. One attempt would be, instead of fixing the order of shear axes, to find for each triplet of rotation angles the (best) order of four shears, thus eliminating the need to consider special angles where the coefficients are large or undefined. This is a future possibility for extending the current work. One of the main advantages of using only four plane shears is that, due to the smaller number of roundings, the rotation error will be smaller than the error using the nine beam shears.

Acknowledgement. This research was partially supported by the Science Fund of the Republic of Serbia, #GRANT No 7632, Project "Mathematical Methods in Image Processing under Uncertainty" - MaMIPU.

References

1. Andres, E.: The quasi-shear rotation. In: Miguet, S., Montanvert, A., Ubéda, S. (eds.) DGCI 1996. LNCS, vol. 1176, pp. 307–314. Springer, Heidelberg (1996). https://doi.org/10.1007/3-540-62005-2_26
2. Andres, E., Dutt, M., Biswas, A., Largeteau-Skapin, G., Zrour, R.: Digital two-dimensional bijective reflection and associated rotation. In: Couprie, M., Cousty, J., Kenmochi, Y., Mustafa, N. (eds.) DGCI 2019. LNCS, vol. 11414, pp. 3–14. Springer, Cham (2019). https://doi.org/10.1007/978-3-030-14085-4_1
3. Andres, E., Largeteau-Skapin, G., Zrour, R.: Shear based bijective digital rotation in hexagonal grids. In: Lindblad, J., Malmberg, F., Sladoje, N. (eds.) DGMM 2021. LNCS, vol. 12708, pp. 217–228. Springer, Cham (2021). https://doi.org/10.1007/978-3-030-76657-3_15
4. Breuils, S., Kenmochi, Y., Andres, E., Sugimoto, A.: Conjecture on characterisation of bijective 3D digitized reflections and rotations. In: Hitzer, E., Papagiannakis, G., Vasik, P. (eds.) Empowering Novel Geometric Algebra for Graphics and Engineering. ENGAGE 2022. LNCS, vol. 13862, pp. 41–53. Springer, Cham (2022). https://doi.org/10.1007/978-3-031-30923-6_4
5. Carstens, H., Deuber, W., Thumser, W., Koppenrade, E.: Geometrical bijections in discrete lattices. Comb. Probab. Comput. **8**(1–2), 109–129 (1999)
6. Chen, B., Kaufman, A.E.: 3D volume rotation using shear transformations. Graph. Model. **62**(4), 308–322 (2000)
7. Chen, B., Kaufman, A.E.: Two-pass image and volume rotation (ST). In: 2nd IEEE TCVG/Eurographics International Workshop on Volume Graphics, VG, pp. 329–344. Eurographics Association (2001)
8. Hanrahan, P.: Three-pass affine transforms for volume rendering. In: Proceedings of the 1990 Workshop on Volume Visualization, VVS, pp. 71–78. ACM (1990)
9. Jacob, M.-A., Andres, E.: On discrete rotations. In: Discrete Geometry for Computer Imagery, pp. 161–174 (1995)
10. Nouvel, B., Rémila, E.: Characterization of bijective discretized rotations. In: Klette, R., Žunić, J. (eds.) IWCIA 2004. LNCS, vol. 3322, pp. 248–259. Springer, Heidelberg (2004). https://doi.org/10.1007/978-3-540-30503-3_19
11. Paeth, A.W.: A fast algorithm for general raster rotation. In: Graphic Interface 86 (reprinted with corrections in Graphic Gems (Glassner Ed.) Academic 1990), pp. 179–195 (1986). pages 77–81
12. Pluta, K., Romon, P., Kenmochi, Y., Passat, N.: Bijectivity certification of 3D digitized rotations. In: Bac, A., Mari, J.-L. (eds.) CTIC 2016. LNCS, vol. 9667, pp. 30–41. Springer, Cham (2016). https://doi.org/10.1007/978-3-319-39441-1_4

13. Richard, A., Fuchs, L., Largeteau-Skapin, G., Andres, E.: Decomposition of nD-rotations: classification, properties and algorithm. Graph. Models **73**(6), 346–353 (2011)
14. Roussillon, T., Coeurjolly, D.: Characterization of bijective discretized rotations by Gaussian integers. Research report, LIRIS UMR CNRS 5205, January 2016
15. Tanaka, A., Kameyama, M., Kazama, S., Watanabe, O.: A rotation method for raster image using skew transformation. In: Proceedings of the IEEE Conference on Computer Vision and Pattern Recognition, pp. 272–277 (1986)
16. Thibault, Y., Sugimoto, A., Kenmochi, Y.: 3D discrete rotations using hinge angles. Theor. Comput. Sci. **412**(15), 1378–1391 (2011)
17. Toffoli, T., Quick, J.: Three-dimensional rotations by three shears. CVGIP Graph. Models Image Process. **59**(2), 89–95 (1997)

New Characterizations of Full Convexity

Fabien Feschet[1] and Jacques-Olivier Lachaud[2](✉)

[1] Université Clermont Auvergne, CNRS, ENSMSE,
LIMOS, 63000 Clermont-Ferrand, France
`fabien.feschet@uca.fr`
[2] Université Savoie Mont Blanc, CNRS, LAMA, 73000 Chambéry, France
`jacques-olivier.lachaud@univ-smb.fr`

Abstract. *Full convexity* has been recently proposed as an alternative definition of digital convexity. In contrast to classical definitions, fully convex sets are always connected and even simply connected whatever the dimension, while remaining digitally convex in the usual sense. Several characterizations were proposed in former works, either based on lattice intersection enumeration with several convex hulls, or using the idempotence of an envelope operator. We continue these efforts by studying simple properties of real convex sets whose digital counterparts remain largely misunderstood. First we study if we can define full convexity through variants of the usual continuous convexity via *segments* inclusion, i.e. "for all pair of points of X, the straight segment joining them must lie within the set X". We show an equivalence of full convexity with this segment convexity in dimension 2, and counterexamples starting from dimension 3. If we consider now d-simplices instead of a segment (2-simplex), we achieve an equivalence in arbitrary dimension d. Secondly, we exhibit another characterization of full convexity, which is recursive with respect to the dimension and uses simple axis projections. This latter characterization leads to two immediate applications: a proof that digital balls are indeed fully convex, and a natural progressive measure of full convexity for arbitrary digital sets.

Keywords: Digital geometry · Digital convexity · Full convexity

1 Introduction

Convexity is a fundamental tool for analyzing functions and shapes. For digital spaces $\mathbb{Z}^d$, digital convexity was first defined as the intersection of real convex sets of $\mathbb{R}^d$ with $\mathbb{Z}^d$ (e.g. see survey [11]). Although easy to formulate, resulting digital convex sets may not be digitally connected in general. This deficiency prevents local shape analysis, so many works have tried to constrain the connectedness of such sets, for instance by relying on digital lines [1,5] or extensions of digital functions [6,7]. Most works are limited to 2D, and 3D extensions do not solve

This work was partly funded by STABLEPROXIES ANR-22-CE46-0006 research grant.

S. Brunetti et al. (Eds.): DGMM 2024, LNCS 14605, pp. 41–53, 2024.
https://doi.org/10.1007/978-3-031-57793-2_4

all geometric issues [4]. It is also possible to define digital convexity from the intersection of Euclidean convex sets with the lattice cubical complex [12], but it remains unclear how to determine if a given cell complex is indeed convex. We take an interest here in a recent alternative definition of digital convexity, called *full convexity* [8,9], which guarantees the connectedness and even the simple connectedness of fully convex sets. It is important to note that classical digital primitives like standard and naive lines and planes are indeed fully convex, so most of the classical tools of digital geometry fall into this setting.

Full convexity has already proven to be a fruitful framework for analyzing the geometry of digital shapes [2,3,9]: local characterization of convex, concave and planar parts, geodesics and visibility problems, reversible and tight reconstruction of triangulated surfaces, digital polyhedral models. The purpose of this paper is mainly to study its core properties and to exhibit new characterizations of full convexity. Such results are important both from a theoretical perspective (new characterizations lead to better understanding of what is digital convexity) and from a practical algorithmic point of view. For instance the characterization by idempotence of some cell operations [3, Theorem 2] lead to an enveloppe operator that builds a fully convex hull. The first morphological characterization [8, Theorem 5] provides an exact algorithm to check full convexity involving $2^d - 1$ convex hull computations and lattice point enumeration; the second characterization in terms of maximal cells of [3, Theorem 5] lead to an exact algorithm that requires only one convex hull computations and lattice point enumeration.

After recalling some essential background related to full convexity (Sect. 2), we study in Sect. 3 if we can define a digital analogue of convexity through its classical formulation of "inclusion of every straight segment". Originally studied by Minsky and Papert [10], their definition was far too unrestrictive since it included many digital weird sets. We propose here a more reasonable analogue (S-convexity) that we prove equivalent to full convexity in $\mathbb{Z}^2$, but not in higher dimension. We then extend this definition to an analogue of "inclusion of every d-simplices" (S^d-convexity) to get another characterization of full convexity in $\mathbb{Z}^d$. Then we propose in Sect. 4 a recursive convexity (P-convexity): a digital set is P-convex whenever it is digitally convex and each one of its projections along axis is P-convex. Quite surprisingly, we show that P-convexity is indeed equivalent to full convexity. This clearly opens new perspective for studying digital sets, and we already provide here two immediate applications in Sect. 5. One is the proof that subsets of the lattice hypercube as well as digital balls are always fully convex, the other is a measure for digital sets that characterizes fully convex sets. We conclude and describe a few perspectives to this work in Sect. 6.

2   Useful Definitions and Properties

We introduce here basic definitions and properties needed in the rest of the paper. The references are [3,9]. In the sequel, $\mathscr{C}^d$ is the cubical cell complex induced by $\mathbb{Z}^d$. Its 0-dimensional cells are identified to points of $\mathbb{Z}^d$. The set $\mathscr{C}^d_k$ is the set of open k-dimensional cells of $\mathscr{C}^d$.

The *(topological) boundary* ∂Y of a subset Y of $\mathbb{R}^d$ is the set of points in its closure but not in its interior. The star of a cell σ in $\mathscr{C}^d$, denoted by $\mathrm{Star}\,(\sigma)$, is the set of cells of $\mathscr{C}^d$ whose boundary contains σ, plus the cell σ itself. The closure $\mathrm{Cl}\,(\sigma)$ of σ contains σ and all the cells in its boundary. In this paper, the cell boundary operator, also denoted by ∂, maps a k-cell to all its proper faces, that is all its k'-cells, $0 \leqslant k' < k$, and not only its $(k-1)$-cells.

A subcomplex K of $\mathscr{C}^d$ with $\mathrm{Star}\,(K) = K$ is *open*, while being *closed* when $\mathrm{Cl}\,(K) = K$. The *body* of a subcomplex K, i.e. the union of its cells in $\mathbb{R}^d$, is written $\|K\|$.

For any real subset Y of $\mathbb{R}^d$, we denote by $\bar{\mathscr{C}}^d_k[Y]$ the set of k-cells whose topological closure intersects Y, i.e. $\bar{\mathscr{C}}^d_k[Y] = \{c \in \mathscr{C}^d_k, \bar{c} \cap Y \neq \emptyset\}$, where $\bar{c} = \|\mathrm{Cl}\,(c)\|$ for any cell c. For any subset $Y \subset \mathbb{R}^d$, it is natural to define $\mathrm{Star}\,(Y) := \bar{\mathscr{C}}^d[Y]$. Last, the set $\mathrm{CvxH}\,(Y)$ is the *convex hull* of Y in $\mathbb{R}^d$.

Definition 1 (Full convexity). *A non empty subset $X \subset \mathbb{Z}^d$ is digitally k-convex for $0 \leqslant k \leqslant d$ whenever*

$$\bar{\mathscr{C}}^d_k[X] = \bar{\mathscr{C}}^d_k[\mathrm{CvxH}\,(X)]. \tag{1}$$

Subset X is fully (digitally) convex *if it is digitally k-convex for all $k, 0 \leqslant k \leqslant d$.*

The following two characterizations will be useful:

Lemma 1 ([2, Lemma 4]). *A digital set X is fully convex iff $\mathrm{Star}\,(X) = \mathrm{Star}\,(\mathrm{CvxH}\,(X))$.*

Lemma 2 ([3, Theorem 2]). *A digital set X is fully convex iff $X = \mathrm{FC}(X)$, with $\mathrm{FC}(X) := \mathrm{Extr}\,(\mathrm{Skel}\,(\mathrm{Star}\,(\mathrm{CvxH}\,(X))))$.*

The Skel operator builds the *skeleton of a set of cells*, which is defined as the intersection of all cell complexes whose star includes the set of cells. The *extreme operator* Extr maps a set of cells to their set of vertices, which is a digital set. The skeleton can be characterized as follows.

Lemma 3 ([3, Lemma 12]). *Let us consider $Y \subset \mathbb{R}^d$. For any $c \in \mathscr{C}^d$, $c \in \mathrm{Skel}\,(\mathrm{Star}\,(Y)) \iff \|c\| \cap Y \neq \emptyset$ and $\|\partial c\| \cap Y = \emptyset$.*

For a cell $c \in \mathscr{C}^d$, we say that a convex set K is *framed* within c if $K \cap \|c\| \neq \emptyset$ and $K \cap \|\partial c\| = \emptyset$. So, considering Lemma 12 in [3] applied to $\mathrm{CvxH}\,(X)$, $c \in \mathscr{C}^d$ belongs to $\mathrm{Skel}\,(\mathrm{Star}\,(\mathrm{CvxH}\,(X)))$ iff $\|c\| \cap \mathrm{CvxH}\,(X) \neq \emptyset$ and $\|\partial c\| \cap \mathrm{CvxH}\,(X) = \emptyset$. In other words, a cell c is in the skeleton of a convex hull if and only if the convex hull is framed within c. Note that if the dimension of c is zero, i.e. c is a lattice point, then ∂c is empty, so a convex set is framed within a lattice point if and only if this convex set contains it.

As $\mathrm{CvxH}\,(\|\partial c\|)$ is the topological closure of $\mathrm{CvxH}\,(\|c\|) = \|c\|$, if K is not framed within c, then K must either intersect only the boundary of c or c and its boundary. Moreover if K is closed, then $K \cap \|c\|$ is closed. The framed properties cannot happen when K is the convex hull of points in $\mathbb{Z}^d$ and $\dim c = d$. Indeed, if K is framed within c then obviously K is entirely included in $\|c\|$. But this is impossible since $\|c\|$ does not contain any points. So, to have the framed property, we must have $\dim c \leq d - 1$.

3   Segment Convexity and Generalizations

In the Euclidean space, convexity is defined through inclusion of every straight line segment joining two points of the set. Minsky and Papert, in their famous book on perceptrons [10], proposed a digital analogue of segment convexity, phrased "A [digital] set X fails to be convex if and only if there exists three [digital] points such that q is in the line segment joining p and r and, $p \in X$, $q \notin X$, $r \in X$." This definition of digital convexity is unfortunately not at all equivalent to the digital convexity (see Fig. 1), and even less to full convexity.

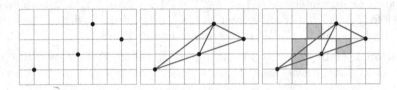

Fig. 1. Minsky-Papert segment convexity versus S-convexity: (a) MP-convex set X, since (b) each segment does not touch any other lattice point. But X is not S-convex, since (c) these segments touch 1-cells and 2-cells that are not in Star (X) (in red). (Color figure online)

Therefore we propose the following digital analogues of "segment" convexity, which are much closer to digital convexity.

Definition 2 (S-convexity and S^k-convexity). *We say that a digital set $X \subset \mathbb{Z}^d$ is S-convex whenever $\forall p \in X, \forall q \in X$, Star $(\mathrm{CvxH}(\{p,q\})) \subset$ Star (X). Furthermore, for $k \geq 2$, the set X is S^k-convex whenever for any k-tuple of points T of X (not necessarily distinct), we have Star $(\mathrm{CvxH}(T)) \subset$ Star (X).*

Otherwise said for S-convexity (resp. S^k-convexity), any pair of points of X (resp. any k-tuple of points of X) must be tangent to X in the terminology of [9]. It is obvious that S^2-convexity is the S-convexity and that S^{k+1}-convexity implies S^k-convexity. We establish the following results in this section.

Theorem 1. *For $d \geqslant 1$, $k \geqslant 2$, full convexity implies S^k-convexity.*

Proof. Let us consider a fully convex set X. If we consider a k-tuple T in X then $\mathrm{CvxH}(T) \subset \mathrm{CvxH}(X)$ because the convex hull operator $\mathrm{CvxH}()$ is increasing. But Star $()$ is also an increasing operator hence Star $(\mathrm{CvxH}(T)) \subset$ Star $(\mathrm{CvxH}(X)) =$ Star (X) (the latter equality given by Lemma 1). So X is S^k-convex. □

Theorem 2. *S-convexity is equivalent to full convexity in $\mathbb{Z}^1$ and $\mathbb{Z}^2$.*

Theorem 3. *S-convexity is not equivalent to full convexity starting from $\mathbb{Z}^3$.*

Theorem 4. *S^d-convexity is equivalent to full convexity in $\mathbb{Z}^d$, for $d \geqslant 2$.*

Some preliminary lemmas will be used to extract impossible configurations for the S^k-convexity. With such situations, we are then able to relate full convexity and S^k-convexity. Theorem 3 is proven by a counter-example.

Lemma 4 (Blocking lemma). *Let us consider an S^k-convex digital set $X \subset \mathbb{Z}^d$. Let us consider $p \in \mathbb{Z}^d$ but $p \notin X$. If Y is a subset of X such that $p \notin$ CvxH(Y) and such that CvxH(Y) has at most $k-1$ vertices, then there exists a pointed convex cone C with apex p such that $C \cap X = \emptyset$.*

Proof. Let us consider an S^k-convex digital set $X \subset \mathbb{Z}^d$. Let us consider $p \in \mathbb{Z}^d \setminus X$, and a subset Y of X such that $p \notin$ CvxH(Y). Since CvxH(Y) is a Euclidean convex set and since p does not belong to it, there exists an hyperplane H_p separating CvxH(Y) from p. We denote by $H_p^{(-)}$ the half-space containing CvxH(Y) and by $H_p^{(+)}$ the half-space containing p. Let us consider any $(k-1)$-tuple T containing the vertices of CvxH(Y) with repetition if needed. We denote by $Y^{(-)}$ the pointed convex cone $Y^{(-)} = $ CvxH$(T \cup \{p\})$. Let us denote by $Y^{(+)}$ the pointed cone obtained by a central symmetry with center p of $Y^{(-)}$. We claim that $Y^{(+)} \cap X = \emptyset$.

Let us hence suppose on the contrary that $Y^{(+)} \cap X \neq \emptyset$ and consider a point x in this intersection. Since T is a $(k-1)$-tuple in X, the set $T_x = T \cup \{x\}$ is a k-tuple in X. Since X is S^k-convex, we have Star$($CvxH$(T_x)) \subset$ Star(X). But by the construction with central symmetry with center p, we have $p \in$ CvxH(T_x). So p is a point in Star(X), which implies $p \in X$. This is a contradiction. $\square$

For any two points $u,\ v \in \mathbb{Z}^d$, let us denote by $(u;v)$ the line passing through u and v. Let us denote by $(\infty;v)_u$ the semi-line in $(u;v)$ containing u and stopping at v. The semi-line $(v;\infty)_u$ is an infinite semi-line in $(u;v)$ containing v but no other point of $(\infty;v)_u$.

Lemma 5 (Line Blocking lemma). *Let us consider an S^k-convex digital set $X \subset \mathbb{Z}^d$. Let us consider $p \in \mathbb{Z}^d$, $p \notin X$. For any $x \in X$, $(p;\infty)_x \cap X = \emptyset$.*

Proof. Let us consider an S^k-convex digital set $X \subset \mathbb{Z}^d$. Let us consider $p \in \mathbb{Z}^d$, $p \notin X$. Let us consider the $(k-1)$-tuple Y by using x, $k-1$ times. Then applying Lemma 4, we get a pointed cone C, which is in fact $(p;\infty)_x$, whose intersection with X is empty. $\square$

Lemma 6 (Grid lemma). *Let us consider a finite S^k-convex digital set $X \subset \mathbb{Z}^d$ with $k \geqslant 2$. Let us denote by $\mathbb{Z}[e_j]$ any line in $\mathbb{Z}^d$ directed by the canonical basis vector e_j. Then, $\mathbb{Z}[e_j] \cap X$ is connected.*

Proof. By using Lemma 5, we known that if we can find a point p outside X then a semi-line will be blocked for X. The same is true for any 1d slice of X, a 1d slice being $\mathbb{Z}[e_j] \cap X$. But since X is finite, there always exists a point p in $\mathbb{Z}[e_j]$ which is outside X. We consider a point p such that one of its neighbor in $\mathbb{Z}[e_j]$ is in X. There are two possible extreme choices for p called $p^{(-)}$ and $p^{(+)}$ and so two neighbors $x^{(-)}$ and $x^{(+)}$. We claim that $\mathbb{Z}[e_j] \cap X = [x^{(-)};x^{(+)}] \cap \mathbb{Z}^d$. Indeed, if a lattice point q is missing in this interval then it belongs to the convex hull of a $k \geq 2$ tuple in X. Hence it must belong to X. $\square$

Lemma 7 (Star lemma). *Let us consider a digital set $X \subset \mathbb{Z}^d$. If X is S^d-convex then $\forall c \in \mathscr{C}^d$, dim $c > 0$, if CvxH(X) is framed within c then $\exists e \in \partial c$, dim $e = 0, e \in X$.*

Proof. Let us consider an S^d-convex digital set $X \subset \mathbb{Z}^d$. Let us consider $c \in \mathscr{C}^d$, dim $c > 0$, $K = \|c\| \cap$ CvxH$(X) \neq \emptyset$. Suppose that CvxH(X) is framed within $\|c\|$. In other words, since CvxH(X) is convex and closed, we have $K \subsetneq \|c\|$. The framed property implies that there exists supporting hyperplanes of CvxH(X) separating $\|c\|$ from $\|\partial c\|$, but touching ∂CvxH(X) too. We can pick one of these supporting hyperplanes that contains a face of ∂CvxH(X): it has an affine dimension $0 \leq k \leq d - 1$. This convex face can be decomposed into a set of k-simplexes $\{S_i\}$. At least one of them, say S_{i_c}, intersects $\|c\|$, otherwise K cannot be a subset of $\|c\|$. So $c \in$ Star(S_{i_c}). Now S_{i_c} is the convex hull of a $k + 1$-tuple of points of X, with $k + 1 \leq d$. By definition of the S^d-convexity, we must have $c \in$ Star(X). But X is a set of points so at least one of the vertices of c must be in X. We have just found some cell $e \in \partial c$, dim $e = 0, e \in X$.

If we assume now that CvxH(X) is not framed within c, we have both CvxH$(X) \cap \|c\| \neq \emptyset$ and CvxH$(X) \cap \|\partial c\| \neq \emptyset$. So, we can consider any cell f in $\|\partial c\|$. If CvxH(X) is framed within f, we got a 0-dimensional cell e, otherwise, we choose a lower dimensional cell in the boundary of f. At each step the dimension decreases. So at the end, either we find a 0-dimensional cell e belonging to CvxH(X) or we find a cell within which CvxH(X) is framed, and we also obtain a 0-dimensional cell e. □

Lemma 8. *Let us consider a digital set $X \subset \mathbb{Z}^d$. If X is S^d-convex then FC$(X) =$ CvxH$(X) \cap \mathbb{Z}^d$.*

Proof. Let us consider an S^d-convex digital set $X \subset \mathbb{Z}^d$. Let us consider a cell c in Skel$($Star$($CvxH$(X)))$. Suppose that dim $c > 0$. Using Lemma 7 since X is S^d-convex, $\exists e \in \partial c$, dim $e = 0, e \in X$. This implies that Star$(c) \subset$ Star(e). Hence, c cannot belong to Skel$($Star$($CvxH$(X)))$, but e does. So we got a contradiction with $\|\partial c\| \cap$ CvxH$(X) = \emptyset$. It follows that c must be a 0-dimensional cell, that is a point. This implies that the Skeleton of CvxH(X) only contains points such that Extr$($Skel$($Star$($CvxH$(X)))) =$ Skel$($Star$($CvxH$(X)))$. But since we only have 0-dimensional cells in the skeleton, we get Skel$($Star$($CvxH$(X))) =$ CvxH$(X) \cap \mathbb{Z}^d$. We conclude that FC$(X) =$ CvxH$(X) \cap \mathbb{Z}^d$. □

Proof (Theorem 2). Using Theorem 1, we only study the case of an S-convex set.

Let us consider an S-convex set X in $\mathbb{Z}^1$. Using Lemma 6, we have that X must be an interval of points. Hence X is fully convex because Star$($CvxH$(X)) =$ Star(X).

Let us consider an S-convex set X in $\mathbb{Z}^2$. Let us suppose that there exists a lattice point z in $($CvxH$(X) \cap \mathbb{Z}^2) \setminus X$. We first remark that z is strictly interior to CvxH(X). Indeed, z cannot be a vertex of the convex hull since the vertices are always in X. Furthermore z cannot be on an edge of CvxH(X) because Lemma 5 would imply a contradiction. We consider the set $L(c)$ of 1-dimensional

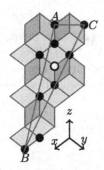

This is piece of the standard digital plane $P = \{(x, y, z) \in \mathbb{Z}^3, 0 \leqslant x + y + 2z < 4\}$. The set X, represented as black disks, is a subset of P. The set Y is the union of X with the point $M = (1, 1, -1)$ represented as a white disk. We have $A = (0, 0, 0), B = (4, 2, -3), C = (-1, 1, 0)$. All four points A, B, C, M have remainder 0 in the digital plane. One can check that $M = \frac{1}{3}(A + B + C)$, hence $M \in \mathrm{CvxH}(X)$. But M does not belong to any straight segment between any pair of points of X.

Fig. 2. Counter-example to S^2-convexity implies full convexity: set Y is S^2-convex and fully convex, while X is S^2-convex but not fully convex (and not digitally 0-convex).

cells $c \in \mathscr{C}^d$ in Star (z). Because z is strictly interior to $\mathrm{CvxH}(X)$, we must have $L(c) \cap \mathrm{CvxH}(X) \neq \emptyset$. So we consider a 1-dimensional cell c in $L(c)$ such that $\|c\| \cap \mathrm{CvxH}(X) \neq \emptyset$. Using Lemma 7, we get a point $e \in \partial c$ with $e \in X$. Let us consider the 1 dimensional cell c_{op} which is on the same axis as z and e but which does not have e on its boundary. We note that $c_{op} = \{f, z\}$. But, $f \notin X$, because of Lemma 6. So we can move on this axis in the direction of f until the 0-cell on the boundary of the 1-cell is not in $\mathrm{CvxH}(X)$. Let us call c_{lim} this last 1-dimensional cell. We have that $\|c_{\mathrm{lim}}\| \cap \mathrm{CvxH}(X) \neq \emptyset$ and no points in $\partial c_{\mathrm{lim}}$ are in X which is in contradiction with Lemma 7. So, $(\mathrm{CvxH}(X) \cap \mathbb{Z}^2) \setminus X = \emptyset$ and it follows that $X = \mathrm{CvxH}(X) \cap \mathbb{Z}^2$.

We now use Lemma 8 to get that $\mathrm{FC}(X) = \mathrm{CvxH}(X) \cap \mathbb{Z}^2$.

We have obtained the equality $\mathrm{FC}(X) = \mathrm{CvxH}(X) \cap \mathbb{Z}^2 = X$ which completes the proof that X is indeed a fully convex set in $\mathbb{Z}^2$.   □

Proof (Theorem 3). A counter-example is given on Fig. 2. We should note that large random constructions of S^2-convex sets by simulation did not lead to any counter-examples. In fact, problematic examples correspond to sets for which there exists an integer point in the relative interior of a maximal face which does not belong to any segments of the face, and are thus very unlikely to be generated randomly.   □

The main result in dimension 2 for S-convexity is that for an S-convex set X, we necessarily have $X = \mathrm{CvxH}(X) \cap \mathbb{Z}^2$. As Theorem 3 states it, this property failed to be true in higher dimension. This explains why we must rely on the more restrictive S^d-convexity when increasing the dimension.

Proof (Theorem 4). Let us consider an S^d-convex set X. Its convex hull $\mathrm{CvxH}(X)$ can be partitioned into a set of d-dimensional simplices C_i in $\mathbb{Z}^d$ for which we can rely on the S^d-convexity property of X. Indeed, if there exists a point $z \in (\text{CvxH}(X) \cap \mathbb{Z}^d) \setminus X$, then $\exists i_z$ such that $z \in C_{i_z}$. Since X is S^d-convex, this means that $z \in \text{Star}(\text{CvxH}(C_{i_z})) \subset \text{Star}(X)$. So since z is a point, this implies that z is in X which is a contradiction. So, $(\text{CvxH}(X) \cap \mathbb{Z}^d) \setminus X = \emptyset$, otherwise said $\text{CvxH}(X) \cap \mathbb{Z}^d = X$, that is X is 0-convex. But applying Lemma 8, we get that $\text{FC}(X) = \text{CvxH}(X) \cap \mathbb{Z}^d$ too. Gathering the two equalities, we obtain $\text{FC}(X) = X$, which means that X is fully convex.   □

4   Projection Convexity

We here study the stability of fully convex sets with respect to orthogonal projections along axes in $\mathbb{R}^d$. We denote by π_j the orthogonal projector associated to the j-th axis, which consists in omitting the j-th coordinates for all points of $\mathbb{Z}^d$. By direct extension, those projectors are defined for cells in $\mathscr{C}^d$. π_j are called *axis projectors*. Those projectors share the property that the image of a cell $c \in \mathscr{C}^d$ is a cell in $\mathscr{C}^{d-1}$ and those projections are the only projections for which this property is true. Moreover, the image of a point in $\mathbb{Z}^d$ by any axis projector is a point in $\mathbb{Z}^{d-1}$. Let us define a convexity by projections as follows.

Definition 3 (*P*-convexity). *Let* $X \subset \mathbb{Z}^d$ *be a digital set. The set* X *is P-convex if and only if* X *is digitally 0-convex (i.e.* $\text{CvxH}(X) \cap \mathbb{Z}^d = X$*) and when* $d > 1$*, for any* j*,* $1 \leqslant j \leqslant d$*,* $\pi_j(X)$ *is P-convex in* $\mathbb{Z}^{d-1}$*.*

Quite surprisingly, we have the equivalence of *P*-convexity with full convexity.

Theorem 5. *For arbitrary dimension* $d \geqslant 1$*, for any* $X \subset \mathbb{Z}^d$*,* X *is fully convex if and only if* X *is P-convex.*

Proof. The fact that a fully convex X is also *P*-convex directly follows from (i) fully convex sets are in particular digitally 0-convex, (ii) projection $\pi_j(X)$ are fully convex in $\mathbb{Z}^{d-1}$ as shown in [3, Lemma 23].

When $d = 1$, 0-convexity is equivalent to full convexity (consequence of [9, Lemma 4] with $d = 1$), so *P*-convexity implies full convexity for this dimension. Let us now show that this implication holds for $d > 1$.

Suppose that $X \subset \mathbb{Z}^d$ is *P*-convex but not fully convex. Since X is 0-convex by definition, we know that $X = \text{CvxH}(X) \cap \mathbb{Z}^d$. So, any 0-dimensional cell of X is in the skeleton $\text{Skel}(\text{Star}(\text{CvxH}(X)))$ and $\text{CvxH}(X)$ does not contain any other points. So since the fact that X is not fully convex implies that $X \neq \text{FC}(X)$, there exists some cell c in $\text{Skel}(\text{Star}(\text{CvxH}(X)))$ with $\dim c > 0$. Indeed, the extreme operator only add points for cells of strictly positive dimension. We can characterize c by the framed property: $\text{CvxH}(X) \cap \|c\| \neq \emptyset$ and $\text{CvxH}(X) \cap \|\partial c\| = \emptyset$.

Let y be some point of $\text{CvxH}(X) \cap \|c\|$, which is also not in X since it is not a lattice point. Being in $\text{CvxH}(X)$, by Carathéodory's convexity theorem, there exists at most $d + 1$ extreme points $v_0, v_1, \ldots, v_d$ of $\text{CvxH}(X)$ such that y is a convex linear combination of these points. We thus have $y = \sum_{i=0}^{d} \lambda_i v_i$, with

$\sum_{i=0}^{d} \lambda_i = 1$ and $\forall i, 0 \leqslant i \leqslant d, \lambda_i \geq 0$. Being extreme points of CvxH (X), every v_i is a lattice point in X.

Let $k := \dim c$. There exists k different directions $J := (j_i)_{i=1,\dots,k}$ such that $\dim(\pi_{j_i}(c)) = k - 1$. Let π_J be the composition of the projections $\pi_{j_1}, \dots, \pi_{j_k}$ (the order is not important since these operators commute). It follows that $\pi_J(c)$ is a lattice point, say z. Since $y \in c$, we have also $z = \pi_J(y)$. It follows that:

$$z = \pi_J \left(\sum_{i=0}^{d} \lambda_i v_i \right) = \sum_{i=0}^{d} \lambda_i (\pi_J(v_i)). \qquad \text{(by linearity of } \pi_j\text{)}$$

Since every $v_i \in X$, we have shown that $z \in$ CvxH $(\pi_J(X))$ and $z \in \mathbb{Z}^{d-k}$. Since X is P-convex, its projection $\pi_J(X)$ is 0-convex, so CvxH $(\pi_J(X)) \cap \mathbb{Z}^{d-k} = \pi_J(X)$ and $z \in \pi_J(X)$.

The last assertion means that there exist a lattice point $x \in X$, such that $z = \pi_J(x)$. Let $C := \pi_J^{-1}(z)$ be the affine k-dimensional space containing z. It contains in particular c, its boundary ∂c, the point y and the lattice point x. We have $y \in c$ while x cannot be in c since it is a lattice point. Furthermore CvxH $(X) \cap \|\partial c\| = \emptyset$ implies also $x \notin \partial c$ (because $x \in X$). Now $y \in$ CvxH (X) by definition, $x \in X \subset$ CvxH (X), so the straight segment $[y; x]$ must be included in CvxH (X). It also included in the k-dimensional space C so it is a connected path from the interior of cell c to the exterior of c in C: it must cross ∂c at some point x'. By convexity we have $x' \in [y; x] =$ CvxH $(\{x, y\}) \subset$ CvxH (X). But we have also $x' \in \partial c$ and CvxH $(X) \cap \|\partial c\| = \emptyset$. This is a contradiction.

Hence $X = FC(X)$ which is equivalent to X fully convex. $\qquad \square$

5   Applications

We present here two quite immediate applications of the previous characterization of fully convex sets.

5.1   New Fully Convex Digital Sets

Proposition 1. *Let A be a digital set with bounding box defined by a lowest point $\mathbf{p}$ and a highest point $\mathbf{q}$, with $\forall i, 1 \leqslant i \leqslant d, |q^i - p^i| \leqslant 1$. Then A is fully convex.*

Proof. If A is empty then the conclusion holds. In dimension 1, it is clear that any subset of the digital set $\{x, x + 1\}$ is 0-convex hence P-convex. Assuming now that the property holds for dimension $d - 1$, let us prove it for $A \subset \mathbb{Z}^d$. Note first that any subset of the hypercube H defined by $\mathbf{p}$ and $\mathbf{q}$ is digitally 0-convex, since any vertex of CvxH (A) must belong to A since it is a vertex of H too. Each projection $\pi_j(A)$ is also a non-empty subset of a $d - 1$-hypercube $\pi_j(H)$, and is thus P-convex by induction hypothesis. The conclusion follows from the equivalence of P-convexity with full convexity (Theorem 5). $\qquad \square$

A *digital ball of* $\mathbb{Z}^d$ is the intersection of *any* Euclidean d-dimensional ball with $\mathbb{Z}^d$. Note that the center of the ball may by any Euclidean point of $\mathbb{R}^d$ and the radius may be any real non negative value.

Proposition 2. *Any digital ball of* $\mathbb{Z}^d$ *is fully convex.*

Proof. We show that this is true by induction on the dimension d. For $d = 1$, full convexity is equivalent to 0-convexity, and a 1-dimensional digital ball is 0-convex. Let us assume that digital balls are fully convex for dimension $d - 1$, $d \geq 2$, and let us prove that this assertion is true for dimension d.

By Theorem 5, it is equivalent to show that d-dimensional digital balls are P-convex. Let X be some digital ball of center $\mathbf{c} \in \mathbb{R}^d$ and radius r, i.e. $X = B_r(\mathbf{c}) \cap \mathbb{Z}^d$. First of all, X is 0-convex since it is the intersection of a real convex set with the grid $\mathbb{Z}^d$. We have to show that, for any axis direction j, $1 \leqslant j \leqslant d$, $\pi_j(X)$ is P-convex.

We write the proof for $j = d$ for simplicity of writings, but the proof is the same for the other directions. The main argument is that the projection of a digital ball is itself a $d - 1$-digital ball but possibly with a slightly lower radius.

For $\mathbf{x} \in X$, we have $\|\pi_d(\mathbf{x}) - \pi_d(\mathbf{c})\|^2 = \|\mathbf{x} - \mathbf{c}\|^2 - |x_d - c_d|^2 \leq r^2 - |x_d - c_d|^2$, where x_d and c_d are the d-th coordinate of their respective point. It is obvious that, for any $z \in \mathbb{Z}$, $|z - c_d| \geq |c_d - \lfloor c_d \rceil| =: \alpha$, where $\lfloor \cdot \rceil$ is the round operator. It follows that $\|\pi_d(\mathbf{x}) - \pi_d(\mathbf{c})\|^2 \leq r^2 - \alpha^2 =: \rho^2$. We have just shown that $\pi_d(X) \subset B_\rho(\pi_d(\mathbf{c})) \cap \mathbb{Z}^{d-1}$.

Reciprocally, let us now pick a point $\mathbf{y} \in B_\rho(\pi_d(\mathbf{c})) \cap \mathbb{Z}^{d-1}$. It follows that $\|\mathbf{y} - \pi_d(\mathbf{c})\|^2 \leq r^2 - \alpha^2$. Let us build a d-dimensional lattice point $\mathbf{z}$ as $\mathbf{z} = (y_1, \ldots, y_{d-1}, \lfloor c_d \rceil)$. We have:

$$\|\mathbf{z} - \mathbf{c}\|^2 = \|\mathbf{y} - \pi_d(\mathbf{c})\|^2 + |\lfloor c_d \rceil - c_d|^2 \leqslant r^2 - \alpha^2 + \alpha^2 = r^2.$$

This proves that $\mathbf{z} \in X$. Since $\pi_d(\mathbf{z}) = \mathbf{y}$, it holds that $\mathbf{y} \in \pi_d(X)$. It follows that $B_\rho(\pi_d(\mathbf{c})) \cap \mathbb{Z}^{d-1} \subset \pi_d(X)$.

So $\pi_d(x)$ is a $d - 1$-dimensional digital ball, hence is fully convex or, equivalently, P-convex. Since all projections are P-convex, it holds that X is P-convex or, equivalently, fully convex. □

Note that the argument does not work for an arbitrary ellipsoid since some projections of digital ellipsoids might not be digital ellipsoids: this is due to possible missing points when the ellipsoid is too thin.

5.2 A Progressive Measure for Full Convexity

Sometimes it is useful to quantify a property over a set in a progressive manner. For instance there exists measures of circularity, convexity, straightness, disconnectedness, and so on (e.g. [13–15]). We would like here to define a *full convexity measure* over a digital set, that has value exactly 1 for fully convex sets, while decreasing to zero as the digital set looks less and less like a fully convex set.

Let $M_d(A)$ be any d-dimensional digital convexity measure of digital set A. A choice could be for instance for finite sets:

$$M_d(A) := \frac{\#(A)}{\#(\mathrm{CvxH}(A) \cap \mathbb{Z}^d)}, \qquad M_d(\emptyset) = 1. \qquad (2)$$

The *full convexity measure* M_d^F for $A \subset \mathbb{Z}^d$, A finite, is then:

$$M_1^F(A) := M_1(A) \qquad\qquad \text{for } d = 1, \qquad (3)$$

$$M_d^F(A) := M_d(A) \prod_{k=1}^{d} M_{d-1}^F(\pi_k(A)) \qquad \text{for } d > 1. \qquad (4)$$

It coincides with the digital convexity measure in dimension 1, but may differ starting from dimension 2.

Theorem 6. *Let $A \subset \mathbb{Z}^d$ finite. Then $M_d^F(A) = 1$ if and only if A is fully convex and $0 < M_d^F(A) < 1$ otherwise. Besides $M_d^F(A) \leqslant M_d(A)$ in all cases.*

Proof. Immediate from the equivalence of P-convexity with full convexity.   □

Figure 3 illustrates the links and the differences between the two convexity measures M_d and M_d^F on simple 2D examples. As one can see, the usual convexity measure may not detect disconnectedness, is sensitive to specific alignments of pixels, while full convexity is globally more stable to perturbation and is never 1 when sets are disconnected.

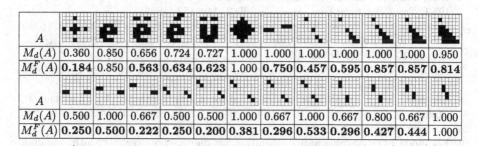

Fig. 3. Common points and differences of convexity measure M_d and full convexity measure M_d^F on small 2D digital sets.

6   Conclusion and Perspectives

We have presented two original characterizations of full convexity. The first one gives a nice analogue of the "segment inclusion" definition of convexity with full convexity in dimension 1 and 2, and it shows also that, in higher dimensional spaces, additional continuity constraints are required. The second characterization tells that full convexity requires the digital convexity of the set and all of its shadows along axes.

Both characterizations shed new light on what is really full convexity. They may provide alternative algorithms to check full convexity, which may have better time complexity. They may help deciding if some digital sets are fully convex, as we show here for digital balls and hypercube subsets. They enable the definition of new measures for digital sets, with a stronger power of categorization. Finally, the characterization of full convexity through projections can be of interest for discrete tomography, as it induces connectedness in a natural way.

References

1. Eckhardt, U.: Digital lines and digital convexity. In: Bertrand, G., Imiya, A., Klette, R. (eds.) Digital and Image Geometry. LNCS, vol. 2243, pp. 209–228. Springer, Heidelberg (2001). https://doi.org/10.1007/3-540-45576-0_13

2. Feschet, F., Lachaud, J.O.: Full convexity for polyhedral models in digital spaces. In: Baudrier, E., Naegel, B., Krahenbuhl, A., Tajine, M. (eds.) Discrete Geometry and Mathematical Morphology. DGMM 2022. LNCS, vol. 13493, pp. 98–109. Springer, Cham (2022). https://doi.org/10.1007/978-3-031-19897-7_9

3. Feschet, F., Lachaud, J.O.: An envelope operator for full convexity to define polyhedral models in digital spaces. J. Math. Imaging Vis. **65**(5), 754–769 (2023). https://doi.org/10.1007/s10851-023-01155-w

4. Kim, C.E., Rosenfeld, A.: Convex digital solids. IEEE Trans. Pattern Anal. Mach. Intell. **6**, 612–618 (1982)

5. Kim, C.E., Rosenfeld, A.: Digital straight lines and convexity of digital regions. IEEE Trans. Pattern Anal. Mach. Intell. **2**, 149–153 (1982)

6. Kiselman, C.O.: Convex functions on discrete sets. In: Klette, R., Žunić, J. (eds.) IWCIA 2004. LNCS, vol. 3322, pp. 443–457. Springer, Heidelberg (2004). https://doi.org/10.1007/978-3-540-30503-3_32

7. Kiselman, C.O.: Elements of digital geometry, mathematical morphology, and discrete optimization. World Scientific (2022). https://doi.org/10.1142/12584

8. Lachaud, J.-O.: An alternative definition for digital convexity. In: Lindblad, J., Malmberg, F., Sladoje, N. (eds.) DGMM 2021. LNCS, vol. 12708, pp. 269–282. Springer, Cham (2021). https://doi.org/10.1007/978-3-030-76657-3_19

9. Lachaud, J.O.: An alternative definition for digital convexity. J. Math. Imaging Vis. **64**(7), 718–735 (2022). https://doi.org/10.1007/s10851-022-01076-0

10. Minsky, M., Papert, S.A.: Perceptrons: An Introduction to Computational Geometry. MIT Press, Cambridge (1969)

11. Ronse, C.: A bibliography on digital and computational convexity (1961–1988). IEEE Trans. Pattern Anal. Mach. Intell. **11**(2), 181–190 (1989)

12. Webster, J.: Cell complexes and digital convexity. In: Bertrand, G., Imiya, A., Klette, R. (eds.) Digital and Image Geometry. LNCS, vol. 2243, pp. 272–282. Springer, Heidelberg (2001). https://doi.org/10.1007/3-540-45576-0_16

13. Zunić, J., Hirota, K., Rosin, P.L.: A Hu moment invariant as a shape circularity measure. Pattern Recognit. **43**(1), 47–57 (2010). https://doi.org/10.1016/j.patcog.2009.06.017

14. Zunić, J., Rosin, P., Ilić, V.: Disconnectedness: a new moment invariant for multi-component shapes. Pattern Recognit. **78**, 91–102 (2018). https://doi.org/10.1016/j.patcog.2018.01.010
15. Zunić, J., Rosin, P.L.: A new convexity measure for polygons. IEEE Trans. Pattern Anal. Mach. Intell. **26**(7), 923–934 (2004)

Decomposition of Rational Discrete Planes

Tristan Roussillon[1]([✉])[iD] and Sébastien Labbé[2][iD]

[1] Univ Lyon, INSA Lyon, CNRS, UCBL, LIRIS, UMR5205, 69622 Villeurbanne,
France
tristan.roussillon@liris.cnrs.fr
[2] Univ Bordeaux, CNRS, Bordeaux INP, LaBRI, UMR5800, 33400 Talence, France
sebastien.labbe@labri.fr

Abstract. This paper is a contribution to the study of rational discrete planes, i.e., sets of points with integer coordinates lying between two parallel planes. Up to translation and symmetry, they are completely determined by a normal vector $a \in \mathbb{N}^3$. Excepted for a few well-identified cases, it is shown that there are two approximations $b, c \in \mathbb{N}^3$ of a, satisfying $a = b + c$, such that the discrete plane of normal a can be partitioned into two sets having respectively the combinatorial structure of discrete planes of normal b and c. Christoffel graphs are used to compactly encode the structure of discretes planes. This result may have practical interest in discrete geometry for the analysis of planar features.

Keywords: Discrete Plane · Christoffel Graph · Approximation

1 Introduction

This paper is a contribution to the study of standard arithmetical rational discrete planes. They are defined from a non-zero normal vector $a \in \mathbb{N}^3$ as follows:

$$\mathcal{P}_a := \{x \in \mathbb{Z}^3 \mid 0 \leq x \cdot a < \|a\|_1\}.$$

Their combinatorial structure has been studied thirty years ago in [3]. The main definitions and results are recalled below. The adjacency graph associated to $\mathcal{P}_a$ is a graph whose vertices are the points of $\mathcal{P}_a$ and that has an edge between two distinct points x and y if and only if $\|x - y\|_1 = 1$. The elementary cycles, which are squares, are called faces. The adjacency graph is connected and, together with the set of faces, defines a two-dimensional combinatorial manifold without boundary [3] (Fig. 1).

In the adjacency graph associated to $\mathcal{P}_a$, the set of edges incident to a given vertex x is determined by the quantity $x \cdot a$, called the height of x (Fig. 1). The arrangement of edges incident to equally high points is thus the same. In addition, there are only eight different arrangements of incident edges in all discrete planes and at most seven in a given one [3] (Fig. 2).

This work has been funded by PARADIS ANR-18-CE23-0007-01 research grant.

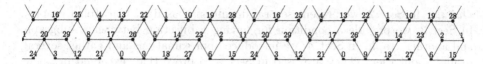

Fig. 1. $\mathcal{P}(4, 9, 17)$. The number displayed close to a point is its height.

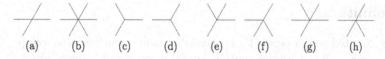

Fig. 2. All arrangements of incident edges – (a) and (b) cannot be in the same discrete plane. This is close to a vertex-atlas in tiling theory [7, section 5.3].

Related Works. In 2d, it is well known that the smallest segment that periodically generates a rational discrete line can be uniquely partitioned into two segments, each of them being the minimal period of another rational discrete line. See, e.g., the splitting formula [8, pp. 153–157] or the standard factorization of Christoffel words [1, pp. 19–22]. This paper aims at extending such decomposition to 3d. However, there are infinitely many sets of faces that can periodically generate the same rational discrete plane and there is no canonical way of decomposing them. See, e.g., [2] for a practical method of generation and decomposition based on a geometrical extension of substitutions.

Contribution. In this paper, we propose another framework based on a symmetric version of Christoffel graphs, introduced in [5] as extensions of Christoffel words. In brief, they describe, for every height h, the arrangement of edges incident to the points of height h in the adjacency graph associated to $\mathcal{P}_a$. They allow us to compare the arrangements of edges in the adjacency graphs of two different sets. We say that two sets have the same combinatorial structure if there is a bijection between their points and the arrangements of edges incident to them. We show that, excepted for a few cases, there are two approximations $b, c \in \mathbb{N}^3$ of a, such that $a = b + c$ and $\mathcal{P}_a$ can be partitioned into two sets having respectively the combinatorial structure of $\mathcal{P}_b$ and $\mathcal{P}_c$. See Fig. 3.

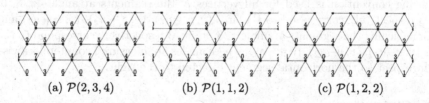

(a) $\mathcal{P}(2, 3, 4)$ (b) $\mathcal{P}(1, 1, 2)$ (c) $\mathcal{P}(1, 2, 2)$

Fig. 3. Decomposition of $\mathcal{P}(2, 3, 4)$ in red and blue sets. The arrangements of edges in the red (resp. blue) set match those of $\mathcal{P}(1, 1, 2)$ (resp. $\mathcal{P}(1, 2, 2)$). (Color figure online)

Outline. In Sect. 2, we present several definitions based on Christoffel graphs. In Sect. 3, we provide a condition for having a bijection between a subgraph and another graph. In Sect. 4, we introduce approximations and gather several results about them. They are used in Sect. 5 to prove that, excepted in a few cases, there is an approximation for which our condition for bijection is true.

2   Definitions

Given a directed graph $\mathcal{G} := (\mathcal{V}, \mathcal{A})$, let us introduce the function $arcs_{\mathcal{G}} : 2^{\mathcal{V}} \mapsto 2^{\mathcal{A}}$ that returns all the arcs emanating from the vertices of a given subset $\mathcal{V}' \subseteq \mathcal{V}$, i.e., such that $arcs_{\mathcal{G}}(\mathcal{V}') = \{(v_1, v_2) \in \mathcal{A} \mid v_1 \in \mathcal{V}'\}$.

For a symmetric graph $\mathcal{G}$, i.e., such that $(v_1, v_2) \in \mathcal{A} \Leftrightarrow (v_2, v_1) \in \mathcal{A}$, a partition of $\mathcal{V}$ induces a partition of $\mathcal{A}$ such that the set of arcs emanating from a given vertex is included in the same subset of $\mathcal{A}$. Indeed, for two subsets $\mathcal{V}', \mathcal{V}''$ such that $\mathcal{V}' \cup \mathcal{V}'' = \mathcal{V}$ and $\mathcal{V}' \cap \mathcal{V}'' = \emptyset$, one has

$$arcs_{\mathcal{G}}(\mathcal{V}') \cup arcs_{\mathcal{G}}(\mathcal{V}'') = \mathcal{A}, \quad arcs_{\mathcal{G}}(\mathcal{V}') \cap arcs_{\mathcal{G}}(\mathcal{V}'') = \emptyset.$$

2.1   Christoffel Graph

Definition 1. *Given $A \in \mathbb{Z}_{>0}$, let us define the set $\mathcal{V}_A := \{0, \ldots, A-1\}$. The symmetric Christoffel graph of normal $\boldsymbol{a} \in \mathbb{N}^3 \setminus \{\boldsymbol{0}\}$, with $\gcd(\boldsymbol{a}) = 1$, is the pair $\mathcal{G}_{\boldsymbol{a}} := (\mathcal{V}_{\|\boldsymbol{a}\|_1}, \mathcal{A}_{\boldsymbol{a}})$, where*

$$\mathcal{A}_{\boldsymbol{a}} := \{(v_1, v_2) \mid v_1, v_2 \in \mathcal{V}_{\|\boldsymbol{a}\|_1}, 1 \leq i \leq 3, v_2 = v_1 \pm a_i\}.$$

We have symmetrized the original definition of Christoffel graph introduced in [5] in order to define partitions such that the set of arcs emanating from a given vertex is always included in the same subset of arcs. Since we only consider symmetric Christoffel graphs in the rest of the paper, we will now omit the term symmetric.

Two representations of the same Christoffel graph are shown in Fig. 4. The two arcs (v_1, v_2) and (v_2, v_1) are merged into one undirected edge in (b), while they are represented by two distinct segments in (c): one incident to v_1 for (v_1, v_2), the other incident to v_2 for (v_2, v_1). In the latter representation, the following convention is used for all vertices v: the segments at angle $4\pi/3$, 0, $2\pi/3$ respectively correspond to the arcs $(v, v + a_1)$, $(v, v + a_2)$, $(v, v + a_3)$ and symmetrically, the segments at angle $\pi/3$, π, $5\pi/3$ respectively correspond to the arcs $(v, v - a_1)$, $(v, v - a_2)$, $(v, v - a_3)$, where angles are measured counterclockwise with respect to the horizontal segment directed to the right.

2.2   Christoffel Subgraph

Given $A \in \mathbb{Z}_{>0}$, a bound $B \in \{0, \ldots, A-1\}$ and an offset $\delta \in \{-B+1, \ldots, A-B\}$ are used to define a subset of $\mathcal{V}_A$:

$$\mathcal{V}_A^{B,\delta} := \{k \in \mathcal{V}_A \mid (kB - \delta) \bmod A < B\}. \tag{1}$$

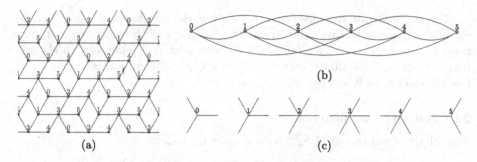

Fig. 4. $\mathcal{P}_{(1,2,3)}$ in (a). Two representations of $\mathcal{G}_{(1,2,3)}$ in (b) and (c). Observe how they encode the arrangements of edges incident to the points of $\mathcal{P}_{(1,2,3)}$.

The following table shows how to compute two subsets of the vertices $\{0, \dots, 8\}$ (in blue and red) thanks to (1), where (B, δ) are equal to $(4, 0)$ (resp. $(5, 1)$) in the second (resp. third) line.

k	0	1	2	3	4	5	6	7	8
$4k \bmod 9$	0	4	8	3	7	2	6	1	5
$(5k-1) \bmod 9$	8	4	0	5	1	6	2	7	3

Then, let us denote by $\mathcal{S}_a^{B,\delta} := \left(\mathcal{V}_{\|a\|_1}^{B,\delta}, arcs_{\mathcal{G}_a}\left(\mathcal{V}_{\|a\|_1}^{B,\delta} \right) \right)$ the subgraph of $\mathcal{G}_a$ of bound B and offset δ. To say it simply, it contains all the arcs of $\mathcal{G}_a$ emanating from at least one vertex of $\mathcal{V}_{\|a\|_1}^{B,\delta}$. Several subgraphs of $\mathcal{G}_{(2,3,4)}$ and $\mathcal{G}_{(1,1,7)}$ are illustrated in Fig. 5.

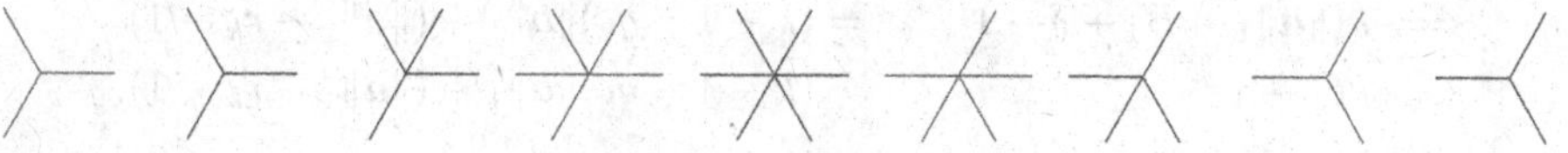

(a) $\mathcal{S}_{(2,3,4)}^{4,0}$ (red) and $\mathcal{S}_{(2,3,4)}^{5,1}$ (blue)

(b) $\mathcal{S}_{(2,3,4)}^{3,1}$ (red) and $\mathcal{S}_{(2,3,4)}^{6,0}$ (blue)

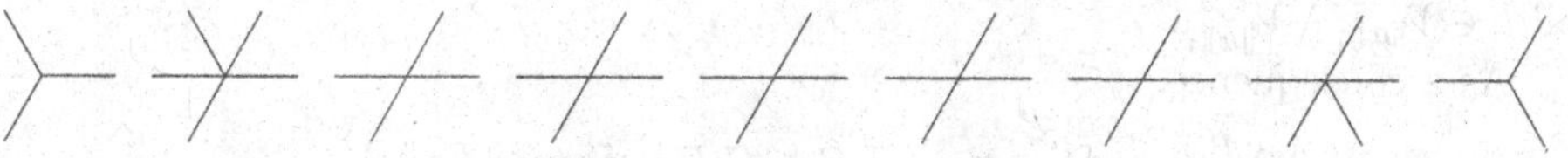

(c) $\mathcal{S}_{(1,1,7)}^{1,2}$ (red) and $\mathcal{S}_{(1,1,7)}^{(8,-1)}$ (blue)

Fig. 5. Subgraphs of $\mathcal{G}_{(2,3,4)}$ (a–b) and $\mathcal{G}_{(1,1,7)}$ (c). Since the vertices $\{0, 1, \dots, 8\}$ are placed from left to right in increasing order, the numbers are omitted. (Color figure online)

The rationale for the offset is two-fold. On the one hand, it allows us to easily describe the relative complement of a subgraph with respect to a graph. See Fig. 5 and the next subsection. On the other hand, it allows us to always find a subgraph such that it corresponds to a Christoffel graph, excepted for a few values of $\boldsymbol{a}$, as it will become clear in Sect. 5.

2.3 Relative Complement

The relative complement of $\mathcal{S}_a^{B,\delta}$ with respect $\mathcal{G}_a$ is defined as

$$\overline{\mathcal{S}_a^{B,\delta}} := \left(\mathcal{V}_{\|a\|_1} \setminus \mathcal{V}_{\|a\|_1}^{B,\delta}, \mathcal{A}_a \setminus arcs_{\mathcal{G}_a}\left(\mathcal{V}_{\|a\|_1}^{B,\delta} \right) \right).$$

Note that $\mathcal{S}_a^{B,\delta} \cup \overline{\mathcal{S}_a^{B,\delta}} = \mathcal{G}_a$ by definition, where the union is done independently on the vertices and arcs.

The relative complement of a subgraph with respect to a graph is also a subgraph in itself with appropriate bound and offset.

Lemma 1. *One has* $\overline{\mathcal{S}_a^{B,\delta}} = \mathcal{S}_a^{C,\gamma}$, *where* $C := \|a\|_1 - B$ *and* $\gamma := -\delta + 1$.

Note that $\gamma \in \{-C+1, \dots, \|a\|_1 - C\}$, see Subsect. 2.2.

Proof. Let us first focus on the vertices. By definition of $\mathcal{V}_{\|a\|_1}$ and $\mathcal{V}_{\|a\|_1}^{B,\delta}$, the set $\mathcal{V}_{\|a\|_1} \setminus \mathcal{V}_{\|a\|_1}^{B,\delta}$ is equal to $\{k \in \mathcal{V}_{\|a\|_1} \mid B \leq (kB - \delta) \bmod \|a\|_1\}$.

Denoting by q_k and r_k, respectively the quotient and remainder of the Euclidean division of $kB - \delta$ by $\|a\|_1$, one has $kB - \delta = q_k\|a\|_1 + r_k$ and $B \leq r_k < \|a\|_1$ for all $k \in \mathcal{V}_{\|a\|_1} \setminus \mathcal{V}_{\|a\|_1}^{B,\delta}$. However, $kB - \delta = q_k\|a\|_1 + r_k$ is equivalent to

$$
\begin{aligned}
(-kB + \delta) + (k\|a\|_1 - 1) &= (-q_k\|a\|_1 - r_k) + (k\|a\|_1 - 1) \\
\Longleftrightarrow k(\|a\|_1 - B) + \delta - 1 \quad &= (k - 1 - q_k)\|a\|_1 + (\|a\|_1 - r_k - 1) \\
\Longleftrightarrow kC - \gamma \qquad\qquad &= (k - 1 - q_k)\|a\|_1 + (\|a\|_1 - r_k - 1).
\end{aligned}
$$

Furthermore, $B \leq r_k < \|a\|_1$ is equivalent to

$$
\begin{aligned}
-1 < (\|a\|_1 - r_k - 1) &\leq \|a\|_1 - B - 1 \\
\Longleftrightarrow \quad 0 \leq (\|a\|_1 - r_k - 1) &< \|a\|_1 - B.
\end{aligned}
$$

The two above results imply that $\|a\|_1 - r_k - 1$ is the remainder of the Euclidean division of $kC - \gamma$ by $\|a\|_1$ and is strictly less than $\|a\|_1 - B = C$ for all $k \in \mathcal{V}_{\|a\|_1} \setminus \mathcal{V}_{\|a\|_1}^{B,\delta}$.

As a consequence,

$$
\begin{aligned}
\mathcal{V}_{\|a\|_1} \setminus \mathcal{V}_{\|a\|_1}^{B,\delta} &= \{k \in \mathcal{V}_{\|a\|_1} \mid B \leq (kB - \delta) \bmod \|a\|_1\} \\
&= \{k \in \mathcal{V}_{\|a\|_1} \mid (kC - \gamma) \bmod \|a\|_1 < C\} = \mathcal{V}_{\|a\|_1}^{C,\gamma}.
\end{aligned}
$$

Finally, by the definitions of $\mathcal{G}_a$ and $arcs_{\mathcal{G}_a}$, $\mathcal{V}_{\|a\|_1} \setminus \mathcal{V}_{\|a\|_1}^{B,\delta} = \mathcal{V}_{\|a\|_1}^{C,\gamma}$ implies $\mathcal{A}_a \setminus arcs_{\mathcal{G}_a}\left(\mathcal{V}_{\|a\|_1}^{B,\delta} \right) = arcs_{\mathcal{G}_a}\left(\mathcal{V}_{\|a\|_1}^{C,\gamma} \right)$, which concludes. $\qquad\square$

2.4   Arc-Preserving Bijection

The goal of the paper is to compare a subgraph with another graph.

Definition 2. *Let $a, b \in \mathbb{N}^3 \setminus \{0\}$, with $\gcd(a) = \gcd(b) = 1$, be such that $a - b \in \mathbb{N}^3 \setminus \{0\}$ and δ be in $\{-\|b\|_1 + 1, \ldots, \|a\|_1 - \|b\|_1\}$. $\mathcal{S}_a^{\|b\|_1, \delta}$ is said to agree with $\mathcal{G}_b$, which is denoted by $\mathcal{S}_a^{\|b\|_1, \delta} \simeq \mathcal{G}_b$, if there exists a bijection $f : \mathcal{V}_{\|a\|_1}^{\|b\|_1, \delta} \mapsto \mathcal{V}_{\|b\|_1}$ such that for all $i \in \{1, 2, 3\}$, $(v, v + a_i) \in \mathcal{S}_a^{\|b\|_1, \delta} \Leftrightarrow (f(v), f(v) + b_i) \in \mathcal{G}_b$ and $(v, v - a_i) \in \mathcal{S}_a^{\|b\|_1, \delta} \Leftrightarrow (f(v), f(v) - b_i) \in \mathcal{G}_b$.*

In other words, $\mathcal{S}_a^{\|b\|_1, \delta} \simeq \mathcal{G}_b$ means that there exists a bijection f, such that, for every vertex v in $\mathcal{S}_a^{\|b\|_1, \delta}$, v and $f(v) \in \mathcal{G}_b$ are surrounded by the same kinds of arcs. You can observe that the blue subgraph in Fig. 5(b) can be mapped to the graph drawn in Fig. 4(c), i.e., $\mathcal{S}_{(2,3,4)}^{6,0} \simeq \mathcal{G}_{(1,2,3)}$.

Note that we do not rely directly on a subgraph isomorphism definition, since it involves a *vertex-induced* subgraph, i.e., such that the endpoints of the edges are all in the vertex subset, which is not the case here.

3   A Criterion for Bijection

In order to compare $\mathcal{S}_a^{\|b\|_1, \delta}$ with $\mathcal{G}_b$, let us consider the following:

Definition 3. *The bijection function $f_A^{B, \delta} : \mathcal{V}_A^{B, \delta} \mapsto \mathcal{V}_B$ is defined such that*

$$f_A^{B, \delta}(k) := \left\lfloor \frac{kB - \delta}{A} \right\rfloor. \tag{2}$$

Corollary 1 provides a simple criterion for $\mathcal{S}_a^{\|b\|_1, \delta} \simeq \mathcal{G}_b$. It depends on b and δ and comes from the two following lemmae:

Lemma 2. *If $k \in \mathcal{V}_A^{B, \delta}$, then $f_A^{B, \delta}(k) \in \mathcal{V}_B$.*

Proof. Using $k \in \mathcal{V}_A^{B, \delta} \subseteq \mathcal{V}_A$ and $\delta \in \{-B + 1, \ldots, A - B\}$ (see Subsects. 2.1 and 2.2), one can show that

$$0 \leq \left\lfloor \frac{kB - \delta}{A} \right\rfloor \leq \frac{kB - \delta}{A} < B.$$

Indeed, the upper bound is implied by $k \leq A - 1$ and $-B < \delta$. The lower bound is trivial if $\delta \leq 0$. Otherwise, it is enough to notice that $k \in \mathcal{V}_A^{B, \delta}$ implies $\frac{\delta}{B} \leq k$. It follows that $0 \leq kB - \delta$, which concludes. $\square$

Lemma 3. *For all $i \in \{1, 2, 3\}$, for all $k \in \mathcal{V}_{\|a\|_1}^{\|b\|_1, \delta}$,*

$(C1)$ *if $b_i \|a\|_1 - a_i \|b\|_1 \leq \mu$,*   *then $(k, k + a_i) \in \mathcal{S}_a^{\|b\|_1, \delta} \Rightarrow (l, l + b_i) \in \mathcal{G}_b$,*

$(C2)$ *if $-(b_i \|a\|_1 - a_i \|b\|_1) \leq \mu$,* *then $(k, k - a_i) \in \mathcal{S}_a^{\|b\|_1, \delta} \Leftarrow (l, l - b_i) \in \mathcal{G}_b$,*

$(C3)$ *if $\nu \leq b_i \|a\|_1 - a_i \|b\|_1$,*   *then $(k, k + a_i) \in \mathcal{S}_a^{\|b\|_1, \delta} \Leftarrow (l, l + b_i) \in \mathcal{G}_b$,*

$(C4)$ *if $\nu \leq -(b_i \|a\|_1 - a_i \|b\|_1)$,*   *then $(k, k - a_i) \in \mathcal{S}_a^{\|b\|_1, \delta} \Rightarrow (l, l - b_i) \in \mathcal{G}_b$,*

where $l = f_{\|a\|_1}^{\|b\|_1, \delta}(k)$, $\mu = \|b\|_1 - 1 + \delta$ and $\nu = -\|a\|_1 + \|b\|_1 + \delta$.

Proof. First, note that the numbers l and $k\|\boldsymbol{b}\|_1 - l\|\boldsymbol{a}\|_1 - \delta$ are respectively the quotient and remainder of the Euclidean division of $k\|\boldsymbol{b}\|_1 - \delta$ by $\|\boldsymbol{a}\|_1$.

Let us denote by P_0 the proposition $\delta \leq k\|\boldsymbol{b}\|_1 - l\|\boldsymbol{a}\|_1 < \delta + \|\boldsymbol{b}\|_1$. Since $k\|\boldsymbol{b}\|_1 - l\|\boldsymbol{a}\|_1 - \delta = (k\|\boldsymbol{b}\|_1 - \delta) \bmod \|\boldsymbol{a}\|_1$, $k \in \mathcal{V}_{\|\boldsymbol{a}\|_1}^{\|\boldsymbol{b}\|_1,\delta}$ implies P_0.

In the four cases, the same proof by contradiction is used. It can be coarsely described as follows: we consider an hypothesis H, and an implication $P \Rightarrow Q$; we show that, assuming H, the negation of the implication, i.e., $(P \wedge \neg Q)$ contradicts P_0, which means that $P \Rightarrow Q$ must be true.

The hypotheses are given after the *if* in the claim of the lemma and involve bounds on the quantity $b_i\|\boldsymbol{a}\|_1 - a_i\|\boldsymbol{b}\|_1$.

The implications and the converse propositions are given in Table 1. We indeed can derive arithmetic constraints from the fact that both v_1 and v_2 must be in $\mathcal{V}_{\|\boldsymbol{a}\|_1}$ for an arc (v_1, v_2) being part of $\mathcal{S}_a^{\|\boldsymbol{b}\|_1,\delta}$. For instance, knowing that $k \in \mathcal{V}_{\|\boldsymbol{a}\|_1}^{\|\boldsymbol{b}\|_1,\delta}$, $(k, k + a_i) \in \mathcal{S}_a^{\|\boldsymbol{b}\|_1,\delta} \subseteq \mathcal{G}_a$ is equivalent to $k < \|\boldsymbol{a}\|_1 - a_i$, while $(k, k - a_i) \in \mathcal{S}_a^{\|\boldsymbol{b}\|_1,\delta} \subseteq \mathcal{G}_a$ is equivalent to $k \geq a_i$. Obviously, the same applies for $\mathcal{G}_b$. Indeed, since $k \in \mathcal{V}_{\|\boldsymbol{a}\|_1}^{\|\boldsymbol{b}\|_1,\delta}$, one has $l \in \mathcal{V}_{\|\boldsymbol{b}\|_1}$ by Lemma 2. Knowing that $l \in \mathcal{V}_{\|\boldsymbol{b}\|_1}$, $(l, l + b_i) \in \mathcal{G}_b$ is equivalent to $l < \|\boldsymbol{b}\|_1 - b_i$, while $(l, l - b_i) \in \mathcal{G}_b$ is equivalent to $l \geq b_i$.

Table 1. The implications to show are on the left. The converse propositions, which contradicts P_0, are on the right.

Case	$P \Rightarrow Q$	$P \wedge \neg Q$
(C1)	$(k < \|\boldsymbol{a}\|_1 - a_i) \Rightarrow (l < \|\boldsymbol{b}\|_1 - b_i)$	$(k < \|\boldsymbol{a}\|_1 - a_i) \wedge (l \geq \|\boldsymbol{b}\|_1 - b_i)$
(C2)	$(l \geq b_i) \Rightarrow (k \geq a_i)$	$(l \geq b_i) \wedge (k < a_i)$
(C3)	$(l < \|\boldsymbol{b}\|_1 - b_i) \Rightarrow (k < \|\boldsymbol{a}\|_1 - a_i)$	$(l < \|\boldsymbol{b}\|_1 - b_i) \wedge (k \geq \|\boldsymbol{a}\|_1 - a_i)$
(C4)	$(k \geq a_i) \Rightarrow (l \geq b_i)$	$(k \geq a_i) \wedge (l < b_i)$

From the constraints given by $P \wedge \neg Q$ (see Table 1), we introduce two integral slack variables to have expressions for k and l. Then, we compute a bound for $k\|\boldsymbol{b}\|_1 - l\|\boldsymbol{a}\|_1$ that contradicts P_0.

In the first two cases, we have an upper bound for k and a lower one for l:

- $\epsilon_A \geq 1$ is such that $k = (\|\boldsymbol{a}\|_1 - a_i) - \epsilon_A$ in (C1) and $k = a_i - \epsilon_A$ in (C2),
- $\epsilon_B \geq 0$ is such that $l = (\|\boldsymbol{b}\|_1 - b_i) + \epsilon_B$ in (C1) and $l = b_i + \epsilon_B$ in (C2).

Now, substituting k and l by their values in $k\|\boldsymbol{b}\|_1 - l\|\boldsymbol{a}\|_1$, we get for (C1):

$$k\|\boldsymbol{b}\|_1 - l\|\boldsymbol{a}\|_1 = (\|\boldsymbol{a}\|_1\|\boldsymbol{b}\|_1 - \|\boldsymbol{b}\|_1\|\boldsymbol{a}\|_1)$$
$$+ \underbrace{(b_i\|\boldsymbol{a}\|_1 - a_i\|\boldsymbol{b}\|_1)}_{\leq \mu = \|\boldsymbol{b}\|_1 - 1 + \delta} \underbrace{- \|\boldsymbol{b}\|_1\epsilon_A - \|\boldsymbol{a}\|_1\epsilon_B}_{\leq -\|\boldsymbol{b}\|_1} \leq \delta - 1,$$

which raises a contradiction, because $k\|\boldsymbol{b}\|_1 - l\|\boldsymbol{a}\|_1 \geq \delta$. We thus conclude that $(k < \|\boldsymbol{a}\|_1 - a_i) \Rightarrow (l < \|\boldsymbol{b}\|_1 - b_i)$ and $(k, k + a_i) \in \mathcal{S}_a^{\|\boldsymbol{b}\|_1,\delta} \Rightarrow (l, l + b_i) \in \mathcal{G}_b$.

Likewise, for (C2),

$$k\|b\|_1 - l\|a\|_1 = \underbrace{-(b_i\|a\|_1 - a_i\|b\|_1)}_{\leq \mu = \|b\|_1 - 1 + \delta} \underbrace{-\|b\|_1 \epsilon_A - \|a\|_1 \epsilon_B}_{\leq -\|b\|_1} \leq \delta - 1,$$

which means that $(k, k - a_i) \in \mathcal{S}_a^{\|b\|_1, \delta} \Leftarrow (l, l - b_i) \in \mathcal{G}_b$.

For the last two cases, (C3) and (C4), we can similarly conclude using ν. $\square$

The following Corollary sums up the previous results:

Corollary 1. *If* $\left\| \|b\|_1 a - \|a\|_1 b \right\|_\infty \leq \min \left(|\|b\|_1 - 1 + \delta|, |\|a\|_1 - \|b\|_1 - \delta| \right)$ *then* $\mathcal{S}_a^{\|b\|_1, \delta} \simeq \mathcal{G}_b$.

In the next section, we introduce the concept of approximation, which will be linked later with Corollary 1.

4 Diophantine Approximation

Definition 4. *Let* $a \in \mathbb{N}^3 \setminus \{0\}$ *such that* $\gcd(a) = 1$. *A vector* $b \in \mathbb{N}^3 \setminus \{0\}$ *is an* approximation *of vector* a *if and only if* $a - b \in \mathbb{N}^3 \setminus \{0\}$ *and*

$$\left\| \frac{a}{\|a\|_1} - \frac{b}{\|b\|_1} \right\|_\infty < \frac{1}{2\|b\|_1}. \tag{3}$$

In addition, an approximation b *is* reduced *if and only if* $\gcd(b) = 1$.

That definition is closely related to the simultaneous approximation of fractions. When the denominator is denoted by q, $\frac{1}{2q}$ is usually considered as a trivial bound, see [4, Section 5.2]. What is relatively uncommon here is that we consider only rationals and that the denominator are sums of numerators: $\|b\|_1 = \sum_i b_i$.

Vectors for which the left-hand side of (3) is strictly less than the trivial bound are called "approximations", while the others do not deserve to be called "approximations". The rationale for the strict inequality sign is technical: it allows us to have an "if and only if" in Lemma 7 of the next section.

Multiplying both sides of (3) by the product $\|b\|_1 \|a\|_1$, one obtains:

$$\left\| \|b\|_1 a - \|a\|_1 b \right\|_\infty < \frac{\|a\|_1}{2}. \tag{4}$$

Note that the condition of Corollary 1 also involves the left-hand side of (4). An approximation b will turn out to be a good candidate to have $\mathcal{S}_a^{\|b\|_1, \delta} \simeq \mathcal{G}_b$. However, it must be proved first that an approximation exists.

4.1   Existence of an Approximation

Lemma 4. *Let $a \in \mathbb{N}^3 \setminus \{0\}$ be such that $\gcd(a) = 1$. There exists at least one approximation of a in $\mathbb{N}^3 \setminus \{0\}$ if and only if a is not a permutation of $(0, 0, 1)$, $(0, 1, 1)$, $(1, 1, 1)$ or $(1, 1, 2)$.*

Proof ($\Longrightarrow$). One can check by enumeration that permutations of $(0, 0, 1)$, $(0, 1, 1)$, $(1, 1, 1)$ and $(1, 1, 2)$ admit no approximation.

($\Longleftarrow$) Suppose that $\|a\|_1 \leq 4$. A convenient permutation of $(0, 0, 1)$ is an approximation of any permutation of $(0, 1, 2)$ and $(0, 1, 3)$. Other cases are permutations of $(0, 0, 1)$, $(0, 1, 1)$, $(1, 1, 1)$ and $(1, 1, 2)$, which are excluded.

Assume now that $\|a\|_1 > 4$. Let us consider the open ball $\mathcal{B} := \{x \in \mathbb{R}^3 \mid \|x\|_\infty < 1/2\}$ and the images $\mathcal{B}_1$ and $\mathcal{B}_2$ of $\mathcal{B}$ under the orthogonal projection onto $(1, 1, 1)$ and onto the orthogonal complementary subspace respectively. Since both images have the same volume [6] and that the first one is trivially equal to $\sqrt{3}$, one has $vol(\mathcal{B}_2) = \sqrt{3}$.

Now, let us consider the open straight segment $\mathcal{S} := \{\lambda a \mid \lambda \in (-1, 1)\}$ and the image $\mathcal{S}_1$ of $\mathcal{S}$ under the orthogonal projection onto $(1, 1, 1)$. It is easy to see that $vol(\mathcal{S}_1) = \frac{2}{\sqrt{3}} \|a\|_1$.

Finally, let us consider the dilation of $\mathcal{S}$ by $\mathcal{B}_2$, i.e., $\mathcal{D} := \mathcal{S} \oplus \mathcal{B}_2$. It is a symmetric and convex region. Note that the volume of $\mathcal{D} = \mathcal{S} \oplus \mathcal{B}_2$ is equal to the volume of $\mathcal{S}_1 \oplus \mathcal{B}_2$, because one can transform one to the other by a volume-invariant shearing. Thus,

$$vol(\mathcal{D}) = vol(\mathcal{B}_2) vol(\mathcal{S}_1) = \sqrt{3} \frac{2}{\sqrt{3}} \|a\|_1 = 2\|a\|_1 > 2 \cdot 4 = 8.$$

By Minkowski's theorem, since the volume of $\mathcal{D}$ is strictly greater than 2^3, $\mathcal{D}$ contains at least a non-zero integer point b.

By definition of $\mathcal{D}$, we have $-\|a\|_1 < b \cdot (1, 1, 1) < \|a\|_1$ and, due to the symmetry, one can assume without loss of generality that $0 < b \cdot (1, 1, 1)$. Furthermore, $\frac{\|b\|_1}{\|a\|_1} a$ is the projection of b onto $\mathcal{S}$ along projecting lines orthogonal to $(1, 1, 1)$. Since both b and its projection are in $\mathcal{D}$ by definition, we have

$$\left\| \frac{\|b\|_1}{\|a\|_1} a - b \right\|_\infty < \frac{1}{2} \quad \text{which implies} \quad \left\| \|b\|_1 a - \|a\|_1 b \right\|_\infty < \frac{\|a\|_1}{2}. \qquad (5)$$

Consequently, b is an approximation of a. $\qquad\qquad\qquad\qquad\qquad\qquad\qquad\square$

4.2   Reduced Approximations

If an approximation exists, then a reduced one exists. That claim is rather obvious but crucial to link approximations with Christoffel subgraphs, because the latter are defined for vectors with coprime coordinates.

Lemma 5. *If b is an approximation of $a \in \mathbb{N}^3 \setminus \{0\}$ such that $\gcd(a) = 1$, then there exists a reduced approximation b^* of a.*

Proof. Let us define $b^\star := b/\gcd(b)$. It is clear that $b^\star, a - b^\star \in \mathbb{N}^3 \setminus \{0\}$ and

$$\max_i \left(|\|b^\star\|_1 a_i - \|a\|_1 b_i^\star| \right) \leq \max_i \left(|\|b\|_1 a_i - \|a\|_1 b_i| \right) < \frac{\|a\|_1}{2},$$

which means that $b^\star$ is a reduced approximation of a. □

Furthermore, in order to be able to deal with the relative complement of a subgraph with respect to a Christoffel graph, the following result is useful.

Lemma 6. *Let b be a reduced approximation of $a \in \mathbb{N}^3 \setminus \{0\}$ such that $\gcd(a) = 1$. Then $a - b$ is also a reduced approximation of a.*

Proof. Let $c := a - b$. From the hypothesis, we have that $c = a - b \in \mathbb{N}^3 \setminus \{0\}$ and $a - c = b \in \mathbb{N}^3 \setminus \{0\}$. On the other hand, one has for all $i \in \{1, 2, 3\}$,

$$\begin{aligned}
|\|c\|_1 a_i - \|a\|_1 c_i| &= |(\|a\|_1 - \|b\|_1) a_i - \|a\|_1 (a_i - b_i)| \\
&= |\|a\|_1 a_i - \|a\|_1 a_i - \|b\|_1 a_i + \|a\|_1 b_i| \\
&= |-(\|b\|_1 a_i - \|a\|_1 b_i)| \leq \left\| \|b\|_1 a - \|a\|_1 b \right\|_\infty < \tfrac{1}{2} \|a\|_1,
\end{aligned}$$

where we used (4). This means that c is a reduced approximations of a. □

5   Existence of a Partition

In this section, we show that there always exist b and δ, such that $\mathcal{G}_a = \mathcal{S}_a^{\|b\|_1, \delta} \cup \overline{\mathcal{S}_a^{\|b\|_1, \delta}}$, $\mathcal{S}_a^{\|b\|_1, \delta} \simeq \mathcal{G}_b$ and $\overline{\mathcal{S}_a^{\|b\|_1, \delta}} \simeq \mathcal{G}_{(a-b)}$. This result is based on the following:

Lemma 7. *Let $a \in \mathbb{N}^3 \setminus \{0\}$ be such that $\gcd(a) = 1$. Then b is an approximation of a if and only if there exists $\delta \in \{-\|b\|_1 + 1, \ldots, \|a\|_1 - \|b\|_1\}$ such that*

$$\left\| \|b\|_1 a - \|a\|_1 b \right\|_\infty \leq \min \left(|\|b\|_1 - 1 + \delta|, |\|a\|_1 - \|b\|_1 - \delta| \right).$$

Proof. For sake of shortness, let us set $q := \left\| \|b\|_1 a - \|a\|_1 b \right\|_\infty$. The goal is to search for δ such that $q \leq \min(|\|b\|_1 - 1 + \delta|, |\|a\|_1 - \|b\|_1 - \delta|)$. However, $\delta \in \{-\|b\|_1 + 1, \ldots, \|a\|_1 - \|b\|_1\}$ implies $0 \leq \|b\|_1 - 1 + \delta$ and $0 \leq \|a\|_1 - \|b\|_1 - \delta$. Therefore, one can equivalently search for δ such that both of these conditions are true:

$$\begin{aligned}
q &\leq \|b\|_1 + \delta - 1 &\Leftrightarrow q - \|b\|_1 + 1 \leq \delta, \\
q &\leq \|a\|_1 - \|b\|_1 - \delta &\Leftrightarrow \delta \leq \|a\|_1 - \|b\|_1 - q.
\end{aligned}$$

We conclude that such δ exists if and only if the lower bound is less than the upper bound, i.e., if and only if

$$q - \|b\|_1 + 1 \leq \|a\|_1 - \|b\|_1 - q \Leftrightarrow q \leq \frac{\|a\|_1 - 1}{2},$$

which is true if and only if b is an approximation of a. □

Based on all previous results, the main result of the paper follows.

Theorem 1. *Let $a \in \mathbb{N}^3 \setminus \{0\}$ be such that $\gcd(a) = 1$. If a is not a permutation of one of the vectors $(0,0,1)$, $(0,1,1)$, $(1,1,1)$ and $(1,1,2)$, then there exist a reduced approximation $b \in \mathbb{N}^3 \setminus \{0\}$ of a and there exists an offset δ such that*

$$\mathcal{G}_a = \mathcal{S}_a^{\|b\|_1,\delta} \cup \overline{\mathcal{S}_a^{\|b\|_1,\delta}}, \quad \mathcal{S}_a^{\|b\|_1,\delta} \simeq \mathcal{G}_b \quad and \quad \overline{\mathcal{S}_a^{\|b\|_1,\delta}} \simeq \mathcal{G}_{(a-b)}.$$

Proof. From Lemma 4 and Lemma 5, there exists a reduced approximation $b \in \mathbb{N}^3 \setminus \{0\}$ of a. There are two consequences. On the one hand, $\gcd(b) = 1$, which means that one can define $\mathcal{G}_b$. On the other hand, Lemma 7 and Corollary 1 together prove that there exists an offset δ such that $\mathcal{S}_a^{\|b\|_1,\delta} \simeq \mathcal{G}_b$ (with bijection function $f_{\|a\|_1}^{\|b\|_1,\delta}$; see Definition 3).

Then, note that $\overline{\mathcal{S}_a^{\|b\|_1,\delta}} = \mathcal{S}_a^{\|c\|_1,\gamma}$ by Lemma 1, where $c := a - b$, $\gamma := -\delta + 1$. In addition, c is also a reduced approximation of a by Lemma 6. Since, $\gcd(c) = 1$, one can define $\mathcal{G}_c$.

For that subgraph, the condition of Corollary 1 writes

$$\left\| \|c\|_1 a - \|a\|_1 c \right\|_\infty \leq \min \left(\left| \|c\|_1 - 1 + \gamma \right|, \left| \|a\|_1 - \|c\|_1 - \gamma \right| \right). \tag{6}$$

Remind that $\left\| \|c\|_1 a - \|a\|_1 c \right\|_\infty = \left\| \|b\|_1 a - \|a\|_1 b \right\|_\infty$ (see the proof of Lemma 6). In addition, note that

(i) $\|c\|_1 - 1 + \gamma = (\|a\|_1 - \|b\|_1) - 1 + (-\delta + 1) = \|a\|_1 - \|b\|_1 - \delta$,
(ii) $\|a\|_1 - \|c\|_1 - \gamma = \|a\|_1 - (\|a\|_1 - \|b\|_1) - (-\delta + 1) = \|b\|_1 - 1 + \delta$.

As a consequence, (6) is equivalent to

$$\left\| \|b\|_1 a - \|a\|_1 b \right\|_\infty \leq \min \left(\left| \|b\|_1 - 1 + \delta \right|, \left| \|a\|_1 - \|b\|_1 - \delta \right| \right),$$

which is true by Lemma 7 and one can conclude as above that $\mathcal{S}_a^{\|c\|_1,\gamma} \simeq \mathcal{G}_c$ □

Remark 1. We have not tried to identify a set Σ such that $f_{\|a\|_1}^{\|b\|_1,\delta}$ makes $\mathcal{S}_a^{\|b\|_1,\delta} \simeq \mathcal{G}_b$ for all $b \in \Sigma$ and *only them*. Indeed, in order to have an "only if" part in Theorem 1, it would be necessary to show that not only $f_{\|a\|_1}^{\|b\|_1,\delta}$, but all bijection functions fail to make $\mathcal{S}_a^{\|b\|_1,\delta} \simeq \mathcal{G}_b$ if $b \notin \Sigma$. That would require other results than those provided above, which are mainly based on $f_{\|a\|_1}^{\|b\|_1,\delta}$.

Remark 2. A few vectors are excluded from Theorem 1 because they do not have an approximation. They are, up to a permutation of the coordinates, $(0,0,1)$, $(0,1,1)$, $(1,1,1)$ and $(1,1,2)$. It is easy to see that the Christoffel graphs defined from those vectors all contain specific local configurations (see Fig. 6). Thus, they cannot contain subgraphs corresponding to a Christoffel graph defined from a shorter vector. That is why it is not possible to have a result similar to Theorem 1 for them.

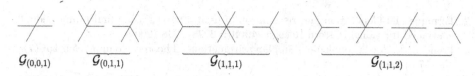

$$\mathcal{G}_{(0,0,1)} \qquad \mathcal{G}_{(0,1,1)} \qquad \mathcal{G}_{(1,1,1)} \qquad \mathcal{G}_{(1,1,2)}$$

Fig. 6. Undecomposable Christoffel graphs.

Remark 3. As shown in Fig. 5 (a–b), there may be several possible b and δ.

6   Conclusion and Perspectives

We have shown in Theorem 1 that for all $a \in \mathbb{N}^3 \setminus \{0\}$ such that $\gcd(a) = 1$ and which is not a permutation of $(0,0,1)$, $(0,1,1)$, $(1,1,1)$ or $(1,1,2)$, there are an approximation b and an offset δ such that $\mathcal{G}_a = \mathcal{S}_a^{\|b\|_1,\delta} \cup \overline{\mathcal{S}_a^{\|b\|_1,\delta}}$, $\mathcal{S}_a^{\|b\|_1,\delta} \simeq \mathcal{G}_b$ and $\overline{\mathcal{S}_a^{\|b\|_1,\delta}} \simeq \mathcal{G}_{(a-b)}$. In other words, $\mathcal{P}_a$ can be partitioned into two parts having respectively the combinatorial structure of $\mathcal{P}_b$ and $\mathcal{P}_{(a-b)}$. Compare Fig. 7 with Fig. 3.

A short-term perspective is to focus on the geometrical aspects of the partition. It seems that a rational discrete plane is decomposed into parallel strips. See Fig. 3(a). Are the subsets in each strip connected? What are the thickness and direction of the strips? Another perspective is to efficiently compute one or all approximations for which a decomposition is possible. Finally, we are also interested in the practical application of these studies in discrete geometry.

$$\mathcal{G}_{(2,3,4)}$$

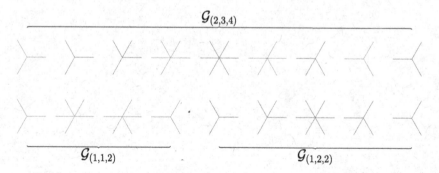

$$\mathcal{G}_{(1,1,2)} \qquad\qquad \mathcal{G}_{(1,2,2)}$$

Fig. 7. $\mathcal{G}_{(2,3,4)} = \mathcal{S}_{(2,3,4)}^{4,0} \cup \mathcal{S}_{(2,3,4)}^{5,1}$, $\mathcal{S}_{(2,3,4)}^{4,0} \simeq \mathcal{G}_{(1,1,2)}$, $\mathcal{S}_{(2,3,4)}^{5,1} \simeq \mathcal{G}_{(1,2,2)}$.

References

1. Berstel, J., Lauve, A., Reutenauer, C., Saliola, F.: Combinatorics on Words: Christoffel Words and Repetition in Words. American Mathematical Society (2008)

2. Fernique, T.: Generation and recognition of digital planes using multi-dimensional continued fractions. Pattern Recogn. **42**(10), 2229–2238 (2009)
3. Françon, J.: Sur la topologie d'un plan arithmétique. Theoret. Comput. Sci. **156**(1), 159–176 (1996)
4. Grötschel, M., Lovász, L., Schrijver, A.: Geometric Algorithms and Combinatorial Optimization. Springer, Heidelberg (1993). https://doi.org/10.1007/978-3-642-78240-4
5. Labbé, S., Reutenauer, C.: A d-dimensional extension of Christoffel words. Discret. Comput. Geom. **54**(1), 152–181 (2015)
6. McMullen, P.: Volumes of projections of unit cubes. Bull. Lond. Math. Soc. **16**(3), 278–280 (1984)
7. Senechal, M.: Quasicrystals and Geometry. Cambridge University Press, Cambridge (1995)
8. Voss, K.: Discrete Images, Objects, and Functions in Zn. Springer, Heidelberg (1993). https://doi.org/10.1007/978-3-642-46779-0

Differential Maximum Euclidean Distance Transform Computation in Component Trees

Dennis J. Silva[1,2]([✉]) [iD], Paulo André Vechiatto Miranda[1] [iD],
Wonder A. L. Alves[3] [iD], Ronaldo F. Hashimoto[1] [iD], Jiří Kosinka[2] [iD],
and Jos B. T. M. Roerdink[2] [iD]

[1] Institute of Mathematics and Statistics, University of São Paulo, R. do Matão,
1010, CEP 05508-090, Butantã, São Paulo, SP, Brazil
{dennis,pmiranda,ronaldo}@ime.usp.br
[2] Bernoulli Institute for Mathematics, Computer Science and Artificial Intelligence,
University of Groningen, Nijenborgh 9, 9747 AG Groningen, The Netherlands
{d.j.da.silva,j.kosinka,j.b.t.m.roerdink}@rug.nl
[3] Informatics and Knowledge Management Graduate Program, Nove de Julho
University, R. Vergueiro, 235/249, CEP 01525-000, Liberdade, São Paulo, SP, Brazil

Abstract. The distance transform is an important binary image transformation that assigns to each foreground pixel the distance to the closest contour pixel. Among other applications, the maximum distance transform (DT) value can describe the thickness of the connected components of the image. In this paper, we propose using the maximum distance transform value as an attribute of component tree nodes. We present a novel algorithm to compute the maximum DT value of all connected components of a greyscale image in a differential way by joining an incremental method for contour extraction in component trees and the Differential Image Foresting Transform (DIFT). We save processing time by reusing the DIFT subtrees rooted at the contour points (DIFT seeds) of a node in its ancestors until those points are not contour points anymore. We experimentally show that we can compute the maximum distance attribute twice as fast as the node-reconstruction approach. Our proposed attribute is increasing and its applicability is exemplified by the design of an extinction value filter. The ability to select thin connected components, like cables, of our filter is compared to filters using other increasing attributes in terms of their parameters and their resulting images.

Keywords: Component tree · Distance transform · Image Foresting Transform

1 Introduction

Component trees are powerful full image representations used in different image processing and analysing tasks, including designing connected operators [14],

S. Brunetti et al. (Eds.): DGMM 2024, LNCS 14605, pp. 67–79, 2024.
https://doi.org/10.1007/978-3-031-57793-2_6

text location [12], eye vessel segmentation [1], interactive image manipulation [19], and many others. In this representation, the connected components
(CCs) of the level-sets are the nodes of the tree and the subset relation of these
CCs is encoded in the parenthood relationship. An important step in processing
images using component trees is describing the nodes through attributes. Different attributes such as area, perimeter, number of holes, height, and volume
have been successfully applied in different applications [1,12,14].

The distance transform is an important binary image transformation that
assigns for each foreground pixel the distance to the closest contour pixel.
Although widely used in image processing [18], the distance transform applied to
component trees has not received much attention in the mathematical morphology community. In particular, we note that the maximum distance transform
value of the CCs describes the thickness of the CC and could be a useful increasing attribute for the nodes of a component tree. Thus, in this paper, we propose
the maximum distance transform value as an attribute of component trees. To
make it feasible, we propose a novel algorithm that combines a recent incremental
approach to compute the contours of component trees with the computation of
the distance transform using the Differential Image Foresting Transform (DIFT),
which can reuse a previously computed IFT by only recomputing the IFT on the
removed and inserted seeds. We experimentally demonstrate that our algorithm
is on average twice as fast as a non-differential approach and that the proposed
attribute can be used in extinction value filters to remove thin objects which
other increasing attributes cannot easily do.

The remainder of this paper is organised as follows. We recall some useful
definitions and notations in Sect. 2. We describe the proposed attribute and the
differential algorithm in Sect. 3. In Sect. 4, we report our run-time experiments
and present a simple example application of the proposed attribute to extinction
filters. We conclude the paper in Sect. 5.

2   Background

2.1   Images

A *greyscale image* is a function $f : D_f \to \mathbb{K}_f$ where $D_f \subseteq \mathbb{Z}^2$ is a regular grid of
pixels and $\mathbb{K}_f = \{0, 1, \ldots, K_f - 1\} \subset \mathbb{Z}$ is the greylevel set. When $K_f = 2$, f is
a *binary image* and we represent it as a set $X = \{p \in D_f : f(p) = 1\}$. We say a
pixel p is a *foreground pixel* if $p \in X$ and a *background pixel* if $p \in \mathbb{Z}^2 \setminus X$. We
can extract binary images from a greyscale image by selecting the pixels whose
grey level is greater or equal to, respectively less than or equal to, a threshold
value $\lambda \in \mathbb{Z}$ as its foreground pixels. This *thresholding* operation is denoted by

$$[f \geq \lambda] = \{p \in D_f : f(p) \geq \lambda\},$$
$$[f \leq \lambda] = \{p \in D_f : f(p) \leq \lambda\}. \tag{1}$$

We call $[f \geq \lambda]$ and $[f \leq \lambda]$), resp.,the *upper* and *lower level-set* of f wrt. λ.

We relate spatially close pixels by an adjacency relation. In particular, we define $\mathcal{N}_4(p) = \{p + q : q \in \{(-1,0),(0,-1),(1,0),(0,1)\}\}$ and $\mathcal{N}_8(p) = \mathcal{N}_4(p) \cup \{p + q : q \in \{(-1,-1),(1,-1),(1,1),(-1,1)\}\}$ as the *4-connected* and *8-connected adjacency relation* (or 4- and 8-connected neighbourhood) of p, resp. We denote an arbitrary neighbourhood of pixel p by $\mathcal{N}(p)$. If $q \in \mathcal{N}(p)$, we say q is a *neighbour* of (or adjacent to) p, and that p and q are *neighbours* (adjacent pixels). Given a set of pixels $X \subseteq \mathbb{Z}^2$, we denote the set of pairs of adjacent pixels by

$$\mathcal{A}(X) = \{(p,q) \in X^2 : q \in \mathcal{N}(p)\}. \tag{2}$$

Given a greyscale image $f : D_f \to \mathbb{K}_f$ and an adjacency relation $\mathcal{A}$, we can enrich the image by representing it by a *vertex-weighted graph* $\mathcal{G}_{f,\mathcal{A}} = (f, D_f, \mathcal{A}(D_f))$, where its vertices are pixels, its edges are defined by the adjacency relation, and the weights of the vertices are the pixel grey levels. Using the set representation of the binary image, we can also create the graph $G_{X,\mathcal{A}} = (X, \mathcal{A}(X))$ that enriches the binary image representation. Using the graph representation of binary images, we define a *path* $\pi(p,q)$ between two pixels $p,q \in X$ as the sequence $(r_0, r_1, \ldots, r_n)$ of pixels in X with $r_i \in \mathcal{N}(r_{i+1})$ for $0 \le i < n$, $r_0 = p$ and $r_n = q$. If there exists such a path $\pi(p,q)$ in X, we say p and q are *connected*, otherwise, we say they are *disconnected*. A subset of pixels $S \subseteq X$ is called *connected* if all pairs of pixels $p,q \in S$ are connected in S. If S is maximally connected, it is called a *connected component* of X. We denote the set of connected components of a graph $G_{X,\mathcal{A}}$ by $CC(G_{X,\mathcal{A}})$, and the connected component containing pixel $p \in \mathbb{Z}^2$ by $CC(G_{X,\mathcal{A}}, p)$, with $CC(G_{X,\mathcal{A}}, p) = \emptyset$ if $p \notin X$. Further, we define the family of connected components of the upper and lower level-sets by

$$\begin{aligned}\mathcal{U}(f,\mathcal{A}) &= \{C \in CC(G_{[f \ge \lambda],\mathcal{A}}) : \lambda \in \mathbb{Z}\}, \\ \mathcal{L}(f,\mathcal{A}) &= \{C \in CC(G_{[f \le \lambda],\mathcal{A}}) : \lambda \in \mathbb{Z}\}. \end{aligned} \tag{3}$$

2.2 Component Trees

A (rooted directed) *tree* T is an acyclic (directed) graph $T = (V(T), E(T))$, where $V(T)$ is the set of vertices/nodes and $E(T)$ is the set of (directed) edges. Given a node $N \in V(T)$, a tree T supports the following operations:

$$\begin{aligned}\texttt{parent}(N,T) = P &\Leftrightarrow (N,P) \in E(T), \\ \texttt{children}(N,T) = \{C &\in V(T) : (C,N) \in E(T)\}, \\ \texttt{rootTree}(T) = N &\Leftrightarrow \nexists P \in V(T) : P = \texttt{parent}(N,T). \end{aligned} \tag{4}$$

A *component tree* $T_f = (V(T_f), E(T_f))$ is a tree such that

$$V(T_f) \text{ is either } \mathcal{U}(f,\mathcal{A}) \text{ or } \mathcal{L}(f,\mathcal{A}),$$
$$E(T_f) = \{(N,P) : N, P \in V(T_f), N \subset P, \texttt{between}(N,P,T_f) = \emptyset\},$$

where

$$\texttt{between}(N,P,T_f) = \{P' \in V(T_f) : N \subset P' \subset P\}. \tag{5}$$

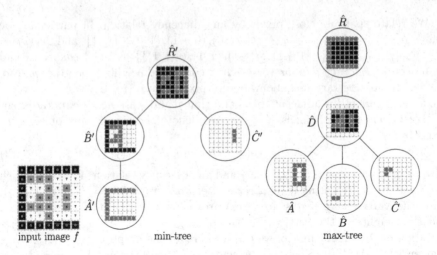

Fig. 1. Max-tree and min-tree of the input image f. The red pixels are the CNPs of the nodes, the black pixels are pixels of the node that are stored in a descendant node and the white pixels are the background pixels. (Color figure online)

We can associate the grey level at which a node N first appears during greylevel decomposition of f with the node in the component tree as follows:

$$\texttt{level}(N, T_f) = \begin{cases} \inf_{p \in N} f(p), & \text{if } V(T_f) = \mathcal{U}(f, \mathcal{A}), \\ \sup_{p \in N} f(p), & \text{if } V(T_f) = \mathcal{L}(f, \mathcal{A}). \end{cases} \tag{6}$$

Naive handling of component trees by computers can be costly due to the pixels that belong to many nodes. To alleviate this, each pixel is stored only at the first node where it becomes a foreground pixel. The full node is then formed by the pixels it stores and the pixels stored by its descendants. Such nodes are called *compact nodes* and are formally defined by $\hat{N} = N \setminus \bigcup_{C \in \texttt{children}(T_f, N)}$. We call a pixel $p \in \hat{N}$ a *compact node pixel* or CNP for short. Since the compact node that stores a pixel p is the smallest CC in the tree containing p, we call it the *small component* of p and denote it by $\mathcal{SC}(T_f, p) = N \Leftrightarrow p \in \hat{N}$. This leads to the definition of the *compact component tree* $\hat{T}_f = (\hat{V}(T_f), \hat{E}(T_f))$:

$$\hat{V}(T_f) = \{\hat{N} : N \in V(T_f)\},$$
$$\hat{E}(T_f) = \{(\hat{A}, \hat{B}) : \hat{A}, \hat{B} \in \hat{V}(T_f), (A, B) \in E(T_f)\}.$$

The compact representations of component trees built using the upper and lower level-sets are called *max-tree* and *min-tree*, respectively; see Fig. 1.

Component trees are useful tools for image processing because we can associate *attributes* to their nodes. Formally, an attribute is a function that maps component tree nodes to a set that describes some feature of the nodes. The function $\texttt{level}$ is an example of an attribute that maps nodes to their associated grey level. Other common examples are area, volume, and perimeter [11].

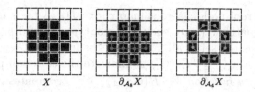

Fig. 2. Contour definition and the impact of the adopted adjacency relation. X is a binary image, $\partial_{\mathcal{A}_8}(X)$ is the contour of X extracted using $\mathcal{A}_8$ and $\partial_{\mathcal{A}_4}(X)$ is the contour of X extracted using $\mathcal{A}_4$. The graph in red (middle) shows that $\partial_{\mathcal{A}_8}(X)$ produces an $\mathcal{A}_4$-contour and the graph in green (right) shows that $\partial_{\mathcal{A}_4}(X)$ produces an $\mathcal{A}_8$-contour. (Color figure online)

Attributes that increase as we move from the leaves to the root are called *increasing*. Formally, an attribute `attr` is increasing if for $A, B \in V(T_f) : A \subset B$ implies $\mathtt{attr}(A) \leq \mathtt{attr}(B)$; e.g. area. Otherwise, the attribute is called *non-increasing*; e.g. perimeter. An *incremental algorithm* is an algorithm that computes an attribute of a node using the node's CNPs and the attribute value of its children [17]. When there exists an incremental algorithm that computes an attribute, we say that the attribute is *incremental*.

2.3 Contours

Given a binary image X and an adjacency relation $\mathcal{A}$, we say that a pixel $p \in X$ is a contour pixel if it is adjacent to a background pixel $q \in \mathbb{Z}^2 \setminus X$. The set of contour pixels of X is the *contour* of X [10]:

$$\partial_{\mathcal{A}}(X) = \{p \in X : \exists q \in \mathcal{N}(p), q \in \mathbb{Z}^2 \setminus X\}. \tag{7}$$

When extracting contours, it is important to note that the connectivity of the adjacency relation used to compute the contour is dual to the connectivity of the resulting contour. That is, $\partial_{\mathcal{A}_8}(X)$ produces $\mathcal{A}_4$-connected contours and $\partial_{\mathcal{A}_4}(X)$ results in $\mathcal{A}_8$-connected contours; see Fig. 2.

2.4 Differential Image Foresting Transform

The *Image Foresting Transform* (IFT) algorithm is a generalization of Dijkstra's algorithm for multiple sources (seeds) and more general connectivity functions. It can be applied to an image graph to develop image operators based on optimum connectivity [7]. In its seeded version, as used in this work, the search for optimal paths is restricted to paths starting in a set of seeds $S \subset D_f$.

For a given image graph $G_{X,\mathcal{A}}$ and seed set $S \subset X \subseteq D_f$, let $\pi^*(s, p)$ with $s \in S$ denote the path computed by IFT ending at pixel $p \in X \subseteq D_f$. Computed paths are stored in a predecessor map $\mathtt{pred} : D_f \to D_f^* = D_f \cup \{\mathbf{NIL}\}$, such that for each pixel $p \in X \setminus S$, $q = \mathtt{pred}(p)$ indicates the path's predecessor node of p in the computed path $\pi^*(s, p)$ and $\mathtt{pred}(s) = \mathbf{NIL}$ indicates that s is the origin of the path (root node). A root map $\mathtt{root} : D_f \to D_f$ is used to explicitly store the origin of paths $\pi^*(s,p)$, such that $\mathtt{root}(p) = s$ and $\mathtt{pred}(s) = \mathbf{NIL}$.

Let $\Pi_q(G_{X,\mathcal{A}})$ be the set of all possible paths in graph $G_{X,\mathcal{A}}$ ending at pixel $q \in X$ and $\Pi(G_{X,\mathcal{A}}) = \bigcup_{q \in X} \Pi_q(G_{X,\mathcal{A}})$ indicate the set of all possible paths in $G_{X,\mathcal{A}}$. A *connectivity function* $\Psi : \Pi(G_{X,\mathcal{A}}) \to \mathbb{R}^+$ computes a cost $\Psi(\pi(p,q))$ for any path $\pi(p,q)$. A path $\pi(p,q)$ is *optimal* if $\Psi(\pi(p,q)) \leq \Psi(\pi'(x,q))$ for any other path $\pi'(x,q) \in \Pi_q(G_{X,\mathcal{A}})$, irrespective of its starting point x. During the IFT calculation, a cost map $\mathtt{cost} : D_f \to \mathbb{R}^+$ is used to store the costs of the computed paths $\pi^*(s,p)$, such that $\mathtt{cost}(p) = \Psi(\pi^*(s,p))$ for all $p \in X$.

In the case of a sequence of IFT applications for different sets of seeds modified by insertion and/or removal of seeds and the same connectivity function, the *Differential Image Foresting Transform* (DIFT) allows the updating of paths stored in $\mathtt{pred}$ and other maps in a time proportional to the size of the modified regions in the image (i.e., in sublinear time) [6]. Let a sequence of IFTs be represented as $\langle IFT_{(\mathcal{S}^1)}, IFT_{(\mathcal{S}^2)}, \dots, IFT_{(\mathcal{S}^n)} \rangle$, where n is the total number of IFT executions on the image. At each execution, the seed set $\mathcal{S}^i$ is modified by adding and/or removing seeds to obtain a new set $\mathcal{S}^{i+1}$. We define a *scene* $\mathcal{G}^i$ as the set of maps $\mathcal{G}^i = \{\mathtt{pred}^i, \mathtt{root}^i, \mathtt{cost}^i\}$, resulting from the *i-th* iteration in a sequence of IFTs. DIFT allows to efficiently compute a scene $\mathcal{G}^i$ from the previous scene $\mathcal{G}^{i-1}$, a set $\Delta_{\mathcal{S}^i}^+ = \mathcal{S}^i \setminus \mathcal{S}^{i-1}$ of new seeds for addition, and a set $\Delta_{\mathcal{S}^i}^- = \mathcal{S}^{i-1} \setminus \mathcal{S}^i$ of seeds marked for removal, by reusing the part of the previous calculation that remains unchanged.

2.5   Distance Transform

The Distance Transform (DT) is a well-known binary image transformation that assigns to each foreground pixel the distance to the closest background/contour pixel. Formally, given a binary image X the DT at a pixel $p \in X$ is defined as

$$\mathtt{edt}(X,p) = \min_{q \in \partial_{\mathcal{A}}(X)} \|q - p\|, \tag{8}$$

where $\| \cdot \|$ denotes the standard Euclidean L_2 norm on $D_f \subset \mathbb{Z}^2$.

DT-based image operators are obtained in the IFT framework by considering the connectivity function

$$\Psi_{\mathrm{Euc}}(\pi(p,q)) = \begin{cases} \|q - p\| & \text{if } p \in \mathcal{S}; \\ +\infty & \text{otherwise.} \end{cases} \tag{9}$$

Its applications involve its use in multi-scale shape skeletonization [8,9] and shape descriptors (e.g., fractal dimensions [5], contour saliences [13], and shape descriptors based on tensor scale [2]).

The distance transform computation in a differential mode requires an updated version of the DIFT algorithm as proposed in [4]. In this work, we adopt this algorithm with some extra modifications, since we also have here the expansion of the graph $G_{X,\mathcal{A}}$ and not just the modification of the set of seeds from $\mathcal{S}^{i-1}$ to $\mathcal{S}^i$, as explained in Sect. 3, given that X grows as we move towards the root of the component tree. Therefore, we have a different X_i for each execution of DIFT with $X_{i-1} \subset X_i$. For each new node of the expanded graph in $X_i \setminus X_{i-1}$, its neighbours in X_{i-1} that already have computed paths, rooted in seeds not marked for removal, must also be inserted into the priority queue used by DIFT to allow their future path extensions.

3   Proposed Method

3.1   Proposed Attribute

We define the *maximum distance transform value* (or maximum distance for short) $\mathtt{maxDist} : V(T_f) \to \mathbb{R}$ as an attribute of component trees for a node $N \in V(T_f)$ as

$$\mathtt{maxDist}(N) = \max_{p \in N} \mathtt{edt}(N, p). \tag{10}$$

Since a node N can never lose pixels when we move to its ancestors, the $\mathtt{edt}$ of the pixels never decreases, making the maximum distance an increasing attribute (see Sect. 2.2). Consequently, it can be used in an extinction value filter (see Sect. 4.2). It describes the thickness of the node in such a way that a thin object independent of its size (area) has a low value of maximum distance. We propose an efficient differential algorithm to compute it.

3.2   Differential Algorithm

Our proposed algorithm processes the max-tree from its leaves to the root. For each level-set, we collect the new seeds by finding new contour pixels and the removed seeds by finding the contour pixels from the previously processed level-set that are not contour pixels on the level-set which is currently being processed. Then, we use DIFT for distance transform computation using the seeds from the previously computed level-set (maintained seeds), the found new seeds (inserted seeds), and removing the seeds (removed seeds) that are not contour pixels in the level-set being processed. The maximum distance transform value for each DIFT root is mapped to a contour pixel of the node. Finally, we scan the contour pixels of the node (computed incrementally) to find the maximum distance transform value mapped to the DIFT root in the previous step.

We summarise these steps in Algorithm 1. In the algorithm, we denote by $\mathbb{Z}^+$ the set of positive integers, $\mathbb{R}^+$ the set of positive real numbers, and $D_f^* = D_f \cup \{\mathbf{NIL}\} : \mathbf{NIL} \notin D_f$, where $\mathbf{NIL}$ denotes an invalid pixel.

Algorithm 1 starts by initialising the $\mathtt{maxDist}$ map, the DIFT variables, and the incremental contour variables in lines 2–7. In line 8, we create a map $\mathtt{levelToNodes}$ that maps to λ all nodes associated with grey level λ such that we can quickly access all nodes associated with λ. Then, we loop over all levels λ of the input image in lines 9–35 and skip processing levels that are not in the image using the **if** statement in line 10. Then, we loop over all nodes N associated with λ in lines 11–32. For each node N, we (*i*) remove the background neighbours by scanning the contour of the children and checking if the contour pixel is a

---

Algorithm 1: Differential algorithm

Input: A greyscale image $f : D_f \to \mathbb{K}_f$ and its max-tree $\hat{T}_f = (\hat{V}(T_f), \hat{E}(T_f))$
Output: A map $\text{maxDist} : V(T_f) \to \mathbb{R}^+$

1   **Function** computeMaximumDistanceDifferential $(f, \hat{T}_f)$
2   Let $\text{maxDist} : V(T_f) \to \mathbb{R}^+$ with $\text{maxDist}[N] = 0, \forall N \in V(T_f)$;
3   Let $\text{bin} = \emptyset$ and $\text{cost} : D_f \to \mathbb{R}^+$ be the cost image;
4   Let $\text{pred} : D_f \to D_f^*$ with $\text{pred}[p] = \textbf{NIL}, \forall p \in D_f$;
5   Let $\text{root} : D_f \to D_f$ with $\text{root}[p] = p, \forall p \in D_f$ and Q be a cost queue;
6   Let $\text{contours} : V(T_f) \to \mathcal{P}(D_f)$ with $\text{contours}[N] = \emptyset, \forall N \in V(T_f)$;
7   Let $\text{ncount} : D_f \to \mathbb{Z}^+$ with $\text{ncount}[p] = 0, \forall p \in D_f$;
8   Let $\text{levelToNodes} : \mathbb{K}_f \to V(T_f)$ with
    $\text{levelToNodes}[\lambda] = \{N \in V(T_f) : \text{level}(T_f, N) = \lambda\}, \forall \lambda \in \mathbb{K}_f$;
9   **foreach** $\lambda \in \max(\mathbb{K}_f)$ *down to* $\min(\mathbb{K}_f)$ **do**
10   **if** $\text{levelToNodes}[\lambda] = \emptyset$ **then continue**;
11   **foreach** $N \in \text{levelToNodes}[\lambda]$ **do**
12   Let $\text{toRemove} = \emptyset$ and $\text{Ncontour} = \emptyset$ be two sets;
13   **foreach** $C \in \text{children}(T_f, N)$ **do**
14   **foreach** $p \in \text{contours}[C]$ **do**
15   **foreach** $q \in \mathcal{N}_4(p)$ **do**
16   **if** $q \in D_f$ *and* $f(q) = \text{level}(T_f, N)$ **then**
17          $\text{ncount}[p]\text{--}$;
18   **if** $\text{ncount}[p] = 0$ **then** $\text{toRemove} \leftarrow \text{toRemove} \cup \{p\}$;
19   **else** $\text{Ncontour} \leftarrow \text{Ncontour} \cup \{p\}$;
20   **if** $\text{toRemove} \neq \emptyset$ **then**
      $\text{treeRemoval}(\text{toRemove}, \text{bin}, Q, \text{root}, \text{pred}, \text{cost})$;
21   **foreach** $p \in \hat{N}$ **do**
22       $\text{bin} \leftarrow \text{bin} \cup \{p\}$;
23   **foreach** $q \in \mathcal{N}_4(p)$ **do**
24   **if** $q \notin D_f$ *or* $f(p) > f(q)$ **then** $\text{ncount}[p]\text{++}$;
25   **if** $\text{ncount}[p] > 0$ **then**
26        $\text{Ncontour} \leftarrow \text{Ncontour} \cup \{p\}$;
27        $\text{root}[p] \leftarrow p, \text{pred}[p] \leftarrow \textbf{NIL}, \text{cost}[p] \leftarrow 0$;
28        $\text{insert}(Q, \text{cost}, p)$;
29   **else**
30        $\text{cost}[p] \leftarrow +\infty$;
31   **foreach** $q \in \mathcal{N}_4(p) : q \in \text{bin}$ **do**
32   **if** $q \notin Q$ *and* $\text{cost}[q] \neq +\infty$ **then** $\text{insert}(Q, \text{cost}, q)$;
33     $\text{Bedt} \leftarrow \text{EDTDiff}(Q, \mathcal{A}_8, \text{bin}, \text{root}, \text{pred}, \text{cost})$
34   **foreach** $N \in \text{levelToNodes}[\lambda]$ **do**
35      $\text{maxDist}[N] \leftarrow \max_{p \in \text{contours}[N]} \text{Bedt}[p]$
36   **return** maxDist;

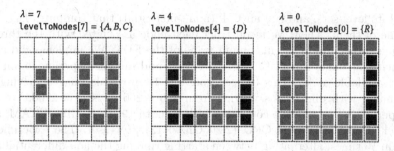

Fig. 3. MIR sets for each level-set of the image in Fig. 1. The pixels in black represent the seeds which are maintained from the previous computed λ, the green pixels represent the inserted seeds at the current level λ, and the red pixels represent the seeds removed from the previous processed λ. (Color figure online)

neighbour of a CNP of N (lines 15–17), (ii) if the analysed contour pixel has no background neighbour anymore it is included in the sets of seeds to be removed, otherwise, it is a maintained seed and is included in the contour of N (lines 18 and 19). In line 20, we remove the collected seeds by calling `treeRemoval` (function as described in [4,6]). Next, we scan the CNPs of N (lines 21–32). For each CNP p, we (i) include it in the current level-set (line 22) such that when finishing the loop of lines 11–32, we have `bin` $= [f \geq \lambda]$, (ii) we count the background neighbours of p (line 24), (iii) if p has at least one background neighbour, we include it in the contour of N and update DIFT variables to include it as a new seed (lines 25–28), otherwise we set the variables of the DIFT to make it a non-seed pixel which needs to be processed (lines 30–32). In line 33, all nodes associated with λ have been processed, the DIFT variables are set and we can run the DIFT to compute the distance transform of the level-set $[f \geq \lambda]$. Then, we call `EDTDiff` in line 33 which computes the distance transform using a DIFT and returns an image `Bedt` containing the maximum distance transform valued mapped to a DIFT's seed (contour pixel) of the node. Finally, we extract the maximum DT value from this boundary image `Bedt` for all nodes associated with λ by scanning their contour pixels in lines 34–35.

The main idea of Algorithm 1 is to incrementally keep the sets of maintained, inserted, and removed seeds (MIR sets for short). It keeps these sets implicitly in lines 18 (removed seeds), 19 (maintained seeds), and 26 (inserted seeds). Using MIR sets we can quickly set up a DIFT for distance transform computation and find its maximum value for each CC. Figure 3 depicts the MIR sets for each level-set of the image in Fig. 1.

4   Experimental Results

4.1   Run Time Analysis

We have performed experiments to compare Algorithm 1 to a non-differential approach. In the non-differential approach, we decompose the input image f

into all its level-sets, and run an IFT (non-differential) that computes EDT for each one of them. Then, we reconstruct each node associated with the threshold value and compute the maximum value of the EDT. We implemented both approaches in single-thread C++14 programs, and ran the experiments on the dataset available at [3]. This dataset contains 18 greyscale images varying on content and with dimensions varying from 256×256 to 8218×2700. We ran the experiments on a laptop computer with Ubuntu 19.10, 16 GiB of RAM, and an Intel®Core™i7-8750H CPU (2.20 Ghz ×12) processor. The experiments were run 10 times, alternating between starting running the non-differential and differential approach. Then, we took the average run-time of the 10 runs for each image. The details of the experiments, including source code, scripts, and results, are available on GitHub [16]. In summary, our differential approach was on average 2.01 times faster than the non-differential approach. In addition, the differential approach was at least 1.04 times faster and was maximally 2.87 times faster than the non-differential approach.

4.2 Extinction Value Filters

If we have an increasing attribute and consider a greyscale image as a topographic surface in which the ground is at level zero and the grey-level of each pixel denotes its altitude, we can associate to the peaks the lowest attribute value such that peak will be extinguished. These attribute values are called *extinction values* and can be associated with leaves of a component tree [15]. Then, we can apply a connected filter to the image by sorting the leaves by their extinction value, selecting a desired number L of leaves, and pruning the branches associated with L lowest extinction values. Height, area, volume and many other increasing attributes are applied in different applications. In particular, moment of inertia is an increasing attribute commonly used in applications that rely on object thickness [20]. Similarly, our proposed attribute is increasing and describes the

Fig. 4. Synthetic figure for attribute illustrations. The max-tree of the image contains three leaves highlighted in red, green, and blue. Their respective moment of inertia extinction values are $+\infty$, 9.51×10^6, and 9.59×10^7, and maximum distance extinction values are $+\infty$, 2350, and 53. Unlike the moment of inertia, the maximum distance extinction value of the spiral is lower than that of the circle. (Color figure online)

thickness of the object. However, our proposed attribute can better describe the thickness of elongated objects as shown in Fig. 4.

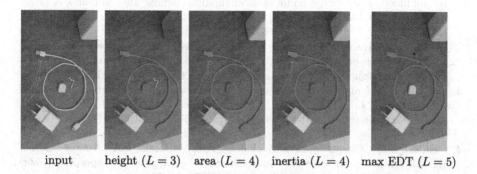

input   height $(L = 3)$ area $(L = 4)$ inertia $(L = 4)$ max EDT $(L = 5)$

Fig. 5. Extinction value filter for different attributes. Each image corresponds to the input image after an extinction value filter applied to keep L leaves for each attribute. The number L was chosen as the highest L-value for which the cable is filtered out.

We demonstrate an extinction value filter to remove a cable on a desk with different objects in Fig. 5. The figure shows that our attribute correctly filters thin objects out before filtering out other thick objects. In this example, only our proposed algorithm kept the semi-rounded eraser.

5 Conclusion

The distance transform of a binary image is an important operator used in many applications. In particular, it can be used to describe the thickness of connected components. In this work, we have proposed using the maximum distance transform value as an attribute of component trees. To do so, we have introduced a novel algorithm that adapts the incremental contour computation and the Differential Image Foresting Transform to quickly compute this attribute. We experimentally show that our algorithm is on average twice as fast as the non-differential approach. We have shown that this attribute is increasing and demonstrated its usage in an extinction value filter to remove thin objects of a greyscale image, also comparing it to other increasing attributes.

Acknowledgment. Dennis J da Silva, Ronaldo F. Hashimoto, Paulo A.V. Miranda and Wonder A.L. Alves acknowledge CNPq - Conselho Nacional de Desenvolvimento Científico e Tecnológico (Proc. 141422/2018-1, Proc. 428720/2018-7, Proc. 407242/2021-0, and Proc. 313087/2021-0) for financial support. Ronaldo F. Hashimoto and Wonder A.L. Alves acknowledge FAPESP - Fundação de Amparo a Pesquisa do Estado de São Paulo (Proc. 2015/22308-2 and 2018/15652-7) for financial support.

References

1. Alves, W.A.L., Gobber, C.F., Araújo, S.A., Hashimoto, R.F.: Segmentation of retinal blood vessels based on ultimate elongation opening. In: Campilho, A., Karray, F. (eds.) ICIAR 2016. LNCS, vol. 9730, pp. 727–733. Springer, Cham (2016). https://doi.org/10.1007/978-3-319-41501-7_81
2. Andalo, F.A., Miranda, P.A.V., da S. Torres, R., Falcão, A.X.: A new shape descriptor based on tensor scale. In: International Symposium on Mathematical Morphology and Its Application to Signal and Image Processing, pp. 141–152 (2007)
3. Carlinet, E., Géraud, T.: A comparison of many max-tree computation algorithms. In: Hendriks, C.L.L., Borgefors, G., Strand, R. (eds.) ISMM 2013. LNCS, vol. 7883, pp. 73–85. Springer, Heidelberg (2013). https://doi.org/10.1007/978-3-642-38294-9_7
4. Condori, M.A., Cappabianco, F.A., Falcão, A.X., Miranda, P.A.: An extension of the differential image foresting transform and its application to superpixel generation. J. Vis. Commun. Image Representation **71**, 102748 (2020)
5. da S. Torres, R., Falcão, A., da F. Costa, L.: A graph-based approach for multiscale shape analysis. Pattern Recogn. **37**(6), 1163–1174 (2004)
6. Falcão, A.X., Bergo, F.P.: Interactive volume segmentation with differential image foresting transforms. IEEE Trans. Med. Imaging **23**(9), 1100–1108 (2004)
7. Falcão, A.X., Stolfi, J., de Alencar Lotufo, R.: The image foresting transform: theory, algorithms, and applications. IEEE Trans. Pattern Anal. Mach. Intell. **26**(1), 19–29 (2004)
8. Falcão, A., da F. Costa, L., da Cunha, B.: Multiscale skeletons by image foresting transform and its application to neuromorphometry. Pattern Recogn. **35**(7), 1571–1582 (2002)
9. Falcão, A., Feng, C., Kustra, J., Telea, A.: Chapter 2 - multiscale 2D medial axes and 3D surface skeletons by the image foresting transform. In: Saha, P.K., Borgefors, G., Baja, G.S.D. (eds.) Skeletonization, pp. 43–70. Academic Press (2017)
10. Gonzalez, R.C., Woods, R.E.: Digital Image Processing, 4th edn. Pearson, London (2018)
11. Najman, L., Couprie, M.: Building the component tree in quasi-linear time. IEEE Trans. Image Process. **15**(11), 3531–3539 (2006)
12. Neumann, L., Matas, J.: Real-time scene text localization and recognition. In: 2012 IEEE Conference on Computer Vision and Pattern Recognition (CVPR), pp. 3538–3545. IEEE (2012)
13. da S. Torres, R., Falcão, A.: Contour salience descriptors for effective image retrieval and analysis. Image Vision Comput. **25**(1), 3–13 (2007). SIBGRAPI
14. Salembier, P., Wilkinson, M.H.: Connected operators. IEEE Signal Process. Mag. **26**(6), 136–157 (2009)
15. Silva, A.G., Lotufo, R.d.A.: New extinction values from efficient construction and analysis of extended attribute component tree. In: SIBGRAPI, pp. 204–211 (2008)
16. Silva, D.J., Miranda, P.A.V., Alves, W.A.L., Hashimoto, R.F., Kosinka, J., Roerdink, J.B.T.M.: Differential maximum Euclidean distance transform value computation in component trees - experiments analysis webpage and source code (2023). https://github.com/dennisjosesilva/max_dist_diff. Accessed 10 Oct 2023
17. Silva, D.J., Alves, W.A., Hashimoto, R.F.: Incremental bit-quads count in component trees: theory, algorithms, and optimization. Pattern Recogn. Lett. **129**, 33–40 (2020)

18. Telea, A.C.: Data Visualization: Principles and Practice, 2 edn., pp. 363–379. CRC Press (2014). Chap. Image visualization

19. Wang, J., Silva, D.J., Kosinka, J., Telea, A., Hashimoto, R.F., Roerdink, J.B.T.M.: Interactive image manipulation using morphological trees and spline-based skeletons. Comput. Graph. **108**, 61–73 (2022)

20. Wilkinson, M.H.F., Roerdink, J.B.T.M.: Fast morphological attribute operations using Tarjan's union-find algorithm. In: Goutsias, J., Vincent, L., Bloomberg, D.S. (eds.) Mathematical Morphology and its Applications to Image and Signal Processing. Computational Imaging and Vision, vol. 18, pp. 311–320. Springer, Boston (2002). https://doi.org/10.1007/0-306-47025-X_34

Construction of Fast and Accurate 2D Bijective Rigid Transformation

Stéphane Breuils[1]([✉])[iD], David Coeurjolly[2][iD], and Jacques-Olivier Lachaud[1][iD]

[1] University of Savoie Mont-Blanc, LAMA Laboratory, Chambéry, France
{stephane.breuils,jacques-olivier.lachaud}@univ-smb.fr
[2] University of Lyon, CNRS, INSA Lyon, UCBL, LIRIS, UMR5205, Lyon, France
david.coeurjolly@lcnrs.fr

Abstract. Preserving surfaces or volumes of digital objects is crucial when applying transformations of 2D/3D digital objects in medical images and computer vision. To achieve this goal, the digital geometry community has focused on characterizing bijective digitized rotations and reflections. However, the angular distribution of these bijective rigid transformations is far from being dense. Other bijective approximations of rigid transformations have been proposed, but the state-of-the-art methods lack the experimental evaluations necessary to include them in real-life applications. This paper presents several new methods to approximate digitized rotations with bijective transformations, including the composition of bijective digitized reflections, bijective rotation by circles and bijective rotation through optimal transport. These new methods and several classical ones are compared both in terms of accuracy with respect to Euclidean rotations, and in terms of computational complexity and practical speed in real-time applications.

1 Introduction

While rotations and translations in $\mathbb{R}^d$ are trivial isometric and bijective transforms, their digitized cousins in $\mathbb{Z}^d$ have attracted a lot more attention as in general, they do not preserve distances and are not bijective. Of course, direct applications of such transformations in $\mathbb{Z}^d$ belong to the image processing or computer vision fields (template matching, object tracking...). However, the study of digitization effects of such rigid motions in $\mathbb{Z}^d$ has led to interesting number theoretic and arithmetical results. For instance, one can characterize the set of angles for which the digitized rotation is bijective in $\mathbb{Z}^2$ [1], in $\mathbb{Z}^3$ [2], on the hexagonal lattice [3]. We can also consider rigid motions from quasi-shear transforms [4–6], or reflections [7,8]. For specific applications, we can even look for an approximation of the rotation preserving the homotopy for subsets of $\mathbb{Z}^2$ [9].

In this article, we follow this line of previous works focusing on a more practical question in $\mathbb{Z}^2$: for a given rotation angle, what is the best discrete bijective transformation we can have. More precisely, we are looking for a bijective transformation from $\mathbb{Z}^2$ to $\mathbb{Z}^2$ (or subsets of $\mathbb{Z}^2$) that minimizes a distance-based metric. In this context, we review existing bijective rotation approaches and propose

S. Brunetti et al. (Eds.): DGMM 2024, LNCS 14605, pp. 80–92, 2024.
https://doi.org/10.1007/978-3-031-57793-2_7

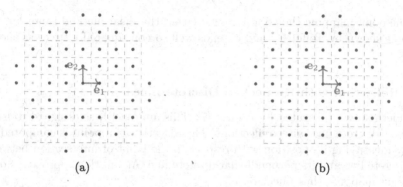

Fig. 1. 2D digitized rotations of points in blue. The digitized rotation of (a) is not bijective since it yields holes and double points whereas the rotation of (b) is bijective. (Color figure online)

two new approaches: the first one relies on composition of discrete reflections following [8], the second one on bijective rotation using circular annulus [10]. We demonstrate that the composition of four discrete reflections leads to the rotation with the lowest metric error. For this last approach, we also provide a lookup table that returns the best sequence of reflections (and their parameter) for a large set of angles.

2   Bijectivity of Digitized Rigid Transformation

Let us consider rigid transformations that act on the integer lattice $\mathbb{Z}^d$. A digitized rigid transformation is the composition of a rigid transformation $T \in E(d)$ (element of the Euclidean group) and a rounding to the nearest integer operator $\mathcal{D} : \mathbb{R}^d \to \mathbb{Z}^d$. The rigid transformation is bijective whereas the rounding operator is not, see Fig. 1.

However, there are two possibilities to ensure bijectivity of digitized rigid transformations, either

(a) define a digitized transformation that leaves invariant lines (reflection) or circles (rotation),
(b) or characterize rigid transformations that are bijective after digitization. Once the characterization is known, it is not difficult to approximate any rigid transformation with the "nearest" bijective one.

In the following, we only consider reflections and rotations that act on the integer lattice $\mathbb{Z}^2$. We start by briefly recalling the state-of-the-art bijective approximation of rigid transformations.

2.1   Quasi-Shears

The quasi-shear approach by Andres [4] consists in the discretization of the continuous horizontal and vertical shears that approximate rotation. This is a one-one point mapping thus bijective. Note that the shear method is not limited to 2D as shown by Toffoli [6] and extends well to non squared lattices as shown by [5].

2.2 Reflection with Respect to Discrete Lines

This method was presented by Andres [7]. This approach was designed to leave invariant discrete lines after reflections. The reflection of a point is computed by simply identifying its position with respect to the point of intersection between the discrete line and its perpendicular discrete line. Again, this approach is one-one point mapping thus bijective.

2.3 Bijective Digitized Reflections and Rotations

We focus in this subsection on the approach that follows (b). First, let us recall the necessary and sufficient condition for a digitized rotation and a digitized reflection to be bijective. This condition provides the subsets

- R^k such that $\forall R_\alpha \in R^k$, $\mathcal{D} \circ R_\alpha$ is bijective for digitized rotations,
- H^k such that $\forall H_\mathbf{m} \in H^k$, $\mathcal{D} \circ H_\mathbf{m}$ is bijective for digitized reflections ($\mathbf{m}$ being the normal vector of the hyperplane used for the reflection).

We start by recalling the characterization of 2D bijective digitized rotation with angle $\alpha \in [0, \frac{\pi}{2}]$ made by Nouvel and Rémila [1]:

$$R^k = \left\{ \cos(\alpha) = \frac{2k + 1}{2k^2 + 2k + 1}, \sin(\alpha) = \frac{2k^2 + 2k}{2k^2 + 2k + 1}, k \in \mathbb{Z}^+ \right\}. \qquad (1)$$

More recently, Roussillon and Coeurjolly [11] expressed the bijectivity condition in the complex plane by the Gaussian integers $\gamma \in \mathbb{Z}[i]$ (ring of Gaussian integers) as follows:

$$R^k = \left\{ \frac{\gamma \cdot \gamma}{\sqrt{(\gamma \cdot \gamma^*)}} \mid \gamma = (k + 1) + ki, k \in \mathbb{Z}^+ \right\}. \qquad (2)$$

In order to express the bijectivity condition of digitized reflections, Breuils et al. [8] used the Geometric Algebra $\mathbb{G}^2$ with basis vectors $\mathbf{e}_1, \mathbf{e}_2$. The resulting subset H^k of bijective digitized reflections is

$$H^k = H_1^k \cup H_2^k \cup H_3^k \cup H_4^k, \qquad (3)$$

where

$$H_1^k = \{\mathbf{m} \in \mathbb{G}^2, \mathbf{m} = -k\mathbf{e}_1 + (k+1)\mathbf{e}_2\}, \ H_2^k = \{\mathbf{m} \in \mathbb{G}^2, \mathbf{m} = -(k+1)\mathbf{e}_1 + k\mathbf{e}_2\},$$
$$H_3^k = \{\mathbf{m} \in \mathbb{G}^2, \mathbf{m} = -\mathbf{e}_1 + (2k+1)\mathbf{e}_2\}, \ H_4^k = \{\mathbf{m} \in \mathbb{G}^2, \mathbf{m} = -(2k+1)\mathbf{e}_1 + \mathbf{e}_2\}.$$
$$(4)$$

Note that the reflection $\mathbf{p}'$ of a point $\mathbf{p} \in \mathbb{R}^2$ with respect to a hyperplane of normal vector $\mathbf{m} \in \mathbb{G}^2$ can be written as

$$\mathbf{p}' = -\mathbf{m}\mathbf{p}\mathbf{m}^{-1} = \mathbf{p} - 2\frac{\mathbf{m} \cdot \mathbf{p}}{\mathbf{m} \cdot \mathbf{m}}\mathbf{m}, \qquad (5)$$

where $\mathbf{mp} = \mathbf{m} \cdot \mathbf{p} + \mathbf{m} \wedge \mathbf{p}$ represents the geometric product between the vector $\mathbf{m}$ and $\mathbf{p}$. Note, that the geometric product acts on basis vectors as follows

$$\mathbf{e}_i \mathbf{e}_j = \begin{cases} 1 & \text{if } i = j \\ -\mathbf{e}_{ji} & \text{otherwise} \end{cases} \quad \text{and} \quad \mathbf{e}_{ij}\mathbf{e}_k = \begin{cases} \mathbf{e}_i & \text{if } j = k \\ -\mathbf{e}_j & \text{if } i = k \end{cases}. \tag{6}$$

where $\mathbf{e}_{ij}$ is called a basis bivector.

In geometric algebra [12], a rotation can be expressed as the composition of two reflections. Assuming $\mathbf{m}, \mathbf{n}$ be the two unit normal vectors of the reflections, the rotation is expressed as

$$\mathbf{x}' = (\mathbf{nm}) \times (\mathbf{nm})^{-1} = (\cos \tfrac{\alpha}{2} + \sin \tfrac{\alpha}{2} \, \mathbf{e}_{12}) \times (\cos \tfrac{\alpha}{2} - \sin \tfrac{\alpha}{2} \, \mathbf{e}_{12}). \tag{7}$$

Furthermore, the composition of reflections with normal vectors $\mathbf{m}_1, \mathbf{m}_2, \cdots, \mathbf{m}_n$ is expressed as the reflection induced by the hyperplane defined by the geometric product of the normal vectors

$$\mathbf{m}_1 \mathbf{m}_2 \cdots \mathbf{m}_n. \tag{8}$$

As a consequence, if n is even and each normal vector is a unit vector then the above geometric product acts as a rotation on a point $\mathbf{x}$.

3   Composition of Bijective Digitized Reflections

These characterisations lead to bijective digitized rigid transformations. However, the resulting angular distribution of both R^k and H^k is far from being dense, see Fig. 2. When computing a rotation by an angle not in R^k and H^k, one option would be to consider the nearest bijective rotation or reflection. However, this leads to low quality transformations (*e.g.* mean squared distance error between rotated grid points and the original subset of the grid). Since the composition of an even number of reflections results to a rotation, an alternative is to compose bijective digitized reflections, for instance 4 of them, to approximate a given target rotation angle. More precisely, we aim at constructing a look-up table that associates, to some prescribed rotation angles, the sequence of reflections that miminises some error metrics.

3.1   Candidate Set Construction and Duplicates

First of all, let us fix $k = k_{\max}$ and compute the composition of the elements of $H^{k_{\max}}$. The set of composition of 4 bijective digitized reflections $C^{k_{\max}}$ is expressed as

$$C^{k_{\max}} = \{(\mathbf{m}_1, \mathbf{m}_2, \mathbf{m}_3, \mathbf{m}_4) \mid \mathbf{m}_1, \mathbf{m}_2, \mathbf{m}_3, \mathbf{m}_4 \in H^{k_{\max}}\}. \tag{9}$$

From our experiments, we do not compose more reflections than 4 since the maximum angular uncertainty is already lower than one degree 0.00015 rad for $k_{\max} = 15$. Figures 2c and 2d show the two normalised angular histograms of $C^{k_{\max}}$ for 2 and 4 reflections.

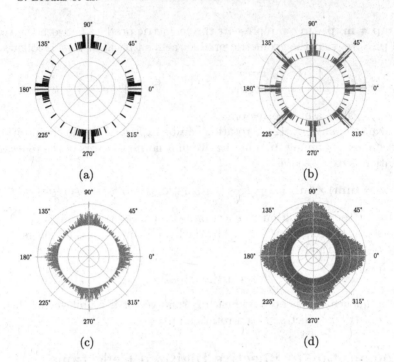

Fig. 2. Angular density distributions of bijective digitized rotations colored in black (a), bijective digitized reflections colored in blue (b), the composition of 2 bijective digitized reflections colored in red (c) and the composition of 4 bijective digitized reflections colored in green (d). (Color figure online)

Remark 1. $card(C^{k_{\max}}) = (4k_{\max})^4$.

Proposition 1. $R^{k_{\max}} \subset C^{k_{\max}}$.

Proof. Let $\gamma = (k+1) + ki \in R^{k_{\max}}$ and since the subalgebra composed of the scalar and bivector $\mathbb{R} \oplus \bigwedge^2 \mathbb{R}^2$ is isomorphic to the complex numbers $((\mathbf{e}_{12})^2 = i^2 = -1)$, then $\exists \mathbf{m}_1, \mathbf{m}_2 \in H^{k_{\max}}, \gamma = \mathbf{m}_1 \mathbf{m}_2$. For instance, choose $\mathbf{m}_1 = \mathbf{e}_1, \mathbf{m}_2 = (k+1)\mathbf{e}_1 + k\mathbf{e}_2$. Their product is

$$\mathbf{m}_1 \mathbf{m}_2 = (\mathbf{e}_1)((k+1)\mathbf{e}_1 + k\mathbf{e}_2) = (k+1) + k\mathbf{e}_{12} \in R^{k_{\max}}.$$

□

We also observe that several compositions of bijective digitized reflections result in the same rotation angle. This becomes critical if the value of $k_{\max}$ increases (with $k_{\max} = 15$, $card(C^{k_{\max}}) = 40.10^6$). In order to reduce overhead associated to the storage of this table, the duplicates must be removed. This involves sorting $C^{k_{\max}}$ by increasing angle of the resulting rotation. Furthermore, it is important to note that two compositions resulting in the same angle might have different digitizations, see Fig. 3. Thus, for each angle α, we choose

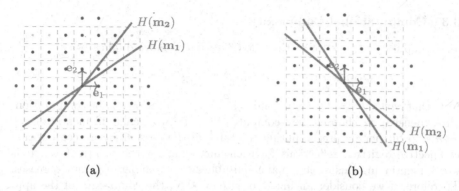

Fig. 3. Composition of 2 bijective digitized reflections $\mathbf{m}_1, \mathbf{m}_2$. (a) $\mathbf{m}_1 = -2\mathbf{e}_1 + 3\mathbf{e}_2$, $\mathbf{m}_2 = -5\mathbf{e}_1 + 4\mathbf{e}_2$. (b) $\mathbf{m}_1 = 4\mathbf{e}_1 + 5\mathbf{e}_2$, $\mathbf{m}_2 = 3\mathbf{e}_1 + 2\mathbf{e}_3$. The angle between the normal vectors in (a) and (b) are the same whereas their digitizations (points in blue) are different, for example $4\mathbf{e}_1 + \mathbf{e}_2$ is in the lattice of (b) but not (a). (Color figure online)

$$\text{RefRot}[\alpha] := \underset{(\mathbf{m}_1, \mathbf{m}_2, \mathbf{m}_3, \mathbf{m}_4) \in C^{k_{\max}}}{\arg\min} \|\Pi_{i=1}^n (\mathscr{D} \circ -\mathbf{m}_i \mathbf{p} \mathbf{m}_i^{-1}) - R_\alpha(\mathbf{p})\|_\infty. \quad (10)$$

Note that $\Pi_{i=1}^n \mathbf{m}_i = \cos\left(\frac{\alpha}{2}\right) + \sin\left(\frac{\alpha}{2}\right)\mathbf{e}_{12}$. Furthermore, in practice, we choose $\mathbf{p} \in D_p$ where $D_p = \mathbb{Z}^2 \cap [-100, 100]^2$. Finally, with $k_{\max} = 15$, we reduce $card(C^{k_{\max}})$ to 10^5 compositions instead of 40.10^6 before.

3.2   Rotation Angle to the most Accurate Bijective Composition

Given a target rotation angle α and $C^{k_{\max}}$ sorted and without duplicates, we seek for $C_i^{k_{\max}} \in C^{k_{\max}}$ that best approximates $R_\alpha(\mathbf{p})$. Firstly, since $C^{k_{\max}}$ is sorted by ascending angle, finding $C_i^{k_{\max}}$ with a resulting angle α is a binary search operation. However, this element is not necessarily the composition that is the most accurate, meaning that minimizes the distance with the Euclidean rotation $R_\alpha(\mathbf{p})$. For instance, let us consider a target rotation angle ϵ near 0, there might be a composition of 4 bijective reflections resulting in ϵ whereas the most accurate one is simply the composition of the two trivials reflections with normal vectors $\mathbf{m}_1 = \mathbf{m}_2 = \mathbf{e}_1$. To address this issue, we start by computing the K compositions of bijective digitized reflections nearest to $C_i^{k_{\max}}$ where $C_i^{k_{\max}}$ is the composition of bijective digitized reflections whose resulting angle is the closest to the target angle. We call this subset $NN(C_i^{k_{\max}}, \alpha)$. We then seek for the composition of digitized reflections that minimizes either

$$\widetilde{R_\alpha} = \underset{(\mathbf{m}_1, \mathbf{m}_2, \mathbf{m}_3, \mathbf{m}_4) \in NN(C_i^{k_{\max}}, \alpha)}{\arg\min} \|\Pi_{i=1}^4 (\mathscr{D} \circ -\mathbf{m}_i \mathbf{p} \mathbf{m}_i^{-1}) - R_\alpha(\mathbf{p})\|_\infty \quad (11)$$

or

$$\widetilde{R_\alpha} = \underset{(\mathbf{m}_1, \mathbf{m}_2, \mathbf{m}_3, \mathbf{m}_4) \in NN(C_i^{k_{\max}}, \alpha)}{\arg\min} \|\Pi_{i=1}^4 (\mathscr{D} \circ -\mathbf{m}_i \mathbf{p} \mathbf{m}_i^{-1}) - R_\alpha(\mathbf{p})\|_2. \quad (12)$$

3.3   Computational Complexity

The computational cost of sorting the $C^{k_{\max}}$ and remove duplicates is

$$O(max(card(D_p), card(C^{k_{\max}}) \log(card(C^{k_{\max}})). \tag{13}$$

Note that $card(D_p) = 201 \times 201$. Since this latter operation can be computationally expensive, we choose to pre-compute $C^{k_{\max}}$. In practical implementation, we go a step further by precomputing the table of the most accurate composition of bijective digitized reflections for each angle and for the points of D_p. It is worth mentioning this table remains reusable as the number of points increases. Therefore, if we consider an image of size $N \times N$, the complexity of the approach is the complexity of applying bijective digitized reflections to each point of the image. Thus, the overall complexity is $O(N^2)$. Figure 6 shows 2 figures of the composition of digitized reflections applied to an image. The resulting implementation of this approach is available in DGtal [13].

4   Bijective Rotation by Circles

We build a bijective approximation of a rotation by decomposing the plane into concentric digital circles around the center of rotation. The points along each digital circle are sorted according to the angle they form with the center of rotation and the x-axis. Then the global transformation is constructed by mapping circles onto themselves, shifting the points according to the desired angle α.

More precisely, assuming the origin of the frame lies at the center of rotation, let $C^r := \{p \in \mathbb{Z}^2, r \leqslant \|p\|_2 < r+1\}$. It is clear that $(C^r)_{r \in \mathbb{Z}, r \geqslant 0}$ forms a partition of $\mathbb{Z}^2$. We then sort the points of each circle C^r according to their angle with the x-axis: let $(C_i^r)_{i=0,\ldots,n^r-1}$ be the induced sequence of points, where n^r is the cardinal of C^r. We have thus $\forall 0 \leqslant i < j < n^r, \angle(C_i^r Ox) < \angle(C_j^r Ox)$. Denoting by $\lceil \cdot \rceil$ the nearest integer rounding operator, we define the *rotation along circles* R_α^C of angle α as:

$$\forall p \in \mathbb{Z}^2, R_\alpha^C(p) = q, \quad \text{with} \quad \begin{cases} p = C_i^r, q = C_j^r \\ \text{and } j = \left(i - \lceil \frac{\alpha}{2\pi} n^r \rceil\right) \mod n^r. \end{cases}$$

This transformation is clearly bijective: it maps the points of a circle onto the same circle, and the shift of indices is a one-one mapping. This transformation also preserves circles and minimizes the radial error in some sense.

From a computational point of view, rotating a whole image of size $N \times N$ takes a time $\Theta(N^2)$. It suffices to proceed circle by circles, each shift takes a time linear in the number of points of the circle. If one is interested in rotating just one point $p = C_i^r$, the complexity is then $O(\log N)$: it takes $O(1)$ to find the correct circle radius r, then $O(\log N)$ worst case to find the index i of p in the sequence, and finally $O(1)$ to get the shifted point.

5   Optimal Transport Method

In recent years, Optimal Transport (OT for short) has become a key mathematical framework for manipulating generalized probability density functions (*e.g.* [14]). The most general way to describe the interest of OT is that it allows quantifying meaningfully how costly it is to move masses from a generalized probability density function to another one, so-called the Wasserstein distance. Depending on the nature of the measures, discrete-to-discrete, semi-discrete, or continuous-to-continuous, a huge literature exists on numerical methods to efficiently solve OT problems [15,16]. When dealing with discrete measures with unit masses, the OT problem boils down to an optimal assignment problem: given two sets of points $X = \{x_i\}_n$ and $Y = \{y_i\}_n$ in $\mathbb{R}^d$, and a cost function $c : \mathbb{R}^d \times \mathbb{R}^d \to \mathbb{R}^+$, we are looking for the permutation σ in $\{1..n\}$ such that

$$\sum_{i=1}^{n} c(x_i, y_{\sigma(i)})$$

is minimal. Back to our setting, if X and Y are two discrete sets and c the squared Euclidean distance, the OT approach allows us to construct a bijective map $X \to Y$ that minimizes the mean squared l_2 error between X and Y. If X is a finite disk of $\mathbb{Z}^d$ and $T \in E(d)$ any continuous rotation, one can define the OT variant of T (*e.g.* OT based rotations) as the optimal assignment between X and $T(X)$ for the (squared) l_2 cost. On the computational side, the Hungarian method can be used to compute the optimal assignment (see for example [15]) with a $O(n^4)$ computational cost, for n the number of pixels. In this paper, we rely on a fast network simplex algorithm [17,18]. The worst-case computational cost remains highly polynomial in n (i.e. $O(N^8)$ for an image $N \times N$), but the bound is not reached in practice. To get an idea of computation times, rotating a 100×100 image takes several minutes on an Apple M2 processor.

6   Optimal Transport by Circles

Since the OT of an image is very costly and impracticable for nowadays image resolutions, we construct a new bijective transformation by mixing rotation by circles and OT. More precisely, for a constant $k \geqslant 2$, we group concentric circles C^r by k-tuples, leaving only C^0 alone. We thus build digital sets D^i that are grouped concentric circles:

$$D^0 := C^0 \qquad \forall i \geqslant 0, D^{i+1} := \bigcup_{j=1}^{k} C^{ki+j}. \qquad (14)$$

Then, given a rotation T_α of angle α, for each circular ring D^i, we perform the optimal transport between $T_\alpha(D^i)$ and D^i to find the best (as of L_2) bijective rotation within each ring. Note that the computational cost is now $N \times O(k^4 N^4)$, since the number of points within a ring is proportionnal to kN. Finally, we can build a look-up table for a fixed number of angles (like 360) that gives the assignment for each ring.

7   Experimental Results and Discussions

7.1   Computational Complexity

We evaluate the computational complexity of rotating a whole image of size $N \times N$. As for the quasi-shear approach, we consider the algorithm of [4] presented in page 313, namely `Final_QSR`. Applying a shift to a point is a constant time algorithm thus the complexity is $O(N^2)$. Concerning the rotation as the composition of discrete line reflections approach, we rely on Algorithm 1 of [7]. The function $X(y)$ defined in line 2 computes a rounding operation and this operation is a constant-time operation. The Table 1 summarizes the complexity of both the image transformation and the table precomputation for the methods presented in this paper.

Table 1. For the main approaches, the first two lines describe the time complexity and the precomputation time complexity to apply each bijective transformation method to a $N \times N$ image. The remaining lines present both the time (ms) to transform a 201×201 image (Image transf.) as well as the precomputation time required for the same image (Precomp.). For CBDR, we choose $kmax = 15$. Methods proposed in this paper are emphasized in bold font. For OTC-k, k stands for the width of each ring (see Eq. (14)).

Method	QSH	CDLR	BROT	CBDR	RBC	OT	OTC-k		
Image transf.	$\Theta(N^2)$	$\Theta(N^2)$	$\Theta(N^2)$	$\Theta(N^2)$	$\Theta(N^2)$	$O(N^8)$	$O(N^2 \log(N))$		
Precomp.	n.a.	n.a.	n.a.	Eq. (13)	$O(N^2)$	n.a.	$O(k^3 N^5)$		
							OTC-2	OTC-3	OTC-4
Image transf. (ms)	2.7	43.5	3.8	3.8	3.5	$>10^5$	15.9	16	16
Precomp. (ms)	0	0	0	5300	13	0	$4 \cdot 10^5$	$7 \cdot 10^5$	$13 \cdot 10^5$

7.2   Accuracy

The accuracy of each method is given in terms of Euclidean distance between the Euclidean rotation and the digital approximation method. More precisely, we compute both the L_2 and L_∞ norms of the error between the Euclidean rotation and the approximation method. As for method CDLR, we improve it by modifying Algorithm 1 [7], where we choose the first reflection such that the L_∞ norm of the error is minimized. Figures 4 and 5 displays respectively the L_2-error and L_∞-error for each method as a function of the angle of rotation.

Overall, bijective rotations (BROT) constitute the worst trade-off, because their angle density is too scarce. Quasi-shears (QSH) have quite regular errors but remain less interesting than a few other methods. Composition of discrete line reflections (CDLR) is among the best methods (both in worst-case or average error). Rotations by circles (RBC) induce quite large errors (especially in worst-case). However their optimal transport extensions (OTC-k) present lower and lower errors as the width k of each ring is increased. Indeed method OTC

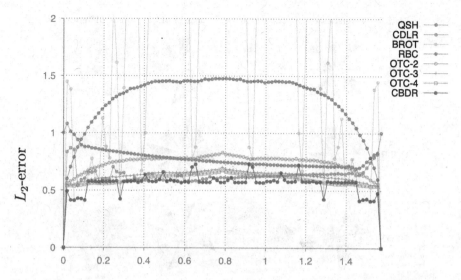

Fig. 4. Plots of L_2-errors for the different bijective transformations as a function of the angle (90 angles between $[0, \frac{\pi}{2}]$), y-range is between 0 and 2 pixels.

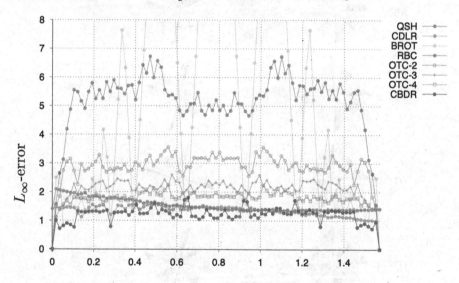

Fig. 5. Plots of L_∞-errors for the different bijective transformations as a function of the angle (90 angles betweeen $[0, \frac{\pi}{2}]$), y-range is between 0 and 8 pixels.

with rings of width 4 is only outperformed by CBDR, and not for all angles. Increasing the ring width would probably induce the method with lowest average error, but its precomputation is very costly (several days). Last, Compositions of Bijective Digitized Reflections (CBDR) provide generally the best results on average and in worst-case, while staying fast to compute. Figure 6 in appendix

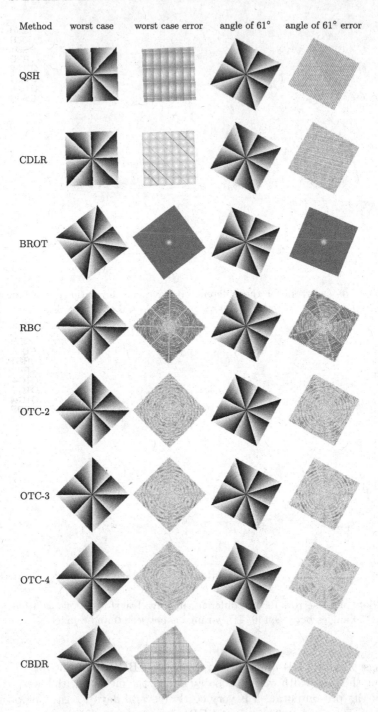

Fig. 6. Comparison of worst case results, norm of the error with respect to the Euclidean rotation (red), and 61-degree rotation for each approximation method (Color figure online)

gives a visual comparison of each approximation method for both the worst case and a fixed angle (61°). For both the worst case and the fixed angle, we also show the L2 norm error field with respect to the Euclidean rotation. Note that all implementations are available in DGtal [13][1].

8   Conclusion

In this paper, we presented multiple approaches for handling bijective rigid transformations and conducted a comparative analysis against state-of-the-art methods. Our experimental results highlight the performances of both optimal transport with circles and the composition of bijective digitized reflections. Extending to 3D the proposed approximation methods is one of our perspectives. A major problem with the extension is the computational complexity. Furthermore, in 3D, the angular density of bijective digitized rotations and reflections is even sparser than in 2D, see [2]. Finally, we are also interested in investigating other mixes of optimal transport and bijective approximation approach.

Acknowledgments. This work is supported by both the Savoie Mont-Blanc University (TRDRECAL project) and the French National Research Agency (StableProxies project, ANR-22-CE46-0006).

References

1. Nouvel, B., Rémila, E.: Characterization of bijective discretized rotations. In: Klette, R., Žunić, J. (eds.) IWCIA 2004. LNCS, vol. 3322, pp. 248–259. Springer, Heidelberg (2004). https://doi.org/10.1007/978-3-540-30503-3_19
2. Breuils, S., Kenmochi, Y., Andres, E., Sugimoto, A.: Conjecture on characterisation of bijective 3D digitized reflections and rotations. In: Hitzer, E., Papagiannakis, G., Vasik, P. (eds.) ENGAGE 2022. LNCS, vol. 13862, pp. 41–53. Springer, Cham (2023). https://doi.org/10.1007/978-3-031-30923-6_4
3. Pluta, K., Roussillon, T., Cœurjolly, D., Romon, P., Kenmochi, Y., Ostromoukhov, V.: Characterization of bijective digitized rotations on the hexagonal grid. J. Math. Imaging Vision **60**(5), 707–716 (2018)
4. Andres, E.: The quasi-shear rotation. In: Miguet, S., Montanvert, A., Ubéda, S. (eds.) DGCI 1996. LNCS, vol. 1176, pp. 307–314. Springer, Heidelberg (1996). https://doi.org/10.1007/3-540-62005-2_26
5. Carstens, H.-G., Deuber, W.A., Thumser, W., Koppenrade, E.: Geometrical bijections in discrete lattices. Combin. Probab. Comput. **8**(1–2), 109–129 (1999)
6. Toffoli, T., Quick, J.: Three-dimensional rotations by three shears. Graph. Models Image Process. **59**(2), 89–95 (1997)
7. Andres, E., Dutt, M., Biswas, A., Largeteau-Skapin, G., Zrour, R.: Digital two-dimensional bijective reflection and associated rotation. In: Couprie, M., Cousty, J., Kenmochi, Y., Mustafa, N. (eds.) DGCI 2019. LNCS, vol. 11414, pp. 3–14. Springer, Cham (2019). https://doi.org/10.1007/978-3-030-14085-4_1

---

[1] https://github.com/DGtal-team/DGtal.

8. Breuils, S., Kenmochi, Y., Sugimoto, A.: Visiting bijective digitized reflections and rotations using geometric algebra. In: Lindblad, J., Malmberg, F., Sladoje, N. (eds.) DGMM 2021. LNCS, vol. 12708, pp. 242–254. Springer, Cham (2021). https://doi.org/10.1007/978-3-030-76657-3_17

9. Passat, N., Ngo, P., Kenmochi, Y., Talbot, H.: Homotopic affine transformations in the 2D cartesian grid. J. Math. Imaging Vision **64**(7), 786–806 (2022)

10. Andres, E.: Cercles discrets et rotations Discrétes. Ph.D. thesis, Université Louis Pasteur, Strasbourg, France (1992)

11. Roussillon, T., Coeurjolly, D.: Characterization of bijective discretized rotations by Gaussian integers. Research report, LIRIS UMR CNRS 5205 (2016)

12. Dorst, L., Fontijne, D., Mann, S.: Geometric Algebra for Computer Science, An Object-Oriented Approach to Geometry. Morgan Kaufmann (2007)

13. DGtal: Digital geometry tools and algorithms library

14. Villani, C.: Optimal Transport: Old and New, vol. 338. Springer, Heidelberg (2009). https://doi.org/10.1007/978-3-540-71050-9

15. Peyré, G., Cuturi, M., et al.: Computational optimal transport: with applications to data science. Found. Trends Mach. Learn. **11**(5–6), 355–607 (2019)

16. Flamary, R., et al.: POT: python optimal transport. J. Mach. Learn. Res. **22**(78), 1–8 (2021)

17. Bonneel, N., van de Panne, M., Paris, S., Heidrich, W.: Displacement interpolation using Lagrangian mass transport. ACM Trans. Graph. (SIGGRAPH ASIA 2011) **30**(6) (2011)

18. Bonneel, N.: Fast network simplex for optimal transport (2018). https://github.com/nbonneel/network_simplex

Digital Calculus Frameworks and Comparative Evaluation of Their Laplace-Beltrami Operators

Colin Weill–Duflos[1]($\boxtimes$) (iD), David Coeurjolly[2]($\boxtimes$) (iD),
and Jacques-Olivier Lachaud[1]($\boxtimes$) (iD)

[1] Université Savoie Mont Blanc, CNRS, LAMA, 73000 Chambéry, France
{colin.weill-duflos,jacques-olivier.lachaud}@univ-smb.fr
[2] Univ Lyon, CNRS, Lyon1, INSA, LIRIS, Lyon, France
david.coeurjolly@cnrs.fr

Abstract. Defining consistent calculus frameworks on discrete meshes is useful for processing the geometry of meshes or model numerical simulations and variational problems onto them. However digital surfaces (boundary of voxels) cannot benefit directly from the classical mesh calculus frameworks, since their vertex and face geometry is too poor to capture the geometry of the underlying smooth Euclidean surface well enough. This paper proposes two new calculus frameworks dedicated to digital surfaces, which exploit a *corrected normal field*, in a manner similar to the recent digital calculus of [3]. First we build a *corrected interpolated calculus* by defining inner products with position and normal interpolation in the Grassmannian. Second we present a *corrected finite element method* which adapts the standard Finite Element Method with a corrected metric per element. Experiments show that these digital calculus frameworks seem to converge toward the continuous calculus, offer a valid alternative to classical mesh calculus, and induce effective tools for digital surface processing tasks.

Keywords: Digital calculus · Laplacian operator · Differential operators

1 Introduction

When solving differential equations on a mesh, it is often required to build a set of differential operators for this mesh. Perhaps the most commonly found is the Laplace-Beltrami operator as it is used in a wide variety of applications such as mesh editing [14,17], mesh smoothing [15] or geodesic path approximation [5]. Building a simple graph Laplacian or discrete Laplacian does not suffice, since the mesh geometry must be taken into account. Using a subdivision scheme and building the operators on it (as done in [7]) do not suffice either, as the limit

This work was partly funded by STABLEPROXIES ANR-22-CE46-0006 research grant.

surface does not solve the metric issues (staircase effects induced by the grid). On triangular and polygonal surfaces, several calculus frameworks produce these differential operators, such as the Finite Element Method (FEM) [16], Discrete Exterior Calculus (DEC) [9], the Virtual Element Method [18], etc (see [1] for a comparative evaluation).

Usually these frameworks operate under the assumption that the mesh interpolates the underlying "true" smooth geometry. In the case of digital surfaces made of surfels (boundary of voxels), which are frequent when processing 3D images, this assumption is false, and these frameworks fail at yielding convergent operators. However several geometric quantities can be evaluated with convergence properties, such as surface area [13], or the normal field and the curvature tensor [10,11] on digital surfaces. We are aware of only two digital analogues to differential operators: Caissard *et al.* [2] proposed a digital Laplacian based on the heat kernel, while two of the authors have adapted in [3] the polygonal calculus of [6], by correcting its normal vector field. The digital *"Heat kernel"* Laplacian of [2] is the only one that is proven convergent and its convergence is observed through experiments. The digital *"Projected PolyDEC"* Laplacian of [3] is not pointwise convergent, but yet provide meaningful results in variational problems.

This paper proposes two new digital calculus frameworks, that are constructed with a tangent space corrected by a prescribed normal vector field (e.g., the II normal estimator [10]). Tangent space correction has proven to be effective for tasks such as estimating curvatures [11] and reconstructing a piecewise smooth surface from a digital surface [4]. The first one, called *"interpolated corrected calculus"*, embeds the digital surface into the Grassmannian with a vertex-interpolated corrected normal vector field: the resulting surface is thus continuous in positions and normals. It is thus more consistent than the Projected PolyDEC, whose embedding is discontinuous between surfels. The second one, called *"corrected FEM"*, adapts the Finite Element Method with metrics tailored to a constant corrected normal vector per element. Both constructions are consistent with classical calculus constructions, and we hope they will allow to prove the convergence of operators. For now, we conducted experiments which show that these frameworks build a consistent Laplacian, convergent when slightly diffused. We achieve results on par with [2] while retaining the ease of build and sparsity from [3].

2   Digital Calculus with Corrected Tangent Space

We demonstrate here that the same approach of corrected lengths and areas used in [3] can be used to build differential operators with other methods. The approach can be summarized as a correction of lengths and areas based on how orthogonal they are to the true normal. Assuming we have a vector $\mathbf{v}$ and a normal $\mathbf{u}$, the corrected length of $\mathbf{v}$ is given by $\|\mathbf{v} \times \mathbf{u}\|$. To correct the area of a parallelogram defined by two vectors $\mathbf{v}$ and $\mathbf{w}$ with normal $\mathbf{u}$ is given by $\det(\mathbf{u}, \mathbf{v}, \mathbf{w})$. These can be seen as the length/areas of the projected vector/parallelogram onto the tangent plane.

Our data here will be defined by values at vertices, meaning that each face has 4 degrees of freedom. We use these degrees of freedom to build a base of functions on the mesh. These functions are bilinear on mesh elements here but other basis functions are possible (e.g. the Virtual Elements Method [18] requires solely the behavior of functions on edges). The methods we present, similarly to [3], use a per face construction of sparse operators.

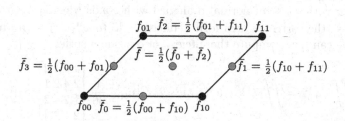

Fig. 1. Notations for the interpolations of function f values on a surfel.

Notation. The parameter space of each surfel is a unit square $\square := [0,1]^2$ parameterized by s and t. We denote by $\mathbf{n}$ the natural or naive normal of a surfel σ, that can be computed with a cross product of two consequent edges. The corrected normal field will be denoted $\mathbf{u}$. Inside a surfel, we can decompose this surfel in the natural base of the surfel into $\mathbf{u} = (\mathbf{u}^x, \mathbf{u}^y, \mathbf{u}^z)$.

A function f in a surfel σ is assumed to be bilinearly interpolated. We denote then by $[f_\square(\sigma)] := [f_{00}(\sigma), f_{10}(\sigma), f_{11}(\sigma), f_{01}\sigma)]^\mathsf{T}$ the degrees of freedom of f, corresponding to its values at each vertex when circulating around σ. We will often write simply $[f_\square]$ when the surfel is obvious from the context. We sometimes use averages of these values, whose notations are illustrated in Fig. 1.

2.1 Interpolated Corrected Calculus

We propose here a calculus where the corrected normal vector field $\mathbf{u}(x)$ is continuous over the mesh: corrected normal vectors are given at vertices; these vectors are bilinearly interpolated within each face. Hence, within a surfel, $\mathbf{u}(s,t) = \mathbf{u}_{00}(1-s)(1-t) + \mathbf{u}_{10}s(1-t) + \mathbf{u}_{01}(1-s)t + \mathbf{u}_{11}st$. Although this naive bilinear interpolation does not respect the condition that normals need to be unitary vectors, it yields much simpler formulas in calculation. Furthermore, experiments show that a more complex interpolation yielding almost unit normals does not improve the results, while increasing the complexity of formulas.

The construction of the calculus is similar to the polygonal calculus of [6], building inner products, sharp and flat operators on a per face basis. However the correction of the geometry does not follow [3], but instead use an embedding of the mesh into the Grassmannian to correct the area/length measures. The Grassmannian is a way to represent affine subspaces, hence tangent spaces here. Within this space, one can define differential forms that are invariant to rigid motions (Lipschitz-Killing forms). We exploit here the *corrected area 2-form* (see

[11,12]): $\omega_0^{\mathbf{u}}(\mathbf{x})(\mathbf{v}, \mathbf{w}) := \det(\mathbf{u}(\mathbf{x}), \mathbf{v}, \mathbf{w})$, for $\mathbf{v}$ and $\mathbf{w}$ tangent vectors. As one can see, thanks to the embedding in the Grassmannian, the corrected area form can be expressed as a simple volume form (i.e. a determinant). Note that it falls back to the usual area measure $\|\mathbf{v} \times \mathbf{w}\|$ when $\mathbf{v}$ and $\mathbf{w}$ are indeed orthogonal to a unit normal vector $\mathbf{u}(\mathbf{x})$, while it gets smaller if there is a mismatch between tangent and normal information.

We first define how we integrate a quantity g defined at vertices. In the case of a surfel σ with constant normal $\mathbf{n}$ aligned with z-axis wlog, and with $\mathbf{v} = \frac{\partial \mathbf{x}}{\partial s}$ and $\mathbf{w} = \frac{\partial \mathbf{x}}{\partial t}$, the corrected area form reduced on $\square$ to $\omega_0^{\mathbf{u}}(s,t) = \langle \mathbf{n} \mid \mathbf{u}(s,t) \rangle = \mathbf{u}^z(s,t)$. We can now compute the integral of g inside a surfel:

$$\iint_\square g\omega_0^{\mathbf{u}} := \iint_\square g(s,t)\mathbf{u}^z(s,t)dsdt = \left[\mathbf{u}_\boxplus^z\right]^\mathsf{T} \frac{1}{36} \begin{bmatrix} 4 & 2 & 1 & 2 \\ 2 & 4 & 2 & 1 \\ 1 & 2 & 4 & 2 \\ 2 & 1 & 2 & 4 \end{bmatrix} \left[g_\boxplus\right].$$

We study now the integral quantity $\iint \nabla\Phi\omega_0^{\mathbf{u}}$, which is an integrated gradient corrected by the normal vector field $\mathbf{u}$. First of all, the scalar field Φ will generally be defined as the bilinear interpolation of a scalar field ϕ defined over the domain. Thus $\phi(s,t) = \Phi(\mathbf{x}(s,t))$. We relate the gradient of Φ with the partial derivatives of ϕ by writing the standard chain rule with Jacobian matrices:

$$J_\phi(s,t) = J_\Phi(\mathbf{x}(s,t))J_\mathbf{x}(s,t) \Leftrightarrow \left[\frac{\partial \phi}{\partial s} \; \frac{\partial \phi}{\partial t}\right](s,t) = (\nabla\Phi)^T(\mathbf{x}(s,t))\left[\frac{\partial \mathbf{x}}{\partial s} \; \frac{\partial \mathbf{x}}{\partial t}\right](s,t).$$

We quite naturally extend ϕ as constant along the $\mathbf{u}$ direction. The preceding relation can now be inverted given that $(\frac{\partial \mathbf{x}}{\partial s} = \begin{bmatrix} 1 & 0 & 0 \end{bmatrix}^T, \frac{\partial \mathbf{x}}{\partial t} = \begin{bmatrix} 0 & 1 & 0 \end{bmatrix}^T, \mathbf{u} = \begin{bmatrix} \mathbf{u}^x & \mathbf{u}^y & \mathbf{u}^z \end{bmatrix}^T)$ forms a basis ((s,t) is omitted for conciseness):

$$\nabla\Phi(\mathbf{x}) = \underbrace{\begin{bmatrix} \mathbf{u}^z & 0 & 0 \\ 0 & \mathbf{u}^z & 0 \\ -\mathbf{u}^x & -\mathbf{u}^y & 1 \end{bmatrix}}_{C} \begin{bmatrix} \frac{\partial \phi}{\partial s} \\ \frac{\partial \phi}{\partial t} \\ 0 \end{bmatrix}.$$

It follows that $\iint_\square \nabla\Phi\omega_0^{\mathbf{u}} = \iint_\square C \left[\frac{\partial \phi}{\partial s} \; \frac{\partial \phi}{\partial t} \; 0\right]^\mathsf{T} \mathbf{u}^z dsdt$. Below, we explicit the vector $\left[\frac{\partial \phi}{\partial s} \; \frac{\partial \phi}{\partial t} \; 0\right]^\mathsf{T}$ involving derivatives of ϕ as

$$\begin{bmatrix} (1-t)(\phi_{10} - \phi_{00}) + t(\phi_{11} - \phi_{01}) \\ (1-s)(\phi_{01} - \phi_{00}) + s(\phi_{11} - \phi_{10}) \\ 0 \end{bmatrix} = \underbrace{\begin{bmatrix} 1-t & 0 & -t & 0 \\ 0 & s & 0 & s-1 \\ 0 & 0 & 0 & 0 \end{bmatrix}}_{B} \underbrace{\begin{bmatrix} -1 & 1 & 0 & 0 \\ 0 & -1 & 1 & 0 \\ 0 & 0 & -1 & 1 \\ 1 & 0 & 0 & -1 \end{bmatrix}}_{D_0} \left[\phi_\boxplus\right].$$

Matrix D_0 is the differential operator, and is common to all quad faces. We get:

$$\iint_\square \nabla\Phi\omega_0^{\mathbf{u}} = \underbrace{\iint_\square CB\mathbf{u}^z dsdt}_{\mathcal{G}_\sigma} D_0 \left[\phi_\boxplus\right],$$

where $\mathscr{G}_\sigma$ is a 3×4 matrix whose expression is (note the use of averages):

$$\mathscr{G}_\sigma = \frac{1}{3} \begin{bmatrix} \bar{\mathbf{u}}^z & 0 & -\bar{\mathbf{u}}^z & 0 \\ 0 & \bar{\mathbf{u}}^z & 0 & -\bar{\mathbf{u}}^z \\ -\bar{\mathbf{u}}^x & -\bar{\mathbf{u}}^y & \bar{\mathbf{u}}^x & \bar{\mathbf{u}}^y \end{bmatrix} + \frac{1}{6} \begin{bmatrix} \bar{\mathbf{u}}_0^z & 0 & -\bar{\mathbf{u}}_2^z & 0 \\ 0 & \bar{\mathbf{u}}_1^z & 0 & -\bar{\mathbf{u}}_3^z \\ -\bar{\mathbf{u}}_0^x & -\bar{\mathbf{u}}_1^y & \bar{\mathbf{u}}_2^x & \bar{\mathbf{u}}_3^y \end{bmatrix}.$$

The (corrected) area a_σ of such a surfel σ has a simple expression, while a pointwise expression of the gradient $\mathbf{G}_\sigma$ is obtained by normalizing $\mathscr{G}_\sigma$ by the corrected area leading to:

$$a_\sigma := \iint_\square \omega_0^{\mathbf{u}} = \iint_\square \mathbf{u}^z ds dt = \bar{\mathbf{u}}^z, \qquad \mathbf{G}_\sigma := \frac{1}{a_\sigma} \mathscr{G}_\sigma D_0.$$

Sharp and Flat Operators. The sharp operator transform a 1-form into a vector field. We use the expression of the pointwise gradient to raise any 1-form as a representative vector per surfel. Within a surfel, a 1-form associates a scalar value to each (oriented) edge. Let β be 1-form, and $[\beta_\circledast(\sigma)] := [\beta_0 \ \beta_1 \ \beta_2 \ \beta_3]^\mathsf{T}$ its values on the 4 edges of σ. Omitting the differential operator D_0 in the pointwise gradient gives the representative 3D vector of β on surfel σ:

$$\beta^\sharp(\sigma) := \frac{1}{a_\sigma} \mathscr{G}_\sigma [\beta_\circledast(\sigma)].$$

The discrete sharp operator on σ is thus the 3×4 matrix $U_\sigma := \frac{1}{a_\sigma}\mathscr{G}_\sigma$.

The flat operator projects a vector field onto the tangent plane and computes its circulation along each edge. The 1-form $\mathbf{v}^\flat$ associated with vector $\mathbf{v}$ is thus:

$$[\mathbf{v}_\circledast^\flat] := \oint_{\partial f} \mathbf{t}^T (I - \mathbf{u}\mathbf{u}^T)\mathbf{v} = \int_0^1 \begin{bmatrix} [1 \ 0 \ 0] \, (I - \mathbf{u}(r,0)\mathbf{u}^T(r,0))\mathbf{v} \\ [0 \ 1 \ 0] \, (I - \mathbf{u}(1,r)\mathbf{u}^T(1,r))\mathbf{v} \\ [-1 \ 0 \ 0] \, (I - \mathbf{u}(r,1)\mathbf{u}^T(r,1))\mathbf{v} \\ [0 \ -1 \ 0] \, (I - \mathbf{u}(0,r)\mathbf{u}^T(0,r))\mathbf{v} \end{bmatrix} dr.$$

By linearity, the flat operator V_σ is a 4×3 matrix (see appendix for details).

Inner Products for Discrete Forms (i.e. Metrics). The inner product between 0-forms is simply the integration of their product on the surfel σ. For any bilinearly interpolated functions ϕ, ψ, we obtain on the surfel σ the scalar:

$$\langle \phi \mid \psi \rangle_0 (\sigma) := \iint_\sigma \phi\psi\omega_0^{(\mathbf{u})} = [\phi_\boxplus(\sigma)]^\mathsf{T} M_{0,\sigma} [\phi_\boxplus(\sigma)].$$

The associated metric matrix is a 4×4 symmetric matrix, called *mass matrix*, whose expression is given in the appendix. If the corrected normal vector $\mathbf{u}$ is consistent with the naive surfel normal $\mathbf{n}$ (i.e. $\langle \mathbf{u}(s,t) \mid \mathbf{n} \rangle > 0$), then $M_{0,\sigma}$ is positive definite.

We would like the inner product between 1-forms β and γ to be defined by emulating the continuous case. We integrate the scalar product between the vectors associated with the 1-forms on the surfel σ:

$$\langle \beta \mid \gamma \rangle_1 (\sigma) := \iint_\square \langle \beta^\sharp \mid \gamma^\sharp \rangle \omega_0^{(\mathbf{u})} = [\beta_\circledast(\sigma)]^\mathsf{T} M_{1,\sigma}^{\mathrm{naive}} [\gamma_\circledast(\sigma)].$$

Using above relations we have:

$$\langle \beta \mid \gamma \rangle_1(\sigma) = a_\sigma(U_\sigma\,[\beta_\circledast(\sigma)])^{\mathsf{T}}(U_\sigma\,[\gamma_\circledast(\sigma)]\,\gamma) = [\beta_\circledast(\sigma)]^{\mathsf{T}}\left(\frac{1}{a_\sigma}\mathscr{G}_\sigma^{\mathsf{T}}\mathscr{G}_\sigma\right)[\gamma_\circledast(\sigma)].$$

Hence $M_{1,\sigma}^{\mathrm{naive}} = \frac{1}{a_\sigma}\mathscr{G}_\sigma^{\mathsf{T}}\mathscr{G}_\sigma$; it is a symmetric matrix. It can be verified that, if $\mathbf{u}$ is a unit constant vector over the surfel σ and $\langle \mathbf{u} \mid \mathbf{n} \rangle > 0$, then this matrix is symmetric positive. However, it is not definite. To remedy this, we follow [6] and complement the definition to get the *stiffness matrix* as

$$M_{1,\sigma} := \frac{1}{a_\sigma}\mathscr{G}_\sigma^{\mathsf{T}}\mathscr{G}_\sigma + \lambda(I - U_\sigma V_\sigma). \tag{1}$$

Calculus on the Whole Mesh. Let n, m and k be respectively the number of vertices, edges and faces of the mesh. Let $\mathscr{V}$ be the space of all sampled functions (an n-dimensional vector space), and $\mathscr{E}$ be the space of all discrete 1-forms (an m-dimensional vector space). Global operators sharp U (size $3k \times m$), flat V (size $m \times 3k$), mass matrix M_0 (size $n \times n$) and stiffness matrix M_1 (size $m \times m$), differential D_0 (size $m \times n$) are obtained by merging the corresponding local operators $U_\sigma, V_\sigma, M_{0,\sigma}, M_{1,\sigma}, D_0$ on the corresponding rows and columns.

Codifferentials and Laplacian. We build the 1-*codifferential* $\delta_1 : \mathscr{E} \to \mathscr{V}$ by adjointness in our inner products.

$$\forall f \in \mathscr{V}, \forall \alpha \in \mathscr{E}, \langle D_0 f \mid \alpha \rangle_1 = -\langle f \mid \delta_1 \alpha \rangle_0 \Leftrightarrow (D_0 f)^{\mathsf{T}} M_1 \alpha = -f^{\mathsf{T}} M_0 \delta_1 \alpha.$$

Being true for all pairs (f, α), it follows that $\delta_1 := -M_0^{-1} D_0^{\mathsf{T}} M_1$. The *Laplacian operator* Δ_0 is the composition of the codifferential and the differential, i.e.

$$\Delta_0 := \delta_1 D_0 = -M_0^{-1} D_0^{\mathsf{T}} M_1 D_0.$$

Since it is very costly to build matrix M_0^{-1}, we will generally not use the two operators δ_1 and Δ_0 as is when solving numerical problems, but we will rather work with their "integrated" version ($M_0 \delta_1$ and $M_0 \Delta_0$). We can now see another approach to compute a Laplace-Beltrami operator coming from the Finite Elements framework.

2.2   Generalization to Finite Element Method

We show here how to adapt the standard Finite Element Method (FEM), e.g. see [16], in order to solve a Poisson problem. The method builds a stiffness matrix L and a mass matrix M to transform the Poisson problem into a linear problem. We will see also that a Laplace operator can be obtained with the same method. Our adaptation consist in correcting the metric used, changing the formulas used for derivatives and dot products. While we only demonstrate here how to correct FEM on a Poisson problem, other problems can also be corrected with the same metric.

The Poisson problem is formulated as solving for g in $\Delta g = f$, with a given border constraint for g if the domain has a boundary, or with a fixed value somewhere if the domain has no boundary. The weak formulation of this problem is given by: solve for f

$$\int_\Omega \nabla g . \nabla \Phi = -\int_\Omega f\Phi + \int_{\partial\Omega} \Phi \langle \nabla g, \mathbf{n} \rangle, \tag{2}$$

for any Φ. In our case, we will evaluate against Φ the functions locally bilinear inside each element. The third term is dependent on the boundary condition, and we will make it vanish here for now.

The FEM approach consists in discretizating the problem at nodes and splitting the domain into elements bordered by nodes (quads here): functions g (say) are discretized at these nodes as vectors $\mathbf{g}$ of their values at nodes. FEM assumes bilinear interpolation of functions within elements. It builds a stiffness matrix L and a mass matrix M such that $L\mathbf{g} = M\mathbf{b}$. This corresponds to the first two terms in (2). Boundary constraints are integrated in this linear problem, either by removing rows and columns or by setting equalities. We can then solve the Poisson problem by solving the linear system $L\mathbf{g} = M\mathbf{b}$, but we can also deduce a Laplacian operator $\Delta := M^{-1}L$.

The matrices are built quad by quad, so here per surfel. We start by defining a metric G per surfel, since it depends on the corrected normal $\mathbf{u}$, then using this metric in the formulas for derivatives and scalar products when building the matrices. Our reference element is a unit square in the plane $\square$. We obtain:

$$G = \begin{bmatrix} 1 - (\mathbf{u}^x)^2 & -\mathbf{u}^x\mathbf{u}^y \\ -\mathbf{u}^x\mathbf{u}^y & 1 - (\mathbf{u}^y)^2 \end{bmatrix}.$$

Since we assume now that our corrected normal field is constant on the surfel, the metric is also constant. This is an arbitrary choice we make in order to keep formulas simple. It becomes easy to compute the gradient and Laplacian. We use the formula $df(\mathbf{w}) = \langle \nabla f, \mathbf{w} \rangle_G$ for any vector $\mathbf{w}$, with $\langle \cdot, \cdot \rangle_G$ the inner product. It follows that $\nabla f = G^{-1} \left[\frac{\partial f}{\partial s} \frac{\partial f}{\partial t} \right]^T$. For the Laplacian, since the metric is constant, we use $\Delta f = \nabla \cdot \nabla f$. We write them more explicitly as:

$$\nabla f = \frac{1}{(\mathbf{u}^z)^2} \begin{bmatrix} (1 - u_y^2)\frac{\partial f}{\partial s} + \mathbf{u}^x\mathbf{u}^y\frac{\partial f}{\partial t} \\ \mathbf{u}^x\mathbf{u}^y\frac{\partial f}{\partial s} + (1 - (\mathbf{u}^x)^2)\frac{\partial f}{\partial t} \end{bmatrix} \tag{3}$$

$$\Delta f = \frac{1}{(\mathbf{u}^z)^2} \left((1 - (\mathbf{u}^y)^2)\frac{\partial^2 f}{\partial s^2} + 2\mathbf{u}^x\mathbf{u}^y\frac{\partial^2 f}{\partial s\partial t} + (1 - (\mathbf{u}^x)^2)\frac{\partial^2 f}{\partial t^2} \right) \tag{4}$$

We choose a basis of bilinear functions on the square as our basis functions. This choice can be disputed: while linear functions are still harmonic regarding to the Laplacian in (4), bilinear functions are no longer harmonics in this setting. However, finding a way to build hat functions that stay harmonic in this setting is not obvious. Yet bilinear functions are still used on quad meshes that are not rectangular and where the same reasoning can be applied to show that they are not harmonic. We define the four basis functions as $f_0 = (1 - s)(1 - t)$, $f_1 = s(1 - t)$, $f_2 = st$, $f_3 = (1 - s)t$.

In order to build our stiffness matrix we evaluate: $\int_\square \langle \nabla f, \nabla p \rangle_G$ for any f and p bilinear. The local stiffness matrix is then:

$$L_M = \frac{1}{6u^z} \begin{bmatrix} 3u^x u^y + 2 + 2(u^z)^2 & 2(u^y)^2 - 1 - (u^x)^2 & 1 - 3u^x u^y - (u^z)^2 & 2(u^x)^2 - 1 - (u^y)^2 \\ 2(u^y)^2 - 1 - (u^x)^2 & 2 - 3u^x u^y + 2(u^z)^2 & 2(u^x)^2 - 1) - (u^y)^2 & 3u^x u^y + 1 - (u^z)^2 \\ 2 - 3u^x u^y - (u^z)^2 & 2(u^x)^2 - 1 - (u^y)^2 & 3u^x u^y + 2 + 2(u^z)^2 & 2(u^y)^2 - 1 - (u^x)^2 \\ 2(u^x)^2 - 1 - (u^y)^2 & 3u^x u^y + 1 + (u^z)^2 & 2(u^y)^2 - 1 - (u^x)^2 & 2 - 3u^x u^y + 2(u^z)^2 \end{bmatrix}. \tag{5}$$

The global stiffness matrix is then obtained by summing over all the local stiffness matrices. The mass matrix is computed from $\int_\Omega fp$, with f and p bilinear:

$$M_M = \frac{u^z}{36} \begin{bmatrix} 4 & 2 & 1 & 2 \\ 2 & 4 & 2 & 1 \\ 1 & 2 & 4 & 2 \\ 2 & 1 & 2 & 4 \end{bmatrix}. \tag{6}$$

We recognize the standard mass matrix for quad mesh with a factor correcting the area of the surfel. It is the same as M_0 for constant $\mathbf{u}$.

We now have two ways of building a sparse Laplace-Beltrami operator. We see now how they compare against operators from previous works.

3 Evaluations and Comparisons

We compare the resulting operators and the ones from previous works on several use cases: first on the sphere, with forward evaluation (compute the laplacian of a function), backward evaluation (solve a Poisson problem to get a function back from its laplacian), eigenvalue comparisons, and then on a standard mesh by comparing with the results obtained on an underlying triangle mesh. Plots related to digitized spheres are the means of the results of 32 computations for each step, each conducted with a different center to better take into account the variability in sphere discretizations.

3.1 Forward Evaluation

Several previous works tried to evaluate the quality and convergence of the Laplacian operator when used in a forward manner: from f defined on the mesh, we compute Δf both analytically and with a discrete Laplacian, then compare the two results. In other words, if our stiffness matrix is called L and our mass matrix M, we solve the equation $LF = MX$ where X is the unknown.

A naive approach consists in computing $X = M^{-1}LF$. This is the one that was used for evaluation in previous works, and was not convergent when using sparse operators. We reproduce this behavior by computing the Laplacian of $f(x) = e^x$ using various methods, none of which seem to converge (see Fig. 2). This is disappointing since we expect the Laplacian operator to have linear convergence when evaluated in forward manner, as observed in [2] and proven for the mesh Laplacian on triangle mesh.

Our idea for improving the convergence consists of adding a small diffusion step to the result. It suffices to replace the mass matrix M by $M - dtL$. In other words, instead of evaluating $X = M^{-1}LF$, we evaluate $X = (M - dtL)^{-1}LF$. The result depends on the choice of parameter dt: we found that we approach linear convergence when dt is in the order of h, and the best quality for $dt = 0.035h$. This means that we add a diffusion with a characteristic length of order $h^{\frac{1}{2}}$, which is coherent with results from other works. Using this method, we achieve what seems to be linear convergence on different functions (Fig. 2), with results comparable or even higher quality than in [2]. We run the same experiment as Fig. 2, with evaluated normals (using Integral Invariant [10]) instead of ground truth. Results are shown in Fig. 3, and also approach linear convergence.

3.2   Backward Evaluation

A Laplacian is often built to solve a Poisson problem. We evaluate a function on our digital surface, we also evaluate its Laplacian using an exact formula, then we compute an approximation of the original function that we compare to the exact original. It is a criterion used for Laplacian evaluation (see [1]), which has not yet been done for digital Laplacians. It also makes more sense to evaluate the Finite Element Methods in this case than in forward evaluation, as this is the problem the operator is built for and is proven (in the case of standard

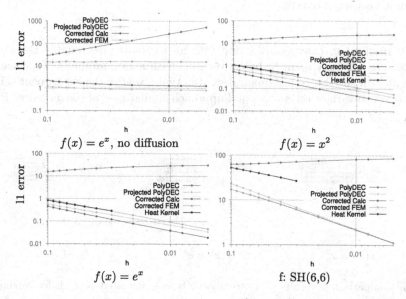

Fig. 2. Forward evaluation of the mesh Laplacian on quadratic, exponential functions and the sixth spherical harmonic. Adding diffusion significantly improves the results, achieving linear convergence on the sphere. We achieve similar rates of convergence as Heat Kernel [2], with a better quality on less smooth functions. Note that the Heat Kernel method is limited to a gridstep of $h \geq 0.03$, due to its enormous memory usage.

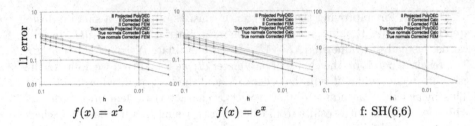

$$f(x) = x^2 \qquad\qquad f(x) = e^x \qquad\qquad \text{f: SH(6,6)}$$

Fig. 3. Comparison on evaluation of the Laplacian using the true normals and using the Integral Invariant estimators. Estimated normals give slightly worse results than with true normals, but still seem to converge. The estimated normals values are limited to a gridstep of $h \geq 0.008$ due to their time to evaluate.

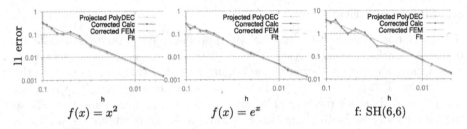

$$f(x) = x^2 \qquad\qquad f(x) = e^x \qquad\qquad \text{f: SH(6,6)}$$

Fig. 4. Error when solving a Poisson problem with different Laplacians. We approach a quadratic convergence rate.

regular meshes) to converge. We find that all methods give roughly the same results (Fig. 4). They seem to be convergent, with a rate around $h^{1.9}$, which is coherent with the theoretical quadratic rate. Again, we run the experiment using the Integral Invariant estimators and approach similar rate of convergence (Fig. 5).

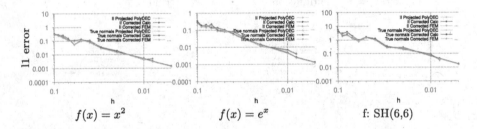

$$f(x) = x^2 \qquad\qquad f(x) = e^x \qquad\qquad \text{f: SH(6,6)}$$

Fig. 5. Comparison of results using true normals and estimated normals for solving a Poisson problem.

3.3   Eigenvalues

We follow the evaluation of eigenvalues on the spherical harmonics used in [1]. Since the spherical harmonics have analytic expressions, we can compare the eigenvalues of our operator to the exact eigenvalues of the Laplace-Beltrami operator on the sphere. To obtain these eigenvalues, we solve for λ in the following generalized eigenvalue problem $L\mathbf{u} = \lambda M \mathbf{u}$. Figure 6 shows the first eigenvalues of our Laplacians on the unit sphere with discretization steps $h = 0.1$ and $h = 0.01$. The PolyDEC method [6] is not accurate, but corrected methods are, with accuracy increased at finer resolutions.

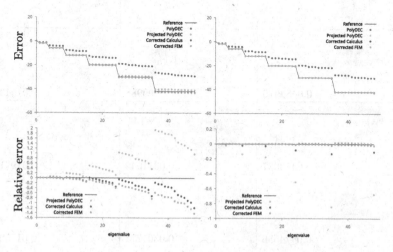

Fig. 6. Smallest 49 eigenvalues of the Laplacian on unit sphere with discretization step 0.1 (left) and 0.01 (right). Relative error is given by $\hat{\lambda} - \lambda$, where $\hat{\lambda}$ is the approximated eigenvalue and λ the correct value.

3.4   Comparison to the Cotan Laplacian

Until now all our comparisons were made on a digitized sphere: this is because there are some closed form expressions of Laplacians, and its eigen decomposition is well studied. However, the sphere is a very specific case, and our evaluations may not reflect well more general cases. We compare here our operators to the results obtained on a regular, high quality triangle mesh with the standard cotan Laplacian. To do so, we use a refined version of a triangle mesh (at 100000 vertices), and equivalent digital surfaces at different resolutions (128^3, 256^3, 512^3). Then we built a projection operator allowing us to map values on the high resolution mesh to the digital one (orthogonal projection and linear interpolation). We also use this projection operator to map the normal vector field computed on the mesh to the surfels or vertices in the calculus frameworks. We then compute a function on the triangle mesh as well as its Laplacian using the cotan Laplacian on the triangle mesh [14], and then their projection on the digital surface, which we use as "ground truth". Forward evaluation results are shown on Fig. 7. We use the same diffusion constant as previously (0.035 h). Error is significantly reduced with the discretization step of the digital surface, suggesting that our operator is convergent toward the result given by the cotan Laplacian.

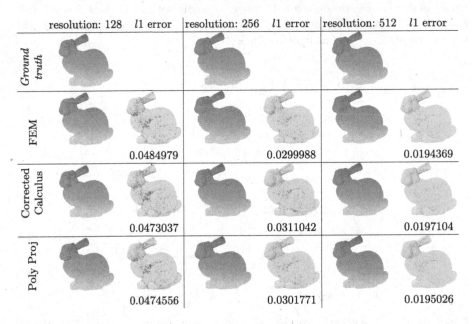

Fig. 7. Comparison between classical cotan Laplacian of a function defined on a triangle mesh and the digital Laplacian operators. All three operators have comparable performances. The error decreases as the resolution increases, suggesting that all three operators are convergent.

4   Conclusion

We show that, similarly to the corrected PolyDEC method [3], a corrected normal field can be inserted within discrete calculus frameworks yielding different Laplace-Beltrami operators. All these operators seem to converge when solving Poisson problems, and when used in a forward evaluation, the addition of a slight diffusion also seems to make them convergent. A limit of our study is that these results are only experimental: only the Heat kernel Laplacian of [2] is yet proven to converge on digital surfaces (strong consistency). However, since results on a common framework (the finite element method) seem promising, it may be interesting to see if the proof of convergence from this framework can be adapted to digital surfaces. The same type of calculus construction could also be tested outside digital surfaces, for instance on a triangle mesh with a corrected normal field or a normal field of much higher resolution such as a normal map (as done in [12]). The idea of adding diffusion and modifying the mass matrix can be seen as similar to the approach of [8]. However Caissard *et al.* [2] were not able to reproduce their experiments and expected convergence, probably because digital surfaces do not have the mesh regularity required by the proof. Indeed, from our metric G it is easy to find that mesh regularity means that $\frac{1}{|\mathbf{u}^z|}$ is bounded. Such a condition can be fulfilled for some specific meshes (such as a digital plane), but is not guaranteed on surface digitizations in general (such as a sphere).

Acknowledgments. This work is supported by the French National Research Agency in the framework of the «France 2030» program (ANR-15-IDEX- 0002), by the LabEx PERSYVAL-Lab (ANR-11-LABX-0025-01) and by the StableProxies project (ANR-22-CE46-0006).

A   Details on the Interpolated Corrected Calculus

Let σ be a surfel aligned with x and y and with normal aligned with z. The flat operator has the following expression:

$$V_\sigma := \frac{1}{6}$$

$6 - 2((u_{00}^x)^2 + u_{00}^x u_{10}^x + (u_{10}^x)^2)$	$-(2u_{00}^x + u_{10}^x)u_{00}^y - (u_{00}^x + 2u_{10}^x)u_{10}^y$	$-(2u_{00}^x + u_{10}^x)u_{00}^z - (u_{00}^x + 2u_{10}^x)u_{10}^z$
$-(2u_{10}^x + u_{11}^x)u_{10}^y - (u_{10}^x + 2u_{11}^x)u_{11}^y$	$6 - 2((u_{10}^y)^2 + u_{10}^y u_{11}^y + (u_{11}^y)^2)$	$(2u_{10}^y + u_{11}^y)u_{10}^z - (u_{10}^y + 2u_{11}^y)u_{11}^z$
$2((u_{01}^x)^2 + u_{01}^x u_{11}^x + (u_{11}^x)^2) - 6$	$(2u_{01}^x + u_{11}^x)u_{01}^y + (u_{01}^x + 2u_{11}^x)u_{11}^y$	$(2u_{01}^x + u_{11}^x)u_{01}^z + (u_{01}^x + 2u_{11}^x)u_{11}^z$
$(2u_{00}^x + u_{01}^x)u_{00}^y + (u_{00}^x + 2u_{01}^x)u_{01}^y$	$2((u_{00}^y)^2 + u_{00}^y u_{01}^y + (u_{01}^y)^2) - 6$	$(2u_{00}^y + u_{01}^y)u_{00}^z + (u_{00}^y + 2u_{01}^y)u_{01}^z$

The metric matrix for 0-forms is defined as the matrix such that, for any bilinearly interpolated functions ϕ, ψ, we obtain on surfel σ the scalar:

$$\langle \phi \mid \psi \rangle_0 (\sigma) := \iint_\sigma \phi\psi\omega_0^{(\mathbf{u})} = [\phi_\boxplus(\sigma)]^\mathsf{T} M_0 [\phi_\boxplus(\sigma)].$$

Let us now define weighted sums for components of $\mathbf{u}$ over the quad. We number the edges when turning along the boundary of the surfel σ from 0 to 3, such that edges 0, 1, 2, 3 connect vertex pairs $(\mathbf{x}_{00}, \mathbf{x}_{10})$, $(\mathbf{x}_{10}, \mathbf{x}_{11})$, $(\mathbf{x}_{01}, \mathbf{x}_{11})$, $(\mathbf{x}_{01}, \mathbf{x}_{00})$, respectively. We define

$$\bar{\mathbf{u}}_{00} := 9\mathbf{u}_{00} + 3\mathbf{u}_{10} + \mathbf{u}_{11} + 3\mathbf{u}_{01} \qquad \bar{\mathbf{u}}_{10} := 3\mathbf{u}_{00} + 9\mathbf{u}_{10} + 3\mathbf{u}_{11} + \mathbf{u}_{01}$$
$$\bar{\mathbf{u}}_{11} := \mathbf{u}_{00} + 3\mathbf{u}_{10} + 9\mathbf{u}_{11} + 3\mathbf{u}_{01} \qquad \bar{\mathbf{u}}_{01} := 3\mathbf{u}_{00} + \mathbf{u}_{10} + 3\mathbf{u}_{11} + 9\mathbf{u}_{01}$$
$$\bar{\mathbf{u}}_{00,10} := 3\mathbf{u}_{00} + 3\mathbf{u}_{10} + \mathbf{u}_{11} + \mathbf{u}_{01} \qquad \bar{\mathbf{u}}_{10,11} := \mathbf{u}_{00} + 3\mathbf{u}_{10} + 3\mathbf{u}_{11} + \mathbf{u}_{01}$$
$$\bar{\mathbf{u}}_{11,01} := \mathbf{u}_{00} + \mathbf{u}_{10} + 3\mathbf{u}_{11} + 3\mathbf{u}_{01} \qquad \bar{\mathbf{u}}_{01,00} := 3\mathbf{u}_{00} + \mathbf{u}_{10} + \mathbf{u}_{11} + 3\mathbf{u}_{01}$$

By integration of left-hand side, we obtain for a surfel with normal z:

$$\mathbf{M}_0 = \frac{1}{144} \begin{bmatrix} \bar{u}_{00}^z & \bar{u}_{00,10}^z & 4\bar{u}^z & \bar{u}_{01,00}^z \\ \bar{u}_{00,10}^z & \bar{u}_{10}^z & \bar{u}_{10,11}^z & 4\bar{u}^z \\ 4\bar{u}^z & \bar{u}_{10,11}^z & \bar{u}_{11}^z & \bar{u}_{11,01}^z \\ \bar{u}_{11,01}^z & 4\bar{u}^z & \bar{u}_{11,01}^z & \bar{u}_{01}^z \end{bmatrix}.$$

References

1. Bunge, A., Botsch, M.: A survey on discrete Laplacians for general polygonal meshes. Comput. Graph. Forum **42**(2), 521–544 (2023)
2. Caissard, T., Coeurjolly, D., Lachaud, J.O., Roussillon, T.: Laplace-Beltrami operator on digital surfaces. J. Math. Imaging Vision **61**(3), 359–379 (2019)
3. Coeurjolly, D., Lachaud, J.O.: A simple discrete calculus for digital surfaces. In: Baudrier, É., Naegel, B., Krähenbühl, A., Tajine, M. (eds.) DGMM 2022. LNCS, vol. 13493, pp. 341–353. Springer, Cham (2022). https://doi.org/10.1007/978-3-031-19897-7_27
4. Coeurjolly, D., Lachaud, J.O., Gueth, P.: Digital surface regularization with guarantees. IEEE Trans. Vis. Comput. Graph. **27**(6), 2896–2907 (2021)
5. Crane, K., Weischedel, C., Wardetzky, M.: The heat method for distance computation. Commun. ACM **60**(11), 90–99 (2017)
6. De Goes, F., Butts, A., Desbrun, M.: Discrete differential operators on polygonal meshes. ACM Trans. Graph. (TOG) **39**(4), 110–1 (2020)
7. de Goes, F., Desbrun, M., Meyer, M., DeRose, T.: Subdivision exterior calculus for geometry processing. ACM Trans. Graph. **35**(4), 1–11 (2016)
8. Hildebrandt, K., Polthier, K.: On approximation of the Laplace-Beltrami operator and the Willmore energy of surfaces. In: Computer Graphics Forum (2011)
9. Hirani, A.N.: Discrete exterior calculus. Ph.D. thesis, USA (2003). aAI3086864
10. Lachaud, J.-O., Coeurjolly, D., Levallois, J.: Robust and convergent curvature and normal estimators with digital integral invariants. In: Najman, L., Romon, P. (eds.) Modern Approaches to Discrete Curvature. LNM, vol. 2184, pp. 293–348. Springer, Cham (2017). https://doi.org/10.1007/978-3-319-58002-9_9
11. Lachaud, J.O., Romon, P., Thibert, B.: Corrected curvature measures. Discrete Comput. Geom. **68**(2), 477–524 (2022)
12. Lachaud, J.O., Romon, P., Thibert, B., Coeurjolly, D.: Interpolated corrected curvature measures for polygonal surfaces. Comput. Graph. Forum **39**(5), 41–54 (2020)
13. Lachaud, J.O., Thibert, B.: Properties of Gauss digitized shapes and digital surface integration. J. Math. Imaging Vision **54**(2), 162–180 (2016)
14. Lévy, B., Zhang, H.: Spectral mesh processing. In: ACM SIGGRAPH 2010 Courses, pp. 1–312 (2010)
15. Nealen, A., Igarashi, T., Sorkine, O., Alexa, M.: Laplacian mesh optimization. In: Proceedings of the 4th International Conference on Computer Graphics and Interactive Techniques in Australasia and Southeast Asia, GRAPHITE 2006, pp. 381–389. Association for Computing Machinery, New York (2006)
16. Reddy, J.N.: Introduction to the Finite Element Method. McGraw-Hill Education (2019)
17. Sorkine, O.: Laplacian mesh processing. In: Chrysanthou, Y., Magnor, M. (eds.) Eurographics 2005 - State of the Art Reports. The Eurographics Association (2005)
18. Veiga, L., Brezzi, F., Cangiani, A., Manzini, G., Marini, L., Russo, A.: Basic principles of virtual element methods. Math. Models Methods Appl. Sci. **23**, 199–214 (2012)

Fitting Egg-Shapes to Discretized Object Boundaries

Jiří Hladůvka[(✉)] [iD] and Walter G. Kropatsch [iD]

Institute of Visual Computing and Human-Centered Technology, TU Wien,
Vienna, Austria
jiri@prip.tuwien.ac.at

Abstract. This paper presents a novel method for fitting egg-shapes to
discrete sets of boundary points. Egg-shapes extend ellipses by assigning
a positive weight to one of the two focal points. Fitting of egg-shapes thus
requires optimization of 6 parameters. Our approach simplifies this to a
1D parameter space exploration. First, we utilize a least square algorithm
to fit an ellipse to the boundary. While the desired egg-shape shares the
orientation of the major axis and to a certain extent also the size of
the ellipse, its fine-tuning to the boundary is more involved than merely
adjusting the focal weight. To this end, we establish a relation between
the eccentricity of the ellipse and the two shape-defining parameters
of the closest egg-shapes. Subsequently, we utilize this relationship to
iterate over a 1D space of closest egg-shape candidates while assessing
their fitness to the boundary. Our results underscore the benefits of using
egg-shapes over ellipses for representing a spectrum of real-world objects.

Keywords: egg-shape · fitting · discrete shape · eccentricity · ellipse ·
generalized conics

1 Introduction

The core task in any computer vision problem is to define and efficiently express
the essential object characteristics [11]. In order to represent a shape, one strategy is to approximate it by fitting a geometric primitive to the set of its boundary
points. For example, an ellipse enables describing the 2D shape by elongation
and orientation and requires 5 parameters. This significantly simplifies the processing and memory costs compared to the entire collection of pixels [1]. As a
result, ellipse fitting is an area of extensive research [2,5,12,17] and finds its use
in a multitude of applications [7,9,10,14].

This paper presents an advancement in the field of shape representation by
exploring the potential of an egg-shape [3]. This generalization of the ellipse
introduces a positive weight to one of the focal points. Despite the broadened
scope of objects that might be described by egg-shapes, there are two key areas
that have not been addressed to the best of our knowledge. Firstly, there is a lack
of real-world examples demonstrating the benefits of egg-shapes. Secondly, it is

S. Brunetti et al. (Eds.): DGMM 2024, LNCS 14605, pp. 107–119, 2024.
https://doi.org/10.1007/978-3-031-57793-2_9

the absence of a method for fitting egg-shapes to object boundaries. To this end, existing literature provides fitting methods for a different generalization, i.e., the super-ellipse [8,18], that yields shapes from four-armed stars with concave sides to rectangles thru an additional parameter.

Addressing these gaps, we have identified various classes of real-world objects, such as chicken eggs, avocados, leaves, spoons, and rackets, that demonstrably benefit from egg-shape representation. We have assembled a collection of 1337 such objects, complete with photographs and boundaries, and have made this collection publicly accessible [15] in conjunction with the publication of this paper. A selection of four boundaries from this collection is presented in the results Sect. 6.

Crucially, we have developed an algorithm, detailed in Sect. 4, that is capable of fitting egg-shapes to the discretized boundaries of these objects. The research code for this algorithm has also been made publicly available [6].

The primary contribution of our work, detailed in Sect. 3, is the reduction of the search space from six parameters to a single one. This was enabled by our key discovery, i.e., the relationship between egg-shape parameters and the eccentricity of its best-fitting ellipse. This contribution was facilitated by addressing two crucial aspects of the egg-shape: its *explicit*, polar-form representation (sec. 2.1) and the derivation of its arc-length sampling (sec. 2.2).

2   Egg-Shapes

Conics have been generalized by assigning real-valued weights w_i to multiple focal points F_i in higher-dimensional metric spaces [4]. These generalizations are elegantly encompassed by $\Sigma_i w_i \|P - F_i\| = c \in \mathbb{R}$, which implicitly defines the conic by its points P.

In this work, we focus on shapes in the Euclidean plane induced by two foci and two positive weights. After a weight normalization [3], we are left with a single weight μ from the unit interval.

Definition 1 (Egg-shape). *Given two distinct foci $F_0 \neq F_1$ in the Euclidean plane, weight $0 \leq \mu \leq 1$, and scaling factor $c \geq 1$. Egg-shape is a set of points P in the Euclidean plane fulfilling*

$$d_0 + \mu d_1 = c \cdot f \tag{1}$$

where $d_i = \|P - F_i\|$ are the distances of P from foci F_i and $f = \|F_1 - F_0\|$ is the focal distance.

Wherever context permits, we may use the term *egg* as a shorthand for egg-shape. Definition 1 encompasses circles ($\mu = 0$), ellipses ($\mu = 1$), as well as a spectrum of egg-like shapes with various sharpness (Fig. 5), including those with sharp corner [3] at $c = 1$ (Fig. 3).

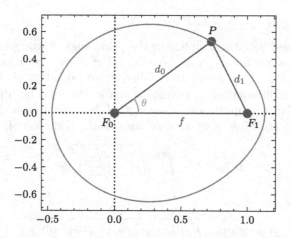

Fig. 1. Deriving polar coordinates. Unit egg U_{01} parametrized by $c = 1.2$, $\mu = 0.5$.

2.1 Polar Coordinates

To explicitly express the egg in polar coordinates, we consider the two foci at the origin $F_0 = [0,0]^\top$ and on the positive horizontal axis, $F_1 = [f,0]^\top$. Setting the focal distance $f = 1$ yields what we refer to as the unit egg U_{01} (Fig. 1). Later, we will also refer to its left-oriented counterpart U_{10} flipped about $x = 1/2$.

For any point P at angle θ on the egg (1) holds. Taking the square results in:

$$d_1^2 = \left(\frac{cf - d_0}{\mu}\right)^2 \tag{2}$$

Furthermore, the law of cosines yields:

$$d_1^2 = d_0^2 + f^2 - 2d_0 f \cos(\theta) \tag{3}$$

Subtracting (3) from (2) results in quadratic equation in d_0:

$$d_0^2 \left(\mu^2 - 1\right) + d_0 \left(2f(c - \mu^2 \cos(\theta))\right) + f^2 \left(\mu^2 - c^2\right) = 0. \tag{4}$$

For $\mu = 1$ the quadratic term vanishes yielding the polar equation of an ellipse. For $\mu = 0$, $d_0 = cf$ implicitly defines points P on a circle centered at F_0. For $0 < \mu < 1$ and fixed θ, the two solutions of (4), i.e.,

$$d_0^+, d_0^- = f \cdot \frac{\mu^2 \cos(\theta) - c \pm \mu \sqrt{c^2 - 2c \cos(\theta) - \mu^2 \sin^2(\theta) + 1}}{\mu^2 - 1} \tag{5}$$

correspond to distances (from F_0) of two points: P on the egg ($\mu > 0 \Rightarrow d_0^+$), and P^- on a generalized hyperbola ($\mu < 0 \Rightarrow d_0^-$) [3], which is further not discussed. Putting $r(\theta) = d_0^+$ yields the polar coordinates $(r(\theta), \theta)$ of the egg-shape.

2.2 Arc Length

Similar to ellipses, uniform sampling of the polar angle θ will generally lead to uneven distribution of points on the boundary (Fig. 3, left). Such sampling would likely introduce bias when fitting to the discretized boundary of an object.

To address undersampled segments, we aim at the arc-length parameterization (Fig. 3, right). Computing the arc-length L between two angles $\theta_0 < \theta_1$ in the polar parametrization (r, θ) involves numerical integration of:

$$L(\theta_0, \theta_1) = \int_{\theta_0}^{\theta_1} \sqrt{r^2 + \left(\frac{dr}{d\theta}\right)^2} \, d\theta \tag{6}$$

Putting

$$R = R(\theta) = \sqrt{c^2 - 2c\cos(\theta) - \mu^2 \sin^2(\theta) + 1} \tag{7}$$

$$M = M(\theta) = \mu^2 \cos(\theta) - c \tag{8}$$

$$A = \mu^2 - 1 \tag{9}$$

simplifies the terms for r and its derivative w.r.t θ to:

$$r = f \cdot \frac{M + \mu R}{A} \tag{10}$$

$$\frac{dr}{d\theta} = -f \cdot \frac{M + \mu R}{A} \cdot \mu \cdot \frac{\sin(\theta)}{R} = -r \cdot \mu \cdot \frac{\sin(\theta)}{R} \tag{11}$$

Equation (11) confirms the discovery regarding the sharp corners [3] in eggs with $c = 1$ (Fig. 3). Specifically, when $c = 1$, $R(\theta)$ vanishes at $\theta = 2k\pi \; \forall k \in \mathbb{Z}$, which results in discontinuities in $dr/d\theta$ and leaves the derivatives undefined. However, $dr/d\theta$ is an odd function with finite and opposing left/right limits in $2k\pi$:

$$\lim_{\substack{c=1 \\ \theta \to 2k\pi^-}} -r\mu \frac{\sin(\theta)}{R} = \frac{r\mu}{\sqrt{1 - \mu^2}} = - \lim_{\substack{c=1 \\ \theta \to 2k\pi^+}} -r\mu \frac{\sin(\theta)}{R} \tag{12}$$

Therefore, the square $(dr/d\theta)^2$ in (6) exists in the limit (from either side):

$$\lim_{\substack{c=1 \\ \theta \to 2k\pi}} \left(\frac{dr}{d\theta}\right)^2 = \frac{r^2 \mu^2}{1 - \mu^2} \tag{13}$$

Equations (10)–(13) result, for given θ, in a recipe to reuse the radius r to compute the square of the derivative. Having the ingredients, the values of arc-length $L(0, \theta)$ can be utilized to sample the egg points at regular distances as exemplified in Figs. 2 and 3.

3   Egg-Shapes and Best-Fitting Ellipses

We aim to establish the correspondence between egg parameters and the eccentricity $\varepsilon = \varepsilon(c, \mu)$ of the best-fitting ellipse [5]. Two bounding cases are apparent from the equation for egg-shape (cf. Fig. 4):

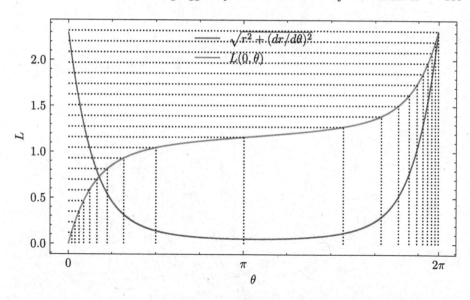

Fig. 2. Uniform sampling along the arc length L (horizontal lines) leads to samples of θ (vertical lines), which correspond to the equidistant sampling of the egg-shape.

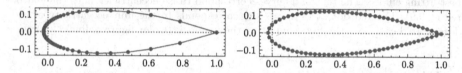

Fig. 3. A corner-egg ($c = 1$) with θ sampled linearly (left) and by arc length (right).

$\varepsilon(c, 0) = 0/c$: egg-shape, and thus also its best-fitting ellipse, is a circle implicitly given by $d_0 = cf$.

$\varepsilon(c, 1) = 1/c$: egg-shape, and thus also its best-fitting ellipse, is an ellipse implicitly given by $d_0 + d_1 = cf$, i.e., one with focal distance f, the length of its main axis cf, and eccentricity $\varepsilon = f/(cf) = 1/c$.

To model $\varepsilon(c, \mu)$ between, we fitted ellipses by the robust algorithm [5] to the sampled boundaries of unit eggs parametrized by $\mu \in (0, 1)$ and $c \in \langle 1, 5\rangle$.[1] This way we densely sampled the parameter space of egg-shapes and tracked the fitting-ellipse eccentricities.

Interestingly, this process reveals a power-rule increase of ε w.r.t. μ. This is illustrated in Fig. 4 by the two dotted lines at $c = 1$ and $c = 5$. The power-rule behavior becomes apparent after the scaled $\varepsilon \cdot c$ curves map to lines through the origin in a log-log plot. Moreover, it can be observed that the slopes S of the log-log lines are inversely proportional to c, $S \propto 1/c$.

---

[1] The upper bound for c was determined by the expected lowest eccentricity of ellipses fitted to modeled objects. For our collection [15]: $c_{\max} = 1/\varepsilon_{\min} \leq 1/0.2 = 5$.

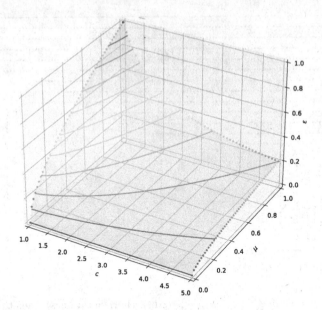

Fig. 4. Relationship between egg-shape parameters c, μ, and the eccentricity of the best-fitting ellipse. The ten solid curves correspond to iso-eccentricity lines (blue at 0, red at 0.9). The red dot at $(c,\mu) = (1,1)$ corresponds to the unit eccentricity of the degenerated case, i.e., the line connecting the two foci. (Color figure online)

This encourages to model the eccentricity by

$$\varepsilon(c, \mu) \approx \frac{1}{c} \cdot \mu^{S(c)} \tag{14}$$

where

$$S(c) \approx S(c; \alpha, \beta, \gamma) = \beta + \frac{\alpha}{c + \gamma} \tag{15}$$

Fitting the three parameters results in $\alpha = 0.01994611$, $\beta = 0.49646579$, $\gamma = -0.7994735$ and yields an error in eccentricity of 0.0000 ± 0.0013.

It is worth mentioning that this modeling is performed only once, before fitting egg-shapes to object boundaries.

4   Fitting Egg-Shapes via an Ellipse Proxy

Equation (14) relates the eccentricity of best-fitting ellipse given egg-shape parameters. More importantly, we can constrain the inverse, egg-to-ellipse fitting problem to the iso-eccentricity line in the (c, μ) space (Fig. 5).

Given an ellipse of eccentricity ε, the parameter c is restricted to the interval $\langle 1, 1/\varepsilon \rangle$. Rewriting (14) allows, for a fixed c, computation of the corresponding μ:

$$\mu = \mu(c; \varepsilon) = (c \cdot \varepsilon)^{1/S(c)} \approx (c \cdot \varepsilon)^{\frac{(c+\gamma)}{\beta(c+\gamma)+\alpha}} \tag{16}$$

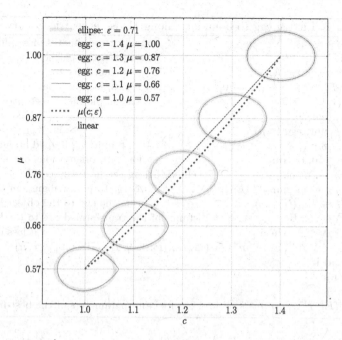

Fig. 5. Sampled best-fitting eggs to an ellipse of fixed eccentricity on the curve given by Eq. (16) in the (c, μ) parameter space. For clarity, the unit eggs are scaled down.

Our objective is to fit an egg-shape to a segmented object represented by boundary points B. It is important to note that, unlike the unit egg, boundaries B can be arbitrarily placed and oriented within the Euclidean plane.

The steps of Algorithm 1 proceed as follows: We begin by fitting a *proxy* ellipse to the boundary points B and recording its eccentricity. It is worth noting that while the proxy provides the angle of the object's main axis, it does not disambiguate between the two possible orientations of the object's tip. Next, we perform an iterative search along the iso-eccentricity curve. During each iteration, we fit ellipses to both the arc-sampled unit egg U_{01} and its horizontally flipped counterpart U_{10}. Since both ellipses share the same eccentricity as the proxy, they naturally determine transformations that align U_{01} and U_{10} with the boundary B. Finally, we resolve the object's tip orientation by assessing both transformed eggs by fitness to the boundary B using a metric Δ. In the context of this paper, Δ represents the average Hausdorff distance defined later in Sect. 5.

Algorithm 1. Fitting egg to boundary along 1D iso-eccentricity curve

---

Require: B ▷ boundary coordinates of egg-like object
1:   $E_B \leftarrow$ fit_ellipse (B) ▷ fit a proxy ellipse to the boundary
2:   $\varepsilon_B \leftarrow$ eccentricity (E_B) ▷ eccentricity to restrict the search
3:   $(\Delta^*, F_0^*, F_1^*, c^*, \mu^*) \leftarrow (\infty, [0,0]^\top, [1,0]^\top, 1, 0)$ ▷ loss, foci, and params to optimize
4:   **for** $c \in \langle 1, 1/c \rangle$ **do**   ▷ sample the iso-eccentricity curve
5:      $\mu \leftarrow \mu(c; \varepsilon_B)$ ▷ equation (16)
6:      $U_{01} \leftarrow$ unit_egg (c, μ) ▷ unit egg given by c, μ
7:      $U_{10} \leftarrow$ flip (U_{01}) ▷ unit egg flipped about $x = 1/2$
8:      **for** $o \in \{0, 1\}$ **do**   ▷ for both orientations of the unit egg
9:          $E_U \leftarrow$ fit_ellipse $(U_{o,1-o})$ ▷ ellipse fitting the unit egg, $ecc(E_U) \approx ecc(E_B)$
10:         $T_{UB} \leftarrow \arg\min_T |T(E_U), E_B|$ ▷ E_U to proxy alignment transform
11:         $G \leftarrow T_{UB}(U_{o,1-o})$ ▷ bring the egg to the object boundary
12:         $\Delta \leftarrow |G, B|$ ▷ distance of the transformed egg to the boundary
13:         **if** $\Delta < \Delta^*$ **then**   ▷ if loss decreased
14:            $(\Delta^*, F_0^*, F_1^*, c^*, \mu^*) \leftarrow (\Delta, T_{UB}([o,0]^\top), T_{UB}([1-o,0]^\top), c, \mu)$ ▷ update
15:         **end if**
16:      **end for**
17: **end for**
18: **return** $(F_0^*, F_1^*, c^*, \mu^*)$ ▷ foci and parameters of the best-fitting egg

---

5   Validation

Our methodology is verified using both overlap-centric and distance-centric metrics [16] on a purposely created collection of boundaries [15] segmented from images of both biological and common objects. These include both deformable and rigid items captured in arbitrary orientations and scales. Each segmentation is represented by a binary mask and its boundary thus as a set of integer coordinates.

5.1   Datasets

Whole eggs: Boundaries of 1,100 photographed eggs [13].

Boiled eggs: Images of longitudinally halved, hard-boiled eggs found on the internet were manually segmented, yielding 12 boundaries each for the egg whites and yolks.

Avocados: Images of longitudinally halved avocados were found on the internet. The shells of these avocados were manually segmented, resulting in 6 boundaries. Some of them are slightly deformed.

Leaves: Tree and plant leaves were deliberately selected and photographed in line with this study. The criteria included being longitudinally symmetrical, egg-shaped, elongated, and possibly pointed. Manual segmentation excluded the stems and produced 23 boundaries.

Cells: Palisade cells of Arabidopsis thaliana in a micro-CT cross-section slice were manually segmented, resulting in 159, mostly elliptic boundaries.

Household items: 11 spoon heads and 2 toilet seats segmented from photos.

Rackets: Images of tennis, badminton, and squash racket heads sourced from the internet were segmented, resulting in 12 boundaries. The outer shell of the squash head is noted to be pointed. One of the tennis heads is elliptic.

5.2   Validation Metrics

The geometric alignment of a model with the object boundary is evaluated using metrics that take into account either the boundary point sets M and B, or the corresponding polygons $\mathbf{M}$ and $\mathbf{B}$.

IoU (also referred to as the Jaccard index) [16], defined as the area of the intersection divided by the area of the union of polygons $\mathbf{M}$ and $\mathbf{B}$:

$$IoU(M, B) = \frac{|\mathbf{M} \cap \mathbf{B}|}{|\mathbf{M} \cup \mathbf{B}|} \tag{17}$$

ranges from 0 to 1, with 1 indicating perfect alignment.

The Average Hausdorff Distance ($\overline{HD}$) is used to assess the fitness for its decreased sensitivity to outliers [16] when compared to the usual Hausdorff distance. $\overline{HD}$ is defined by:

$$\overline{HD}(M, B) = \max\left(d(M, B), d(B, M)\right) \tag{18}$$

where d(X, Y) is the *directed* Average Hausdorff distance [16] given by:

$$d(X, Y) = \frac{1}{|X|} \sum_{x \in X} \min_{y \in Y} ||x - y||. \tag{19}$$

To assess improvements of egg models M_G over elliptic models E_L across contours of different lengths we further introduce the following normalization:

Normalized Improvement (NI) in average Hausdorff distance $\overline{HD}$ of egg model M_G over elliptical model M_L when fitted to boundary B. We define $NI = 0$ if models align. Otherwise:

$$NI(M_G, M_L | B) = \frac{\overline{HD}(M_L, B) - \overline{HD}(M_G, B)}{\max\left(\overline{HD}(M_L, B)\,,\,\overline{HD}(M_G, B)\right)}. \tag{20}$$

Being already normalized, IoU is naturally suited to assess improvements by:

$$\Delta IoU(M_G, M_L | B) = IoU(M_G, B) - IoU(M_L, B). \tag{21}$$

Both improvement metrics fall in the closed interval $\langle -1, 1 \rangle$. Positive values indicate an improvement due M_G, while negative values signify a decline.

6   Results

Figure 7 showcases a selection of top-improvement examples. To avoid clutter, Fig. 6 summarizes both improvement metrics (20), (21) in two separate scatter plots. 94.68% of data points indicating improvements in both metrics are located in the 1st quadrants. The ties in the 2nd and 4th quadrants correspond to cases improved in only one of the metrics. They account for $10+11$ out of 159 cells and $9+12$ out of 1100 whole eggs. Finally, the 3rd quadrant shows minimal declines in both metrics: less than 4% of cells (6 out of 159) and less than 2% of whole eggs (20 out of 1100). The leftmost point of the upper part is the elliptic tennis racket that can not be improved by the egg-shape.

In the marginal-improvement rectangle $(0, 0.1) \times (0, 0.02)$ in the upper part, a cluster of 9 boiled-egg yolks can be found. They are almost elliptical making it difficult for the egg-shapes to improve them. This explains the almost identical IoU and only marginally improved NI values. The enclosed badminton racket already deviates from an ellipse and was slightly improved by an egg-shape.

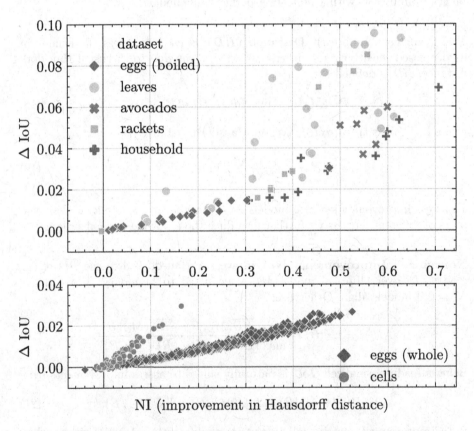

Fig. 6. Normalized improvements in average Hausdorff distance (NI) and in IoU.

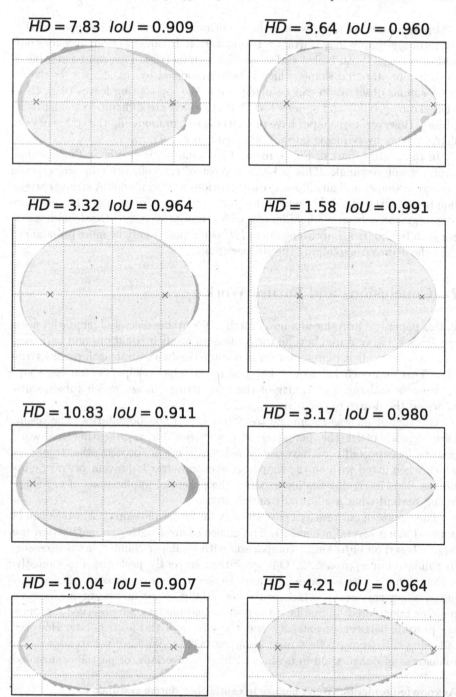

Fig. 7. Improvements by egg-shape fit (right column) over elliptic fit (left column) for selected contours. Top to bottom: avocado ($NI = 0.54$), whole egg ($NI = 0.52$), spoon head ($NI = 0.71$), and leaf ($NI = 0.58$). Gray and dark green areas are not explained well by the respective model. The blue crosses show the foci. (Color figure online)

Upon examining areas with higher improvements, it becomes apparent that the term *egg*-shapes may warrant reconsideration. In our selection, avocados outperformed eggs (both boiled and whole) in both metrics. This can be attributed to their more tapered shape, which is better modeled by egg-shapes.

A similar observation can be made about the top-scoring leaves (e.g., those with $\Delta IoU \gtrsim 0.06$) and spoons ($NI \gtrsim 0.6$), which are difficult to model using ellipses. However, egg-shapes have proven effective in modeling the spikes or even sharp corners by reducing c towards 1 (refer to Fig. 7).

In the bottom part of Fig. 6, most of the plant cells cluster in the marginal improvement rectangle. This is because most of the cells are elliptical or even circular in shape, and an ellipse already provides a good model. A strong correlation between the metrics is evident for both datasets. Unlike eggs, the cells have been segmented from a single image, which results in more jagged boundaries. For such boundaries, improvements in IoU could potentially be more pronounced than those in NI, explaining the steeper trend.

7   Conclusions and Future Work

In this paper, we introduced a novel method for fitting discrete boundaries using egg-shapes. We ventured beyond the confines of implicit equations and harnessed the power of explicit representation in conjunction with arc-length parametrization. This approach led to our key finding: a relationship between egg-shape parameters and the eccentricity of the best-fitting ellipse, which subsequently informed the design of our fitting algorithm.

Our algorithm was rigorously tested on over 1,000 boundaries. The results have demonstrated the potential of egg-shapes for representing real-world objects. Unexpectedly, we have discovered that there are objects other than *eggs*, which when fitted with an *egg*-shape, show even better improvements over elliptic fits. This intriguing finding expands the potential applications of egg-shape fitting beyond what was initially anticipated.

The results of our research pave the way for future development, which can be pursued along several avenues: 1. The choice of arc-length parametrization was largely based on intuition. A comparison with its linear counterpart is necessary to validate our approach. 2. Our algorithm currently performs an exhaustive search along the iso-eccentricity curve. However, the existence of a global minimum along this curve, which could potentially speed up the fitting process, remains unexplored. 3. Similar to egg-shape parameters, we observed a promising pattern between eccentricity and the offsets of the focal points. However, modeling this seems to be a more complex task. 4. We have yet to address the robustness of our method in dealing with noise, overlaps, or partial occlusions.

Acknowledgements. We would like to express our sincere gratitude to Aysylu Gabdulkhakova for sharing her invaluable insights and knowledge on egg-shapes.

This project was supported by the Vienna Science and Technology Fund (WWTF) project LS19-013 and the Vienna Scientific Cluster (VSC).

References

1. Costa, L.D.F., Cesar, R.M., Jr.: Shape Analysis and Classification: Theory and Practice. CRC Press, Boca Raton (2000)
2. Fitzgibbon, A., Pilu, M., Fisher, R.: Direct least square fitting of ellipses. IEEE Trans. Pattern Anal. Mach. Intell. **21**(5), 476–480 (1999)
3. Gabdulkhakova, A., Kropatsch, W.G.: Generalized conics with the sharp corners. In: Progress in Pattern Recognition, Image Analysis, Computer Vision, and Applications, pp. 419–429 (2021)
4. Groß, C., Strempel, T.K.: On generalizations of conics and on a generalization of the Fermat-Torricelli problem. Am. Math. Mon. **105**(8), 732–743 (1998)
5. Halíř, R., Flusser, J.: Numerically stable direct least squares fitting of ellipses. In: 6th International Conference in Central Europe on Computer Graphics and Visualization (WSCG), pp. 125–132 (1998)
6. Hladůvka, J.: Eggfit: an iterative algorithm to fit egg-shapes to object boundaries. Software stack, TU Wien (2024). https://doi.org/10.48436/d17s6-p5d44
7. Kanatani, K., Sugaya, Y., Kanazawa, Y.: Ellipse Fitting for Computer Vision: Implementation and Applications. Springer, Heidelberg (2016). https://doi.org/10.1007/978-3-031-01815-2
8. Köhntopp, D., Lehmann, B., Kraus, D., Birk, A.: Segmentation and classification using active contours based superellipse fitting on side scan sonar images for marine demining. In: IEEE International Conference on Robotics and Automation, pp. 3380–3387 (2015)
9. Kothari, R.S., Chaudhary, A.K., Bailey, R.J., Pelz, J.B., Diaz, G.J.: Ellseg: an ellipse segmentation framework for robust gaze tracking. IEEE Trans. Visual Comput. Graphics **27**(5), 2757–2767 (2021)
10. Liao, M., et al.: Automatic segmentation for cell images based on bottleneck detection and ellipse fitting. Neurocomputing **173**, 615–622 (2016)
11. Loncaric, S.: A survey of shape analysis techniques. Pattern Recogn. **31**(8), 983–1001 (1998)
12. Long, C., Hu, Q., Zhao, M., Li, D., Ouyang, Z., Yan, D.M.: A triple-stage robust ellipse fitting algorithm based on outlier removal. IEEE Trans. Instr. Meas. **72**, 1–14 (2023)
13. Nho, T.: Egg-segmentation dataset (2023). https://universe.roboflow.com/the-nho/egg-segmentation. Accessed 26 July 2023
14. Panagiotakis, C., Argyros, A.: Region-based fitting of overlapping ellipses and its application to cells segmentation. Image Vis. Comput. **93**, 103810 (2020)
15. Pelletier, V., Hladůvka, J.: Eggshapes: a collection of egg-shaped objects and their boundaries. Dataset, TU Wien (2023). https://doi.org/10.48436/de66n-2pj41
16. Taha, A.A., Hanbury, A.: Metrics for evaluating 3D medical image segmentation: analysis, selection, and tool. BMC Med. Imaging **15**, 29 (2015)
17. Wang, W., Wang, G., Hu, C., Ho, K.C.: Robust ellipse fitting based on maximum correntropy criterion with variable center. IEEE Trans. Image Process. **32**, 2520–2535 (2023)
18. Zhang, X., Rosin, P.L.: Superellipse fitting to partial data. Pattern Recogn. **36**(3), 743–752 (2003)

Computational Aspects of Discrete Structures and Tilings

Construction of Tilings with Transitivity Properties on the Square Grid

Mark D. Tomenes[ID] and Ma. Louise Antonette N. De Las Peñas[✉][ID]

Ateneo de Manila University, Quezon City, Philippines
mdelaspenas@ateneo.edu

Abstract. Tilings have been a subject of interest in discrete geometry for their algebraic and geometric properties. Their applications extend to other areas such as crystallography and mathematical morphology. In this paper, we study a class of tilings called (a, b, c) tilings where their respective symmetry groups form a orbits of vertices, b orbits of edges and c orbits of tiles. A method is presented where (a, b, c) tilings are derived and enumerated from the square grid using groups of symmetries involving 4-fold rotations. Conditions on a, b and c are provided for (a, b, c) tilings that arise. Using these conditions, examples of $(a, a-1, c)$ tilings are obtained.

Keywords: Tilings · Vertex-edge-tile transitivity · Square grid · Square lattice

1 Introduction

Classification of tilings has been carried out in discrete geometry based on transitivity properties of vertices [12, 13, 19], edges [10, 11, 13, 16] or tiles [5, 7–9, 13, 15]. In this paper, we consider a class of tilings that have simultaneous transitivity properties of vertices, edges and tiles. These tilings, where their symmetry groups form a, b and c orbits of vertices, edges and tiles, respectively, in the tilings are called (a, b, c) tilings. These were first introduced by Chavey [1] and have been studied in recent work [20, 21]. These tilings provide structures that have applications in other fields such as crystallography and mathematical morphology.

In crystallography, (a, b, c) tilings find relevance within the context of reticular chemistry. In their work, Delgado-Friedrichs and O'Keeffe [6] noted that $(a, a-1, c)$ tilings are rare and were able to enumerate seven $(3, 2, c)$ tilings of $\mathbb{E}^2$. These tilings also serve as ideal blueprints for the design and construction of metal–organic frameworks (MOFs), with specific emphasis on tilings with low values of b [2].

The central idea of mathematical morphology is the analysis of the geometrical structure of an image [14]. In connection with the current paper, this image can be a tiling pattern and the investigation of its geometrical structures. The analysis involves matching small patterns within the tiling at various locations, giving information on the different types of vertices, edges, and tiles present in the tiling, as well as their interrelations. The different types of vertices, edges and tiles are linked to a group action on the tiling.

© The Author(s), under exclusive license to Springer Nature Switzerland AG 2024
S. Brunetti et al. (Eds.): DGMM 2024, LNCS 14605, pp. 123–136, 2024.
https://doi.org/10.1007/978-3-031-57793-2_10

This paper focuses on families of (a, b, c) tilings, derived from the square grid or square lattice. The approach used in this paper capitalizes on the subgroup relations and orbit-stabilizer conditions to construct these tilings. By examining the growth of fundamental regions of subgroups of varying indices within the symmetry group of the square grid, we gain valuable insights into how symmetries affect the number of orbits of vertices, edges, and tiles within the tiling. This exploration provides a comprehensive understanding of the transitivity properties exhibited by (a, b, c) tilings originating from the square grid.

In particular, we present (a, b, c) tilings obtained using subgroups that contain 4-fold rotational symmetries of the symmetry group G of the square grid, $G \cong *442$ (plane crystallographic group in Conway notation). There are five types of subgroups satisfying this condition, of plane crystallographic types $*442$, $4*2$ and 442. These yield different (a, b, c) tilings described in Propositions 4–8.

In Sect. 2, we lay the foundation by presenting essential concepts that will enable the reader to comprehend the results that follow. Section 3 presents a method of constructing (a, b, c) tilings and a formula for the permissible values of a, b and c for the realization of (a, b, c) tilings with symmetry groups containing 4-fold rotations. In Sect. 4, we apply the results from Sect. 3 by providing examples. Finally, Sect. 5 gives the conclusion and future outlook of the study.

2 Preliminary Concepts

A *tiling* of the Euclidean plane $\mathbb{E}^2$ is a countable collection of closed topological disks called *tiles* that is a covering as well as a packing. Two distinct tiles of $\mathcal{T}$ may intersect in an arc or a point. These intersections are called *edges* or *vertices* of $\mathcal{T}$, respectively. The *symmetry group* G of $\mathcal{T}$ is the group of all isometries of $\mathbb{E}^2$ that leave $\mathcal{T}$ invariant. The elements of G are called *symmetries* of $\mathcal{T}$. Associated with $\mathcal{T}$ and its symmetry group G is the *fundamental region* $\mathcal{R}$ which is an open and connected subset of $\mathbb{E}^2$ such that (i) $\mathbb{E}^2 = \bigcup_{g \in G} g\overline{\mathcal{R}}$; and (ii) if $g_1, g_2 \in G$ with $g_1 \neq g_2$ then $g_1 \mathcal{R} \cap g_2 \mathcal{R} = \varnothing$. It is the smallest region that generates $\mathcal{T}$ when acted on by G.

Consider a vertex V, edge E or tile T of $\mathcal{T}$ and let H be a subgroup of G. The *stabilizer* of V under H consists of all symmetries of H that fix V, that is, $Stab_H(V) = \{h \in H : hV = V\}$. The set $HV = \{hV : h \in H\}$ is called the *orbit* of V under H. The stabilizer and orbit of E or T is defined similarly. The stabilizer of a vertex, edge or tile of a tiling can either be a cyclic group of order n denoted by C_n or a dihedral group of order $2n$ denoted by D_n [17].

If a tiling $\mathcal{T}$ forms a orbits of vertices under its symmetry group G, $\mathcal{T}$ is called an *a-isogonal* tiling. If G forms b orbits of edges, then $\mathcal{T}$ is called *b-isotoxal*. Finally, $\mathcal{T}$ is called *c-isohedral* if G forms c orbits of tiles. If $\mathcal{T}$ satisfies all these transitivity properties simultaneously, $\mathcal{T}$ is called an (a, b, c) tiling.

The *square grid* or *square tiling* is a tiling of $\mathbb{E}^2$ where the square tiles appear four times around a vertex. It has symmetry group G generated by three reflections and has the presentation $G = \langle P, Q, R | P^2 = Q^2 = R^2 = (PQ)^4 = (QR)^4 = (PR)^2 = e \rangle \cong *442$ which is used throughout the paper. Here, $*442$ is the Conway's orbifold notation [3, 4] of G, which indicates that there are two distinct 4-fold rotations (PQ and QR) and a twofold rotation (PR) in the fundamental region $\mathcal{R}$ that lies on a reflection axis (Fig. 1a). In [18], Rapanut determined the subgroups of plane crystallographic groups of finite index and the number of conjugacy classes of these subgroups. We state one of the results pertaining to the subgroups of $*442$ as follows.

Lemma 1. Let G be a plane crystallographic group of type $*442$. Then G has the following subgroups containing 4-fold rotational symmetries for all positive integer n:
i) $H_n \cong *442$ of index n^2 with conjugacy class/es 1 or 2 if n is odd or even, respectively;
ii) $H_n \cong *442$ of index $2n^2$ with 2 conjugacy classes; iii) $H_n \cong 4*2$ of index $2n^2$ with 2 conjugacy classes; iv) $H_n \cong 4*2$ of index $4n^2$ with 2 conjugacy classes; and
v) $H_n \cong 442$ of index $2n$ with conjugacy classes $\theta(n)$ or $2\,\theta(n)$ if n is odd or even, respectively.

where $\theta(n) = ((r_1 + 1) \cdots (r_k + 1) + 1)/2$ if $n = 2^r p_1^{r_1} \cdots p_k^{r_k} q_1^{s_1} \cdots q_m^{s_m}$ with $p_i \equiv 1(\mathrm{mod}4)$, $q_i \equiv 3(\mathrm{mod}4)$ and s_i even.

The Conway notation $4*2$ indicates a 4-fold rotation not lying on a reflection axis and a twofold rotation lying on a reflection axis in the fundamental region $\mathcal{R}$. For example, G has subgroup $H_2 \cong 4*2$ of index 8 generated by R, a reflection, and $QPRQPQ$, a 4-fold rotation not lying on a reflection axis. $(QPRQPQ)R(QPRQPQ)^{-1}R$ is a twofold rotation in H_2 lying on a reflection axis (Fig. 1b). The Conway notation 442 indicates two distinct 4-fold rotations and a twofold rotation in $\mathcal{R}$ not lying on a reflection axis. For example, G has subgroup $H_2 \cong 442$ of index 8 generated by QR and $PQRQPRQR$, 4-fold rotations, and $RQPRQRQPQR$, a twofold rotation (Fig. 1c).

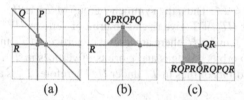

(a)   (b)   (c)

Fig. 1. Location of the centers of twofold rotations (circle), 4-fold rotations (square) and axes of reflections relative to the square grid and the fundamental region (shaded region) of (a) G; (b) $H_2 = \langle R, QPRQPQ \rangle \cong 4*2$; and (c) $H_2 = \langle QR, RQPRQRQPQR \rangle \cong 442$

3 Main Results

In this section, we present a method of constructing (a, b, c) tiling from the square grid using a subgroup H of G, the symmetry group of the square grid.

Theorem 2. Let $\mathcal{T}$ be the square grid with symmetry group G. Consider a subgroup H of G that forms a orbits of tiles, b orbits of edges and c orbits of vertices in $\mathcal{T}$. Then there exists an (a, b, c) tiling $\mathcal{T}^*$ with symmetry group H.

Proof. Consider the square grid $\mathcal{T}$ with symmetry group G and let $H \le G$. Suppose H forms a orbits of tiles, b orbits of edges and c orbits of vertices in $\mathcal{T}$. Let $T_1, \ldots, T_a$ be the tile orbit representatives for each orbit of tiles. Construct points V_i^* such that there is only one point in each T_i. To do this, we consider the following:

a. V_i^* is the incenter of T_i if $Stab_H(T_i) \cong C_d$ or D_d, $d \geq 2$;
b. V_i^* lies in the reflection axis but not in the incenter of T_i if $Stab_H(T_i) \cong D_1$;
c. V_i^* is any point in T_i not lying in any reflection axis if $Stab_H(T_i) \cong C_1$.

Then, form $V = HV_1^* \cup HV_2^* \cup \cdots \cup HV_a^*$ to get the set of vertices of the tiling that we are constructing.

Now, let $E_1, \ldots, E_b$ be the edge orbit representatives for each orbit of edges. For each edge orbit representative E_j construct an edge E_j^* connecting the points in the interior of tiles adjacent to E_j such that the following are satisfied:

a. E_j^* is a straight line segment if $Stab_H(E_j) \cong D_2$ or if $Stab_H(E_j) \cong D_1$ and the reflection axis is perpendicular to E_j (Fig. 2a);
b. E_j^* is an arc with stabilizer D_1 if $Stab_H(E_j) \cong D_1$ and the reflection axis is parallel to E_j (Fig. 2b);
c. E_j^* is an arc with stabilizer C_2 if $Stab_H(E_j) \cong C_2$ (Fig. 2c); and
d. E_j^* is an arc with stabilizer C_1 if $Stab_H(E_j) \cong C_1$ (Fig. 2d).

These edges should also satisfy the condition that $gE_i^* \neq E_j^*$ for all $g \in G\backslash H$, $i \neq j$. These conditions ensures that every pair of points belonging to adjacent tiles are connected by only one edge and that no two edges from different orbits will be sent to each other. Finally, form $T^* := HE_1^* \cup HE_2^* \cup \cdots \cup HE_b^*$. The vertices of T^* are elements of V and its tiles are 4-gons. Each vertex V_k ($k = 1, \ldots, c$) of T give rise to a tile T_k^* in T^*, $V_k \in Int(T_k^*)$.

Since $HT^* = T^*$, its symmetry group contains H. But since $gE_i^* \neq E_j^*$ for all $g \in G\backslash H$, then $gT^* \neq T^*$. Thus, the symmetry group of T^* is H.

Each orbit of tiles HT_i of T give rise to an orbit of vertices HV_i^* in T^*. Since there are a orbits of tiles in T under H, there are a orbit of vertices in T^* under H. In a similar way, since there are b orbits of edges and c orbits of vertices in T under H, there are b orbits of edges and c orbits of tiles in T^* under H. Therefore, T^* is an (a, b, c) tiling. ∎

(a)   (b)   (c)   (d)

Fig. 2. Different types of edges used in the construction of (a, b, c) tilings: (a) straight line segment; and arcs with stabilizer (b) D_1; (c) C_2; and (d) C_1

The following lemma from [21] will be very helpful in proving the succeeding theorems regarding the permissible values of a, b and c.

Lemma 3. Suppose G acts transitively on the set $\mathcal{O}$ of vertices, edges or tiles of a tiling. Let $H \leq G$. If H forms m orbits of vertices, edges or tiles in $\mathcal{O}$, then

$$[G : H] = \sum_{i=1}^{m} \frac{|Stab_G(x)|}{|Stab_H(x_i)|}$$

where $[G : H]$ is the index of H in G, x_i, $i = 1, \ldots, m$, is an orbit representative in each orbit of vertices, edges or tiles under H and $x \in \mathcal{O}$.

Let G be the symmetry group of the square grid. The results that follow were obtained by applying Theorem 2 and considering different families of subgroups of G containing 4-fold rotational symmetries given in Lemma 1 in constructing (a, b, c) tilings. The index of these subgroups in G play an important role in the (a, b, c) tiling that will be constructed as this will indicate the location of the centers of rotations and axes of reflections. These centers and axes will then determine the types of vertices, edges and tiles present in the (a, b, c) tiling.

Proposition 4. Let T be the square grid with symmetry group G. Suppose T_n^* is a tiling obtained from T using a subgroup $H_n \cong *442$ of G of index n^2, n a positive integer, then T_n^* is an (a, b, c) tiling where

$$(a, b, c) = \begin{cases} \left(\frac{n^2+4n+3}{8}, \frac{n^2+2n+1}{4}, \frac{n^2+4n+3}{8} \right) & \text{if } n \text{ is odd} \\ \left(\frac{n^2+6n+8}{8}, \frac{n^2+2n}{4}, \frac{n^2+2n}{8} \right) \text{ or } \left(\frac{n^2+2n}{8}, \frac{n^2+2n}{4}, \frac{n^2+6n+8}{8} \right) & \text{if } n \text{ is even} \end{cases}$$

Proof. Consider the square grid T with symmetry group $G = \langle P, Q, R \rangle \cong *442$. Suppose T_n^* is a tiling obtaining from T using a subgroup $H_n \cong *442$ of G of index n^2, n a positive integer. We consider two cases, i) n is odd and ii) n is even.

Case 1. Suppose n is odd. The subgroups of G of type $*442$ and index n^2 are $H_n = \langle P, Q, (QPQR)^{-(n-1)/2} R(QPQR)^{(n-1)/2} \rangle$. The generators are shown in Fig. 3a for $n = 1, 3, 5, 7$. Each H_n yields an (a, b, c) tiling T_n^* with symmetry group H_n as discussed in Theorem 2. We want to know what kinds of (a, b, c) tilings will be obtained by determining the number of orbits of vertices, edges and tiles of T under H_n.

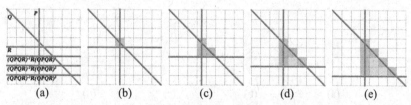

Fig. 3. (a) Axes of reflections of subgroups $H_n = \langle P, Q, (QPQR)^{-(n-1)/2} R(QPQR)^{(n-1)/2} \rangle$, $n = 1, 3, 5, 7$; and (b)-(e) tile orbit representatives that intersect the fundamental region $\mathcal{R}_n$ under H_n. Red tiles have stabilizer D_4, blue tiles have stabilizer D_1 and green tiles have stabilizer C_1 (Color figure online)

We first determine the number of orbits of tiles of T under H_n. To do this, we count the number of tiles of certain types of stabilizers that intersect the fundamental region $\mathcal{R}_n$ corresponding to H_n. This is because all other tiles of T are images of these tiles under some symmetry in H_n. Observe that there are three types of tiles that intersect $\mathcal{R}_n$. These are tiles having stabilizer D_4, D_1 or C_1. Under H_n, these are the tiles that intersect $\mathcal{R}_n$: one tile with stabilizer D_4 which is the tile where the axes of P and Q intersects; $n - 1$ tiles with stabilizer D_1 which are the tiles that intersect either the axis of P or Q but not both; and $a - n$ tiles with stabilizer C_1 which are the tiles that is not intersected by any reflection in H_n (Fig. 3b–3e). We compute a using the formula provided in Lemma 3.

$$[G : H_n] = \sum_{i=1}^{m} \frac{|Stab_G(T)|}{|Stab_{H_n}(T_i)|} \Rightarrow n^2 = \sum_{i=1}^{m} \frac{8}{|Stab_{H_n}(T_i)|} \Rightarrow \frac{n^2}{8}$$

$$= \sum_{i=1}^{m} \frac{1}{|Stab_{H_n}(T_i)|} \Rightarrow \frac{n^2}{8} = \frac{1}{8} + (n-1)\left(\frac{1}{2}\right) + (a - n) \Rightarrow a = \frac{n^2 + 4n + 3}{8}$$

To compute b, we determine the number of orbits of edges of T under H_n. This is done in a similar manner as in above. We count the number of edges of certain types of stabilizers that intersect $\mathcal{R}_n$. In $\mathcal{R}_n$, the types of edges are the following: one edge with stabilizer D_2 which is the edge where the axes of P and $(QPQR)^{-(n-1)/2}R(QPQR)^{(n-1)/2}$ intersect; $n - 1$ edges with stabilizer D_1 which are the edges that intersect the axis of P or $(QPQR)^{-(n-1)/2}R(QPQR)^{(n-1)/2}$ but not both; and $b - n$ edges with stabilizer C_1 which are the edges that is not intersected by any reflection in H_n (Fig. 4). We compute b using Lemma 3.

$$[G : H_n] = \sum_{i=1}^{m} \frac{|Stab_G(E)|}{|Stab_{H_n}(E_i)|} \Rightarrow n^2 = \sum_{i=1}^{m} \frac{4}{|Stab_{H_n}(E_i)|} \Rightarrow \frac{n^2}{4}$$

$$= \sum_{i=1}^{m} \frac{1}{|Stab_{H_n}(E_i)|} \Rightarrow \frac{n^2}{4} = \frac{1}{4} + (n-1)\left(\frac{1}{2}\right) + (b - n) \Rightarrow b = \frac{n^2 + 2n + 1}{4}$$

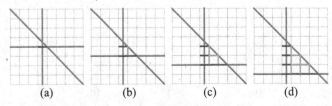

(a)   (b)   (c)   (d)

Fig. 4. Edge orbit representatives that intersect the fundamental region $\mathcal{R}_n$ under the subgroups $H_n, n = 1, 3, 5, 7$. Red edges have stabilizer D_2, blue edges have stabilizer D_1 and green edges have stabilizer C_1 (Color figure online)

We use a similar approach to compute c. In $\mathcal{R}_n$, the types of vertices are the following: one vertex with stabilizer D_4 which is the vertex where the axes of Q and $(QPQR)^{-(n-1)/2}R(QPQR)^{(n-1)/2}$ intersects; $n-1$ vertices with stabilizer D_1 which are the vertices that intersects Q or $(QPQR)^{-(n-1)/2}R(QPQR)^{(n-1)/2}$ but not both; and $c-n$ vertices with stabilizer C_1 which are the edges that is not intersected by any reflection in H_n (Fig. 5). We compute c using Lemma 3.

$$[G:H_n] = \sum_{i=1}^{m} \frac{|Stab_G(V)|}{|Stab_{H_n}(V_i)|} \Rightarrow n^2 = \sum_{i=1}^{m} \frac{8}{|Stab_{H_n}(V_i)|} \Rightarrow \frac{n^2}{8}$$

$$= \sum_{i=1}^{m} \frac{1}{|Stab_{H_n}(V_i)|} \Rightarrow \frac{n^2}{8} = \frac{1}{8} + (n-1)\left(\frac{1}{2}\right) + 7(c-n) \Rightarrow c = \frac{n^2 + 4n + 3}{8}$$

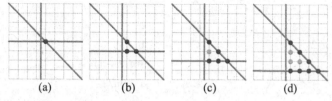

(a)   (b)   (c)   (d)

Fig. 5. Vertex orbit representatives that intersect the fundamental region $\mathcal{R}_n$ under the subgroups H_n, $n = 1, 3, 5, 7$. Red vertices have stabilizer D_4, blue vertices have stabilizer D_1 and green vertices have stabilizer C_1 (Color figure online)

Thus, when n is odd, the resulting tiling T_n^* is a $((n^2 + 4n + 3)/8, (n^2 + 2n + 1)/4, (n^2 + 4n + 3)/8)$ tiling.

Case 2. Suppose n is even. A similar approach can be done as in the previous case. Here, there are two conjugacy classes of subgroups of index n^2. Subgroups coming from a conjugacy class are $H_n^{(1)} = \langle P, Q, (QPQR)^{-n/2}QPQ(QPQR)^{n/2}\rangle$ and $H_n^{(2)} = \langle Q, R, (QRQP)^{-n/2}QRQ(QRQP)^{n/2}\rangle$. For each case, we put the relevant information in Table 1 to compute the values of a, b and c.

Using Lemma 3 and computing the values of a, b and c in the same manner as in Case 1, we see that the subgroups $H_n^{(1)}$ results to $((n^2 + 6n + 8)/8, (n^2 + 2n)/4, (n^2 + 2n)/8)$ tilings while the subgroups $H_n^{(2)}$ results to $((n^2 + 2n)/8, (n^2 + 2n)/4, (n^2 + 6n + 8)/8)$ tilings. ∎

The subsequent propositions can be proven using a similar method employed in Proposition 4, wherein we examine subgroups of G, each characterized by a specified symmetry group type and index. We then proceed to identify and analyze the distinct types of vertices, edges and tiles present on $\mathcal{R}_n$ based on their respective stabilizers under these subgroups. The values of a, b, and c are obtained through the application of Lemma 3.

Table 1. Number of vertices, edges and tiles in $\mathcal{R}_n$ with different stabilizers under subgroups $H_n^{(1)} = \langle P, Q, (QPQR)^{-n/2}QPQ(QPQR)^{n/2}\rangle$ or $H_n^{(2)} = \langle Q, R, (QRQP)^{-n/2}QRQ(QRQP)^{n/2}\rangle$ of types *442 and index n^2 in G, where n is a positive even integer.

	$H_n^{(1)}$		$H_n^{(2)}$	
	Number of T_i, E_i or V_i in $\mathcal{R}_n$	Stabilizer	Number of T_i, E_i or V_i in $\mathcal{R}_n$	Stabilizer
Tiles	2	D_4	$n/2$	D_1
	1	D_2	$c-n/2$	C_1
	$3(n/2-1);$	D_1	-	-
	$a-3n/2$	C_1	-	-
Edges	n	D_1	n	D_1
	$b-n$	C_1	$b-n$	C_1
Vertices	$n/2$	D_1	2	D_4
	$c-n/2$	C_1	1	D_2
	-	-	$3(n/2-1)$	D_1
	-	-	$a-3n/2$	C_1

Proposition 5. Let T be a square grid with symmetry group G. Suppose T_n^* is a tiling obtained from T using a subgroup $H_n \cong *442$ of G of index $2n^2$, n a positive integer, then T_n^* is a $((n^2+2n+1)/4, (n^2+n)/2, (n^2+4n+3)/4)$ or $((n^2+4n+3)/4, (n^2+n)/2, (n^2+2n+1)/4)$ tiling if n is odd, and $((n^2+2n)/4, (n^2+n)/2, (n^2+4n+4)/4)$ or $((n^2+4n+4)/4, (n^2+n)/2, (n^2+2n)/4)$ tiling if n is even (Table 2).

Proposition 6. Let T be a square grid with symmetry group G. Suppose T_n^* is a tiling obtained from T using a subgroup $H_n \cong 4*2$ of G of index $2n^2$, n a positive integer, then T_n^* is a $((n^2+3)/4, (n^2+n)/2, (n^2+2n+1)/4)$ or $((n^2+2n+1)/4, (n^2+n)/2, (n^2+3)/4)$ tiling if n is odd and an $(n^2/4, (n^2+n)/2, (n^2+2n+4)/4)$ or $((n^2+2n+4)/4, (n^2+n)/2, n^2/4)$ tiling if n is even (Table 3).

Table 2. Number of vertices, edges and tiles in $\mathcal{R}_n$ with different stabilizers under subgroups $H_n^{(1)}$ $= \langle R, PQP, (RQPQ)^{-(n-1)}Q(RQPQ)^{n-1}\rangle$ or $H_n^{(2)} = \langle QPQ, PQPRQRPQP, (RQPQ)^{-(n-1)}$ $Q(RQPQ)^{n-1}\rangle$ of types $*442$ and index $2n^2$ in G

	$H_n^{(1)}$		$H_n^{(2)}$	
	Number of T_i, E_i or V_i in $\mathcal{R}_n$	Stabilizer	Number of T_i, E_i or V_i in $\mathcal{R}_n$	Stabilizer
Tiles	1 if n is odd 0 if n is even	D_2	2	D_4
	$n-1$ if n is odd n if n is even	D_1	0 if n is odd 1 if n is even	D_2
	$a-n$	C_1	$2n-2$ if n is odd $2n-3$ if n is even	D_1
	-	-	$a-2n$	C_1
Edges	n	D_1	n	D_1
	$b-n$	C_1	$b-n$	C_1
Vertices	2	D_4	1 if n is odd 0 if n is even	D_2
	0 if n is odd 1 if n is even	D_2	$n-1$ if n is odd n if n is even	D_1
	$2n-2$ if n is odd $2n-3$ if n is even	D_1	$c-n$	C_1
	$c-2n$	C_1	-	-

Proposition 7. Let $\mathcal{T}$ be a square grid with symmetry group G. Suppose $\mathcal{T}_n^*$ is a tiling obtained from $\mathcal{T}$ using a subgroup $H_n \cong 4*2$ of G of index $4n^2$, n a positive integer, then $\mathcal{T}_n^*$ is an $((n^2+n)/2, n^2, (n^2+n+2)/2)$ or an $((n^2+n+2)/2, n^2, (n^2+n)/2)$ tiling (Table 3).

Proposition 8. Let $\mathcal{T}$ be a square grid with symmetry group G. Suppose $\mathcal{T}_n^*$ is a tiling obtained from $\mathcal{T}$ using a subgroup $H_n \cong 442$ of G of index $2n$, $n = \alpha^2 + \beta^2$, α, β are nonnegative integers but not both zero, then $\mathcal{T}_n^*$ is one of the following tilings: $((\alpha^2+\beta^2+2)/4, (\alpha^2+\beta^2)/2, (\alpha^2+\beta^2+6)/4)$, $((\alpha^2+\beta^2+6)/4, (\alpha^2+\beta^2)/2, (\alpha^2+\beta^2+2)/4)$, $((\alpha^2+\beta^2)/4, (\alpha^2+\beta^2)/2, (\alpha^2+\beta^2+8)/4)$, $((\alpha^2+\beta^2+8)/4, (\alpha^2+\beta^2)/2, (\alpha^2+\beta^2)/4)$ or $((\alpha^2 + \beta^2 + 3)/4, (\alpha^2 + \beta^2 + 1)/2, (\alpha^2 + \beta^2 + 3)/4)$ tiling (Table 4).

Table 3. Number of vertices, edges and tiles in $\mathcal{R}_n$ with different stabilizers under subgroups $H_n^{(1)} = \langle R, (RPQPRQ)^{-\frac{(n-1)}{2}} PQ(RPQPRQ)^{\frac{(n-1)}{2}}\rangle$, $H_n^{(2)} = \langle R, (RPQPRQ)^{-\frac{n}{2}} PQRP(RPQPRQ)^{n/2}\rangle H_n^{(1)} = \langle R, (RPQPRQ)^{-\frac{(n-1)}{2}} PQ(RPQPRQ)^{\frac{(n-1)}{2}}\rangle$, $H_n^{(2)} = \langle R, (RPQPRQ)^{-\frac{n}{2}} PQRP(RPQPRQ)^{n/2}\rangle$, $H_n^{(3)} = \langle QPQ, (RPQPRQ)^{-(n+1)/2} PQRP(RPQPRQ)^{(n+1)/2}\rangle$, or $H_n^{(4)} = \langle QPQ(RPQPRQ)^{-n/2}, PQ(RPQPRQ)^{n/2}\rangle$ of types $4*2$ and index $2n^2$ in G; and subgroups $\overline{H}_n^{(1)} = \langle PQRP, (PQRQ)^{-(n-1)} Q(PQRQ)^{n-1}\rangle$ or $\overline{H}_n^{(2)} = \langle PQ, (PQRQ)^{-n} Q(PQRQ)^n\rangle$ of types $4*2$ and index $4n^2$ in G

	$H_n^{(1)}$ or $H_n^{(2)}$		$H_n^{(3)}$ or $H_n^{(4)}$		$\overline{H}_n^{(1)}$		$\overline{H}_n^{(2)}$	
	Number of T_i, E_i or V_i in $\mathcal{R}_n$	Stabilizer	Number of T_i, E_i or V_i in $\mathcal{R}_n$	Stabilizer	Number of T_i, E_i or V_i in $\mathcal{R}_n$	Stabilizer	Number of T_i, E_i or V_i in $\mathcal{R}_n$	Stabilizer
Tiles	1, n odd 0, n even	C_4	0, n odd 1, n even	C_4	n	D_1	1	C_4
	$a-1$, n odd a, n even	C_1	1	D_2	$a-n$	C_1	1	D_2
	-	-	$n-1$	D_1	-	-	$n-1$	D_1
	-	-	$a-n$, n odd $a-n-1$, n even	C_1	-	-	$a-n-1$	C_1
Edges	n	D_1	n	D_1	b	C_1	b	C_1
	$b-n$	C_1	$b-n$	C_1	-	-	-	-
Vertices	0, n odd 1, n even	C_4	1, n odd 0, n even	C_4	1	C_4	n	D_1
	1	D_2	$c-1$, n odd c, n even	C_1	1	D_2	$c-n$	C_1
	$n-1$	D_1	-	-	$n-1$	D_1	-	-
	$c-n$, n odd $c-n-1$, n even	C_1	-	-	$c-n-1$	C_1	-	-

Table 4. Number of vertices, edges and tiles in $\mathcal{R}_n$ with different stabilizers under subgroups $H_n^{(1)} = \langle QR, y^{(\beta+1)/2}x^{-(\alpha+1)/2}PQPQx^{(\alpha+1)/2}y^{-(\beta+1)/2}\rangle$, $H_n^{(2)} = \langle PQ, y^{(\beta-1)/2}x^{-(\alpha-1)/2}QRQ \; Rx^{(\alpha-1)/2}y^{-(\beta-1)/2}\rangle$, $H_n^{(3)} = \langle QR, y^{\beta/2}x^{-\alpha/2} \; QRQRx^{\alpha/2}y^{-\beta/2}\rangle$, $H_n^{(4)} = \langle PQy^{\beta/2}x^{-\alpha/2}PQPQ \; x^{\alpha/2}y^{-\beta/2}\rangle$ or $H_n^{(5)} = \langle QR, y^{\beta/2}x^{-(\alpha+1)/2} \; PRx^{(\alpha+1)/2}y^{-\beta/2}\rangle$ of types 442 and index $2n$ in G, $n = \alpha^2 + \beta^2$, $x = PQRQ$ and $y = RQPQ$.

	$H_n^{(1)}$		$H_n^{(2)}$		$H_n^{(3)}$		$H_n^{(4)}$		$H_n^{(5)}$	
	Number of T_i, E_i or V_i in $\mathcal{R}_n$	Stabilizer	Number of T_i, E_i or V_i in $\mathcal{R}_n$	Stabilizer	Number of T_i, E_i or V_i in $\mathcal{R}_n$	Stabilizer	Number of T_i, E_i or V_i in $\mathcal{R}_n$	Stabilizer	Number of T_i, E_i or V_i in $\mathcal{R}_n$	Stabilizer
Tiles	1	C_2	2	C_4	a	C_1	2	C_4	1	C_4
	$a-1$	C_1	$a-2$	C_1	-	-	1	C_2	$a-1$	C_1
	-	-	-	-	-	-	$a-3$	C_1	-	-
Edges	b	C_1	b	C_1	b	C_1	b	C_1	1	C_2
	-	-	-	-	-	-	-	-	$b-1$	C_1
Vertices	2	C_4	1	C_2	2	C_4	c	C_1	1	C_4
	$c-2$	C_1	$c-1$	C_1	1	C_2	-	-	$c-1$	C_1
	-	-	-	-	$c-3$	C_1	-	-	-	-

4 Examples

In this section, we present the application of Theorem 2 by showing an example of a construction of a $(3, 4, 3)$ tiling with symmetry group of type $*442$. Consider the square grid $\mathcal{T}$ having symmetry group $G = \langle P, Q, R \rangle \cong *442$ (Fig. 6a). Take the subgroup $H_3 = \langle P, Q, RQPQRQPQR \rangle \cong *442$ where $[G : H_3] = 9$ (Fig. 6b). Under H_3, there are three orbits of tiles with orbit representatives T_1, T_2 and T_3 (Fig. 6c). Take points x_1 in the incenter of T_1; x_2 lying in the axis of P but not in the incenter of T_2; and x_3 lying in the axis of Q but not in the incenter of T_3 (Fig. 6d). This is because $Stab_H(T_1) = \langle P, Q \rangle \cong D_4$, $Stab_H(T_2) = \langle P \rangle \cong D_1$, and $Stab_H(T_3) = \langle Q \rangle \cong D_1$. Then we form $\mathcal{V} = Hx_1 \cup Hx_2 \cup Hx_3$ (Fig. 6e).

Now, there are four orbits of edges under H with orbit representatives E_1, E_2, E_3 and E_4 (Fig. 6f). Since $Stab_H(E_1) = \langle P \rangle \cong D_1$, the points in the interior of tiles incident to E_1 are connected by a straight line segment E_1^*. In a similar manner, since $Stab_H(E_2) \cong C_1$, $Stab_H(E_3) = \langle P, RQPQRQPQR \rangle \cong D_2$ and $Stab_H(E_4) = \langle RQPQRQPQR \rangle \cong D_1$, the edges E_2^*, E_3^* and E_4^* are constructed such that $Stab_H(E_2^*) \cong C_1$, $Stab_H(E_3^*) \cong D_2$ and $Stab_H(E_4) \cong D_1$ as shown in Fig. 6g. Then we form $\mathcal{T}^* = HE_1^* \cup HE_2^* \cup HE_4^* \cup HE_4^*$ which is a $(3, 4, 3)$ tiling with symmetry group $H \cong *442$ (Fig. 6h).

Now, Proposition 4 tells us the types of (a, b, c) tilings that we can obtain by using subgroups H_n having symmetry group type $*442$ and index n^2 in G. In particular, an $((n^2 + 4n + 3)/8, (n^2 + 2n + 1)/4, (n^2 + 4n + 3)/8)$ tiling if n is odd, and $((n^2 + 6n + 8)/8, (n^2 + 2n)/4, (n^2 + 2n)/8)$ or $((n^2 + 2n)/8, (n^2 + 2n)/4, (n^2 + 6n + 8)/8)$ tiling if n is even with symmetry groups of type $*442$. For $n = 1, \ldots, 6$, we obtain the

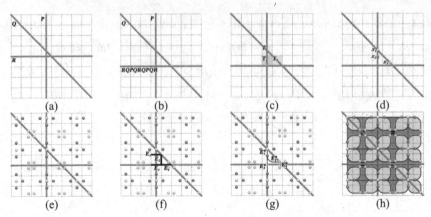

Fig. 6. Construction of a $(3, 4, 3)$ tiling with symmetry group $H \cong *442$ from the square grid $\mathcal{T}$

following tilings shown in Fig. 7 following the construction provided in Theorem 2 and the subgroups provided in Lemma 3. When n is even we only illustrate the case for a $((n^2 + 6n + 8)/8, (n^2 + 2n)/4, (n^2 + 2n)/8)$ tiling.

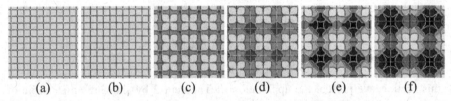

Fig. 7. (a, b, c) tilings with symmetry groups of type $*442$: (a) $(1,1,1)$ tiling; (b) $(3,2,1)$ tiling; (c) $(3,4,3)$ tiling; (d) $(6,6,3)$ tiling; (e) $(6,9,6)$ tiling; and (f) $(10,12,6)$ tiling

5   Conclusion and Future Outlook

In this paper, we present a method of constructing (a, b, c) tiling derived from the square grid utilizing the subgroup structure of its symmetry group and orbit-stabilizer conditions. We also provide a formula for the possible values for a, b and c such that an (a, b, c) tiling with symmetry group of type $*442, 4 * 2$ or 442 is realized. In characterizing the (a, b, c) tilings, the subgroups used in the construction play an important role since these give the precise location of the symmetries and hence have an impact on the types of vertices, edges and tiles present in the tiling. It is interesting as a future study to consider other symmetry group types of (a, b, c) tilings from the square grid. Further research can also be done by considering (a, b, c) tiling derived from the hexagonal and triangular tiling or other lattices and providing a formula for the possible values of a, b and c.

In [6], it is noted that $(a, a - 1, c)$ tilings are rare. Using the propositions presented in Sect. 3, we can find out all $(a, a - 1, c)$ tilings derived from a square grid having symmetry group of type $*442$, $4 * 2$ or 442. For instance, from Proposition 4 we can obtain an $((n^2 + 6n + 8)/8, (n^2 + 2n)/4, (n^2 + 2n)/8)$ tiling. If we consider the equation $b = a - 1$ and solve for n, we get $n = 2$, which results to a $(3, 2, 1)$ tiling. This is illustrated in Fig. 8a. Similarly, from Proposition 5, we get a $(2, 1, 1)$ and a $(4, 3, 2)$ tiling (Fig. 8b–8c); from Proposition 7, we get a $(2, 1, 1)$ tiling (Fig. 8d); and from Proposition 8, we get a $(2, 1, 1)$ and a $(3, 2, 1)$ tiling (Fig. 8e–8f).

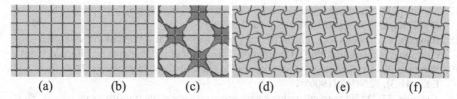

(a)   (b)   (c)   (d)   (e)   (f)

Fig. 8. $(a, a - 1, c)$ tilings derived from the square grid: (a) (3,2,1) tiling; (b) (2,1,1) tiling; (c) (4,3,2) tiling; (d) (2,1,1) tiling; (e) (2,1,1) tiling; and (f) (3,2,1) tiling

Another problem that may be worth exploring is analyzing the geometrical structure of tiling patterns with transitivity properties through mathematical morphology. Image analysis can give insights on the number of different types of vertices, edges and tiles that can be found in a given tiling. This can be very useful specially when the underlying tiling is no longer the square grid but a tiling with a much more complicated structure.

References

1. Chavey, D.P.: Periodic tilings and tilings by regular polygons. Ph.D. Dissertation, University of Wisconsin – Madison (1984)
2. Chen, Z., Jiang, H., O'Keeffe, M., Eddaoudi, M.: Minimal edge-transitive nets for the design and construction of metal–organic frameworks. Faraday Discuss. **201**, 127–143 (2017)
3. Conway, J. H., Burgiel, H., Goodman-Strauss, C.: The Symmetries of Things. A K Peters, Natick (2008)
4. Conway, J. H.: The orbifold notation for surface groups. Groups Comb. Geom. **165**, 438–447 (1992)
5. Delgado-Friedrichs, O., Huson, D., Zamorzaeva, E.: The classification of 2-isohedral tilings of the plane. Geom. Dedicata **42**, 43–117 (1992)
6. Delgado-Friedrichs, O., O'Keeffe, M.: Edge-2-transitive trinodal polyhedra and 2-periodic tilings. Acta Crystallogr. A **73**, 227–230 (2017)
7. Dress, A.W.M., Huson, R.: On tilings of the plane. Geom. Dedicata **24**, 295–310 (1987)
8. Grünbaum, B., Löckenhoff, H.-D., Shephard, G.C., Temesvári, Á.H.: The Enumeration of normal 2-homeohedral tilings. Geom. Dedicata **19**, 109–173 (1985)
9. Grünbaum, B., Shephard, G.C.: The eighty-one types of isohedral tilings in the plane. In: Mathematical Proceedings of the Cambridge Philosophical Society vol. 82, no. 2, pp. 177–196 (1977)
10. Grünbaum, B., Shephard, G.C.: The 2-homeotoxal tilings of the plane and the 2-sphere. J. Combin. Theory Ser. B **34**, 113–150 (1983)
11. Grünbaum, B., Shephard, G.C.: Isotoxal tilings. Pacific J. Math. **76**, 407–430 (1978)
12. Grünbaum, B., Shephard, G.C.: The ninety-one types of isogonal tilings in the plane. Trans. Amer. Math. Soc. **242**, 335–353 (1978)

13. Grünbaum, B., Shephard, G. C.: Tilings and Patterns. W. H Freeman and Company, New York (1987)
14. Heijmans, H.J.A.M.: Mathematical morphology: a modern approach in image processing based on algebra and geometry. SIAM Rev. **37**, 1–36 (1995)
15. Huson, D.H.: The generation and classification of tile-k-transitive tilings of the Euclidean plane, the sphere and the hyperbolic plane. Geom. Dedicata **47**, 269–296 (1993)
16. Lučić, Z., Molnár, E., Stojanović, M.: The 14 infinite families of isotoxal tilings in the planes of constant curvature. Period. Math. Hungar. **29**, 177–195 (1994)
17. Martin, G.E.: Transformation Geometry. An Introduction to Symmetry. Springer, New York (1982).https://doi.org/10.1007/978-1-4612-5680-9
18. Rapanut, T.A.: Subgroups, conjugate subgroups and n-color groups of the seventeen plane crystallographic groups. Ph.D. Dissertation, University of the Philippines, Diliman (1988)
19. Sommerville, D.: Semi-regular networks of the plane in absolute geometry. Trans. Roy. Soc. Edin. **41**(3), 725–747 (1906)
20. Tomenes, M. D.: Edge-to-edge tilings with vertex, edge and tile transitivity properties. Ph.D. Dissertation, Ateneo de Manila University, Quezon City, Philippines (2021)
21. 21. Tomenes, M. D., De Las Peñas, M. L. A. N.: Construction of (a, b, c) tilings of the Euclidean plane, hyperbolic plane and the sphere. Contrib. Discrete Math. (in press)

Some Geometric and Tomographic Results on Gray-Scale Images

Michela Ascolese[1], Paolo Dulio[2], and Silvia M.C. Pagani[3]

[1] Dipartimento di Matematica e Informatica "U. Dini", Università di Firenze, Firenze, Italy
michela.ascolese@unifi.it
[2] Dipartimento di Matematica, Politecnico di Milano, Milano, Italy
paolo.dulio@polimi.it
[3] Dipartimento di Matematica e Fisica "N. Tartaglia", Università Cattolica del Sacro Cuore, Brescia, Italy
silvia.pagani@unicatt.it

Abstract. Discrete tomography deals with the reconstruction of images from a (usually small) set of X-ray projections. This is achieved by modeling the tomographic problem as a linear system of equations and then applying a suitable discrete reconstruction algorithm based on iterations. In this paper we adopt the well-known grid model and prove some geometric properties of integer solutions consisting of $p \geq 2$ gray levels. In particular, we show that all gray-scale solutions having the same two-norm belong to a same hypersphere, centered at the uniform image related to the data and having radius ranging in an interval whose bounds are explicitly computed.

Moving from a uniqueness theorem for gray-scale images, we compute special sets of directions that guarantee uniqueness of reconstruction and exploit them as the input of the Conjugate Gradient Least Squares algorithm. Then we apply an integer rounding to the resulting output and, basing on previously described geometric parameters, we test the quality of the obtained reconstructions for an increasing number of iterations, which leads to a progressive improvement of the percentage of correctly reconstructed pixels, until perfect reconstruction is achieved. Differently, using sets of directions which are classically employed, but far from being sets of uniqueness, only partial reconstructions are obtained.

Keywords: Discrete tomography · Gray-scale image · Grid model · Minimum norm solution · Projection · Unique reconstruction

1 Introduction

Recovering a function from the knowledge of its line sums is the aim of discrete tomography. When consistent, the reconstruction task is usually ill-posed, so an active area of research seeks uniqueness conditions. In this work we consider functions whose co-domain is a finite subset of $\mathbb{Z}$.

© The Author(s), under exclusive license to Springer Nature Switzerland AG 2024
S. Brunetti et al. (Eds.): DGMM 2024, LNCS 14605, pp. 137–149, 2024.
https://doi.org/10.1007/978-3-031-57793-2_11

The present paper reflects a general two-step approach to the tomographic reconstruction problem. The first step looks for some kind of geometric conditions that guarantee uniqueness of reconstruction. The second step assumes the resulting geometric constraints as prior information, trying to incorporate them in some tomographic reconstruction algorithm. By exploiting such extra knowledge, there is the hope of reducing the complexity of the tomographic problem, so to get the unique existing solution in polynomial time. Here, Theorem 1 (proven in [5]) provides the first theoretical step that, combined with the considered additional geometrical results, leads to the proposed polynomial-time reconstruction based on the integer rounding of the Conjugate Gradient Least Squares (CGLS) algorithm.

We report the above uniqueness result for gray-scale images with a couple of examples in Sect. 2, while Sect. 3 explores the geometric properties of the solutions of a tomographic problem, particularly focusing on gray-scale images. Experiments are reported in Sect. 4, where outputs are obtained as a kind of integer rounding of the minimum norm solution, computed by CGLS algorithm. We show that sets of directions ensuring unique reconstruction behave better than classically employed sets. Section 5 resumes the obtained results and shows possible extensions.

For a point $\mathbf{x} = (x_1, \ldots, x_n) \in \mathbb{R}^n$ we denote by $\|\mathbf{x}\|_2$ the Euclidean norm of the corresponding vector, while $\|\mathbf{x}\|_1 = \sum_{i=1}^{n} |x_i|$. For a finite set E, $|E|$ means the number of elements in E. The *norm* of a real-valued function f is $|f| = \max_{x \in \mathrm{Dom} f} |f(x)|$. The *Frobenius norm* $\|A\|_F$ of an $m \times n$ matrix $A = [a_{ij}]$ is $\|A\|_F = \left(\sum_{i=1}^{m} \sum_{j=1}^{n} |a_{ij}|^2 \right)^{\frac{1}{2}}$.

A vector $u = (a, b) \in \mathbb{Z}^2$ such that $a \geq 0$ and $\gcd(a, b) = 1$ is said to be a *direction*. A *line with direction* (a, b) is the set of points satisfying the equation $ay = bx + t$ for $t \in \mathbb{Z}$.

Consider a rectangular subset of the lattice $\mathbb{Z}^2$ of size $M \times N$ with bottom-left corner in the origin, namely, $\mathcal{A} = \{(i, j) \in \mathbb{Z}^2 : 0 \leq i < M, 0 \leq j < N\}$. Points of $\mathcal{A}$ are also called *pixels*. A function $f : \mathcal{A} \longrightarrow \mathbb{R}$ is called an *image*. In the following, we will restrict the co-domain to a finite set. The *projection* of f along a line $\ell : ay = bx + t$ with direction (a, b) is $\sum_{(i,j) \in \mathcal{A} \cap \ell} f(i, j)$.

Let $S = \{(a_r, b_r) : r = 1, \ldots, d\}$ be a set of d lattice directions, and

$$h = \sum_{r=1}^{d} a_r \qquad k = \sum_{r=1}^{d} |b_r|.$$

The so-called Katz condition states that $h \geq M$ or $k \geq N$. When it is fulfilled, uniqueness of reconstruction is guaranteed inside the grid $\mathcal{A}$ [12]. If conversely $h < M$ and $k < N$, S is said to be a *valid* set of directions for $\mathcal{A}$ [11]. When removing the Katz condition, uniqueness of reconstruction is not achievable without extra information (see for instance [2,7,8]).

The *tomographic problem* is as follows: given a grid $\mathcal{A}$, a set S of directions and a set of projections, find an image $f : \mathcal{A} \longrightarrow \mathbb{R}$ which is consistent with the projections. The pixels of $\mathcal{A}$, labeled in some order, can be encoded as a column vector, so that the problem is naturally reinterpreted as a linear system

$$A\mathbf{x} = \mathbf{p}, \tag{1}$$

where $\mathbf{x} \in \mathbb{R}^n = \mathbb{R}^{MN}$ contains the pixels of the image to be reconstructed, the vector $\mathbf{p} \in \mathbb{R}^m$ collects the projections and the matrix A is the binary incidence matrix between pixels and lines having directions in S and non-trivial intersection with $\mathcal{A}$. In other words, in this paper we adopt the so-called *grid model*, which is largely employed in discrete tomography. For example, it is the basic model of the Mojette Transform [10,13], it represents the ideal theoretical framework for various uniqueness results [3,4,7,9] and, also, it can be used to investigate nanocrystals, since these consist of discrete atoms positioned in a regular grid [14]. We also assume that $\mathbf{p} \in \mathbb{Z}^m$, which implies that, if solutions exist (namely, there are no inconsistent projections), then at least one of them is integer [6]. Note that the space of solutions of (1) constitutes an affine subspace $\mathcal{L}$ of $\mathbb{R}^n$ of dimension $(M - h) \times (N - k)$, whose explicit determination is equivalent to exactly solving (1). Due to large dimensions involved, this is not feasible in general and iterative algorithms are employed. In what follows, we will determine hyperplanes and hypersurfaces containing $\mathcal{L}$.

We call *central solution* the minimum norm solution $\mathbf{x}^*$ of (1), obtainable by singular value decomposition. We call *uniform image* associated to (1) the $(M \times N)$-sized image having all its entries equal to $\frac{\|\mathbf{p}\|_1}{\|A\|_F^2}$. It can be immediately computed from the projections, even though, in general, it is not a solution of the given tomographic problem.

2   Some Sets of Uniqueness for Gray-Scale Tomography

For an integer number $p > 1$, a function $f : \mathcal{A} \to \{0, 1, \ldots, p - 1\}$ is said to be an *image with p gray levels*, or simply *gray-scale image* if the number of gray levels is obvious from the context. If $p = 2$, we speak of *binary* images. Given a set of directions S, an *S-weakly bad configuration* is a pair (Z, W) of multisets of lattice points in $\mathbb{Z}^2$, containing the same number of lattice points not necessarily distinct (i.e., counted with multiplicity), such that, for each direction $(a, b) \in S$ and for each $z \in Z$, the line through z in direction (a, b) contains a point $w \in W$. If both Z and W are sets (i.e., all points have multiplicity 1), we call (Z, W) an *S-bad configuration*.

To each finite set S of lattice directions we can associate an S-weakly bad configuration as follows. For each $I \subseteq S$, let $u(I) = \sum_{u \in I} u$, with $u(\emptyset) = \mathbf{0} \in \mathbb{Z}^2$, and define the multisets $E_S = \{u(I) : I \subseteq S, |I| \text{ even}\}$, $O_S = \{u(I) : I \subseteq S, |I| \text{ odd}\}$, where the points $u(I)$ are counted with the appropriate multiplicities, whenever there exist distinct subsets I, J with the same parity such that $u(I) = u(J)$. For every $v \in S$, each point $u(I)$ such that $v \in I$ can be paired to the point

$u(I) - v$. This shows that the pair (E_S, O_S) is an S-weakly bad configuration, and it is an S-bad configuration if all the points $u(I)$, $I \subseteq S$, are distinct.

The idea behind uniqueness results for images with p gray levels, under valid sets of directions, is that one must ensure that any weakly bad configuration that can be built inside the grid requires at least one point whose multiplicity is at least p. The number p is assumed to be a prior information.

The following result [5, Theorem 5.1] provides a uniqueness condition for images with p gray levels.

Theorem 1. *For $p \geq 3$, let $S = \{u_s : s = 1, \ldots, 2p\}$ be a valid set of lattice directions for a given $M \times N$ grid $\mathcal{A}$. Suppose that $u_i + u_{p+i} = u_j + u_{p+j}$, and both $u_i \in S$ and $u_{p+i} \in S$ can not be written as sum of directions of $S \setminus \{u_i, u_{p+i}\}$, for $i, j = 1, \ldots, p$. Set $(\alpha, \beta) = u_i + u_{p+i}$ and*

$$D = \{\pm v : v = u(X) - (\alpha, \beta) \neq \mathbf{0}, \ X \subseteq S\}.$$

If, for each $v = (\mu, \nu) \in D$, $|\mu| \geq M - h$ or $|\nu| \geq N - k$ holds, then each $g : \mathcal{A} \to \mathbb{Z}$ with null projections along the lines with direction in S and $|g| < p$ is identically zero.

The next two examples show that the sufficient condition of Theorem 1 is not necessary and can be refined.

Example 1. Let $p = 3$, and consider the set $S_1 = \{u_1 = (1, 2), u_2 = (1, 3), u_3 = (1, 8), u_4 = (1, -1), u_5 = (1, -2), u_6 = (1, -7)\}$. Then, $u_1 + u_4 = u_2 + u_5 = u_3 + u_6$ and no other subsets I, J of S_1 different from $\{u_1, u_4\}$, $\{u_2, u_5\}$, $\{u_3, u_6\}$ are such that $u(I) = u(J)$. The corresponding set D is

$$D = \pm \{(0, 4), (0, 9), (0, 10), (0, 1), (0, 6), (0, 5), (1, 12), (1, \pm 3), (1, \pm 2), (1, \pm 8), (1, \pm 7),$$
$$(1, 9), (1, -1), (1, -6), (1, 4), (1, -11), (2, 11), (2, 10), (2, \pm 5), (2, 1), (2, -4), (2, 6), (2, 0),$$
$$(2, -9), (2, 7), (2, 2), (2, -8), (2, -3), (3, 9), (3, 4), (3, 0), (3, -1), (3, 3), (3, -6), (4, 2)\}.$$

Note that the assumptions of Theorem 1 are not fulfilled, for instance because of the presence of the vector $v = (2, 0)$. Anyway, the set D does not coincide with the set of translations which move a pair of points of the weakly bad configuration to the two triple points, so lowering their multiplicity. An easy but long computation shows that the set of prohibited translations is the subset

$$D' = \pm\{(1, \pm 2), (1, 3), (1, 8), (1, -1), (1, -7), (0, 1), (0, 4), (0, 6), (0, 9), (0, 5), (0, 10)\} \cup \{(2, 1)\}.$$

Being $h = 6$ and $k = 23$, the set S_1 ensures uniqueness of reconstruction for images with 3 gray levels defined on a grid of size $M \times 24$, M a positive integer.

We also note that Theorem 1 does not provide all sets of directions which uniquely reconstruct gray-scale images. In the following example we provide a set of uniqueness with an odd number of directions.

Example 2. Let $p = 3$. The set $S_2 = \{u_1 = (1, 1), u_2 = (1, -2), u_3 = (3, 4), u_4 = (5, 3), u_5 = (1, 6), u_6 = (1, 7), u_7 = (3, -10)\}$ is such that $u_1 + u_2 + u_3 = u_4 =$

$u_5 + u_6 + u_7$. Therefore, there are three mutually disjoint subsets of S_2 with the same parity and having the same sum. Being $h = 15$ and $k = 33$, S_2 is valid for the grid $\mathcal{A}$ of size 16×34 depicted in Fig. 1. Only one weakly bad configuration can be built inside $\mathcal{A}$ (see [11] for details) and it is reported in the figure. No translation can be done without exceeding the grid sides, so the multiplicity of triple points cannot decrease. This means that S_2 uniquely reconstructs any image with 3 gray levels defined on $\mathcal{A}$.

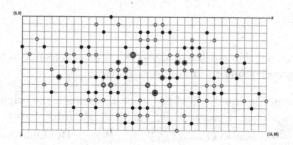

Fig. 1. Points of the only weakly bad configuration related to S_2 in Example 2. The x-axis is oriented downwards, while the y-axis goes from left to right. Full (resp., empty) points belong to O_{S_2} (resp., E_{S_2}). Doubly and triply circled points represent double and triple points, respectively.

3   Geometric Properties of Gray-Scale Solutions

Gray-scale solutions of the tomographic problem are of special interest in discrete tomography (see e.g. [3,4,6]). In view of the explicit reconstruction of possibly existing gray-scale images, it is worth investigating the geometry of all solutions to retrieve parameters which will be useful in the experiments.

Let $\mathcal{S}(S, m, n)$ be the set of all real-valued solutions of a tomographic problem in the grid model concerning images of n pixels, depending on m measurements collected along a set S of lattice directions.

From now on we assume that a solution with p gray levels exists. This implies that the projections are non-negative integers. The following property reproduces [1, Lemma 1], but, for a better understanding of the related results, we prefer to give the proof explicitly.

Lemma 1. *All* $\mathbf{x} \in \mathcal{S}(S, m, n)$ *belong to the hyperplane* $\pi : x_1 + \ldots + x_n = \frac{\|\mathbf{p}\|_1}{|S|}$.

Proof. Let $\mathbf{x} \in \mathcal{S}(S, m, n)$. Then $A\mathbf{x} = \mathbf{p}$ and, being $p_i \geq 0$ for all $i = 1, \ldots, m$, we have

$$\|\mathbf{p}\|_1 = \sum_{i=1}^{m} p_i = \sum_{i=1}^{m} \left(\sum_{j=1}^{n} a_{ij} x_j \right) = \sum_{i=1}^{m} a_{i1} x_1 + \sum_{i=1}^{m} a_{i2} x_2 + \ldots + \sum_{i=1}^{m} a_{in} x_n.$$

Since A is an incidence matrix, there exist precisely $|S|$ entries equal to 1 in each column of A, so $\sum_{i=1}^{m} a_{ij} = |S| \; \forall j \in \{1, \ldots, n\}$, and consequently we get $\|\mathbf{p}\|_1 = |S| \sum_{j=1}^{n} x_j$, which proves the statement. $\square$

Note that $\mathcal{L} \subseteq \pi$ and $\mathbf{x}^* \in \pi$.

Lemma 2. *It results* $\|A\|_F = \sqrt{|S|n}$.

Proof. As noted in the proof of Lemma 1, each column of A has $|S|$ ones. Then

$$\|A\|_F^2 = \sum_{i=1}^m \sum_{j=1}^n |a_{ij}|^2 = \sum_{i=1}^m \sum_{j=1}^n a_{ij} = |S|n.$$

$\square$

Lemma 3. *The line through the origin of* $\mathbb{R}^n$ *and orthogonal to* π *intersects* π *in a point* C *corresponding to the uniform image.*

Proof. The parametric equations of the line ℓ_{OC} through the origin and orthogonal to π are $x_i = \lambda$, $i = 1, \ldots, n$, for $\lambda \in \mathbb{R}$. Intersecting ℓ_{OC} with π, we get $\lambda = \frac{\|\mathbf{p}\|_1}{n|S|}$, hence $C = \left(\frac{\|\mathbf{p}\|_1}{n|S|}, \ldots, \frac{\|\mathbf{p}\|_1}{n|S|} \right)$. By Lemma 2, it results $C = \left(\frac{\|\mathbf{p}\|_1}{\|A\|_F^2}, \ldots, \frac{\|\mathbf{p}\|_1}{\|A\|_F^2} \right)$, so C is the uniform image. $\square$

It was shown in [1, Lemma 3] that all binary images in $\mathcal{S}(S, m, n)$ belong to a same hypersphere $\Sigma(\mathbf{x}^*, R)$, centered at $\mathbf{x}^*$ and with radius

$$R = \sqrt{\frac{\|\mathbf{p}\|_1}{|S|} - \|\mathbf{x}^*\|_2^2}.$$

We can now prove the following generalization of [1, Lemma 3] to all real-valued solutions.

Theorem 2. *All* $\mathbf{x} \in \mathcal{S}(S, m, n)$, $n \geq 2$, *having the same Euclidean norm belong to a same* $(n-2)$-*dimensional hypersphere contained in* π, *centered at the uniform image* C. *Conversely, all* $\mathbf{x} \in \mathcal{S}(S, m, n)$ *that belong to such* $(n-2)$-*dimensional hypersphere have the same Euclidean norm.*

Proof. Let $\mathbf{x}, \mathbf{y} \in \mathcal{S}(S, m, n)$ be two solutions with the same Euclidean norm. Then $\mathbf{x}, \mathbf{y}$ belong to both π and the $(n-1)$-dimensional hypersphere $\Sigma(O, \|\mathbf{x}\|_2)$, centered at the origin O and having radius $\|\mathbf{x}\|_2$, so belong to their intersection $\Sigma(O, \|\mathbf{x}\|_2) \cap \pi$, which is an $(n-2)$-dimensional hypersphere. Its center lies where the line through the origin and orthogonal to π intersects π, so, by Lemma 3, it is the uniform image. Denote by $\Sigma(C, r_{\mathbf{x}})$ the $(n-2)$-dimensional hypersphere $\Sigma(O, \|\mathbf{x}\|_2) \cap \pi$, with radius $r_{\mathbf{x}}$. The second part of the statement follows immediately by Pythagoras' theorem, since $\|\mathbf{x}\|_2^2 = \|C\|_2^2 + r_{\mathbf{x}}^2$. $\square$

Theorem 2 shows that all the solutions are distributed on concentric hyperspheres, centered at the uniform image C, which is not necessarily a solution, but it is immediately available from the data of the tomographic problem. Also, the result provided by [1, Lemma 3] can be further detailed.

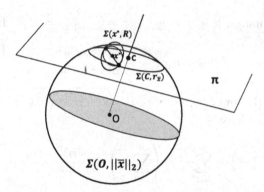

Fig. 2. All solutions having the same 2-norm belong to the $(n-2)$-dimensional hypersphere $\Sigma(C, r_{\mathbf{x}})$. Possible binary solutions belong to the $(n-3)$-dimensional hypersphere (here represented by a pair of points) $\Sigma(\mathbf{x}^*, R) \cap \Sigma(C, r_{\overline{\mathbf{x}}})$.

Corollary 1. *For $n \geq 3$, if $C \neq \mathbf{x}^*$ then binary solutions of the reconstruction problem belong to an $(n-3)$-dimensional sub-sphere of π.*

Proof. By [1, Lemma 3] and Theorem 2, any binary solution $\overline{\mathbf{x}}$ belongs to $\Sigma(\mathbf{x}^*, R) \cap \Sigma(C, r_{\overline{\mathbf{x}}})$, which is a sub-sphere of π of dimension $n-3$.   □

The result in Corollary 1 is shown in Fig. 2.

The assumption $C \neq \mathbf{x}^*$ in Corollary 1 cannot be removed, as shown in the following example.

Example 3. Assume that a 2×2 image $\mathbf{x} = [x_1, x_2, x_3, x_4]^t$ is scanned along the horizontal and the vertical directions ($n = 4$, $|S| = 2$), so collecting a projection array $\mathbf{p} = [h_1, h_2, v_1, v_2]^t$, where $h_1 + h_2 = v_1 + v_2$. The minimum condition on $\|\mathbf{x}\| = \sqrt{x_1^2 + x_2^2 + x_3^2 + x_4^2}$ provides $x_1^* = \frac{h_1}{2} + \frac{v_1 - v_2}{4}$, $x_2^* = \frac{h_1}{2} - \frac{v_1 - v_2}{4}$, $x_3^* = -\frac{h_1}{2} + \frac{3v_1 + v_2}{4}$, $x_4^* = -\frac{h_1}{2} + \frac{v_1 + 3v_2}{4}$. Therefore, in case $h_1 = h_2 = v_1 = v_2$, we get $x_1^* = x_2^* = x_3^* = x_4^* = \frac{v_2}{2} = \frac{\|\mathbf{p}\|_1}{n|S|}$, so the central solution coincides with the uniform image.

When gray-scale images are considered, the bounds of the radius of $\Sigma_{\mathbf{x}}$ can be explicitly estimated.

Corollary 2. *If $\mathbf{x} \in \mathcal{S}(S, m, n)$ is a gray-scale image and $\overline{\mathbf{x}} \in \mathcal{S}(S, m, n)$ is a binary image, then*

$$\sqrt{q\left(1 - \frac{q}{n}\right)} \leq r_{\mathbf{x}} \leq \sqrt{\frac{n-1}{n}} \, q,$$

where $q = \|\overline{\mathbf{x}}\|_2^2$ and $r_{\mathbf{x}}$ is the radius of $\Sigma(C, r_{\mathbf{x}})$.

Proof. By Theorem 2, and since binary solutions are the shortest (w.r.t. the Euclidean norm) among all integer elements of $\mathcal{S}(S, m, n)$ [1, Lemma 2], it results

$$r_{\mathbf{x}} = \sqrt{\|\mathbf{x}\|_2^2 - \|C\|_2^2} \geq \sqrt{\|\overline{\mathbf{x}}\|_2^2 - \|C\|_2^2}.$$

Being $\overline{\mathbf{x}}$ a binary image, it is $q = \sum_{i=1}^{n} \overline{x}_i^2 = \sum_{i=1}^{n} \overline{x}_i = \frac{\|\mathbf{p}\|_1}{|S|}$. By Lemma 2, it results

$$\|C\|_2^2 = n \frac{\|\mathbf{p}\|_1^2}{\|A\|_F^2} = \frac{\|\mathbf{p}\|_1^2}{n|S|^2} = \frac{q^2}{n},$$

and consequently the lower bound of $r_{\mathbf{x}}$ is $\sqrt{q - \frac{q^2}{n}} = \sqrt{q\left(1 - \frac{q}{n}\right)}$.

By Lemma 1, we have

$$\|\mathbf{x}\|_2^2 = \sum_{i=1}^{n} x_i^2 \leq \left(\sum_{i=1}^{n} x_i\right)^2 = \frac{\|\mathbf{p}\|_1^2}{|S|^2} = q^2,$$

for all gray-scale images $\mathbf{x} \in \mathcal{S}(S, m, n)$. By Theorem 2, we get

$$r_{\mathbf{x}} \leq \sqrt{\max_{\mathbf{x} \in \mathcal{S}(S,m,n)} \|\mathbf{x}\|_2^2 - \|C\|_2^2} \leq \sqrt{q^2 - \frac{q^2}{n}} = \sqrt{\frac{n-1}{n}}\, q,$$

and also the upper bound is obtained. □

Denote by H^n the n-dimensional hypercube having a vertex at the origin, sides of length 1 parallel to the coordinate axes, and all of whose vertices have non-negative coordinates.

Lemma 4. *If $\mathcal{S}(S, m, n)$ contains a binary solution, then $\frac{\|\mathbf{p}\|_1}{|S|} \in \{0, \ldots, n\}$.*

Proof. Each binary image corresponds to a vertex $\mathbf{v}$ of H^n, with $\|\mathbf{v}\|_2^2 = \|\mathbf{v}\|_1 \in \{0, \ldots, n\}$. Consequently, $\frac{\|\mathbf{p}\|_1}{|S|} = \|\mathbf{v}\|_2^2 \in \{0, \ldots, n\}$. □

Remark 1. The lower bound $r_{\overline{\mathbf{x}}}$ can be reached by possibly existing binary elements of $\mathcal{S}(S, m, n)$, since these are the shortest integer solutions [1]. However, such a lower bound could also be obtained by solutions consisting of $p > 2$ gray levels, and having Euclidean norm equal to $\frac{\|\mathbf{p}\|_1}{|S|} \leq n$. Differently, according to Lemma 4, if $\frac{\|\mathbf{p}\|_1}{|S|} > n$ then no binary solution exists. Also, the lower bound in Corollary 2 becomes meaningless and cannot be obtained by any element of $\mathcal{S}(S, m, n)$. In this case, the upper bound holds once q is replaced by $\frac{\|\mathbf{p}\|_1}{|S|}$, as one can immediately deduce from the proof of Corollary 2.

Theorem 3. *Let $n \geq 2$ and $0 \neq q = \frac{\|\mathbf{p}\|_1}{|S|} < n$. Then π is the unique affine hyperplane of $\mathbb{R}^n$ containing the set of all vertices of H^n having two-norm $\sqrt{q}$.*

Proof. Let H_q^n be the set of all vertices of H^n whose Euclidean norm equals $\sqrt{q}$. Each $\mathbf{v} \in H_q^n$ has $q = \frac{\|\mathbf{p}\|_1}{|S|}$ entries equal to 1, so that $\mathbf{v}$ belongs to π. Therefore, we must prove that H_q^n is not contained in any other affine hyperplane of $\mathbb{R}^n$. Being $q \neq 0, n$, we have $|H_q^n| = \binom{n}{q} \geq n$.

Consider the vectors $\mathbf{e}_\kappa = (\mathbf{0}_{\kappa-1}, 1, \mathbf{0}_{n-\kappa})$. For $s \in \{1, \ldots, q+1\}$ let $\mathbf{v}_s^1 = (\mathbf{1}_{s-1}, 0, \mathbf{1}_{q-s+1}, \mathbf{0}_{n-q-1})$, namely, $\mathbf{v}_s^1$ has 0 in the s-th entry, 1 in the remaining

q entries of the first $q+1$ entries, and the last $n-q-1$ entries equal to 0. Then $\mathbf{v}_s^1 \in H_q^n$ for all $s \in \{1, \ldots, q+1\}$, and

$$\mathbf{e}_1 = -\frac{q-1}{q}\mathbf{v}_1^1 + \frac{1}{q}\mathbf{v}_2^1 + \frac{1}{q}\mathbf{v}_3^1 + \ldots + \frac{1}{q}\mathbf{v}_{q+1}^1,$$

so that $\mathbf{e}_1$ is a linear combination of the vectors $\mathbf{v}_s^1$, $s \in \{1, \ldots, q+1\}$. Analogously, for $\kappa = 2, \ldots, n$ we can write

$$\mathbf{e}_\kappa = -\frac{q-1}{q}\mathbf{v}_1^\kappa + \frac{1}{q}\mathbf{v}_2^\kappa + \frac{1}{q}\mathbf{v}_3^\kappa + \ldots + \frac{1}{q}\mathbf{v}_{q+1}^\kappa,$$

where $\mathbf{v}_s^\kappa$ is obtained by taking the entries of $\mathbf{v}_s^1 \bmod(n)$ and moving them of $\kappa - 1$ steps ahead. Therefore H_q^n spans $\mathbb{R}^n$, and consequently H_q^n contains n linearly independent vectors. This implies that all elements of H_q^n belong to a same affine hyperplane. $\qquad\square$

Remark 2. Theorem 3 points out that π is precisely the hyperplane spanned by the considered vertices of the hypercube, so each vertex of $H_q^n \setminus \{\mathbf{0}, (1, \ldots, 1)\}$ is a possible candidate for a binary solution of the tomographic problem.

4   Experiments on Reconstructing Gray-Scale Images

According to [6], all integer solutions are vertices of parallelepipeds covering the whole space of solutions. By the previous section, the vertices that can be really involved must belong to $(n-2)$-dimensional concentric sub-spheres of the hyperplane π, centered at the uniform image C. When binary solutions exist, these are vectors of π having $q \in \{0, \ldots, n\}$ entries equal to 1, so these are vertices of H^n.

In [8], the algorithm BRA (Binary Reconstruction Algorithm) has been given such that, using as input a set of four directions that guarantee unique binary solution, it returns the correctly reconstructed image. The output is obtained thanks to a rounding theorem [8, Theorem 13] applied to the central solution $\mathbf{x}^*$. Even if a corresponding algorithm is not yet available, we are induced to apply the same approach to find possibly existing gray-scale solutions having $p > 2$ gray levels. To this, we propose to focus on integer images that can be reached by some combination of elements of H_q^n, so assuming them as candidates to approximate the tomographic reconstructions. In fact, due to Theorem 3, each point of π, and in particular each solution of the tomographic problem, can be obtained as a linear combination of n independent elements of H_q^n. Since $\mathbf{x}^*$ is a solution, it is such a linear combination, so that variations of the corresponding coefficients can be regarded as variations of $\mathbf{x}^*$. In this experimental section we investigate what happens when such variations consist of rounding the entries of $\mathbf{x}^*$ to their nearest integer falling in the interval $\{0, \ldots, p-1\}$. It turns out that, as in the binary case, perfect reconstructions are still achieved.

We focus in particular on two images with $p = 3$ gray levels, so encoded by matrices whose entries are equal to $0, 1$ and 2. Inspired by BRA, our strategy consists in the computation of the central solution using the Conjugate Gradient Least Squares method (CGLS), and then of its integer rounding. Note that, in general, the integer rounding of the central solution turns out to be an image with $p' > 3$ gray levels; to avoid this inconvenience, we lower to 2 all its values greater than 2, as well as all the negative entries are updated to 0. The resulting image, R_p, will be the candidate solution for the reconstruction.

In order to investigate the pertinence of our reconstruction, we evaluate the following parameters: the number of wrongly reconstructed pixels, the percentage of exact reconstruction, the distance of R_p from the hyperplane π and the radius $r_{\mathbf{x}}$ of the corresponding hypersphere. For simplicity, we work with two phantom images of small size, respectively 24×24 and 34×34 pixels, as shown in Fig. 3.

(a)   (b)

Fig. 3. The images with 3 gray values used to test the sets of directions. The phantom (a) has size 24×24 pixels, while the phantom (b) has size 34×34 pixels. (Color figure online)

We consider the following (valid) sets of directions, used for our tests:

$$S_1 = \{(1,2), (1,-1), (1,3), (1,-2), (1,8), (1,-7)\},$$
$$S_2 = \{(1,1), (1,-2), (3,4), (5,3), (1,6), (1,7), (3,-10)\},$$
$$S_3 = \{(1,2), (1,-1), (1,3), (1,-2), (1,6), (1,-5)\},$$
$$S_4 = \{(1,0), (2,1), (1,1), (1,2), (0,1), (1,-1)\},$$
$$S_5 = \{(1,0), (2,1), (1,1), (1,2), (0,1), (1,-2)\}.$$

According to Examples 1 and 2, the sets S_1 and S_2 ensure uniqueness of reconstruction of any image with 3 gray values of size $M \times 24$ and $M \times 34$, respectively, where M is a positive integer. S_3 is a valid set in a grid of size 24×24, but is not a set of uniqueness: we will use it for the reconstruction of our phantoms in order to compare the improvement of the reconstruction results when choosing an appropriate set of directions. Finally, the sets S_4 and S_5 are not sets of uniqueness even in a 24×24 sized grid; we consider them in our experiments since these are classically employed in this field (see [9]).

4.1   Discussion of the Results

We have computed, by means of the CGLS algorithm, the real-valued solution of the tomographic problem having minimum Euclidean norm, when the input is a set of directions that ensures uniqueness of reconstruction. Then we have reshaped the output by means of rounding steps, assuming the finally resulting image as a candidate solution for the tomographic problem.

As a first remark, we focus on the progressive increasing of the percentage of correctly reconstructed pixels, as shown in Tables 1 and 2. Actually, we did not expect to reach perfect reconstruction, since the adopted algorithm, though running under the use of sets of uniqueness, exploits a rounding method that is driven by geometrical and empirical considerations, not supported by rigorous theoretical results. Nevertheless, and rather surprisingly, perfect reconstructions of both phantoms are obtained indeed. Therefore, it seems that the choice of a set of uniqueness already allows the use of the CGLS, properly modified through rounding methods, to detect gray-scale images. Of course, a deeper investigation must be carried out on phantoms of larger size and having $p > 3$ gray levels, which could lead to a refinement of the adopted approach. However, it is remarkable that running the same algorithm with sets of directions not ensuring uniqueness leads to a huge decreasing of the percentage of correctly reconstructed pixels, even if the sets are well selected, as in the cases of S_4 or S_5, where the spacing between two consecutive directions is almost the same (so mimicking the usual continuous choice of sets of equispaced directions). Indeed, using such sets of directions, none of the phantoms is perfectly reconstructed. In addition, the comparison of the output provided by the different sets of directions shows a greater variance in the case of the larger phantom (see Table 2, where it is shown that only about 74% of the pixels are in fact correctly reconstructed by means of non-uniqueness sets).

Table 1. Parameters related to the reconstruction of the 24×24 phantom using the set of uniqueness S_1 and the set of non-uniqueness S_3. The percentage of correct reconstruction for the set S_4 (resp., S_5) is 73.09% (resp., 74.83%) only

# iter	# wrong pixels		% reconstruction		distance from π		r_x	
	S_1	S_3	S_1	S_3	S_1	S_3	S_1	S_3
10	92	100	84.03	82.64	1.17	0.58	12.35	11.99
20	71	85	87.67	85.24	0.71	0.29	12.43	11.99
30	58	81	89.93	85.94	1.17	0.46	12.81	12.17
40	51	69	91.15	88.02	0.88	0.46	12.76	12.25
50	48	61	91.67	89.41	0.92	0.46	12.86	12.40
100	22	43	93.18	92.53	0.50	0.38	13.11	12.79
150	4	20	99.31	96.53	0.17	0.17	13.23	13.06
200	0	19	100.0	96.70	0.00	0.13	13.24	13.06

Table 2. Parameters related to the reconstruction of the 34×34 phantom using the set of uniqueness S_2 and the sets of non-uniqueness S_1 and S_3. The percentage of correct reconstruction for the set S_4 (resp., S_5) is 73.44% (resp., 73.18%) only.

# iter	# wrong pixels			% reconstruction			distance from π			r_x		
	S_2	S_1	S_3	S_2	S_1	S_3	S_2	S_1	S_3	S_2	S_1	S_3
10	165	346	384	85.73	70.07	66.78	0.27	0.62	0.06	15.38	14.69	13.11
20	124	301	371	89.27	73.96	67.91	0.41	0.35	0.12	16.11	16.00	14.09
30	111	282	356	90.40	75.61	69.20	0.38	0.32	0.09	16.36	15.86	14.27
50	83	276	356	92.82	76.13	69.20	0.09	0.21	0.27	17.01	15.82	14.50
100	41	269	346	96.45	76.73	70.07	0.09	0.21	0.24	17.25	16.10	14.52
150	19	269	331	98.36	76.73	71.37	0.03	0.03	0.38	17.48	16.02	14.23
200	5	251	312	99.57	78.29	73.01	0.03	0.18	0.24	17.52	15.88	14.22
250	1	241	304	99.91	79.15	73.70	0.03	0.24	0.12	17.56	15.86	14.25
300	0	239	303	100.0	79.33	73.79	0.00	0.18	0.09	17.62	15.78	14.09

5   Conclusions and Comments

In this paper we have moved from a uniqueness result concerning tomographic images with $p \geq 2$ gray levels and have constructed sets of directions ensuring uniqueness of reconstruction, in view of exploiting them in explicit reconstruction algorithms, so to compare the corresponding outputs with the reconstructions obtained by means of other sets of directions. The proposed algorithm is obtained by generalizing a similar algorithm that works well for the reconstruction of binary images [8]. Gray-scale reconstructions are provided by some kind of approximation of the central solution, which can be numerically computed by means of the CGLS algorithm, and assumed, in case of integer projections, as an approximation of the shortest integer solution (always existing, due to [6]). Maybe, a better performance could be obtained by replacing CGLS with some different approach, but we wish to emphasize that the optimization of the reconstruction procedure is beyond the scope of the present paper, that aims in showing how uniqueness results can support a polynomial time algorithm and exact reconstruction of tomographic images.

In order to find a good strategy of approximation, we have investigated the geometry of gray-scale solutions, showing that these are contained in hyperspheres centered at the uniform image and providing bounds for their radius. In the experimental section we have considered small images in order to explicitly show all details of the achieved geometric characterizations, but the possible real applications are independent of the size. We have evaluated the quality of the reconstructed images according to geometric parameters, and compared the output with the reconstructions obtained by exploiting sets of directions that do not match the uniqueness assumptions. This shows a huge difference in the percentage of the correctly reconstructed pixels as the number of iterations increases.

As a conclusion, the obtained results seem to be very promising, and encourage us to deepen the theoretical properties that support our proposal.

References

1. Batenburg, K., Fortes, W., Hajdu, L., Tijdeman, R.: Bounds on the quality of reconstructed images in binary tomography. Disc. Appl. Math. **161**(15), 2236–2251 (2013). https://doi.org/10.1016/j.dam.2012.11.010
2. Brunetti, S., Dulio, P., Peri, C.: Characterization of (-1,0,+1) valued functions in discrete tomography under sets of four directions. In: Debled-Rennesson, I., Domenjoud, E., Kerautret, B., Even, P. (eds.) DGCI 2011. LNCS, vol. 6607, pp. 394–405. Springer, Heidelberg (2011). https://doi.org/10.1007/978-3-642-19867-0_33
3. Brunetti, S., Dulio, P., Peri, C.: Discrete tomography determination of bounded lattice sets from four X-rays. Disc. Appl. Math. **161**(15), 2281–2292 (2013). https://doi.org/10.1016/j.dam.2012.09.010
4. Brunetti, S., Dulio, P., Peri, C.: Discrete tomography determination of bounded sets in $\mathbb{Z}^n$. Disc. Appl. Math. **183**, 20–30 (2015). https://doi.org/10.1016/j.dam.2014.01.016
5. Brunetti, S., Dulio, P., Peri, C.: Uniqueness results for grey scale digital images. Fund. Inf. **172**(2), 221–238 (2020). https://doi.org/10.3233/fi-2020-1902
6. Dalen, B.V., Hajdu, L., Tijdeman, R.: Bounds for discrete tomography solutions. Indag. Math. **24**(2), 391–402 (2013). https://doi.org/10.1016/j.indag.2012.12.005
7. Dulio, P., Frosini, A., Pagani, S.: A geometrical characterization of regions of uniqueness and applications to discrete tomography. Inverse Prob. **31**(12), 125011 (2015). https://doi.org/10.1088/0266-5611/31/12/125011
8. Dulio, P., Pagani, S.: A rounding theorem for unique binary tomographic reconstruction. Disc. Appl. Math. **268**, 54–69 (2019). https://doi.org/10.1016/j.dam.2019.05.005
9. Gardner, R., Gritzmann, P.: Discrete tomography: determination of finite sets by X-rays. Trans. Am. Math. Soc. **349**(6), 2271–2295 (1997). https://doi.org/10.1090/S0002-9947-97-01741-8
10. Guédon, J.P., Normand, N.: The Mojette transform: the first ten years. In: Andres, E., Damiand, G., Lienhardt, P. (eds.) DGCI 2005. LNCS, vol. 3429, pp. 79–91. Springer, Heidelberg (2005). https://doi.org/10.1007/978-3-540-31965-8_8
11. Hajdu, L., Tijdeman, R.: Algebraic aspects of discrete tomography. J. Reine Angew. Math. **534**, 119–128 (2001). https://doi.org/10.1515/crll.2001.037
12. Katz, M.: Questions of Uniqueness and Resolution in Reconstruction from Projections/Myron Bernard Katz. Springer, New York (1978). https://doi.org/10.1007/978-3-642-45507-0
13. Normand, N., Kingston, A., Évenou, P.: A geometry driven reconstruction algorithm for the Mojette transform. In: Kuba, A., Nyúl, L.G., Palágyi, K. (eds.) DGCI 2006. LNCS, vol. 4245, pp. 122–133. Springer, Heidelberg (2006). https://doi.org/10.1007/11907350_11
14. Van Aert, S., Batenburg, K., Rossell, M., Erni, R., Van Tendeloo, G.: Three-dimensional atomic imaging of crystalline nanoparticles. Nature **470**(7334), 374–377 (2011). https://doi.org/10.1038/nature09741

A Khalimsky-Like Topology
on the Triangular Grid

Benedek Nagy[1,2(✉)]

[1] Department of Mathematics, Faculty of Arts and Sciences, Eastern Mediterranean
University, Famagusta, North Cyprus, Mersin-10, Turkey
nbenedek.inf@gmail.com
[2] Department of Computer Science, Institute of Mathematics and Informatics,
Eszterházy Károly Catholic University, Eger, Hungary

Abstract. It is well known that there are topological paradoxes in digital geometry and in digital image processing. The most studied such
paradoxes are on the square grid, causing the fact that the digital version of the Jordan curve theorem needs some special care. In a nutshell,
the paradox can be interpreted by lines, e.g., two different color diagonals of a chessboard that go through each other without sharing a pixel.
The triangular grid also has a similar paradox, here diamond chains of
different directions may cross each other without having an intersection
trixel (triangle pixel). In this paper, a new topology is offered on the
triangular grid, which gives a solution to the topological problems in the
triangular grid analogous to the Khalimsky's solution on the square grid.

Keywords: Digital geometry · Nontraditional grids · Digital
topology · Topological paradoxes · Adjacency relations · Jordan curve
theorem

1 Introduction

With this introduction, our aim is twofold. On the one hand, we would like to
recall some topological paradoxes appearing on some digital grids, and thus, to
give our problem statement. On the other hand, we are also giving a brief and
concise introduction with the known possible "solutions" of the problem.

The digital planes (based on various grids, i.e., tessellations of the plane)
and spaces have some different properties than the Euclidean plane and space
have. In this paper, we consider the plane. For instance, there are infinitely
many points between two distinct points in their connecting line segment in the
Euclidean plane. On the other hand, in the digital scenarios, the neighborhood
relation is essential, and there is no pixel between the closest neighbor pixels.
Such differences may lead to some paradoxical situations on the digital planes
[16,17,19]. The Jordan curve theorem states that in the Euclidean plane a simple closed curve separates the plane into exactly two connected components: the
interior and the exterior. Unfortunately, this theorem may not be trivially trans-
ferred to the digital scenarios as, e.g., none of the usual neighborhood relations
of the square grid support this theorem [15,18].

S. Brunetti et al. (Eds.): DGMM 2024, LNCS 14605, pp. 150–162, 2024.
https://doi.org/10.1007/978-3-031-57793-2_12

There are various types of solutions to overcome these paradoxical situations. The first solution may work on binary images, where two different adjacencies (neighborhoods) are used for the two basic colors (usually black and white). In digital image processing, one can use object (or foreground) pixels and background pixels, and, e.g., on the square grid 4-8 or 8-4 adjacency, meaning that black pixels form connected components according to the first type of mention connectedness, while white pixels form connected components according to the second type of mentioned connectedness. Similar solution also exists on the triangular grid, by using the closest neighborhood (the 3 side-neighbors, or type-1 neighbors) and the neighborhood containing all 12 corner neighbors (type-3 neighbors) for the black and white trixels (triangle pixels), respectively, or vice-versa. This solution will be recalled also in the beginning of Sect. 3. Because of this solution, the type-1 and type-3 neighborhoods are referred as Jordan type neighborhood in [30]. One of the main drawbacks of the previous approaches is that it supports only binary images, with more than two colors, it does not work, two colors using the same adjacency cause violations of the Jordan curve theorem. A possible solution is based on a topological modification of the square grid, by alternating use of the 4 and 8-adjacency. This topology is called Khalimsky-topology [13,14]. There are various related topologies, where some additional aim is also fulfilled, see, e.g., [39–42]. We should also mention and emphasize here that the underlying grid plays a crucial role in digital geometry and in digital image processing [34]. For instance, the third regular grid, the hexagonal grid does not suffer from the mentioned topological paradoxes, there is only one type of neighborhood relation, all neighbor hexels (hexagonal pixels) share a full side. This fortunate fact is applied in some table games and also in some computer games when hexagonal boards are used. There are also eight semi-regular grids, and some of them also avoid these paradoxes (see [34]). The grid built by alternating use of octagons and squares, the truncated quadrille (also called truncated square tiling) shows exactly the Khalimsky structure and studied e.g., in [21,22].

Generalising the idea of [25,26] to the triangular grid, in [31,32] topological (also called combinatorial) coordinate system for the triangular grid is shown that resolve the topological problem using cell complexes containing not only trixels, but some of the grid edges and also some of the grid points [32,44]. However, in this paper, we shall give a solution using only the trixels, but playing with the neighborhood (adjacency). As both the hexagonal (honeycomb) grid and the triangular grid become more and more popular in image processing and related fields, based, e.g., on their good symmetrical properties, there are more and more applications and studies concerning them, e.g., in graphics [4], in mathematical morphology [1,2], in thinning with topology preserving [12] and in diagrams [33]. We also recall some related topological studies on the triangular grid, e.g., the numbers of concave and convex vertices of a closed curve are counted in [5] and well-composedness and gaps are analyzed in [6].

The structure of this paper is as follows. In the next section, we formally describe the triangular grid, as this grid is our main focus. Then, we display the topological problems of the square grid and analogous topological problems on the triangular grid. In Sect. 3, we show our proposed solution. Finally, Conclusions close the paper.

2   Preliminaries

On the one hand, the geometry of every "digital plane/world" differs from the Euclidean geometry [16,17], as e.g., in any discrete space based on a grid, the neighborhood relation plays an essential role. In contrast, there are no neighbor points in the Euclidean plane, but for any $\varepsilon > 0$, there are continuum many points that have distance less than ε from any point of the plane. On the other hand, there are various digital geometries based on the chosen grid.

The grid-points or vertices of a grid are those points where some gridlines meet. The regions of the planar graph, the tiles, as we mainly refer to imaging, are also called pixels.

There are three regular grids in the plane and eight semi-regular grids [7,9, 34,37]. All these grids are also referred to as Archimedean, they are built up by regular polygons in an isogonal way (the grid-points of the tiling are identical, isometric transformations may map any of them to any other). Already Kepler studied and described them. The dual (by planar duality) of a grid is obtained by inverting the roles of the pixels and gridpoints (corners): by putting a point to the midpoint of each tile/pixel and connecting those which points representing side-neighbor pixels, the dual grid is obtained [38].

As one may see that only the square (or generally, rectangular) and hexagonal grids are point lattices, i.e., they are discrete subgroups of the 2-dimensional Euclidean space, meaning that any of their pixels has exactly the same role, every vector connecting the midpoints of any two pixels translate grid into itself. The other grids do not have a similar property, and thus, the triangular grid may share various properties with the dual grids of the semi-regular ones. Although in this paper we concentrate on the triangular grid, other grids are also used, e.g., the square grid and its specific modification, the Khalimsky grid (i.e., the truncated quadrille or truncated square tiling).

2.1   Description of the Triangular Grid

As grids can be seen as graphs, they can be bipartite and non-bipartite. In bipartite grids, based on steps/moves on side neighbors, every cycle has an even length, while at non-bipartite grids there are paths between the same two tiles such that the difference of their lengths is odd [20]. In fact, both the square and triangular grids are bipartite: in the square grid, the usual coloring of a chessboard gives the partitions, while in the triangular grid, the orientation of the trixels defines it.

The hexels (hexagon pixels) of the hexagonal grid are addressed by 0-sum integer triplets in [11] which suggests to use a similar coordinate system for the triangular grid based on their related symmetry. Thus, the trixels are addressed by coordinate triplets [27,29,43] with sum 0 and 1 according to the two types of pixels (in this way also related to a subspace of the cubic grid [28,36]). The 0-sum integer triplets are used to address the even, the 1-sum integer triplets to address the odd trixels. As every even pixel has three odd side neighbors, and vice versa, the names even and odd are very apt for these tiles. A part of the grid with this symmetric coordinate system is shown in Fig. 1.

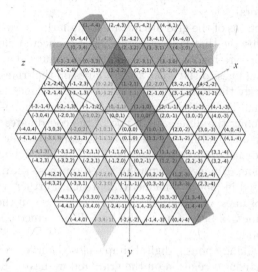

Fig. 1. The coordinate system. A set of trixels form a lane with a fixed value of a coordinate (see the red trixels with $x = 1$ and the blue ones with $y = -3$). A set of trixels is a diamond chain where the difference of two of the coordinates is fixed (see yellow trixels with $x - z = -2$ and green trixels with $y - z = -2$). (Color figure online)

The formal, mathematical description of the neighbor relations is also nice and reflects the symmetry caught by the 3-valued coordinate systems. The three types of neighborhood on the triangular grid was already mentioned and used in the 1970's [8] in relation to image processing. They can be written as: the trixels $p(p_1, p_2, p_3)$ and $q(q_1, q_2, q_3)$ are proper type-m neighbors (with $m \in \{1, 2, 3\}$) if and only if

- $|p_i - q_i| \leq 1$ for each $i \in \{1, 2, 3\}$, and
- $\sum_i |p_i - q_i| = m$.

We say that a trixel p is type-m neighbor of q if it is a proper type-k neighbor for a value $1 \leq k \leq m$, in this way all type-1 neighbors are also type-2 neighbors and all type-2 neighbors are also type-3 neighbors of q. Observe that these conditions are pretty much the same for the two types of neighborhood on the square grid (where $m \in \{1, 2\}$ the number of changing coordinates of a cityblock or a chessboard move). As the triangular grid is not a point lattice, moves or translations may have some interesting or unexpected properties [3]. We note that in [10], to avoid some of these difficulties, the moves for the type-1 neighbors are called 'half moves', while moves to type-2 neighbors are called 'full moves', and type-3 neighbors are not used there. However, for us, all the three types of neighborhoods will play important roles. By the other usual naming convention, where the number of neighbors is used, addition to the cityblock 4-neighbors, and chessboard 8-neighbors, in the triangular grid we have 3-neighbors (for the closest, i.e., type-1, side-neighbors), 9-neighbors (for type-2) and 12-neighbors (for type-3 neighbors).

The main directions of the grid are its lanes, similarly as rows and columns in the square grid. Observe that lane segments are 3-connected, i.e., for each trixel its closest neighbors are involved if the segment continues. On the other hand, since the orthogonal lines to the lanes are not lanes on the triangular grid, but in fact, they are exactly the bisectors of the angles of two lanes, they play similar roles as the diagonal lines on the square grid. These *diamond-chain* directions are described by fixing the difference of two of the coordinates (see also Fig. 1). Observe that these diagonal line segments are 12-connected, in the sense that in a diamond lane for each of its trixel exactly two of its 12-neighbors are belonging to that diamond lane. Binary tomography based on these six natural directions of the grid were used in [35] in a memetic approach. Further, we may also have 9-connected digital lines, or generally curves, when, e.g., all the even or all the odd trixels of a lane are belonging to the line. This connectedness is based on type-2 neighborhood.

When we are talking about a digital simple closed curve, we are fixing one of the defined connectedness relation on the grid, and we have a finite set of pixels such that for each pixel of the curve exactly two of its given type of neighbors belong also to the curve. (For technical reasons, we also assume that the curve contains at least four pixels).

2.2 Topological Problems

Recall that the Jordan curve theorem states that in the Euclidean plane a simple closed curve separates the plane into exactly two connected components: the interior and the exterior.

The square grid has the following topological paradoxes: There are lines that go through on each other without a common tile, e.g., think about the two diagonal lines of a usual chessboard: the black and white diagonals connect opposite corners of the board, but there is no shared place/pixel. In this way, the inner and outer parts of a closed curve can also be connected (violating the Jordan curve theorem). On the other hand, if one uses the city-block connectedness, it may happen that, e.g., the interior is not connected according to this connectedness.

Similar paradox also exists on the triangular grid as we show some non-intersecting diamond chains in Fig. 2. Observe that this phenomenon is closely connected to the previous example on the square grid, however, in the triangular grid three diamond chains may go through on each other at the same point, not only two as in the square grid. Furthermore, similar paradox occur with 9-connected lines, e.g., the line consist of the even trixels of the lane $x = 1$

and the line consists of the odd trixels of the lane $y = 1$ "crosses" each other without having a common, shared trixel. Finally, by using only 3-connectedness, it is easy to draw simple closed curve with a shape that surrounds not only one 3-connected region, but more (see, e.g., Fig. 3, left); in this way violating also the Jordan curve theorem.

Fig. 2. Various diamond chains are crossing each other without sharing any trixel.

In contrast, to the previous two grids, as it is shown in [34], the hexagonal, the truncated hexagonal, the truncated trihexagonal and the Khalimsky grids are the regular and semi-regular grids that do not have topological paradoxes, as in these grids the number of all neighbors of each tile is the same as the number of the side-neighbors. Maybe this is one of the main reasons why they are already involved in digital geometric studies. Actually, this property also gives the importance of the Khalimsky grid in various applications [13,14]. Furthermore, the Khalimsky grid shares the symmetric properties with the square grid, and thus, it is relatively easy to convert images and algorithms from the traditional square grid to this grid.

3   Proposed Solution to the Topological Problems on the Triangular Grid

As we have already mentioned, for binary image processing, the 3-12 and 12-3 topologies could work on the triangular grid. This alternative is closely connected to the usual 4-8 and 8-4 solutions on the square grid, as 3 means only the closest (type-1 or side-neighbors); while 12 represents type-3, i.e., all corner neighbors.

More precisely, in the triangular grid, if a digital closed curve is 3-connected, i.e., for each trixel of the curve there are exactly 2 of its closest neighbors are also members of the curve, then this curve separates the grid into two disjoint 12-connected regions, the interior and exterior regions. Figure 3 (left) presents an example. Notice that the interior (green color) is 12-connected but may not

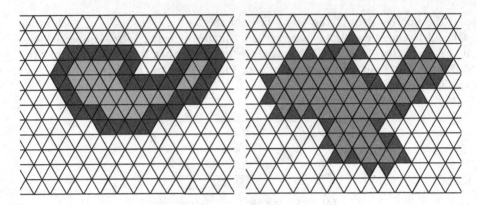

Fig. 3. A 3-connected closed curve separates the grid into two 12-connected regions (left) and a 12-connected closed curve separates the grid into two 3-connected regions (right).

be 3-connected, while the curve itself is 3-connected (representing also one of the earlier mentioned paradoxes).

On the other hand, the "dual" solution also works, if a digital closed curve is 12-connected, i.e., for each trixel of the curve there are exactly 2 of its 12 neighbors are also members of the curve, then this curve separates the grid into two disjoint 3-connected regions, the interior and exterior regions. Figure 3 (right) shows an example. Based on these results, one may feel that only the closest (side) neighbors (equivalent to city block neighbors in square grid) and the set of all 12 corner neighbors (can be seen equivalent to all corner neighbors, i.e., the chessboard neighborhood of the square grid) play importance in the triangular grid from topological point of view: One important thing to observe here is that the 9-connectedness seems not to be useful for topological studies, as if we use this type of connectedness for one of the colors, none of the connectedness match with it (as we have described 9-connected lines may also cross each other without intersecting).

However, as we have already seen, having more than two colors, we need to find other, newer solution to resolve the topological problems. In this section, now, we present our proposed general solution.

The dual of the square grid is also a square grid, but instead of the pixels, the grid points are considered. The Khalimsky topology is usually displayed in the dual representation of the square grid, by connecting neighbor grid points (see Fig. 4 (left), as the square grid is a bipartite grid, the alternate use of the two basic neighborhood provides this solution). Formally, we can give that for points with even (or odd, resp.) coordinate sum the 8-neighborhood (adjacency) is used and for odd (even, resp.) coordinate sum the 4-neighborhood is used.

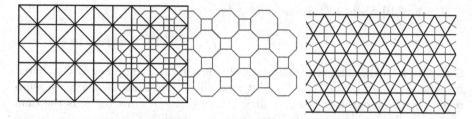

Fig. 4. Left: The usual representation of the Khalimsky topology, and its dual representation, the Khalimsky grid (middle). Right: the dual of the triangular grid is the honeycomb (hexagonal) grid.

Considering the triangular grid, our aim is similar: to use various connectedness of the gridpoints of the dual (honeycomb) representation (see Fig. 4, right) such that the representation is planar, i.e., no edges cross each other.

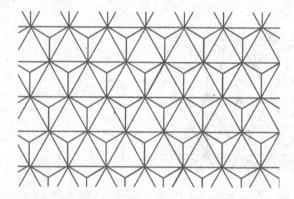

Fig. 5. A Khalimsky-type topology solution on the triangular grid by alternate use of 3-neighborhood and 9-neighborhood.

The triangular grid is also a bipartite grid suggesting the alternate usage of two types of neighborhood. Actually, we may use the closest neighbors at the odd (even, resp.) gridpoints, i.e., grid points representing the odd (or even, resp.) trixels; and the 9-neighborhood for the odd (even, resp.) gridpoints. Figure 5 shows this solution, that can be seen as a variant of the Khalimsky-type solution on the triangular grid. (Since our grids are bipartite we may have two equally good solutions by flipping the roles of even and odd points.) Thus, we can see that, from this point of view the type-2 neighborhood (the 9-neighbors) plays the role of the chessboard neighbors on the triangular grid. The neighborhood of this grid can be written formally in the following way: the pixels $p(p_1, p_2, p_3)$ and $q(q_1, q_2, q_3)$ are neighbors if and only if one of the following conditions hold:

- $q_i - p_i \in \{0,1\}$ for $i \in \{1,2,3\}$ and $\sum_i (q_i - p_i) = 1$ if $p_1 + p_2 + p_3 = 0$ and

- $|q_i - p_i| \leq 1$ for $i \in \{1,2,3\}$ and $\sum_i |q_i - p_i| \in \{1,2\}$ if $p_1 + p_2 + p_3 = 1$.

Going back to the dual representation of this solution, the plane is tessellated by enneagons and triangles as Fig. 6 shows. This is nor a regular neither a semi-regular grid, moreover it is not a dual of any of those grids (as it is the dual of our Khalimsky-like connected triangular based grid). The enneagons are not regular, as in some places three of them meet, and thus they have angles of size 120°, while in some other corners (grid point) two enneagons and a triangle meet and the angle of the enneagons is 150° in these corners. This dual grid is based on our solution and it has no topological problems since all neighbor pixels are side-neighbors, see Fig. 6 (right). Figure 7 (left) also shows the neighborhood relation: The yellow triangle has exactly three neighbors: the purple enneagons marked by 'X'. The enneagon marked by 'O' has three triangle neighbors (blue color) and the six green enneagon neighbors (marked by 'I'). All neighbors of a pixel are side neighbors of the given pixel.

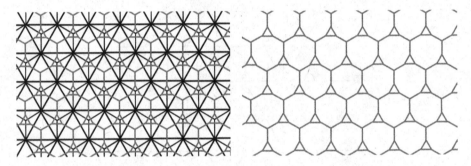

Fig. 6. The dual grid of our proposal, the grid based on our solution without topological problems, i.e., it has the triangular symmetry and all neighbor pixels are side-neighbors.

Actually, the simple closed curves on this grid built up by only enneagons (see Fig. 7, right), since if a triangle would be involved, two of its neighbors, two enneagons must also be involved, however, they are also neighbors of each other. Therefore, the only digital curve that includes a triangle, and every pixel has exactly two of their neighbors involved, consists of a triangle and two of its enneagon neighbors. However, this special case is not really a curve as its every pixel is neighbor to every other pixel (and there is no interior, similarly, if on the square grid we take 3 pairwise neighbor points, a 4-adjacent point with two of its neighbor 8-adjacent ones), thus we can exclude it from our studies (and we have defined our curves including at least four pixels, as usual). Furthermore, considering only the enneagons, their positions and structure (half of the points of the triangular grid, one of the partitions of the bipartite grid) gives the hexagonal structure, and it is already mentioned that the hexagonal grid is without these topological paradoxes. Indeed, for any closed simple curve in which every enneagon has exactly two of its neighbors involved, separates the grid into two connected segments (as we have only side-neighbors, only one type of connectedness is used in this grid, no choice, no paradox) as the Jordan curve theorem states. One may also observe, that the curves in this topology could have only mild turns, i.e., the direction may continue (straight way, 180°), or turn to left or right meaning 120 and 240°, respectively to the direction of the previous step, this fact also shows the close relationship to the hexagonal structure. Finally, we note that the grid obtained in this way is closely connected to other triangular based grids, namely, to the trihexagonal [23] and to the truncated hexagonal grids [24]. To describe a more formal connection between those left to some future task.

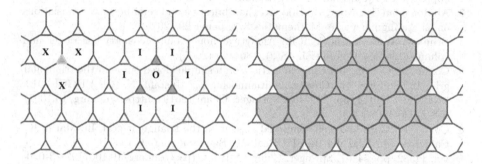

Fig. 7. Left: The neighborhood relation: every triangle has three enneagon neighbors and every enneagon has three triangle and six enneagon neighbors. Right: An example for a simple closed curve (in yellow) in the proposed grid (the interior is light blue). (Color figure online)

4   Conclusions

In this paper, we give a convenient structure to the digital triangular plane (i.e., triangular grid) for digital image analysis and processing that allows to study topological and geometric properties of digital images. Convenient structure, in this term, means that analogously to the Euclidean plane, a digital version of Jordan curve theorem works. Our result is somewhat analogous to the Khalimsky type topology of the square grid, indeed, in our solution the closest neighbors (type-1 neighborhood) is used for one type of trixels, while the type-2 neighborhood is used for the other type of trixels, similarly to the Khalimsky's solution. However, in the triangular grid, this type-2 neighborhood does not consists all the 12 neighbors, as there is a larger corner neighborhood, opposite to the square grid, where the type-2 neighborhood coincide to the set of all corner neighbors. We believe that our result can play a similar role in digital image processing and analysis on the triangular grid, as the Khalimsky topology plays on the square grid.

The new grid, we have prosed as solution, is neither regular or semi-regular, as it contains non-regular enneagons. Further analysis of this grid is a topic of future research including a deeper analysis of this topology and developing its digital geometry, like (weighted) distances, shortest paths, disks, lines.

References

1. Abdalla, M., Nagy, B.: Dilation and erosion on the triangular tessellation: an independent approach. IEEE Access **6**, 23108–23119 (2018)
2. Abdalla, M., Nagy, B.: Mathematical morphology on the triangular grid: the strict approach. SIAM J. Imaging Sci. **13**, 1367–1385 (2020)
3. Abuhmaidan, K., Nagy, B.: Bijective, non-bijective and semi-bijective translations on the triangular plane. Mathematics 8/1, paper 29 (2020)
4. Brimkov, W.E., Barneva, R.P.: Analytical honeycomb geometry for raster and volume graphics. Comput. J. **48**(2), 180–199 (2005)
5. Čomić, L.: Convex and concave vertices on a simple closed curve in the triangular grid. In: Couprie, M., Cousty, J., Kenmochi, Y., Mustafa, N. (eds.) DGCI 2019. LNCS, vol. 11414, pp. 397–408. Springer, Cham (2019). https://doi.org/10.1007/978-3-030-14085-4_31
6. Čomić, L.: Gaps and well-composed objects in the triangular grid. In: Marfil, R., Calderón, M., Díaz del Río, F., Real, P., Bandera, A. (eds.) CTIC 2019. LNCS, vol. 11382, pp. 54–67. Springer, Cham (2019). https://doi.org/10.1007/978-3-030-10828-1_5
7. Conway, J.H., Burgiel, H., Goodman-Strauss, C.: The Symmetries of Things. AK Peters (2008)
8. Deutsch, E.S.: Thinning algorithms on rectangular, hexagonal and triangular arrays. Commun. ACM **15**(3), 827–837 (1972)
9. Grünbaum, B., Shephard, G.C.: Tilings by regular polygons. Math. Mag. **50**(5), 227–247 (1977)
10. Hartman, N.P., Tanimoto, S.L.: A hexagonal pyramid data structure for image processing. IEEE Trans. Syst. Man Cybern. **14**(2), 247–256 (1984)
11. Her, I.: A symmetrical coordinate frame on the hexagonal grid for computer graphics and vision. ASME J. Mech. Design **115**(3), 447–449 (1993)
12. Kardos, P., Palágyi, K.: On topology preservation of mixed operators in triangular, square, and hexagonal grids. Discret. Appl. Math. **216**, 441–448 (2017)
13. Khalimsky, E.D., Kopperman, R., Meyer, P.R.: Computer graphics and connected topologies on finite ordered sets. Topol. Appl. **36**, 1–17 (1990)
14. Khalimsky, E.D., Kopperman, R., Meyer, P.R.: Boundaries in digital planes. J. Appl. Math. Stoch. Anal. **3**, 27–55 (1990)
15. Kiselman, C.O.: Digital Jordan curve theorems. In: Borgefors, G., Nyström, I., di Baja, G.S. (eds.) DGCI 2000. LNCS, vol. 1953, pp. 46–56. Springer, Heidelberg (2000). https://doi.org/10.1007/3-540-44438-6_5
16. Kiselman, C.O.: Elements of Digital Geometry, Mathematical Morphology, and Discrete Optimization. World Scientific, Singapore (2022)
17. Klette, R., Rosenfeld, A.: Digital Geometry - Geometric Methods for Digital Picture Analysis. Morgan Kaufmann, Elsevier Science B.V (2004)

18. Kong, T.Y., Kopperman, R., Meyer, P.R.: A topological approach to digital topology. Am. Math. Mon. **98**, 902–917 (1991)
19. Kong, T.Y., Rosenfeld, A. (eds.): Topological Algorithms for Digital Image Processing. Elsevier, Amsterdam (1996)
20. Kovács, G., Nagy, B., Stomfai, G., Turgay, N.D., Vizvári, B.: On chamfer distances on the square and body-centered cubic grids: an operational research approach. Math. Probl. Eng. **2021**, 1–9 (2021). Article ID 5582034
21. Kovács, G., Nagy, B., Vizvári, B.: On weighted distances on the Khalimsky grid. In: Normand, N., Guédon, J., Autrusseau, F. (eds.) DGCI 2016. LNCS, vol. 9647, pp. 372–384. Springer, Cham (2016). https://doi.org/10.1007/978-3-319-32360-2_29
22. Kovács, G., Nagy, B., Vizvári, B.: Weighted distances and digital disks on the Khalimsky grid - disks with holes and islands. J. Math. Imag. Vis. **59**, 2–22 (2017)
23. Kovács, G., Nagy, B., Vizvári, B.: Weighted distances on the trihexagonal grid. In: Kropatsch, W.G., Artner, N.M., Janusch, I. (eds.) DGCI 2017. LNCS, vol. 10502, pp. 82–93. Springer, Cham (2017). https://doi.org/10.1007/978-3-319-66272-5_8
24. Kovács, G., Nagy, B., Vizvári, B.: Weighted distances on the truncated hexagonal grid. Pattern Recognit. Lett. **152**, 26–33 (2021)
25. Kovalevsky, V.: Algorithms in digital geometry based on cellular topology. In: Klette, R., Žunić, J. (eds.) IWCIA 2004. LNCS, vol. 3322, pp. 366–393. Springer, Heidelberg (2004). https://doi.org/10.1007/978-3-540-30503-3_27
26. Kovalevsky, V.A.: Geometry of Locally Finite Spaces (Computer Agreeable Topology and Algorithms for Computer Imagery). Editing House Dr. Bärbel Kovalevski, Berlin (2008)
27. Nagy, B.: Finding shortest path with neighborhood sequences in triangular grids. In: ISPA 2001: 2nd IEEE R8-EURASIP International Symposium, Pula, Croatia, pp. 55–60 (2001)
28. Nagy, B.: A family of triangular grids in digital geometry. In: ISPA 2003: 3rd International Symposium on Image and Signal Processing and Analysis, Rome, Italy, pp. 101–106 (2003)
29. Nagy, B.: A symmetric coordinate frame for hexagonal networks. In: Theoretical Computer Science - Information Society (ACM Conference), Ljubljana, Slovenia, pp. 193–196 (2004)
30. Nagy, B.: Optimal neighborhood sequences on the hexagonal grid. In: ISPA 2007: 5th International Symposium, Istanbul, Turkey, pp. 310–315. IEEE (2007)
31. Nagy, B.: Cellular topology on the triangular grid. In: Barneva, R.P., Brimkov, V.E., Aggarwal, J.K. (eds.) IWCIA 2012. LNCS, vol. 7655, pp. 143–153. Springer, Heidelberg (2012). https://doi.org/10.1007/978-3-642-34732-0_11
32. Nagy, B.: Cellular topology and topological coordinate systems on the hexagonal and on the triangular grids. Ann. Math. Artif. Intell. **75**(1–2), 117–134 (2015)
33. Nagy, B.: Diagrams based on the hexagonal and triangular grids. Acta Polytech. Hung. **19**(4), 27–42 (2022)
34. Nagy, B.: Non-traditional 2D grids in combinatorial imaging - advances and challenges. In: Barneva, R.P., Brimkov, V.E., Nordo, G. (eds.) IWCIA 2022. LNCS, vol. 13348, pp. 3–27. Springer, Cham (2022). https://doi.org/10.1007/978-3-031-23612-9_1
35. Nagy, B., Moisi, E.V.: Memetic algorithms for reconstruction of binary images on triangular grids with 3 and 6 projections. Appl. Soft Comput. **52**, 549–565 (2017)
36. Nagy, B., Strand, R.: A connection between $\mathbb{Z}^n$ and generalized triangular grids. In: Bebis, G., et al. (eds.) ISVC 2008. LNCS, vol. 5359, pp. 1157–1166. Springer, Heidelberg (2008). https://doi.org/10.1007/978-3-540-89646-3_115

37. Radványi, A.G.: On the rectangular grid representation of general CNN networks. Int. J. Circuit Theory Appl. **30**(2–3), 181–193 (2002)
38. Saadat, M., Nagy, B.: Digital geometry on the dual of some semi-regular tessellations. In: Lindblad, J., Malmberg, F., Sladoje, N. (eds.) DGMM 2021. LNCS, vol. 12708, pp. 283–295. Springer, Cham (2021). https://doi.org/10.1007/978-3-030-76657-3_20
39. Slapal, J.: A digital analogue of the Jordan curve theorem. Discret. Appl. Math. **139**(1–3), 231–251 (2004)
40. Slapal, J.: A Jordan curve theorem with respect to a pretopology on Z^2. Int. J. Comput. Math. **90**(8), 1618–1628 (2013)
41. Slapal, J.: Convenient adjacencies for structuring the digital plane. Ann. Math. Artif. Intell. **75**(1–2), 69–88 (2015)
42. Slapal, J.: Alexandroff pretopologies for structuring the digital plane. Discret. Appl. Math. **216**, 323–334 (2017)
43. Stojmenovic, I.: Honeycomb networks: topological properties and communication algorithms. IEEE Trans. Parallel Distrib. Syst. **8**, 1036–1042 (1997)
44. Wiederhold, P., Morales, S.: Thinning on quadratic, triangular, and hexagonal cell complexes. In: Brimkov, V.E., Barneva, R.P., Hauptman, H.A. (eds.) IWCIA 2008. LNCS, vol. 4958, pp. 13–25. Springer, Heidelberg (2008). https://doi.org/10.1007/978-3-540-78275-9_2

Learning Based Morphology

Group Equivariant Networks Using Morphological Operators

Valentin Penaud--Polge[(✉)], Santiago Velasco-Forero[(✉)],
and Jesus Angulo-Lopez

Mines Paris, PSL University, Center for Mathematical Morphology (CMM),
Fontainebleau, France
{valentin.penaud_polge,santiago.velasco,jesus.angulo}@minesparis.psl.eu

Abstract. With the increase of interest upon rotation invariance and
equivariance for Convolutional Neural Network (CNN), a fair amount
of papers have been published on the subject and the literature keeps
increasing. This paper aims to fill the lack of morphological approaches
on the matter. We propose a set of group equivariant layers using morphological operators, several model configurations are tested and compared with a convolutional equivalent network. The results show that
the proposed morphological networks are capable of classifying rotated
images even when trained only with upright samples.

Keywords: Mathematical Morphology · Deep Learning · Equivariant
Network

1 Introduction

In the intense interest towards deep learning in the computer vision community
and the plethora of papers coming with it, one particular contribution remains
the cornerstone of most proposed methods: the convolutional layer. Through this
contribution in [16], LeCun *et al.* have shown that using translation equivariant
transformations offers great improvements to neural networks for image processing and analysis. This powerful property allows to process in the same way (or
at least predictively) an object in an image independently of its position. Yet,
other types of fluctuation may arise in the pose of an object in an image: orientations or scales, for example, may also change between two images or between two
parts of a same image. Equivariance or invariance to certain transformations can
also be beneficial to process objects presenting symmetries. A popular method
generalizing regular convolution to group convolution have shown interesting
capacities [10]. Even though a theory of Group Morphology has been proposed
before the deep learning era [18] and the idea of using group equivariant morphological operators in deep learning has been suggested by Angulo [2], there

---

V. Penaud--Polge—This work was granted access to the HPC resources of IDRIS under
the allocation 2023-[AD011013367R1] made by GENCI.

S. Brunetti et al. (Eds.): DGMM 2024, LNCS 14605, pp. 165–177, 2024.
https://doi.org/10.1007/978-3-031-57793-2_13

have not been any morphological version of the *Group Equivariant Convolutional Network* (GECN) [10]. We propose in this paper to combine the theory of Group Morphology with GECN to fill this gap. The paper is organized in the following way. Section 2 recalls definitions and mathematical results of the fields of group action and mathematical morphology. Section 3 reviews the most related work concerning group morphology and group equivariant neural networks. Then Sect. 4 introduces the contribution of the paper: sixteen group equivariant layers based on mathematical morphology operators. Section 5 describes the preliminary results obtained with the proposed layers on the Fashion-MNIST dataset. Finally Sect. 6 concludes the paper and gives future perspectives.

2   Mathematical Preliminaries

2.1   Group Action

This section begins by giving a general definition of the notion of equivariance and in-variance. We provide definitions inspired by the ones in [14] and slightly modified. We will consider that a transformation is a bijective map between a set and itself. In this case, we say that the transformation acts on the set. For any set X, we will denote by $\mathcal{T}(X)$, the set of all transformations acting on X. In the following, $\mathcal{F}(X, \mathbb{R})$ will denote the space of maps from X to $\mathbb{R}$.

Definition 1 (Equivariant and Invariant Mapping to a Transformation). *Given a set X and a transformation $\pi \in \mathcal{T}(X)$ acting on it $(\pi : X \to X)$, π can also act on $\mathcal{F}(X, \mathbb{R})$ through an induced transformation $\Pi : \mathcal{F}(X, \mathbb{R}) \to \mathcal{F}(X, \mathbb{R})$ defined as follows :*

$$\forall f \in \mathcal{F}(X, \mathbb{R}), \forall x \in X, \Pi(f)(x) := f\left(\pi^{-1}(x)\right).$$

Given two sets X, Y, a map $\Phi : \mathcal{F}(X, \mathbb{R}) \to \mathcal{F}(Y, \mathbb{R})$, two transformations $\pi \in \mathcal{T}(X)$ and $\psi \in \mathcal{T}(Y)$ acting respectively on X and Y and their induced transformations $\Pi \in \mathcal{T}(\mathcal{F}(X, \mathbb{R}))$ and $\Psi \in \mathcal{T}(\mathcal{F}(Y, \mathbb{R}))$:

– Φ is said to be (Π, Ψ)-equivariant if and only if

$$\forall f \in \mathcal{F}(X, \mathbb{R}), \ \Phi(\Pi(f)) = \Psi(\Phi(f)).$$

When the context is clear and without doubts on the nature of Ψ, we can abusively say that Φ is Π-equivariant.
– Φ is said to be Π-invariant if and only if it is $(\Pi, id_{\mathcal{F}(Y, \mathbb{R})})$-equivariant:

$$\forall f \in \mathcal{F}(X, \mathbb{R}), \ \Phi(\Pi(f)) = \Phi(f).$$

When considering specific kinds of transformations, e.g. rotations or translations, one can sometimes benefit from a group structure if the transformations behave accordingly. The scope of this paper is limited to such situations.

Definition 2 (Group Actions). *Given a set X and a group $(\mathcal{G}, \cdot)$, $\mathcal{G}$ is acting on X if it exists a subgroup $\mathcal{T}_\mathcal{G}(X)$ of $(\mathcal{T}(X), \circ)$, where $\circ$ is map composition, and a homomorphic mapping*

$$T : (\mathcal{G}, \cdot) \to (\mathcal{T}_\mathcal{G}(X), \circ)$$
$$g \mapsto T_g$$

Such transformations T_g are called $\mathcal{G}$-actions. It can be noted that if a group $\mathcal{G}$ acts on X, then $\mathcal{G}$ acts on $\mathcal{F}(X, \mathbb{R})$ through the induced transformations from $\mathcal{T}_\mathcal{G}(X)$ with the induced $\mathcal{G}$-actions defined as:

$$\forall f \in \mathcal{F}(X, \mathbb{R}), \forall x \in X, \mathbb{T}_g(f)(x) = f\left(T_g^{-1}(x)\right) = f\left(T_{g^{-1}}(x)\right)$$

Definition 3 (Group Equivariant and Group Invariant Mapping). *Given a group $\mathcal{G}$ acting on two sets X and Y with $\mathcal{G}$-actions $\mathcal{T}_\mathcal{G}(X) = \{T_g^X \in \mathcal{T}(X) \mid g \in \mathcal{G}\}$ and $\mathcal{T}_\mathcal{G}(Y) = \{T_g^Y \in \mathcal{T}(Y) \mid g \in \mathcal{G}\}$. Let us consider the induced $\mathcal{G}$-actions on $\mathcal{F}(X, \mathbb{R})$ and $\mathcal{F}(Y, \mathbb{R})$, denoted as $\mathbb{T}_g^X$ and $\mathbb{T}_g^Y$ for all $g \in \mathcal{G}$.*
 Given a mapping $\Phi : \mathcal{F}(X, \mathbb{R}) \to \mathcal{F}(Y, \mathbb{R})$,

- *Φ is said to be $\mathcal{G}$-equivariant (or equivariant to the action of $\mathcal{G}$) if for all $g \in \mathcal{G}$, Φ is $(\mathbb{T}_g^X, \mathbb{T}_g^Y)$-equivariant.*
- *Φ is said to be $\mathcal{G}$-invariant (or invariant to the action of $\mathcal{G}$) if for all $g \in \mathcal{G}$, Φ is $\mathbb{T}_g^X$-invariant.*

Definition 4 (Transitivity and Homogeneous Space). *Given a set X and group $\mathcal{G}$, $\mathcal{G}$ is said to act transitively on X if for all $x, y \in X$ it exists a group element $g \in \mathcal{G}$ such that $y = T_g(x)$. In this case, X is said to be a homogeneous space of $\mathcal{G}$. The existence of the group element g is not necessarily unique. If a point x_o is chosen to be the origin of the space X, the set of group elements mapping the origin to itself is called the stabilizer of x_o and will be written Σ_{x_o} and by an abuse of notation, Σ will be used.*

Remark 1. The stabilizer of the origin Σ is a subgroup of $\mathcal{G}$. For every element $y \in X$, if X is a homogeneous space of $\mathcal{G}$, there is at least one $g \in \mathcal{G}$ such that $y = T_g(x_o)$. It can be shown that every other group element mapping x_o to y is an element of the left-coset $g\Sigma = \{gh \in \mathcal{G} \mid h \in \Sigma\}$. The set of all the left-cosets, called the left quotient-space $\mathcal{G}/\Sigma$, is then isomorphic to X.

Example 1. Considering $X = \mathbb{R}^2$ and the group of 2-dimensional translations and rotations $\mathcal{G} = \mathbb{R}^2 \rtimes [-\pi, \pi]$. $\mathcal{G}$ acts on X with the following group action:

$$\forall (\tau, \theta) \in \mathcal{G}, \forall x \in \mathbb{R}^2 \mid T_{(\tau, \theta)}(x) = R_\theta x + \tau,$$

where R_θ is a rotation matrix. The stabilizer $\Sigma = \{(0, \theta) \in \mathcal{G} \mid \theta \in [-\pi, \pi]\}$ when setting $x_o = (0, 0)$.

In practice, to obtain group equivariance, a commonly used strategy is to lift the input image to the desired group, apply an operator or a filter defined on the group and project it back on the original space. For a map $f : X \to \mathbb{R}$, the lifting of f to $\mathcal{G}$ is defined as follow:

$$\forall g \in \mathcal{G}, \; f^{X \to \mathcal{G}} (g) := f \left(T_g \left(x_o \right) \right). \tag{1}$$

One possibility proposed in [14], in the case of discrete groups, for the projection of a map $f : \mathcal{G} \to \mathbb{R}$ is the following one

$$\forall x \in X, \; \exists g_x \in \mathcal{G} \text{ such that } x = T_g \left(x_o \right), \text{ then } f_{\mathcal{G} \to X} \left(x \right) = \frac{1}{\# \left(\Sigma \right)} \sum_{h \in g_x \Sigma} f \left(h \right), \tag{2}$$

where $\#$ gives the cardinality of a set.

2.2   Basis of Mathematical Morphology

This section aims to recall the theoretical foundation of Mathematical Morphology (MM) in order to enhance the self-contained nature of the paper. The subsequent content is strongly inspired by [12] and interested readers are invited to consult this reference for a more detailed overview of MM. Originally defined using set theory to process and analyse binary images, the main idea of MM was to use probes, defined as sets and called *Structuring Elements*, to describe and modify binary shapes. Two operators called *Dilation* and *Erosion* based on the Minkowski addition and subtraction are essential to this end. Given an Euclidean space $\mathcal{E}$ and two subsets X and B of $\mathcal{E}$, where X plays the role of the image/shape to study and B is the structuring element, the dilation and erosion of X using B are defined by the following equations

$$\delta_B \left(X \right) := X \oplus B = \cup_{x \in X} B_x = \cup_{b \in B} X_b. \tag{3}$$

$$\varepsilon_B \left(X \right) := X \ominus B = \{ y \in \mathcal{E} \, | B_y \in X \} = \cap_{b \in B} X_{-b}. \tag{4}$$

For any $B \subset \mathcal{E}$ and any $x \in \mathcal{E}$, the notation B_x denotes the set $\{ x + b \in \mathcal{E} \, | \, b \in B \}$.

MM has then been generalized to other mathematical objects of various natures (e.g. bounded functions on Euclidean spaces [21] or Riemannian manifolds [3], graphs [13], etc) under a common framework of complete lattices defined as follow,

Definition 5 (Complete Lattice). *A complete lattice is defined to be a set $\mathcal{L}$ equipped with a partial ordering $\leq$ satisfying the following properties: for any family of elements $(X_i)_{i \in I} \in \mathcal{L}^I$, it exists a supremum $\bigvee_{i \in I} X_i$ and an infimum $\bigwedge_{i \in I} X_i$ such that:*

- $\forall i \in I, \; \bigwedge_{i \in I} X_i \leq X_i \leq \bigvee_{i \in I} X_i$
- $\forall Y \in \mathcal{L} \; (\forall i \in I, X_i \leq Y) \implies \left(\bigvee_{i \in I} X_i \leq Y \right)$
- $\forall Y \in \mathcal{L} \; (\forall i \in I, Y \leq X_i) \implies \left(Y \leq \bigwedge_{i \in I} X_i \right)$

Two important elements arise from this definition: the supremum and the infimum of the entire complete lattice. They will be denoted $\perp_{\mathcal{L}}$ for the infimum of $\mathcal{L}$ and $\top_{\mathcal{L}}$ for the supremum of $\mathcal{L}$.

Remark 2 Using the reverse partial ordering $\geq$, if $(\mathcal{L}, \leq)$ is a complete lattice, then $(\mathcal{L}, \geq)$ is a complete lattice were the supremum (resp. infimum) of $(\mathcal{L}, \geq)$ is the infimum (resp. supremum) of $(\mathcal{L}, \leq)$. The two lattices are said to be *dual lattices*. This notion of duality is recurrent in MM.

Under the framework of complete lattices, erosions and dilations are defined to be the mappings commuting with supremums and infimums.

Definition 6 (Erosion and Dilation). *Given a complete lattice $(\mathcal{L}, \leq)$, a map $\delta : \mathcal{L} \to \mathcal{L}$ is called a dilation if for any family of elements $(X_i)_{i \in I} \in \mathcal{L}^I$, $\delta\left(\bigvee_{i \in I} X_i\right) = \bigvee_{i \in I} \delta(X_i)$. In the same manner, a map $\varepsilon : \mathcal{L} \to \mathcal{L}$ is called an erosion if for any family of elements $(X_i)_{i \in I} \in \mathcal{L}^I$, $\varepsilon\left(\bigwedge_{i \in I} X_i\right) = \bigwedge_{i \in I} \varepsilon(X_i)$.*

Following the notations of [12], the set of maps $\mathcal{L}^{\mathcal{L}}$ between a complete lattice and itself will be denoted $\mathcal{O}$.

Remark 3. The notions of dilation and erosion are dual and it can be shown (Proposition 2.3 of [12]) that the set of dilations and the set of erosions are both complete lattices with the following partial ordering: Given $\Psi, \Phi \in \mathcal{O}$

$$\Psi \leq \Phi \iff \forall X \in \mathcal{L}, \ \Psi(X) \leq \Phi(X).$$

Definition 7 (Adjunction). *Given two mappings $\delta, \varepsilon \in \mathcal{O}$, the pair (ε, δ) is said to be an adjunction if*

$$\forall X, Y \in \mathcal{L}, \ \delta(X) \leq Y \iff X \leq \varepsilon(Y).$$

Proposition 1. *Proposition 2.5 of [12] Given $\delta, \varepsilon \in \mathcal{O}$, (ε, δ) being an adjunction implies that δ is a dilation and ε is an erosion.*

Definition 8. *For any $\Psi \in \mathcal{O}$, the two dual mappings $\Psi^{\bullet}$ and $\Psi_{\bullet}$ can be defined in the following way: $\forall Y \in \mathcal{L}$*

$$\Psi^{\bullet}(Y) := \bigvee \{Z \in \mathcal{L} \mid \Psi(Z) \leq Y\} \ and \ \Psi_{\bullet}(Y) := \bigwedge \{Z \in \mathcal{L} \mid Y \leq \Psi(Z)\}.$$

Remark 4. Heijmans and Ronse showed in Proposition 2.6 of [12] that for any adjunction $(\varepsilon, \delta) \in \mathcal{O}^2$, the following relations between ε and δ holds:

$$\varepsilon = \delta^{\bullet} \ \text{and} \ \delta = \varepsilon_{\bullet}. \tag{5}$$

This remark goes even further with the following theorem

Theorem 1 (Theorem 2.7 of [12]). *The set of adjunctions forms a dual isomorphism between the complete lattice of erosions and the complete lattice of dilations. Meaning that for every dilation δ (resp. erosion ε), it exists a unique erosion ε (resp. dilation δ) such that (ε, δ) is an adjunction and thus respects (5).*

The concepts of dilations, erosions and more generally the concept of adjunction can be generalized to mappings between two distinct complete lattices $\mathcal{L}_1$ and $\mathcal{L}_2$. In this case, it is possible to have an adjunction (ε, δ) such that $\delta : \mathcal{L}_1 \to \mathcal{L}_2$ and $\varepsilon : \mathcal{L}_2 \to \mathcal{L}_1$.

Definition 9. (Sup-Generating family). *Given a complete lattice $\mathcal{L}$ and a subset $l \in \mathcal{L}$, l is said to be sup-generating if every element $X \in \mathcal{L}$ can be associated with a subset of l, written $l(X)$, defined as $l(X) = \{x \in l \mid x \leq X\}$ and satisfying $X = \bigvee l(X)$. In other words, every element of the complete lattice can be expressed as the supremum of a collection of elements of l.*

When studying gray level images, it comes naturally to consider the image as functions to real numbers, maybe bounded [5], on a Euclidean space. More generally, a gray level image can be seen as a map between a set (or a Euclidean space) $\mathcal{E}$ and a complete lattice $\mathcal{L}$. The space $\mathcal{F}(\mathcal{E}, \mathcal{L})$ of such maps is again a complete lattice where the partial ordering, the infimum and the supremum are defined using the partial ordering, the infimum and the supremum of $\mathcal{L}$: Considering any collection $(f_i)_{i \in I} \in \mathcal{F}(\mathcal{E}, \mathcal{L})^I$

- $\forall f, g \in \mathcal{F}(\mathcal{E}, \mathcal{L})$, $f \leq g \iff \forall x \in \mathcal{E}$, $f(x) \leq g(x)$
- $\forall x \in \mathcal{E}$, $\left(\bigvee_{i \in I} f_i\right)(x) = \bigvee_{i \in I} f_i(x)$ and $\left(\bigwedge_{i \in I} f_i\right)(x) = \bigwedge_{i \in I} f_i(x)$

We will consider two sets $\mathcal{E}_1$ and $\mathcal{E}_2$ and a complete lattice $\mathcal{L}$, the notion of generating family can be useful to describe erosions and dilations on the complete lattices $\mathcal{F}(\mathcal{E}_1, \mathcal{L})$ and $\mathcal{F}(\mathcal{E}_2, \mathcal{L})$ by looking at the dilations (or erosions) of the elements of the generating family. More precisely, we will consider a dilation $\delta : \mathcal{F}(\mathcal{E}_1, \mathcal{L}) \to \mathcal{F}(\mathcal{E}_2, \mathcal{L})$ and for the complete lattice $\mathcal{F}(\mathcal{E}_1, \mathcal{L})$ we will consider the sup-generating family $l_{\mathcal{F}(\mathcal{E}_1, \mathcal{L})} = \{f_{x,t} \in \mathcal{F}(\mathcal{E}_1, \mathcal{L}) \mid x \in \mathcal{E}_1, t \in \mathcal{L}\}$ with

$$\forall y \in \mathcal{E}_1, \; f_{x,t}(y) = \begin{cases} t, & \text{if } y = x \\ \bot_{\mathcal{L}}, & \text{if } y \neq x \end{cases}$$

as for any $F \in \mathcal{F}(\mathcal{E}_1, \mathcal{L})$, $F = \bigvee_{x \in \mathcal{E}_1} f_{x, F(x)}$. In a similar way that a sup generating family can be used to describe an element of the complete lattice, a dilation can be described by its effect on the sup generating family. The set of mappings $\delta_{y,x} : \mathcal{L} \to \mathcal{L}$ are defined using the elements of $l_{\mathcal{F}(\mathcal{E}_1, \mathcal{L})}$ and δ:

$$\forall x \in \mathcal{E}_1, \; \forall y \in \mathcal{E}_2, \; \forall t \in \mathcal{L}, \; \delta_{y,x}(t) = \delta(f_{x,t})(y)$$

Using the assumption that δ is a dilation, it can be shown that $\delta_{y,x}$ is a dilation for all $x \in \mathcal{E}_1$, $y \in \mathcal{E}_2$. The dilations $\delta_{y,x}$ define the dilation δ which gives the following proposition.

Proposition 2 (Proposition 2.10 of [12]). *Given two sets $\mathcal{E}_1$, $\mathcal{E}_2$ and a complete lattice $\mathcal{L}$. The map $\delta : \mathcal{F}(\mathcal{E}_1, \mathcal{L}) \to \mathcal{F}(\mathcal{E}_2, \mathcal{L})$ is a dilation if and only if for every $x \in \mathcal{E}_1$ and $y \in \mathcal{E}_2$, there exists a dilation $\delta_{y,x} : \mathcal{L} \to \mathcal{L}$ such that for all $F \in \mathcal{F}(\mathcal{E}_1, \mathcal{L})$ and for all $y \in \mathcal{E}_2$, it holds*

$$\delta\left(F\right)\left(y\right) = \bigvee_{x \in \mathcal{E}_1} \delta_{y,x}\left(F\left(x\right)\right).$$

The adjoint erosion $\varepsilon : \mathcal{F}(\mathcal{E}_2, \mathcal{L}) \to \mathcal{F}(\mathcal{E}_1, \mathcal{L})$ *is such that, for all* $F \in \mathcal{F}(\mathcal{E}_2, \mathcal{L})$ *and for all* $x \in \mathcal{E}_1$:

$$\varepsilon\left(F\right)\left(x\right) = \bigwedge_{y \in \mathcal{E}_2} \varepsilon_{x,y}\left(F\left(y\right)\right),$$

with every $\varepsilon_{x,y}$ *being the adjoint of* $\delta_{y,x}$ *for* $\mathcal{L}$.

To make the transition to group morphology, Heijmans and Ronse [12] pointed out that by taking $\mathcal{E}_1 = \mathcal{E}_2 = \mathbb{R}^d$, $\mathcal{L} = \mathbb{R}$ and by imposing $\delta_{y,x} = \delta_{y-x,0}$ to respect translation equivariance, then, using a structuring function G, one can choose to consider

$$\forall h \in \mathbb{R}^d, \forall t \in \mathbb{R}, \ \delta_{h,0}(t) = t + G(h). \tag{6}$$

3   Related Work

3.1   Group Morphology

Under the name of invariance, equivariance to translation has always been a concern in MM. As MM is usually applied to image analysis and processing using "small" structuring elements running spatially through the entire image of study, it is natural to be concerned with translation equivariance. Some authors extended MM to general groups [18] or gave fairly general results applicable to any group when studying translations equivariance [12]. This subsection focuses on the main results of these two references. The first and intuitive approach was to generalize the Minkowski addition and subtraction. Roerdink [18] proposed such a generalization for groups that are not necessarily Abelian. Given a group $\mathcal{G}$ and two of its subsets H and G, Roerdink defined the left-equivariant erosion and left-equivariant dilation on the complete lattice $\mathcal{P}\left(\mathcal{G}\right)$ as follows:

$$\delta_H^L\left(G\right) := G \oplus_\mathcal{G} H = \cup_{h \in H} Gh = \cup_{g \in G} gH, \tag{7}$$

$$\varepsilon_H^L\left(G\right) := G \ominus_\mathcal{G}^L H = \cap_{h \in H} Gh^{-1} = \{g \in \mathcal{G} \mid gH \subset G\}. \tag{8}$$

The right-equivariant dilation and erosion are defined in the same manner:

$$\delta_H^R\left(G\right) := H \oplus_\mathcal{G} G = \cup_{h \in H} hG = \cup_{g \in G} Hg, \tag{9}$$

$$\varepsilon_H^R\left(G\right) := G \ominus_\mathcal{G}^R H = \cap_{h \in H} h^{-1}G = \{g \in \mathcal{G} \mid Hg \subset G\}. \tag{10}$$

In practice, a binary image is defined on a Euclidean space $\mathcal{E}$. In this case, the considered complete lattice is the set of subsets of $\mathcal{E}$, written $\mathcal{P}\left(\mathcal{E}\right)$ with the inclusion as partial ordering. If the space $\mathcal{E}$ is a homogeneous space of $\mathcal{G}$, then, Roerdink [18] proposed the following lifting ϑ and projections ϖ, ϖ_Σ in order to apply the erosions and dilations defined by (7), (8) or (9), (10).

$$\forall X \in \mathcal{P}\left(\mathcal{E}\right), \ \vartheta\left(X\right) := \{g \in \mathcal{G} \mid T_g\left(x_o\right) \in X\}, \tag{11}$$

$$\forall G \in \mathcal{G}, \; \varpi(G) := \{T_g(x_o) \mid g \in G\} \text{ and } \varpi_\Sigma(G) := \{T_g(x_o) \mid g\Sigma \subset G\}, \quad (12)$$

where x_o is a chosen origin for $\mathcal{E}$. Using these maps, Roerdink characterized the adjunctions on $\mathcal{P}(\mathcal{E})$ using (7), (8). They are called left $\mathcal{G}$-adjunctions and they take the following form:

$$\delta_H^{\mathcal{G}}(G) := \varpi(\vartheta(G) \oplus_{\mathcal{G}} \vartheta(H)) \text{ and } \varepsilon_H^{\mathcal{G}}(G) := \varpi_\Sigma(\vartheta(G) \ominus_{\mathcal{G}}^L \vartheta(H)). \quad (13)$$

The name left $\mathcal{G}$-adjunction comes from the fact that given such an adjunction $(\varepsilon_H^{\mathcal{G}}, \delta_H^{\mathcal{G}})$, the dilation $\delta_H^{\mathcal{G}}$ and the erosion $\varepsilon_H^{\mathcal{G}}$ are left $\mathcal{G}$-equivariant. A similar definition can be given for right $\mathcal{G}$-adjunctions using (9) and (10). When the image of study is not boolean, it was proposed to use a sup-generative family as an intermediate step. The principal requirement is that the group $\mathcal{G}$ acts transitively on the sup-generative family, allowing to use the lifting and projections defined by the equations (11), (12) and therefore the $\mathcal{G}$-adjunction $(\varepsilon_H^{\mathcal{G}}, \delta_H^{\mathcal{G}})$.

3.2   Group Equivariant Networks

Conceptually there are two different approaches to the construction of invariant and equivariant models [23]: *symmetrization* based one and the *intrinsic* one. In the first case, one starts with an non-invariant model and symmetrizing it by a group averaging. In the second case, the intrinsic approach consists of imposing prior structural constraints on the model that guarantee its invariance.

In Deep Learning, both approaches can justify practices that are common to induce equivariance and invariances. For instance, a) *data augmentation* techniques advocates to augment the available set of training samples $(x, f(x))$ by new ones of the form $(T_g(x), f(x))$, then loss values of theses realizations are average on batches during training [9]. b) *Invariance by regularization* uses differences of $f(x)$ and many realization of $f(T_g(x))$ are computed to regularize the model during training [6, 9, 22]. Both approaches can be seen as exemplifications of the symmetrization-based approach. On the other hand, the *weight sharing* mechanism specially used in convolutional networks [15], and other approaches as PDE based CNNs [20], Elementary Symmetric Polynomials based CNNs [17], Moving Frame based CNNs [19], Deep Scattering CNNs [1], Steerable CNNs [7] and Group Convolutional networks [10, 14] are manifestation of intrinsic approach. The method proposed in [10] will be recalled in more details as the contribution of the paper uses a similar approach. The main idea is to generalize the convolution to any compact group. In the usual convolution, a filter runs through the entire input image to locally match its pattern with the image, i.e., the filter is translated to every position in the domain of the image. When considering other transformations forming a different group than the translations, Cohen and Welling proposed to apply all the transformations of the group to the filter. For example, if the considered group is composed of all the translations with also a finite number of rotations discretizing the circle (subgroup of the circle). Then the proposition would be to transform the filter using all the possible translations and all the rotations. In practice, two types of layers are proposed: group lifting layers and group convolutional layers. The first one plays the role of (1) with in addition a learnable filtering part. To be more precise, copies of the input image (or tensor) are created, which correspond to (1) and rotated copies of a learned two-dimensional filter are used to perform regular convolutions. This layer can be summarized in one formula. Given a discrete group $\mathcal{G}$, an input image f composed with N feature maps and a kernel ϕ, the lifting layer performs the following formula

$$\forall g \in \mathcal{G}, \ \left(f \star^{\mathbb{Z}^2 \to \mathcal{G}} \phi\right)(g) := \sum_{k=1}^{N} \sum_{x \in \mathbb{Z}^2} f_k(x) \, \mathbb{T}_g(\phi_k)(x), \tag{14}$$

followed by a bias addition and an activation function. The output of this operation is a map whose domain is the group $\mathcal{G}$. The group layers, also followed by a bias addition and an activation function, apply the group convolution in the following way

$$\forall g \in \mathcal{G}, \ \left(f \star^{\mathcal{G}} \phi\right)(g) := \sum_{k=1}^{N} \sum_{h \in \mathcal{G}} f_k(h) \, \mathbb{T}_g(\phi_k)(h). \tag{15}$$

4 Proposed Layers

This section aims to define group-equivariant morphological layers. These layers will be named after classical morphological operators due to their evident parallels. Nevertheless, it is important to note that no formal proofs are provided regarding their inherent properties. As an illustration, layers labeled as *dilation* can be characterized by a formula closely resembling that of an established dilation operation. We introduce a total of sixteen layers, with eight of them serving as nonlinear counterparts to the lifting layer outlined in (14). The other half are nonlinear counterparts of the group convolutional layer described by (15). We denote by $\bar{\mathbb{R}} = \mathbb{R} \cup \{-\infty, +\infty\}$ the extended real number line and by $\mathbb{R}_+^*$ the real strictly positive numbers. The proposed layers are based on two types of adjunction on $\bar{\mathbb{R}}$. Given a group $\mathcal{G}$, each element $h \in \mathcal{G}$ gives rise to an adjunction on $\bar{\mathbb{R}}$. Such an adjunction can be denoted by (e_h, d_h). Following the work of Heijmans in [11], we propose to use the following forms for e_h and d_h

$$d_h^\star(t) = A(h) t + G(h) \text{ and } e_h^\star(t) = (t - G(h)) / A(h), \tag{16}$$

where $A(\cdot) : \mathcal{G} \mapsto \mathbb{R}_+^*$ and $G(\cdot) : \mathcal{G} \mapsto \mathbb{R}$ play the role of structuring functions. An easier alternative (see (6) for $\mathcal{G} = \mathbb{R}^d$) consists of using

$$d_h^\dagger(t) = t + \hat{G}(h) \text{ and } e_h^\dagger(t) = t - G(h). \tag{17}$$

From these adjunctions on $\bar{\mathbb{R}}$, we define the following operators on $\mathcal{G}$. Given $F : \mathcal{G} \to \bar{\mathbb{R}}$,

$$D(F)(g) := \bigvee_{h \in \mathcal{G}} d_h\left(F\left(gh^{-1}\right)\right) \text{ and } E(F)(g) := \bigwedge_{h \in \mathcal{G}} e_h\left(F(gh)\right) \tag{18}$$

where the pair (d_g, e_g) can be (16) or (17). We use (18) to propose nonlinear equivariant layers. The eight *lifting* layers can be described by the following formulas. Given a discrete group $\mathcal{G}$, an input image $f : \mathbb{Z}^2 \to \bar{\mathbb{R}}^N$ composed of N feature maps and N adjunctions $(e_{k,x}, d_{k,x})_{k \in [1,N]}$ defined for all $x \in \mathbb{Z}^2$ following either eq. (16) or (17), then for all $g \in \mathcal{G}$

$$\varepsilon_\uparrow (f) (g) := \sum_{k=1}^{N} \bigwedge_{x \in \mathbb{Z}^2} e_{k, T_{g^{-1}(x)}} (f_k (x)), \qquad (19)$$

$$\eth_\uparrow (f) (g) := \sum_{k=1}^{N} \bigvee_{x \in \mathbb{Z}^2} d_{k, -T_{g^{-1}(x)}} (f_k (x)), \qquad (20)$$

$$o_\uparrow (f) (g) := \sum_{k=1}^{N} \bigvee_{y \in \mathbb{Z}^2} d_{k, -T_{g^{-1}(y)}} \left(\bigwedge_{x \in \mathbb{Z}^2} e_{k, T_{g^{-1}(x)} - T_{g^{-1}(y)}} (f_k (x)) \right), \qquad (21)$$

$$c_\uparrow (f) (g) := \sum_{k=1}^{N} \bigwedge_{y \in \mathbb{Z}^2} e_{k, T_{g^{-1}(y)}} \left(\bigvee_{x \in \mathbb{Z}^2} d_{k, -T_{g^{-1}(x)} + T_{g^{-1}(y)}} (f_k (x)) \right), \qquad (22)$$

where $\varepsilon, \eth, o$ and c stand respectively for erosion, dilation, opening and closing. In the same manner, we define nonlinear counterparts of (15). Given a map $F : \mathcal{G} \to \bar{\mathbb{R}}$:

$$\varepsilon_\mathcal{G} (F) (g) := \sum_{k=1}^{N} \bigwedge_{h \in \mathcal{G}} e_{k,h} (F (gh)) \qquad (23)$$

$$\eth_\mathcal{G} (F) (g) := \sum_{k=1}^{N} \bigvee_{h \in \mathcal{G}} d_{k,h} (F (gh^{-1})) \qquad (24)$$

$$o_\mathcal{G} (F) (g) := \sum_{k=1}^{N} \bigvee_{h \in \mathcal{G}} d_{k,h} \left(\bigwedge_{m \in \mathcal{G}} e_{k,m} (F (gh^{-1}m)) \right) \qquad (25)$$

$$c_\mathcal{G} (F) (g) := \sum_{k=1}^{N} \bigwedge_{h \in \mathcal{G}} e_{k,h} \left(\bigvee_{m \in \mathcal{G}} d_{k,m} (F (ghm^{-1})) \right) \qquad (26)$$

These layers are followed by a bias addition and an activation function.

5   Experiments

In practice, we used the group $P4$ defined to be the group of translations and rotations of $\frac{\pi}{2}$ [10]. Several configurations[1] have been tested on a classification task using the Fashion-MNIST dataset. The configurations use either the adjunctions defined by equations (16) or (17) and use different types of layers among those presented in Sect. 4. We use the notation † in superscript to denote the

---

[1] https://github.com/Penaud-Polge/Group_Equivariant_Morphological_Layers.

use of the adjunctions of (17) and $\star$ to denote the use of the adjunctions of (16). The type of layer used will be given by their names. Therefore, an architecture denoted "$\partial_{\mathcal{G}}^{\dagger}\ \partial_{\uparrow}^{\dagger}$" stands for a two layer network where the lifting is given by (20) and the group layer, i.e. the second one, is given by (24) and with adjunctions (e_g, d_g) given by (17). For every configuration, a dense layer is used after a global average pooling to predict the class for each orientation. A max-pooling operation is then applied between the different orientations to obtain invariance to rotations of $\frac{\pi}{2}$. All the structuring functions were of size $(3, 3)$ for the lifting layers and of size $(3, 3, 3)$ for the group layers. For all the configurations, the network has been trained using only original (untransformed) images. On the other hand, rotated copies of the test dataset have been used to determine if the networks were able to classify images having orientations unseen during the training process. An equivalent network using group convolution and an equivalent CNN have been used as references for comparison. The results obtained are presented in Table 1. The results show that all networks except the regular CNN were able to generalize to unseen orientations of $P4$. All the networks are less efficient when evaluating at an orientation not contained in the group $P4$. Results also highlight the fact that for dilation and erosion networks, using adjunctions defined by equation (16) offers better performances. Using the opening and closing layers seems to lower the performances. And finally, except $e_{\mathcal{G}}^{\star}\ e_{\uparrow}^{\star}$ that tends to have performances closer to the convolutional reference, it seems that morphological layers, on this task, are less efficient. It is important to highlight that CNNs have benefited from extensive and in-depth studies by the research community regarding their appropriate optimization, initialization, and other aspects. This is not the case for morphological counterparts [4,8]. We hope that the presented results related to the morphological layers will improve with future research on the subject.

Table 1. Test accuracies in percentage of several configurations of $P4$ equivariant morphological networks depending on the rotations applied to the test dataset.

Networks	Acc_0	$Acc_{\frac{\pi}{6}}$	$Acc_{\frac{\pi}{4}}$	$Acc_{\frac{\pi}{2}}$	Acc_{π}	$Acc_{\frac{3\pi}{2}}$
CNN	**87.13** %	**38.41** %	23.30 %	7.28 %	34.76 %	10.24%
CNN P4	80.16 %	23.08 %	27.17 %	**80.16** %	**80.16** %	**80.16** %
$c_{\mathcal{G}}^{\dagger}\ c_{\uparrow}^{\dagger}$	62.75 %	17.68 %	17.26 %	62.73 %	62.73 %	62.73 %
$\partial_{\mathcal{G}}^{\dagger}\ \partial_{\uparrow}^{\dagger}$	64.90 %	26.63 %	22.46 %	64.90 %	64.90 %	64.90 %
$e_{\mathcal{G}}^{\dagger}\ e_{\uparrow}^{\dagger}$	69.83 %	25.54 %	23.31 %	69.83 %	69.83 %	69.83 %
$o_{\mathcal{G}}^{\dagger}\ o_{\uparrow}^{\dagger}$	70.23 %	30.44 %	27.70 %	70.23 %	70.23 %	70.23 %
$c_{\mathcal{G}}^{\star}\ c_{\uparrow}^{\star}$	62.71 %	22.71 %	19.97 %	62.71 %	62.71 %	62.71 %
$\partial_{\mathcal{G}}^{\star}\ \partial_{\uparrow}^{\star}$	71.64 %	23.67 %	23.89 %	71.64 %	71.64 %	71.64 %
$e_{\mathcal{G}}^{\star}\ e_{\uparrow}^{\star}$	76.21 %	21.96 %	15.76 %	76.21 %	76.21 %	76.21 %
$o_{\mathcal{G}}^{\star}\ o_{\uparrow}^{\star}$	57.66 %	34.51 %	**33.60** %	57.66 %	57.66 %	57.66 %

6   Conclusion

This paper is a first approach to generalize morphological layers to group morphology. Multiple layers, incorporating morphological operators, have been proposed and tested on the Fashion-MNIST dataset. The networks exhibited almost perfect invariance to the chosen group but offered lower performances compared to their convolutional counterpart. The morphological operators employed in these layers have not been theoretically described yet; therefore, providing a theoretical foundation is an essential aspect of future work.

References

1. Andén, J., Mallat, S.: Deep scattering spectrum. IEEE Trans. Signal Process. **62**(16), 4114–4128 (2014)
2. Angulo, J.: Some open questions on morphological operators and representations in the deep learning era. In: Lindblad, J., Malmberg, F., Sladoje, N. (eds.) DGMM 2021. LNCS, vol. 12708, pp. 3–19. Springer, Cham (2021). https://doi.org/10.1007/978-3-030-76657-3_1
3. Angulo, J., Velasco-Forero, S.: Riemannian mathematical morphology. Pattern Recogn. Lett. **47**, 93–101 (2014)
4. Blusseau, S.: Training morphological neural networks with gradient descent: some insights. In: Rinaldi, S. (ed.) DGMM 2024. LNCS, vol. 14605, pp. 229–241. Springer, Heidelberg (2024)
5. Blusseau, S., Velasco-Forero, S., Angulo, J., Bloch, I.: Morphological adjunctions represented by matrices in max-plus algebra for signal and image processing. In: Baudrier, É., Naegel, B., Krähenbühl, A., Tajine, M. (eds.) DGMM 2022. LNCS, vol. 13493, pp. 206–218. Springer, Cham (2022). https://doi.org/10.1007/978-3-031-19897-7_17
6. Botev, A., et al.: Regularising for invariance to data augmentation improves supervised learning. arXiv preprint arXiv:2203.03304 (2022)
7. Cesa, G., Lang, L., Weiler, M.: A program to build E(N)-equivariant steerable CNNs. In: International Conference on Learning Representations (2022)
8. Charisopoulos, V., Maragos, P.: Morphological perceptrons: geometry and training algorithms. In: Angulo, J., Velasco-Forero, S., Meyer, F. (eds.) ISMM 2017. LNCS, vol. 10225, pp. 3–15. Springer, Cham (2017). https://doi.org/10.1007/978-3-319-57240-6_1
9. Chen, W., et al.: Augmentation invariant training. In: Proceedings of the IEEE/CVF ICCV Workshops (2019)
10. Cohen, T., Welling, M.: Group equivariant convolutional networks. In: International conference on machine learning, pp. 2990–2999. PMLR (2016)
11. Heijmans, H.J.A.M.: Theoretical aspects of gray-level morphology. IEEE Trans. Pattern Anal. Mach. Intell. **13**(06), 568–582 (1991)
12. Heijmans, H.J.A.M., Ronse, C.: The algebraic basis of mathematical morphology I. Dilations and erosions. Comput. Vision Graph. Image Process. **50**(3), 245–295 (1990)
13. Heijmans, H.J.A.M., et al.: Graph morphology. J. Vis. Commun. Image Represent. **3**(1), 24–38 (1992)

14. Kondor, R., Trivedi, S.: On the generalization of equivariance and convolution in neural networks to the action of compact groups. In: International Conference on Machine Learning, pp. 2747–2755. PMLR (2018)
15. Laptev, D., et al.: TI-POOLING: transformation-invariant pooling for feature learning in convolutional neural networks. In: IEEE-CVPR, pp. 289–297 (2016)
16. LeCun, Y., et al.: Handwritten digit recognition with a back-propagation network. In: Advances in Neural Information Processing Systems, vol. 2 (1989)
17. Penaud-Polge, V., et al.: GenHarris-ResNet: a rotation invariant neural network based on elementary symmetric polynomials. In: Calatroni, L., Donatelli, M., Morigi, S., Prato, M., Santacesaria, M. (eds.) SSVM 2023. LNCS, vol. 14009, pp. 149–161. Springer, Cham (2023). https://doi.org/10.1007/978-3-031-31975-4_12
18. Roerdink, J.B.: Group morphology. Pattern Recogn. **33**(6), 877–895 (2000)
19. Sangalli, M., et al.: Moving frame net: SE(3)-equivariant network for volumes. In: NeurIPS Workshop, pp. 81–97. PMLR (2023)
20. Smets, B.M., et al.: PDE-based group equivariant convolutional neural networks. J. Math. Imaging Vision **65**(1), 209–239 (2023)
21. Sternberg, S.R.: Grayscale morphology. Comput. Vision Graph. Image Process. **35**(3), 333–355 (1986)
22. Velasco-Forero, S.: Can generalised divergences help for invariant neural networks? In: Nielsen, F., Barbaresco, F. (eds.) GSI 2023. LNCS, vol. 14071, pp. 82–90. Springer, Cham (2023). https://doi.org/10.1007/978-3-031-38271-0_9
23. Yarotsky, D.: Universal approximations of invariant maps by neural networks. Constr. Approx. **55**(1), 407–474 (2022)

An Algorithm to Train Unrestricted Sequential Discrete Morphological Neural Networks

Diego Marcondes[1,2(✉)] [ID], Mariana Feldman[1] [ID], and Junior Barrera[1] [ID]

[1] Department of Computer Science, Institute of Mathematics and Statistics, University of São Paulo, São Paulo, Brazil
dmarcondes@ime.usp.br
[2] Department of Electrical and Computer Engineering, Texas A&M University, College Station, USA

Abstract. There have been attempts to insert mathematical morphology (MM) operators into convolutional neural networks (CNN), and the most successful endeavor to date has been the morphological neural networks (MNN). Although MNN have performed better than CNN in solving some problems, they inherit their black-box nature. Furthermore, in the case of binary images, they are approximations that loose the Boolean lattice structure of MM operators and, thus, it is not possible to represent a specific class of W-operators with desired properties. In a recent work, we proposed the Discrete Morphological Neural Networks (DMNN) for binary image transformation to represent specific classes of W-operators and estimate them via machine learning. We also proposed a stochastic lattice descent algorithm (SLDA) to learn the parameters of Canonical Discrete Morphological Neural Networks (CDMNN), whose architecture is composed only of operators that can be decomposed as the supremum, infimum, and complement of erosions and dilations. In this paper, we propose an algorithm to learn unrestricted sequential DMNN, whose architecture is given by the composition of general W-operators. We illustrate the algorithm in a practical example.

Keywords: discrete morphological neural networks · image processing · mathematical morphology · U-curve algorithms · stochastic lattice descent algorithm

1 Introduction

Mathematical morphology is a theory of lattice mappings which can be employed to design nonlinear mappings, the morphological operators, for image processing and computer vision. It originated in the 60 s, and its theoretical basis was

---

D. Marcondes was funded by grants #22/06211-2 and #23/00256-7, São Paulo Research Foundation (FAPESP), and J. Barrera was funded by grants #14/50937-1 and #2020/06950-4, São Paulo Research Foundation (FAPESP).

S. Brunetti et al. (Eds.): DGMM 2024, LNCS 14605, pp. 178–191, 2024.
https://doi.org/10.1007/978-3-031-57793-2_14

developed in the 70s and 80s [24,33,34]. From the established theory followed the proposal of many families of operators, which identify or transform geometrical and topological properties of images. Their heuristic combination permits to design methods for image analysis. We refer to [14] for more details on practical methods and implementations of the design of mathematical morphology operators.

Since combining basic morphological operators to form a complex image processing pipeline is not a trivial task, a natural idea is to develop methods to automatically design morphological operators based on machine learning techniques [9], what has been extensively done in the literature with great success on solving specific problems. The problems addressed by mathematical morphology concern mainly the processing of binary and gray-scale images, and the learning methods are primarily based on discrete combinatorial algorithms. More details about mathematical morphology in the context of machine learning may be found in [7,19].

Mathematical morphology methods have also been studied in connection with neural networks. The first papers about morphological neural networks (MNN), such as [10–12,32], proposed neural network architectures in which the operation performed by each neuron is either an erosion or a dilation. MNN usually have the general structure of neural networks, and their specificity is on the fact that the layers realize morphological operations. Many MNN architectures and training algorithms have been proposed for classification and image processing problems [4,13,17,26,38]. Special classes of MNN such as morphological/rank neural networks [27,28], the dendrite MNN [5,31,36]; and the modular MNN [3,37] have been proposed. We refer to [35] and the references therein for a review of the early learning methods based on MNN.

More recently, mathematical morphology methods have been studied in connection with deep neural networks, by either combining convolutional neural networks (CNN) with a morphological operator [21], or by replacing the convolution operations of CNN with basic morphological operators, such as erosions and dilations [16,18,25]. Although it has been seen empirically that convolutional layers could be replaced by morphological layers [16], and MNN have shown a better performance than CNN in some tasks [20], they are not more interpretable than an ordinary CNN. In this context, the interpretability is related to the notion of hypothesis space, which is a key component of machine learning theory.

In [22] we proposed the Discrete Morphological Neural Networks (DMNN) for binary image transformation to represent translation invariant and locally defined operators, i.e., W-operators, and estimate them via machine learning. A DMNN architecture is represented by a Morphological Computational Graph that combines operators via composition, infimum, and supremum to build more complex operators. In [22] we proposed a stochastic lattice descent algorithm (SLDA) to train the parameters of Canonical Discrete Morphological Neural Networks (CDMNN) based on a sample of input and output images under the usual machine learning approach. The architecture of a CDMNN is composed only of canonical operations (supremum, infimum, or complement) and operators that can be decomposed as the supremum, infimum, and complement of erosions and dilations with the same structural element. The DMNN is a true mathematical morphology method since it retains the control over the design and the interpretability of results intrinsic to classical mathematical morphology methods, which is a relevant advantage over CNN.

In this paper, we propose an algorithm to train unrestricted sequential DMNN (USDMNN), whose architecture is given by the composition of general W-operators. The algorithm considers the representation of a W-operator by its characteristic Boolean function, which is learned via a SLDA in the Boolean lattice of functions. This SLDA differs from that of [22] which minimizes an empirical error in a lattice of sets and intervals that are the structural elements and intervals representing operators such as erosions, dilations, openings, closings and sup-generating.

Unlike CDMNN, which can be designed from prior information to represent a constrained class of operators, USDMNN are subject to overfitting the data, since it can represent a great class of operators, that may fit the data, but not generalize to new examples. In order to address this issue, we control the complexity of the architecture by selecting from data the window of each W-operator in the sequence (layer). By restricting the window, we create equivalence classes on the characteristic function domain that constrain the class of operators that each layer can represent, decreasing the overall complexity of the operators represented by the architecture and mitigating the risk of overfitting. We propose a SLDA to select the windows by minimizing a validation error in a lattice of sets within the usual model selection framework in machine learning.

We note that both the CDMNN and the USDMNN are fully transparent, the properties of the operators represented by the trained architectures are fully known, and their results can be interpreted. The advantage of USDMNN is that they are not as dependent on prior information as the CDMNN, which require a careful design of the architecture to represent a *simple* class of operators with properties necessary to solve the practical problem, as was illustrated in the empirical application in [22]. When there is no prior information about the problem at hand, an USDMNN may be preferable.

In Sect. 2, we present some notations and definitions, and in Sect. 3 we formally define the USDMNN that is a particular example of the DMNN proposed by [22]. In Sect. 4, we present the SLDAs to learn the windows and the characteristic functions of the W-operators in a USDMNN, and in Sect. 5 we apply the USDMNN to the same dataset used in [22] in order to compare the results. In Sect. 6, we present the next steps of this research.

2   W-Operators

Let $E = \mathbb{Z}^2$ and denote by $\mathcal{P}(E)$ the collection of all subsets of E. Denote by $+$ the vector addition operation. We denote the zero element of E by o. A *set operator* is any mapping defined from $\mathcal{P}(E)$ into itself. We denote by Ψ the collection of all the operators from $\mathcal{P}(E)$ to $\mathcal{P}(E)$. Denote by ι the identity set operator: $\iota(X) = X, X \in \mathcal{P}(E)$.

For any $h \in E$ and $X \in \mathcal{P}(E)$, the set $X + h := \{x \in E : x - h \in X\}$ is called the translation of X by h. We may also denote this set by X_h. A set operator ψ is called *translation invariant* (t.i) if, and only if, $\forall h \in E$, $\psi(X + h) = \psi(X) + h$ for $X \in \mathcal{P}(E)$.

Let W be a finite subset of E. A set operator ψ is called *locally defined within a window W* if, and only if, $\forall h \in E$, $h \in \psi(X) \iff h \in \psi(X \cap W_h)$ for $X \in \mathcal{P}(E)$. The collection Ψ_W of t.i. operators locally defined within a window $W \in \mathcal{P}(E)$ inherits the complete lattice structure of $(\mathcal{P}(E), \subseteq)$ by setting, $\forall \psi_1, \psi_2 \in \Psi_W$,

$$\psi_1 \le \psi_2 \iff \psi_1(X) \subseteq \psi_2(X), \forall X \in \mathcal{P}(E). \tag{1}$$

Define by $\Omega = \cup_{W \in \mathcal{P}(E), |W| < \infty} \Psi_W$ the collection of all operators from $\mathcal{P}(E)$ to $\mathcal{P}(E)$ that are t.i. and locally defined within some finite window $W \in \mathcal{P}(E)$. The elements of Ω are called *W-operators*.

Any W-operator can be uniquely determined by its kernel, its basis or its characteristic function (see [8] for more details). In special, denote by $\mathfrak{B}_W := \{0, 1\}^{\mathcal{P}(W)}$ the set of all Boolean functions on $\mathcal{P}(W)$ and consider the mapping T between Ψ_W and $\mathfrak{B}_W$ defined by

$$T(\psi)(X) = \begin{cases} 1, & \text{if } o \in \psi(X) \\ 0, & \text{otherwise.} \end{cases} \qquad \psi \in \Psi_W, X \in \mathcal{P}(W). \tag{2}$$

The mapping T constitutes a lattice isomorphism between the complete lattices $(\Psi_W, \le)$ and $(\mathfrak{B}_W, \le)$, and its inverse T^{-1} is defined by $T^{-1}(f)(X) = \{x \in E : f(X_{-x} \cap W) = 1\}$ for $f \in \mathfrak{B}_W$ and $X \in \mathcal{P}(E)$. We denote by $f_\psi := T(\psi)$ the characteristic function of ψ.

3   Unrestricted Sequential Discrete Morphological Neural Networks

In this section, we formally define the USDMNN. In Sect. 3.1, we define the morphological computational graph of USDMNN. This is a concept introduced in [22] that is related to their architecture, defined in Sect. 3.2. In Sect. 3.3, we propose a representation of USDMNN by the windows and characteristic functions of the operators in their layers.

3.1   Sequential Morphological Computational Graph

Let $\mathcal{G} = (\mathcal{V}, \mathcal{E}, \mathcal{C})$ be a *computational graph*, in which $\mathcal{V}$ is a general set of vertices, $\mathcal{E} \subset \{(\mathfrak{v}_1, \mathfrak{v}_2) \in \mathcal{V} \times \mathcal{V} : \mathfrak{v}_1 \ne \mathfrak{v}_2\}$ is a set of directed edges, and $\mathcal{C} : \mathcal{V} \to \Omega \cup \{\vee, \wedge\}$ is a mapping that associates each vertex $\mathfrak{v} \in \mathcal{V}$ to a *computation* given by either applying a t.i. and locally defined operator $\psi \in \Omega$ or one of the two basic operations $\{\vee, \wedge\}$.

Denoting $\mathcal{V} = \{\mathfrak{v}_i, \mathfrak{v}_1, \dots, \mathfrak{v}_n, \mathfrak{v}_o\}$ for a $n \ge 1$, $\mathcal{G}$ is a sequential morphological computational graph (MCG) if

$$\mathcal{E} = \{(\mathfrak{v}_i, \mathfrak{v}_1), (\mathfrak{v}_n, \mathfrak{v}_o)\} \bigcup \{(\mathfrak{v}_j, \mathfrak{v}_{j+1}) : j = 1, \dots, n - 1\}, \tag{3}$$

with $\mathcal{C}(\mathfrak{v}_i) = \mathcal{C}(\mathfrak{v}_o) = \iota$ and $\mathcal{C}(\mathfrak{v}_j) \in \Omega$, $j = 1, \ldots, n$. This computational graph satisfies the axioms of MCG (see [22] for more details).

In a sequential MCG, the computation of a vertex $\mathfrak{v}_j$ in $\mathcal{G}$ receives as input the output of the computation of the previous vertex $\mathfrak{v}_{j-1}$, and the output of its computation will be used as the input of the computation of the vertex $\mathfrak{v}_{j+1}$. We assume there is an input vertex $\mathfrak{v}_i$, and an' output vertex $\mathfrak{v}_o$, that store the input, which is an element $X \in \mathcal{P}(E)$, and output of the computational graph, respectively, by applying the identity operator. Furthermore, each vertex computes an operator in Ω and there are no vertices computing supremum or infimum operations.

Denote by $\psi_{\mathcal{G}}(X)$ the output of vertex $\mathfrak{v}_o$ when the input of vertex $\mathfrak{v}_i$ is $X \in \mathcal{P}(E)$ and let $\psi_{\mathcal{G}} : \mathcal{P}(E) \to \mathcal{P}(E)$ be the set operator generated by MCG $\mathcal{G}$. The operator $\psi_{\mathcal{G}}$ is actually t.i. and locally defined within a window $W_{\mathcal{G}}$ (cf. Proposition 5.1 in [22]). We define the sequential Discrete Morphological Neural Network represented by $\mathcal{G}$ as the translation invariant and locally defined set operator $\psi_{\mathcal{G}}$.

3.2 Sequential Discrete Morphological Neural Networks Architecture

A triple $\mathcal{A} = (\mathcal{V}, \mathcal{E}, \mathcal{F})$, in which $\mathcal{F} \subseteq \Omega^{\mathcal{V}}$, is a sequential Discrete Morphological Neural Network (SDMNN) architecture if $(\mathcal{V}, \mathcal{E}, \mathcal{C})$ is a sequential MCG for all $\mathcal{C} \in \mathcal{F}$. A SDMNN architecture is a collection of sequential MCG with the same graph $(\mathcal{V}, \mathcal{E})$ and computation map $\mathcal{C}$ in $\mathcal{F}$. Since a SDMNN architecture represents a collection of MCG, it actually represents a collection of t.i. and locally defined set operators that can be represented as the composition of W-operators.

For an architecture $\mathcal{A} = (\mathcal{V}, \mathcal{E}, \mathcal{F})$, let $\mathbb{G}(\mathcal{A}) = \{\mathcal{G} = (\mathcal{V}, \mathcal{E}, \mathcal{C}) : \mathcal{C} \in \mathcal{F}\}$ be the collection of MCG generated by $\mathcal{A}$. We say that $\mathcal{G} \in \mathbb{G}(\mathcal{A})$ is a realization of architecture $\mathcal{A}$ and we define $\mathcal{H}(\mathcal{A}) = \{\psi \in \Omega : \psi = \psi_{\mathcal{G}}, \mathcal{G} \in \mathbb{G}(\mathcal{A})\}$ as the collection of t.i. and locally defined set operators that can be realized by $\mathcal{A}$.

A SDMNN is said unrestricted if its interior vertices $\mathfrak{v}_j$ can compute any W-operator locally defined within a W_j, so it holds $\mathcal{F} = \{\iota\} \times \Psi_{W_1} \times \cdots \times \Psi_{W_n} \times \{\iota\}$. We denote by $n = |\mathcal{V}| - 2$ the depth of an unrestricted sequential DMNN (USDMNN) and by $|W_j|$ the width of the layer represented by vertex $\mathfrak{v}_j, j = 1, \ldots, n$.

As an example, consider an USDMNN with three hidden layers. For fixed windows $W_1, W_2, W_3 \in \mathcal{P}(E)$, this sequential architecture realizes the operators in Ψ_W, with $W = W_1 \oplus W_2 \oplus W_3$, in which $\oplus$ stands for the *Minkowski addition*, that can be written as the composition of operators in Ψ_{W_1}, Ψ_{W_2} and Ψ_{W_3}, that is, $\mathcal{H}(\mathcal{A}) = \{\psi \in \Psi_W : \psi = \psi^{W_3}\psi^{W_2}\psi^{W_1}; \psi^{W_i} \in \Psi_{W_i}, i = 1, 2, 3\}$. For a proof of this fact, see [8].

3.3 Representation of USDMNN

The USDMNN realized by MCG $\mathcal{G} = (\mathcal{V}, \mathcal{E}, \mathcal{C})$ can be represented by a sequence $\{\psi_1, \ldots, \psi_n\}$, $n \geq 2$, of W-operators with windows $W_1, \ldots, W_n$ as the composition

$$\psi_{\mathcal{G}} = \psi_n \circ \cdots \circ \psi_1 \qquad (4)$$

in which $\psi_j = \mathcal{C}(\mathfrak{v}_j), j = 1, \ldots, n$. Based on (4), we propose a representation for the class of operators realized by an USDMNN based on the window and characteristic function of the operators ψ_j.

For each $i = 1, \ldots, n$, we fix a $d_i \geq 3$ odd and assume that each window $W_i, i = 1, \ldots, n$, is a connected subset of $F_{d_i} = \{-(d_i - 1)/2, \ldots, (d_i - 1)/2\}^2$, the square of side d_i centered at the origin of E. This means that, for every $w, w' \in W_i$, there exists a sequence $w_0, \ldots, w_r \in W_i, r \geq 1$, such that $w_0 = w, w_r = w'$ and $\|w_i - w_{i+1}\|_\infty = 1$, for all $i = 0, \ldots, r - 1$. Denoting $\mathcal{C}_d = \{W \subseteq F_d : W \text{ is connected}\}$, we assume that $W_i \in \mathcal{C}_{d_i}$ for all $i = 1, \ldots, n$.

For each $W \in \mathcal{C}_{d_i}$, let $\mathcal{B}_W = \{f : \mathcal{P}(W) \mapsto \{0, 1\}\}$ be the set of all binary functions on $\mathcal{P}(W)$, and define $\mathcal{F}_i = \{(W, f) : W \in \mathcal{C}_{d_i}, f \in \mathcal{B}_W\}$ as the collection of W-operators with window W in $\mathcal{C}_{d_i}$ and characteristic function f in $\mathcal{B}_W$, which are completely defined by a pair (W, f). Finally, let $\Theta = \prod_{i=1}^n \mathcal{F}_i$ be the Cartesian product of $\mathcal{F}_i$. Observe that an element θ in Θ is actually a sequence of n W-operators with windows in $\mathcal{C}_{d_i}, i = 1, \ldots, n$, which we denote by $\theta = \{(W_1, f_1), \ldots, (W_n, f_n)\}$. Denoting the W-operator represented by (W_i, f_i) as ψ_i, a $\theta \in \Theta$ generates a USDMNN ψ_θ via expression (4).

The USDMNN architecture $\mathcal{A} = (\mathcal{V}, \mathcal{E}, \mathcal{F})$ with n hidden layers and $\mathcal{F} = \{\iota\} \times \Psi_{F_{d_1}} \times \cdots \times \Psi_{F_{d_n}} \times \{\iota\}$ is such that $\mathcal{H}(\mathcal{A}) = \{\psi_\theta : \theta \in \Theta\}$, so Θ is a representation for the class of operators generated by the architecture $\mathcal{A}$.

Since an operator locally defined in W is also locally defined in any $W' \supset W$ (cf. Proposition 5.1 in [8]) it follows that Θ is actually an overparametrization of $\mathcal{H}(\mathcal{A})$ since there are many representations (W', f_ψ) for a same W-operator locally defined in $W \subsetneq F_{d_i}$. This overparametrization, discussed more generally in [23], also happens in the representation of CDMNN proposed in [22] and we take advantage of it to propose a SLDA to train USDMNN and mitigate the risk of overfitting.

4   Training USDMNN via the Stochastic Lattice Descent Algorithm

The training of a USDMNN is performed by obtaining a sequence $\hat{\theta} \in \Theta$ of W-operators via the minimization of an empirical error on a sample $\{(X_1, Y_1), \ldots, (X_N, Y_N)\}$, of N input images X and output target transformed images Y. In order to mitigate overfitting the sample, the training of a USDMNN will be performed by a two-step algorithm that, for a fixed sequence of windows, learns a sequence of characteristic functions by minimizing the empirical error L_t in a training sample, and then learns a sequence of windows by minimizing an empirical error L_v in a validation sample over the sequences of windows. More details about the empirical errors L_t and L_v will be given in Sect. 5.

These two steps are instances of an algorithm for the minimization of a function in a subset of a Boolean lattice. On the one hand, the set $\mathcal{C} := \prod_{i=1}^n \mathcal{C}_{d_i}$ of all sequences of n connected windows is a subset of a Boolean lattice isomorphic to $\prod_{i=1}^n \{0, 1\}^{F_{d_i}}$, so minimizing the validation error over the windows means minimizing a function in a subset of a Boolean lattice. On the other hand, for a fixed sequence of windows $(W_1, \ldots, W_n) \in \mathcal{C}$, the set $\mathcal{B}_{W_1} \times \cdots \times \mathcal{B}_{W_n}$ of all sequences of characteristic functions with these windows is a Boolean lattice isomorphic to $\{0,1\}^{\mathcal{P}(W_1)} \times \cdots \times \{0,1\}^{\mathcal{P}(W_n)}$, so minimizing the training error in this space means minimizing a function in a Boolean lattice.

The U-curve algorithms [6,15,29,30] were proposed to minimize a U-curve function on a Boolean lattice. In summary, these algorithms perform a greedy search of a Boolean lattice, at each step jumping to the point at distance one with the least value of the function, and stopping when all neighbor points have a function value greater or equal to that of the current point. This greedy search of a lattice, that at each step goes to the direction that minimizes the function, is analogous to the dynamic of the gradient descent algorithm to minimize a function with domain in $\mathbb{R}^p$. Inspired by the U-curve algorithms and by the success of stochastic gradient descent algorithms for minimizing overparametrized functions in $\mathbb{R}^p$, and following the ideas of [22], we propose a stochastic lattice descent algorithm (SLDA) to train USDMNN. In Fig. 1, we present the main ideas of the algorithm, and in Sects. 4.1 and 4.2 we formally define it.

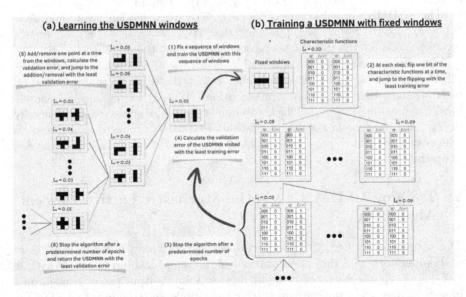

Fig. 1. Illustration of the deterministic version of the SLDA to (a) learn the USDMNN windows and (b) train a USDMNN with fixed windows. To simplify the illustration, we considered only the possibility of adding a point to a window, and flipping a bit from 0 to 1 of a characteristic function, at each step, even though a point can be erased from a window, and a flipping from 1 to 0 may happen, if they have the least respective error. The training error L_t or validation error L_v of each point is on top of it.

4.1   Training a USDMNN with Fixed Windows

For each $\boldsymbol{W} := \{W_1, \ldots, W_n\} \in \mathcal{C}$ let $\Theta_{\boldsymbol{W}} := \{\{(W_1, f_1), \ldots, (W_n, f_n)\} : f_i \in \mathcal{F}_{W_i}, i = 1, \ldots, n\}$ be all sequences of W-operators with windows $\boldsymbol{W}$. Observe there is a bijection between $\Theta_{\boldsymbol{W}}$ and the Boolean lattice $\prod_{i=1}^{n}\{0,1\}^{\mathcal{P}(W_i)}$, and consider the Boolean lattice $(\Theta_{\boldsymbol{W}}, \leq)$.

Denoting by d the distance in the acyclic directed graph $(\Theta_{\boldsymbol{W}}, \leq)$, we define the neighborhood of a $\theta \in \Theta_{\boldsymbol{W}}$ as $N(\theta) = \{\theta' \in \Theta_{\boldsymbol{W}} : d(\theta, \theta') = 1\}$. Observe that $\theta, \theta' \in \Theta_{\boldsymbol{W}}$ are such that $d(\theta, \theta') = 1$ if, and only if, all their characteristic functions but one are equal, and in the one in which they differ, the difference is in only one point of their domain. In other words, θ' is obtained from θ by flipping the image of one point of one characteristic function from 0 to 1 or from 1 to 0.

The SLDA for learning the characteristic functions of a USDMNN with fixed windows is formalized in Algorithm 1. The initial point $\theta \in \Theta_{\boldsymbol{W}}$, a batch size b, the number n of neighbors to be sampled at each step, and the number of training epochs are fixed. For each epoch, the training sample is randomly partitioned in N/b batches, and we denote the training error on batch j by $L_t^{(j)}$. For each batch, n neighbors of θ are sampled and θ is updated to the sampled neighbor with the least training error $L_t^{(j)}$, that is calculated on the sample batch j. Observe that θ is updated at each batch, so during an epoch, it is updated N/b times. At the end of each epoch, the training error $L_t(\theta)$ of θ on the whole training sample is compared with the error of the point with the least training error visited so far at the end of an epoch, and it is stored as this point if its training error is lesser. After the predetermined number of epochs, the algorithm returns the point with the least training error on the whole sample visited at the end of an epoch.

Observe that Algorithm 1 has two sources of stochasticity: the sampling of neighbors and the sample batches. If $n = n(\theta) := |N(\theta)|$ and $b = N$, then this algorithm reduces to the deterministic one illustrated in Fig. 1. Furthermore, the complexity of the algorithm is controlled by the number n of sampled neighbors, the batch size b and the number of epochs. See [22] for a further discussion about a SLDA.

4.2   Learning the Windows of USDMNN Windows

In order to learn the windows of a USDMNN, we apply the SLDA on $\mathcal{C}$, that is a subset of the Boolean lattice $\prod_{i=1}^{n}\{0,1\}^{F_{d_i}}$. For each $\boldsymbol{W} \in \mathcal{C}$, let $L_v(\boldsymbol{W}) := L_v(\theta_{\boldsymbol{W}}^{\mathbb{A}})$ be the validation error of the USDMNN realized by $\theta_{\boldsymbol{W}}^{\mathbb{A}}$, which was learned by Algorithm 1.

We define the neighborhood of $\boldsymbol{W}$ in $\mathcal{C}$ as $N(\boldsymbol{W}) = \{\boldsymbol{W}' \in \mathcal{C}, d(\boldsymbol{W}, \boldsymbol{W}') = 1\}$ in which d means the distance in the acyclic directed graph $(\mathcal{C}, \leq)$. Observe that $\boldsymbol{W}, \boldsymbol{W}' \in \mathcal{C}$ are such that $d(\boldsymbol{W}, \boldsymbol{W}') = 1$ if, and only if, all their windows but one are equal, and in the one in which they differ, the difference is of one point. In other words, $\boldsymbol{W}'$ is obtained from $\boldsymbol{W}$ by adding or removing one point from one of its windows.

Algorithm 1. Stochastic lattice descent algorithm for learning the characteristic functions of a USDMNN with fixed windows $\boldsymbol{W} = \{W_1, \ldots, W_n\}$.

Ensure: $\theta \in \Theta_{\boldsymbol{W}}, n, b, Epochs$
1: $L_{min} \leftarrow L_t(\theta)$
2: $\theta_{\boldsymbol{W}}^{\mathbb{A}} \leftarrow \theta$
3: **for** run $\in \{1, \ldots, \text{Epochs}\}$ **do**
4:    ShuffleBatches(b)
5:    **for** $j \in \{1, \ldots, N/b\}$ **do**
6:       $\tilde{N}(\theta) \leftarrow$ SampleNeighbors(θ, n)
7:       $\theta \leftarrow \theta'$ s.t. $\theta' \in \tilde{N}(\theta)$ and $L_t^{(j)}(\theta') = \min\{L_t^{(j)}(\theta'') : \theta'' \in \tilde{N}(\theta)\}$
8:    **if** $L_t(\theta) < L_{min}$ **then**
9:       $L_{min} \leftarrow L_t(\theta)$
10:       $\theta_{\boldsymbol{W}}^{\mathbb{A}} \leftarrow \theta$
11: **return** $\theta_{\boldsymbol{W}}^{\mathbb{A}}$

The SLDA for learning the windows of a USDMNN is formalized in Algorithm 2. The initial point $\boldsymbol{W} \in \mathcal{C}$, a batch size b, the number n of neighbors to be sampled at each step, and the number of training epochs are fixed. For each epoch, the validation sample is randomly partitioned in N/b batches, and we denote the validation error of any $\boldsymbol{W}$ on batch j by $L_v^{(j)}(\boldsymbol{W}) := L_v^{(j)}(\theta_{\boldsymbol{W}}^{\mathbb{A}})$ which is the empirical error on the j-th batch of the validation sample of the USDMNN realized by $\theta_{\boldsymbol{W}}^{\mathbb{A}}$, learned by Algorithm 1.

For each batch, n neighbors of $\boldsymbol{W}$ are sampled and $\boldsymbol{W}$ is updated to the sampled neighbor with the least validation error $L_v^{(j)}$. At the end of each epoch, the validation error $L_v(\boldsymbol{W})$ of $\boldsymbol{W}$ on the whole validation sample is compared with the error of the point with the least validation error visited so far at the end of an epoch, and it is stored as this point if its validation error is lesser. After the predetermined number of epochs, the algorithm returns the point with the least validation error on the whole sample visited at the end of an epoch.

Algorithms 1 and 2 are analogous and differ only on the lattice ($\Theta_{\boldsymbol{W}}$ and $\mathcal{C}$) they search and the function they seek to minimize (the training and the validation error).

5   Application: Boundary Recognition of Digits with Noise

As an example, we treat the problem of boundary recognition of digits with noise. We consider the USDMNN with two layers, and windows contained in the 3×3 square trained with a training sample of 10, and a validation sample of 10, 56×56 binary images. The training and validation samples are the same used in [22] and the initial windows were considered as the five point cross centered at the origin. The initial characteristic functions for the first sequence of windows were randomly chosen, while the initial characteristic functions of

---

Algorithm 2. Stochastic lattice descent algorithm for learning the windows of a USDMNN.

Ensure: $\boldsymbol{W} \in \mathcal{C}, n, b, Epochs$

1: $L_{min} \leftarrow L_v(\boldsymbol{W})$
2: $\boldsymbol{W}^{\mathbb{A}} \leftarrow \boldsymbol{W}$
3: **for** run $\in \{1, \ldots, Epochs\}$ **do**
4:    ShuffleBatches(b)
5:    **for** $j \in \{1, \ldots, N/b\}$ **do**
6:       $\tilde{N}(\boldsymbol{W}) \leftarrow$ SampleNeighbors($\boldsymbol{W}, n$)
7:       $\boldsymbol{W} \leftarrow \boldsymbol{W}'$ s.t. $\boldsymbol{W}' \in \tilde{N}(\boldsymbol{W})$ and $L_v^{(j)}(\boldsymbol{W}') = \min\{L_v^{(j)}(\boldsymbol{W}'') : \boldsymbol{W}'' \in \tilde{N}(\boldsymbol{W})\}$
8:       **if** $L_v(\boldsymbol{W}) < L_{min}$ **then**
9:          $L_{min} \leftarrow L_v(\boldsymbol{W})$
10:         $\boldsymbol{W}^{\mathbb{A}} \leftarrow \boldsymbol{W}$
11: **return** $\boldsymbol{W}^{\mathbb{A}}$

---

a later sequence of windows were those that minimized the training error of its neighbor visited before, except for the window on which they differ, where the characteristic function is initiated randomly. The algorithms were implemented in **python** and are available at https://github.com/MarianaFeldman/USDMM. The training was performed on a personal computer with processor Intel Core i7-1355U x 12 and 16 GB of RAM.

We consider the intersection of union (IoU) error that is more suitable for form or object detection tasks. The IoU error of ψ_θ in the training sample is

$$L_t(\theta) = 1 - \frac{1}{10} \sum_{k=1}^{10} \frac{|Y_k \cap \psi_\theta(X_k)|}{|Y_k \cup \psi_\theta(X_k)|} \tag{5}$$

that is, the mean proportion of pixels in $Y \cup \psi_\theta(X)$ that are not in $Y \cap \psi_\theta(X)$ among the sample points in the training sample. The respective error on the validation sample is denoted by $L_v(\theta)$.

The results are presented in Table 1 and Fig. 2. We trained the two layer USDMNN with batch sizes 1, 5 and 10, sampling 8 neighbors in the SLDA for the characteristic functions and considering all neighbors in the SLDA for the windows. These batch sizes refer to the SLDA for the characteristic functions, and in all cases we considered a batch size of 10 in the SLDA for the windows. We considered 50 epochs to train the windows and 100 epochs to train the characteristic functions. Each scenario was trained ten times, starting from distinct initial values of the characteristic functions. We present the minimum, mean and standard deviation of the results over the ten repetitions in Table 1.

The best results were obtained with batch size 10, and it seems that, with the initial windows and characteristic functions we are considering, it is not possible to properly train with batch sizes 1 and 5. The training and validation error of the USDMNN trained with batch size 10 did not vary a lot over the repetitions, so the algorithm is not sensible to the initial value for batch size

10. Moreover, with batch size 10, the SLDA for the characteristic functions took in average around 80 epochs to reach the minimum in all repetitions, and the SLDA for the windows took an average of 26 epochs to the minimum over the ten repetitions, although the variation was great and there was a repetition in which the minimum was achieved after only five epochs.

The time it took to train the USDMNN was greater than that it took to train CDMNN in [22], but it can be decreased with a more sophisticated implementation of the algorithms. The minimum training and validation error of the trained operator attained with USDMNN (0.032 and 0.052) was slightly greater than the respective minimum error obtained in [22] (around 0.029 and 0.042). Therefore, the USDMNN have obtained similar empirical results as the CDMNN in [22] without strong prior information.

Table 1. Results of the two layer USDMNN trained with batch sizes $b = 1$, $b = 5$, and $b = 10$ for the characteristic function SLDA. The results are in the form Minimum - Average (Standard Deviation) over the ten repetitions. We present the minimum training error observed during training; the training and validation error of the trained USDMNN; the algorithm total time and time until the minimum validation error; the number of epochs to the minimum validation error; and the average number of epochs until the minimum training error with fixed windows.

b	Min. Train	Learned Train	Min Val	Total time (h)	Time to min. (h)	Epo. to min. (W)	Mean Epo. to min (f)
1	0.442–0.57 (0.07)	0.597–0.677 (0.053)	0.183–0.238 (0.031)	113.4–135.6 (16.1)	11.6–45.7 (37.2)	4–16.8 (15.4)	47.8–49.0 (0.8)
5	0.421–0.49 (0.04)	0.492–0.611 (0.076)	0.127–0.155 (0.015)	78.5–91.9 (20.3)	13.648–32.478 (19.0)	4–18.7 (11.9)	51.4–52.9 (1.3)
10	0.031–0.036 (0.004)	0.032–0.039 (0.006)	0.052–0.056 (0.003)	173.6–182.6 (13.3)	14.262–98.324 (70.9)	5–26.2 (17.7)	80.9–82.6 (1.3)

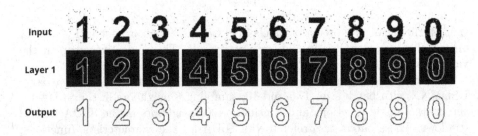

Fig. 2. Result obtained after applying, to the validation sample, each layer of the USDMNN trained with a batch size of 10 that had the least validation error.

6   Next Steps and Future Research

In this paper, we proposed a hierarchical SLDA to train USDMNN and illustrated it in a simple example. We obtained empirical results almost as good as that of [22], but without the necessity to design a DMNN based on prior information. The next step of this work is to develop an efficient implementation of the algorithms and perform a more extensive experimental study to better understand features of them. In particular, it is necessary to better understand the sensitivity to the initial values and the role of the batch size on both the SLDA for the windows and for the characteristic functions. A more thorough comparison with the CDMNN proposed in [22] is also necessary.

We are currently working on an efficient implementation of the hierarchical SLDA that, we believe, will significantly decrease the training time, so methods based on USDMNN may also be competitive with CDMNN from a computational complexity perspective. With a more efficient algorithm, it will be possible to fine train USDMNN and better study the effect of settings such as the number of layers and the window size on the performance. An efficient method to train DMNN without the need of strong prior information may help popularize methods based on them among practitioners which do not have strong knowledge in mathematical morphology to design specific CDMNN architectures.

A promising line of research is to compare the USDMNN and CDMNN with other methods proposed in the literature. Since the DMNN are exact representations of morphological operators, an interesting study would be to compare them with other exact methods, rather than with approximations such as the classical MNN. In particular, it would be interesting to compare the DMNN with the binary morphological neural networks proposed in [1,2]. We leave such a comparison as a topic for future research.

References

1. Aouad, T., Talbot, H.: Binary morphological neural network. In: 2022 IEEE International Conference on Image Processing (ICIP), pp. 3276–3280. IEEE (2022)
2. Aouad, T., Talbot, H.: A foundation for exact binarized morphological neural networks. In: ICCV 2023-International Conference on Computer Vision (2023)
3. Araújo, R.D.A., Madeiro, F., de Sousa, R.P., Pessoa, L.F.: Modular morphological neural network training via adaptive genetic algorithm for designing translation invariant operators. In: 2006 IEEE International Conference on Acoustics Speech and Signal Processing Proceedings, vol. 2, pp. II-II. IEEE (2006)
4. Araújo, R.D.A., Oliveira, A.L., Meira, S.: A morphological neural network for binary classification problems. Eng. Appl. Artif. Intell. **65**, 12–28 (2017)
5. Arce, F., Zamora, E., Sossa, H., Barrón, R.: Differential evolution training algorithm for dendrite morphological neural networks. Appl. Soft Comput. **68**, 303–313 (2018)
6. Atashpaz-Gargari, E., Reis, M.S., Braga-Neto, U.M., Barrera, J., Dougherty, E.R.: A fast branch-and-bound algorithm for u-curve feature selection. Pattern Recogn. **73**, 172–188 (2018)
7. Barrera, J., Hashimoto, R.F., Hirata, N.S., Hirata, R., Jr., Reis, M.S.: From mathematical morphology to machine learning of image operators. São Paulo J. Math. Sci. **16**(1), 616–657 (2022)
8. Barrera, J., Salas, G.P.: Set operations on closed intervals and their applications to the automatic programming of morphological machines. J. Electron. Imaging **5**(3), 335–352 (1996)

9. Barrera, J., Terada, R., Hirata, R., Jr., Hirata, N.S.: Automatic programming of morphological machines by PAC learning. Fund. Inform. **41**(1–2), 229–258 (2000)
10. Davidson, J.L.: Simulated annealing and morphology neural networks. In: Image Algebra and Morphological Image Processing III, vol. 1769, pp. 119–127. SPIE (1992)
11. Davidson, J.L., Hummer, F.: Morphology neural networks: an introduction with applications. Circ. Syst. Sig. Process. **12**(2), 177–210 (1993)
12. Davidson, J.L., Ritter, G.X.: Theory of morphological neural networks. In: Digital Optical Computing II, vol. 1215, pp. 378–388. SPIE (1990)
13. Dimitriadis, N., Maragos, P.: Advances in morphological neural networks: training, pruning and enforcing shape constraints. In: ICASSP 2021–2021 IEEE International Conference on Acoustics, Speech and Signal Processing (ICASSP), pp. 3825–3829. IEEE (2021)
14. Dougherty, E.R., Lotufo, R.A.: Hands-on Morphological Image Processing, vol. 59. SPIE press, Bellingham (2003)
15. Estrela, G., Gubitoso, M.D., Ferreira, C.E., Barrera, J., Reis, M.S.: An efficient, parallelized algorithm for optimal conditional entropy-based feature selection. Entropy **22**(4), 492 (2020)
16. Franchi, G., Fehri, A., Yao, A.: Deep morphological networks. Pattern Recogn. **102**, 107246 (2020)
17. Grana, M., Raducanu, B.: Some applications of morphological neural networks. In: IJCNN 2001. International Joint Conference on Neural Networks. Proceedings (Cat. No. 01CH37222), vol. 4, pp. 2518–2523. IEEE (2001)
18. Groenendijk, R., Dorst, L., Gevers, T.: Morphpool: efficient non-linear pooling & unpooling in CNNs. arXiv preprint arXiv:2211.14037 (2022)
19. Hirata, N.S., Papakostas, G.A.: On machine-learning morphological image operators. Mathematics **9**(16), 1854 (2021)
20. Hu, Y., Belkhir, N., Angulo, J., Yao, A., Franchi, G.: Learning deep morphological networks with neural architecture search. Pattern Recogn. **131**, 108893 (2022)
21. Julca-Aguilar, F.D., Hirata, N.S.: Image operator learning coupled with CNN classification and its application to staff line removal. In: 2017 14th IAPR International Conference on Document Analysis and Recognition (ICDAR), vol. 1, pp. 53–58. IEEE (2017)
22. Marcondes, D., Barrera, J.: Discrete morphological neural networks. arXiv preprint arXiv:2309.00588 (2023)
23. Marcondes, D., Barrera, J.: The lattice overparametrization paradigm for the machine learning of lattice operators. arXiv preprint arXiv:2310.06639 (2023)
24. Matheron, G.: Random Sets and Integral Geometry. Wiley, Hoboken (1974)
25. Mondal, R., Purkait, P., Santra, S., Chanda, B.: Morphological networks for image de-raining. In: Couprie, M., Cousty, J., Kenmochi, Y., Mustafa, N. (eds.) DGCI 2019. LNCS, vol. 11414, pp. 262–275. Springer, Cham (2019). https://doi.org/10.1007/978-3-030-14085-4_21
26. Mondal, R., Santra, S., Mukherjee, S.S., Chanda, B.: Morphological network: how far can we go with morphological neurons? arXiv preprint arXiv:1901.00109 (2019)
27. Pessoa, L.F., Maragos, P.: Morphological/rank neural networks and their adaptive optimal design for image processing. In: 1996 IEEE International Conference on Acoustics, Speech, and Signal Processing Conference Proceedings, vol. 6, pp. 3398–3401. IEEE (1996)
28. Pessoa, L.F., Maragos, P.: Neural networks with hybrid morphological/rank/linear nodes: a unifying framework with applications to handwritten character recognition. Pattern Recogn. **33**(6), 945–960 (2000)

29. Reis, M.S., Estrela, G., Ferreira, C.E., Barrera, J.: Optimal Boolean lattice-based algorithms for the u-curve optimization problem. Inf. Sci. **471**, 97–114 (2018)
30. Ris, M., Barrera, J., Martins, D.C.: U-curve: a branch-and-bound optimization algorithm for u-shaped cost functions on Boolean lattices applied to the feature selection problem. Pattern Recogn. **43**(3), 557–568 (2010)
31. Ritter, G.X., Iancu, L., Urcid, G.: Morphological perceptrons with dendritic structure. In: The 12th IEEE International Conference on Fuzzy Systems, 2003. FUZZ 2003, vol. 2, pp. 1296–1301. IEEE (2003)
32. Ritter, G.X., Sussner, P.: An introduction to morphological neural networks. In: Proceedings of 13th International Conference on Pattern Recognition, vol. 4, pp. 709–717. IEEE (1996)
33. Serra, J.: Image Analysis and Mathematical Morphology. Academic Press, London (1982)
34. Serra, J.: Image Analysis and Mathematical Morphology. Theoretical Advances, vol. 2. Academic Press, London (1988)
35. Monteiro da Silva, A., Sussner, P.: A brief review and comparison of feedforward morphological neural networks with applications to classification. In: Kůrková, V., Neruda, R., Koutník, J. (eds.) Artificial Neural Networks-ICANN 2008, vol. 5164, pp. 783–792. Springer, Heidelberg (2008). https://doi.org/10.1007/978-3-540-87559-8_81
36. Sossa, H., Guevara, E.: Efficient training for dendrite morphological neural networks. Neurocomputing **131**, 132–142 (2014)
37. de Sousa, R.P., de Carvalho, J.M., de Assis, F.M., Pessoa, L.F.: Designing translation invariant operations via neural network training. In: Proceedings 2000 International Conference on Image Processing (Cat. No. 00CH37101), vol. 1, pp. 908–911. IEEE (2000)
38. Sussner, P., Esmi, E.L.: Constructive morphological neural networks: some theoretical aspects and experimental results in classification. Constr. Neural Netw. 123–144 (2009)

Discovering Repeated Patterns from the Onsets in a Multidimensional Representation of Music

Paul Lascabettes[1]([✉])[iD] and Isabelle Bloch[2][iD]

[1] STMS - Sorbonne Université, Ircam, CNRS, Ministère de la Culture, Paris, France
`paul.lascabettes@ircam.fr`
[2] Sorbonne Université, CNRS, LIP6, Paris, France
`isabelle.bloch@sorbonne-universite.fr`

Abstract. This article deals with the discovery of repeated patterns in a multidimensional representation of music using the theory of mathematical morphology. The main idea proposed here is to use the onsets to discover musical patterns. By definition, the morphological erosion of musical data by a musical pattern corresponds to its onsets. However, the erosion of musical data by the onsets is not always equal to the musical pattern. We propose a theorem which guarantees the equality if the musical pattern satisfies a topological condition. This condition is met when the patterns do not intersect, or only slightly, which is coherent in a musical context. Due to the importance of repetition in music, this idea proves to be relevant for the musical pattern discovery task.

Keywords: Pattern discovery · Mathematical morphology · Point-set algorithms · Geometric pattern discovery in music · Music analysis

1 Introduction

While mathematical morphology has been widely applied to image processing, analysis and understanding, there are only few direct applications of this theory, in particular in its algebraic setting, to symbolic representations of music. This is partly due to the different nature of the patterns. On the one hand, the objects associated with digital image processing are defined on a discrete grid, endowed with a discrete connectivity, and are often connected sets. On the other hand, those associated with symbolic music are sparse, in the sense that they are often not connected, according to the underlying connectivity of the space of representation. It is therefore necessary to adapt the morphological tools and the expected results when applying this theory to symbolic representations of music. This has been started by adapting the basic operators of mathematical morphology to find a musical meaning, for example to extract harmonic components or to obtain musical transformations [6,8]. Among the other existing

This work was partly supported by the chair of I. Bloch in Artificial Intelligence (Sorbonne Université and SCAI).

applications, Karvonen et al. have developed automatic methods to discover approximate occurrences of a given pattern in symbolic musical databases [4,5]. We advocate in this article that mathematical morphology can provide relevant tools for the discovery of repeated patterns in a multidimensional representation of music, i.e. in a discrete set of points. Even if we focus on symbolic representations of music, the developed results can be applied to other types of discrete data composed of repeated patterns.

There are many algorithms for discovering repeated patterns from a multidimensional representation of music. Most of these algorithms are based on the SIA algorithm [12,13], which stands for "Structure Induction Algorithm". This algorithm, developed by Meredith et al., consists in discovering the largest translatable patterns in a multidimensional dataset. Among other fundamental algorithms, the SIATEC algorithm [10,12,13] reveals the repetitions of the largest translatable patterns discovered by SIA. However, these algorithms discover too many patterns, some of which are not musically relevant. Therefore, with the aim of improving the precision and efficiency of SIA, Collins et al. proposed the SIAR algorithm [1], which corresponds approximately to using a sliding window of size r in SIA in order to avoid discovering patterns that are too long, and the SIACT algorithm [1], to detect if a sub-pattern of the largest translatable pattern is musically more important. Finally, the method we propose here belongs to the category of algorithms that discover patterns in order to describe musical data, such as COSIATEC [12] (cover the dataset without overlaps), SIATECCompress [11] (cover the dataset with overlaps) or Forth's algorithm [2] (cover not the entire dataset and with overlaps).

The originality of our approach comes from the use of a musical meaning to discover patterns by distinguishing the role of onsets or musical pattern. In particular, we provide various mathematical results related to musical problems. These results optimize the discovery of repeated patterns, and are the foundation for a new approach to discover musical patterns.

This article is organized as follows. Section 2 summarizes the principal operators of mathematical morphology and their properties used in this article. Section 3 presents the problem of discovering the musical patterns from their onsets. Section 4 provides a solution to this problem if the musical patterns satisfy a specific topological condition. Section 5 interprets this result from a musical point of view. Section 6 demonstrates how to use this result to optimize the discovery of musical patterns. Finally, Sect. 7 concludes this article and proposes some future work.

2 Notations and Background: Binary Mathematical Morphology

In this section, we recall the concepts of mathematical morphology used in this article, for it to be self-contained. In particular, we consider here the simple case of binary mathematical morphology, where the set E is equal to $\mathbb{R}^n$. In the remainder of this article, $\mathcal{P}(E)$ is the power set of E. First, let $S \in \mathcal{P}(E)$ and

$t \in E$, we denote the *translate* of S by t as $S_t = \{s + t \mid s \in S\}$. The majority of morphological operators, in the deterministic setting, result from two basic ones: the *dilation* resulting from the *Minkowski addition* $\oplus$ [14] and the *erosion* from the *substraction* $\ominus$ [3].

Definition 1 (Dilation and Erosion). *Let* $S \in \mathcal{P}(E)$, *the dilation* δ_S *and erosion* ε_S *by* S *are defined by:*

$$
\begin{aligned}
\delta_S : \mathcal{P}(E) &\longrightarrow \mathcal{P}(E) \\
X &\longmapsto X \oplus S = \{x + s \mid x \in X, s \in S\}
\end{aligned}
\tag{1}
$$

$$
\begin{aligned}
\varepsilon_S : \mathcal{P}(E) &\longrightarrow \mathcal{P}(E) \\
X &\longmapsto X \ominus S = \{x \in E \mid S_x \subseteq X\}
\end{aligned}
\tag{2}
$$

In this case, S is called a *structuring element*. By composing these two operations, we obtain the two other fundamental operators of mathematical morphology: *opening* and *closing*.

Definition 2 (Opening and Closing). *Given* $S \in \mathcal{P}(E)$, *the opening* γ_S *and the closing* φ_S *are defined by:*

$$
\begin{aligned}
\gamma_S : \mathcal{P}(E) &\longrightarrow \mathcal{P}(E) \\
X &\longmapsto \delta_S(\varepsilon_S(X)) = \bigcup\{S_x \mid x \in E \land S_x \subseteq X\}
\end{aligned}
\tag{3}
$$

$$
\begin{aligned}
\varphi_S : \mathcal{P}(E) &\longrightarrow \mathcal{P}(E) \\
X &\longmapsto \varepsilon_S(\delta_S(X))
\end{aligned}
\tag{4}
$$

These four operations satisfy many different properties. In particular, they are all *increasing* (i.e. $X \subseteq X' \Rightarrow f_S(X) \subseteq f_S(X')$, where f_S denotes one of the four operations). Also, the dilation and the closing are *increasing according to the structuring element* (i.e. $S \subseteq S' \Rightarrow \forall X, \delta_S(X) \subseteq \delta_{S'}(X)$ and $\varphi_S(X) \subseteq \varphi_{S'}(X)$). While the erosion and the opening are *decreasing according to the structuring element* (i.e. $S \subseteq S' \Rightarrow \forall X, \varepsilon_{S'}(X) \subseteq \varepsilon_S(X)$ and $\gamma_{S'}(X) \subseteq \gamma_S(X)$). Moreover, the position of the structuring element S with respect to the origin O_E of the space E has an impact on the result of the dilation or erosion. In particular, if the origin belong to S, the dilation is *extensive* (i.e. $X \subseteq \delta_S(X)$) and the erosion is *anti-extensive* (i.e. $\varepsilon_S(X) \subseteq X$), and conversely, these properties hold only if the origin belong to S. However, the position of S in relation to the origin does not change the result of the opening and the closing, only the shape is involved. In any case, the opening is anti-extensive (i.e. $\gamma_S(X) \subseteq X$) and the closing is extensive (i.e. $X \subseteq \varphi_S(X)$). In addition, it is easy to check that the dilation is *commutative* ($\delta_S(S') = \delta_{S'}(S)$). Finally, if we consider that the origin is included in S and that X is a piece of music, among the four basic operations of mathematical morphology, erosion and opening are *analysis operators* (extracting musical data), while dilation and closing are *generation operators* (enriching the music).

3   Presentation of the Problem of Discovering the Musical Pattern from its Onsets

We present here the problem of *discovering the musical pattern from its onsets*. We define the onsets of a pattern as the origins of its occurrences in musical data. This makes the connection with the morphological erosion.

Definition 3 (Onsets). *Let $P, X \in \mathcal{P}(E)$. The onsets O of P in X are defined by:*

$$O = \varepsilon_P(X) \tag{5}$$

The onsets of a pattern can be interpreted as the beginnings of this pattern in the musical data if the origin of the pattern is placed on its first note, assuming that the dimensions of space are oriented. For example, in Fig. 1(a), musical data are represented by dots, which can be seen as a pattern repeated four times. In this two-dimensional representation, which is used in the rest of this article, the vertical axis refers to note pitch, while the horizontal axis indicates time. Repetitions of the pattern are indicated in the figure by dotted lines. The onsets of this pattern are therefore composed of four dots, which correspond to the beginnings of the repeated pattern. The interesting result is illustrated in Fig. 1(b), by now considering the onsets as a pattern: the onsets of the onsets of the pattern are equal to the pattern. In other words, by computing the points where the onsets occur, we discover the pattern (the repetitions of the onsets are indicated by the dotted lines in the figure). This result is quite surprising, and we can formalize it with morphological operators as follows.

Definition 4 (Discovering the Musical Pattern From Its Onsets). *Let $P, O, X \in \mathcal{P}(E)$, where $O = \varepsilon_P(X)$. The problem of discovering the musical pattern from its onsets is to understand when the following equation is true:*

$$P = \varepsilon_O(X) \tag{6}$$

Let $P, O, X \in \mathcal{P}(E)$ such that $O = \varepsilon_P(X)$. This can be interpreted as X the musical data, P a musical pattern and O its onsets. First of all, the problem of discovering the musical pattern from its onsets, i.e. the equality $P = \varepsilon_O(X)$, is not always true as shown in Fig. 2. However, Lemma 1 allows us to state that the inclusion $P \subseteq \varepsilon_O(X)$ is always true. Therefore, P is always included in the onsets of O. This implies that taking the onsets of the onsets of a pattern enlarges it, as can be seen in Fig. 2 where P, represented in Fig. 2(b), is included in $\varepsilon_O(X)$, represented in Fig. 2(d). Note that the figures are 2D examples, for the sake of clarity, but all theoretical results apply for any finite dimension that can represent note duration, velocity or voice in a musical context.

Lemma 1. *Let $P, O, X \in \mathcal{P}(E)$, such that $O = \varepsilon_P(X)$ and $O \neq \emptyset$. We have:*

$$P \subseteq \varepsilon_O(X) \tag{7}$$

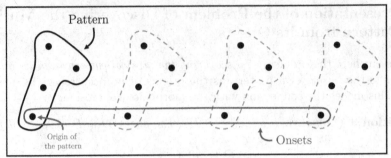

(a) Representation of the pattern and its onsets, which are interpreted as the points where the pattern begins. Pattern repetitions are indicated by the dotted-line border. The origin of the pattern is chosen at the starting point of the pattern (assuming that the time axis is from left to right).

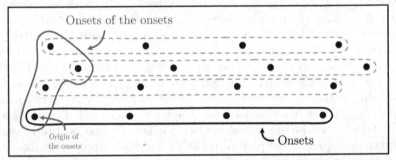

(b) The onsets of the onsets are equal to the pattern. Onsets repetitions are indicated by the dotted-line border. That is, the dots corresponding to where the onsets begin are the same as the pattern dots. As before, the origin of the onsets is chosen at its starting point.

Fig. 1. Presentation of the problem of discovering the pattern from its onsets. In this example, the onsets of the onsets of the pattern are equal to the pattern, which shows that it is sufficient to know the onsets to discover the pattern.

Proof (Lemma 1). Let $p \in P$ and $o \in O$. Because $O = \varepsilon_P(X)$ and $O \neq \emptyset$, $\exists x \in X : p + o = x$. This is true for all $o \in O$, consequently $O_p \subseteq X$, which leads to $p \in \varepsilon_O(X)$. $\qquad\qquad\square$

In the remainder of this article, we prove that the problem of discovering the musical pattern from its onsets can be solved under some assumptions that are musically interpretable.

4   Main Results

Because the equality $P = \varepsilon_O(X)$ is not always true, we need to add an additional condition on P and X to ensure this equality. First, we consider the particular case where the data X is composed of patterns P which are repeated with

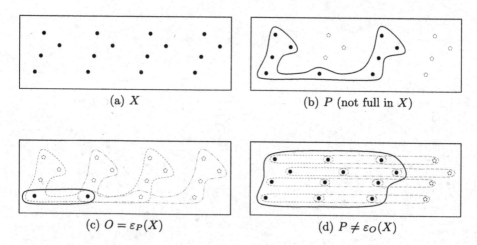

(a) X

(b) P (not full in X)

(c) $O = \varepsilon_P(X)$

(d) $P \neq \varepsilon_O(X)$

Fig. 2. Example where P is not full in X (Definition 5) and $P \neq \varepsilon_O(X)$, however the inclusion $P \subseteq \varepsilon_O(X)$ is satisfied (the origin is the first point on the left).

translations, that is to say $\gamma_P(X) = X$. Moreover, we introduce a new definition, illustrated in Fig. 3, that is, to some extent, the counterpart of the connectivity (without holes) used in image processing. For $P \in \mathcal{P}(E)$, let us note $\mathrm{CH}(P)$ the *convex hull* of P defined by $\mathrm{CH}(P) = \{\sum_{i=1}^{k} \lambda_i p_i \mid k \in \mathbb{N}^* \wedge \sum_{i=1}^{k} \lambda_i = 1 \wedge \forall i \in [\![1, k]\!], p_i \in P, \lambda_i \geq 0\}$, where $[\![1, k]\!]$ is the set of integers from 1 to k included.

Definition 5 (P Full in X). *Let $P, X \in \mathcal{P}(E)$. P is full in X if $P_t \subseteq X$ for any t implies that $\mathrm{CH}(P_t)$ does not contain any point of X other than P_t, i.e.:*

$$\forall t \in E, P_t \subseteq X \Rightarrow \mathrm{CH}(P_t) \cap X = P_t \tag{8}$$

Under the condition that P is full in X, Theorem 1 provides the equality $P = \varepsilon_O(X)$. Therefore, it is possible, under the assumptions of the theorem, to discover the pattern P from its onsets O.

Theorem 1. *Let $P, X \in \mathcal{P}(E)$ such that:*

- *$\gamma_P(X) = X$,*
- *P is full in X,*
- *X is bounded.*

Then, by defining $O = \varepsilon_P(X)$, we have:

$$P = \varepsilon_O(X) \tag{9}$$

We provide another link between P and O using morphological operators with the following lemma, which is useful for the proof of Theorem 1.

Lemma 2. *Let $P, O, X \in \mathcal{P}(E)$, such that $O = \varepsilon_P(X)$ and $\gamma_P(X) = X$. We have:*

$$\varepsilon_O(X) = \varphi_O(P) \tag{10}$$

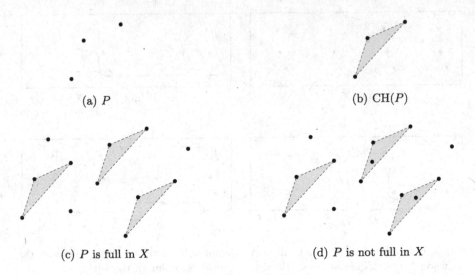

(a) P (b) CH(P)

(c) P is full in X (d) P is not full in X

Fig. 3. Musical pattern P (a) and its convex hull (b). P is full in X (set of dots) in (c) because there are no dots inside the convex hull where it is included in X. However, this is not the case for (d).

Proof (Lemma 2). Under the assumptions of the lemma, we have:

$$\gamma_P(X) = X$$
$$\Rightarrow \quad \delta_P(\varepsilon_P(X)) = X \qquad \text{(definition of } \gamma_P)$$
$$\Rightarrow \quad \delta_P(O) = X \qquad \text{(definition of } O)$$
$$\Rightarrow \quad \delta_O(P) = X \qquad \text{(commutativity of } \oplus)$$
$$\Rightarrow \quad \varepsilon_O(\delta_O(P)) = \varepsilon_O(X) \qquad \text{(composing by } \varepsilon_O)$$
$$\Rightarrow \quad \varphi_O(P) = \varepsilon_O(X) \qquad \text{(definition of } \varphi_O) \qquad \square$$

Proof (Theorem 1). Under the assumptions of the theorem, we can use Lemma 2 and Lemma 1. We prove that $P = \varepsilon_O(X)$ with two inclusions.

$\boxed{\subseteq}$ The first inclusion comes directly from Lemma 1.

$\boxed{\supseteq}$ It has been proved by Serra that for every $C \in \mathcal{P}(E)$ convex and $S \in \mathcal{P}(E)$ bounded, we have: $\varphi_S(C) = C$ (proposition IV-4 in [15]). Since X is bounded, $O = \varepsilon_P(X)$ is also bounded and CH(P) is convex by definition. Consequently, we have:

$$\varphi_O(\text{CH}(P)) = \text{CH}(P) \tag{11}$$

Let $t \in E$ such that $P_t \subseteq X$ (such t exists because $\gamma_P(X) = X$), therefore $t \in O$ and we have the following implications:

$$P \subseteq \mathrm{CH}(P)$$
$$\Rightarrow \quad \varphi_O(P) \subseteq \varphi_O(\mathrm{CH}(P)) \qquad\qquad\qquad (\varphi_O \text{ is increasing})$$
$$\Rightarrow \quad \varphi_O(P) \subseteq \mathrm{CH}(P) \qquad\qquad (\varphi_O(\mathrm{CH}(P)) = \mathrm{CH}(P) \text{ Eq. (11)})$$
$$\Rightarrow \quad \varepsilon_O(X) \subseteq \mathrm{CH}(P) \qquad\qquad\qquad (\varphi_O(P) = \varepsilon_O(X) \text{ Lemma 2})$$
$$\Rightarrow \quad \varepsilon_O(X)_t \subseteq \mathrm{CH}(P)_t \qquad\qquad\qquad\qquad (\text{translation by } t)$$
$$\Rightarrow \quad \varepsilon_O(X)_t \subseteq \mathrm{CH}(P_t) \qquad\qquad\qquad (\mathrm{CH}(P)_t = \mathrm{CH}(P_t))$$
$$\Rightarrow \quad \varepsilon_{O_{-t}}(X) \subseteq \mathrm{CH}(P_t) \qquad\qquad\qquad (\varepsilon_O(X)_t = \varepsilon_{O_{-t}}(X))$$
$$\Rightarrow \quad \varepsilon_{O_{-t}}(X) \cap X \subseteq \mathrm{CH}(P_t) \cap X \qquad\qquad (\text{intersection with } X)$$
$$\Rightarrow \quad \varepsilon_{O_{-t}}(X) \subseteq \mathrm{CH}(P_t) \cap X \qquad (\varepsilon_{O_{-t}}(X) \subseteq X \text{ because } O_E \in O_{-t})$$
$$\Rightarrow \quad \varepsilon_{O_{-t}}(X) \subseteq P_t \qquad (\mathrm{CH}(P_t) \cap X = P_t \text{ because } P \text{ is full})$$
$$\Rightarrow \quad \varepsilon_O(X) \subseteq P \qquad\qquad\qquad\qquad\qquad (\text{translation by } -t)$$

$$\square$$

5 Musical Interpretations of the Theorem

The main result of this article, i.e. Theorem 1, is illustrated in Fig. 4 with a piece of music X composed of a musical pattern P repeated several times (therefore $\gamma_P(X) = X$). The assumptions of Theorem 1 are satisfied because X is bounded and P is full in X. By definition, the morphological erosion of X by the musical pattern P is equal to the onsets O. Moreover, Theorem 1 ensures that the other way around is true: we can obtain the pattern P by applying the erosion by the onsets O. All these morphological links between the piece X, the pattern P and its onsets O are summarized in Fig. 4. In this case, we can describe the piece of music X by a morphological dilation between the pattern and its onsets:

$$X = P \oplus O \tag{12}$$

This last equality increases the relevance of the use of morphological operators applied to music.

Remark 1 (About the assumptions in Theorem 1). In Theorem 1, there are three assumptions which are: P is full in X, $\gamma_P(X) = X$ and X is bounded.

– <u>P is full in X</u>: This assumption is coherent in a musical context. In the *Generative Theory of Tonal Music* (GTTM), the authors state several rules about overlaps in music [9]. In particular, the *Grouping Well-Formedness Rules 4* asserts:

"If a group G_1 contains part of a group G_2 , it must contain all of G_2" ›

This proves that it is very rare for patterns to intersect in music. However, later in GTTM, in Section *3.4 Grouping Overlaps*, it is mentioned that this

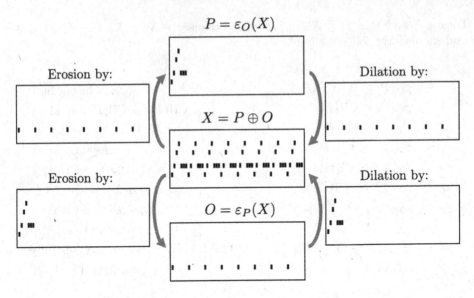

Fig. 4. Illustration using morphological operators of the relations between the musical pattern P, which is full in X, and its onsets O.

rule is not always true and that patterns can overlap in a specific case: if patterns overlap, it is only the first or last note of the pattern that can be part of two patterns. Therefore, if we assume that a pattern P satisfies the GTTM rules (i.e. patterns do not intersect or only share their first or last note) and that the data X is composed of P which is repeated with translations (i.e. $\gamma_P(X) = X$), this leads to having a pattern P that is full in X. This reasoning shows that the property that a pattern is full is perfectly coherent in a musical context. Moreover, this property allows us to obtain the fundamental result of Theorem 1, and if we remove this assumption, the theorem is not always true, as shown in Fig. 2.

– $\gamma_P(X) = X$: In general, there are several musical patterns in a piece of music and we cannot always have $\gamma_P(X) = X$. However, to ensure this equality, we can restrict the analysis to musical sections composed of repeated patterns, and not the whole piece of music. Since musical sections do not intersect, it is therefore possible to apply Theorem 1 to temporal regions separately, and thus to apply Theorem 1 independently to each musical section and not to the whole piece. Therefore, to respect this assumption, we can consider the discovery of each pattern P_i separately and restrict X to $\gamma_{P_i}(X)$.

– X is bounded: In our analysis, the case where X is not bounded never happens because X represents a piece of music which is by definition bounded. There is neither an infinite number of notes, nor an infinite value for pitch or other musical characteristics. Therefore, this assumption is necessary to ensure that Serra's proposition is true but is not a restriction for the musical applications.

6 Optimize the Discovery of Musical Patterns with the Onsets

In this article, we propose the novelty of having two methods for discovering the musical patterns in a multidimensional representation of music: one can discover either the musical patterns P or their onsets O. Using Theorem 1, it is enough to identify only one of the two to obtain the other one (if P is full in X which is usually the case in music). In each case, it is necessary to proceed differently, but the main idea is to use X to discover the musical patterns. The following lemma indicates that musical patterns or the onsets must be composed of sub-patterns in X, i.e. if a sub-pattern P' does not appear in X then it cannot be contained in musical patterns or the onsets.

Lemma 3. *Let $P, P', X \in \mathcal{P}(E)$. If $\gamma_{P'}(X) = \emptyset$ then $\gamma_{P \cup P'}(X) = \emptyset$.*

Proof (Lemma 3). This is due to the decreasingness of the opening with respect to the structuring element, i.e.: $P' \subseteq P \cup P' \Rightarrow \gamma_{P \cup P'}(X) \subseteq \gamma_{P'}(X) = \emptyset$. □

From the previous lemma, we can deduce that the points of the musical patterns or the onsets must belong to the set:

$$\{x_2 - x_1 \mid x_1, x_2 \in X \wedge x_1 \leq x_2\}, \tag{13}$$

where $x_1 \leq x_2$ means that the temporal component of x_1 is less than or equal to the temporal component of x_2. In other words, we can restrict ourselves to this set, rather than trying to cover the whole E set to discover musical patterns. This observation has already been made by Meredith et al. with the SIA algorithm to optimize the discovery of musical patterns [12]. However, we propose here to give a musical meaning to the discovered patterns, distinguishing the role of pattern P and onsets O. Because many elements of this set are not musically interesting, additional conditions have to be added to find the appropriate musical patterns. Depending on whether we want to discover P or O, we then add different constraints.

Learning the Musical Patterns P. To reduce the possibilities to be tested to discover musical patterns P, we need to add more constraints that remove irrelevant candidates. For example, to avoid obtaining musical patterns that are too long in time, Collins et al. proposed to add a constraint on the temporal length of the musical pattern with the SIAR algorithm [1]. However, the possibilities remains very large and we propose another approach to discover musical patterns, using the onsets.

Learning the Onsets O. One of the most important results of this article is the possibility of discovering musical patterns from the onsets. Under the assumptions mentioned in Theorem 1, it is possible to discover the musical pattern from the onsets with the morphological erosion. The major usefulness of this result arises from the algorithmic complexity reduction: it can be faster to discover the onsets than the patterns because there are far fewer possible choices.

In order to discover *exact repetitive patterns*, it is possible to consider time-periodic onsets. For example, if we are looking for musical patterns P that repeat in time at regular intervals of length $L > 0$, the onsets O are:

$$O = \{(iL, 0, \ldots, 0) \mid i \in [\![0, m-1]\!]\}, \tag{14}$$

where the first coordinate iL represents the temporal component. In this case, we are looking for at least m exact repetitions of the musical pattern P. As a result, only one parameter L is required to discover patterns that repeat exactly at regular intervals. However, in the case where the patterns are not repeated exactly with a time translation, they can also be discovered with *transpositions*, which means that a pitch translation is also allowed. Up to an octave transposition up or down is enough to cover most cases, i.e. a pitch translation in $[\![-12, 12]\!]$. Representing the pitch by the second coordinate, the onsets O are of the following form:

$$O = \{(iL, p_i, 0, \ldots, 0) \mid i \in [\![0, m-1]\!] \wedge p_i \in [\![-12, 12]\!]\} \tag{15}$$

We can search for the values of L and p_i that are in the set defined in Eq. 13, choosing those that maximize the number of notes after an opening by the onsets. Therefore, only a few parameters L and p_i are required to discover patterns that repeat at regular intervals with transpositions, where each parameter has a limited number of values.

7   Conclusion and Future Work

In this article, we have proposed an original approach based on mathematical morphology for discovering musical patterns using a multidimensional representation of music. The originality of our method is to provide a musical meaning to the discovered patterns, by distinguishing the role of onsets from the musical patterns. With this characterization, we introduced the problem of *discovering the musical pattern from its onsets*. We then proposed a solution to this problem if the patterns respect a topological property that is musically coherent, because patterns do not intersect in music. This reveals the possibility of discovering musical patterns from their onsets. We have demonstrated the relevance of this result in a musical context, as it optimizes the discovery of musical patterns due to the importance of repetition in musical data. An interesting feature is that we proved a mathematically strong result, with a relevant musical interpretation, based on simple morphological operations. In future work, we intend to apply the method developed in this article to music databases, and compare the obtained results with existing methods. In addition, we aim to determine the links between point-set algorithms and mathematical morphology [7], and produce other mathematical results to characterize the discovered pairs of musical patterns and their onsets. These directions of research are original in that they produce mathematical results that come from a musical motivation, but which can be applied to other types of discrete data, e.g. in topological data analysis. Finally, based on this approach, we hope to extend the application of mathematical morphology to symbolic representations of music, which will allow us to develop additional theories and produce further original mathematical results.

References

1. Collins, T., Thurlow, J., Laney, R., Willis, A., Garthwaite, P.: A comparative evaluation of algorithms for discovering translational patterns in baroque keyboard works. In: 11th International Society for Music Information Retrieval Conference, Utrecht, Netherlands, pp. 3–8 (2010)
2. Forth, J., Wiggins, G.A.: An approach for identifying salient repetition in multidimensional representations of polyphonic music. London Algorithmics 2008: Theory and Practice, pp. 44–58 (2009)
3. Hadwiger, H.: Minkowskische Addition und Subtraktion beliebiger Punkt-mengen und die Theoreme von Erhard Schmidt. Math. Z. **53**, 210–218 (1950)
4. Karvonen, M., Laitinen, M., Lemström, K., Vikman, J.: Error-tolerant content-based music-retrieval with mathematical morphology. In: Ystad, S., Aramaki, M., Kronland-Martinet, R., Jensen, K. (eds.) CMMR 2010. LNCS, vol. 6684, pp. 321–337. Springer, Heidelberg (2010). https://doi.org/10.1007/978-3-642-23126-1_20
5. Karvonen, M., Lemström, K.: Using mathematical morphology for geometric music information retrieval. In: International Workshop on Machine Learning and Music (2008)
6. Lascabettes, P.: Mathematical morphology applied to music. Master's thesis, École Normale Supérieure Paris-Saclay (2019)
7. Lascabettes, P.: Mathematical Models for the Discovery of Musical Patterns and Structures, and for Performances Analysis. Ph.D. thesis, Sorbonne Université (2023)
8. Lascabettes, P., Bloch, I., Agon, C.: Analyse de représentations spatiales de la musique par des opérateurs simples de morphologie mathématique. In: Journées d'Informatique Musicale. Strasbourg, France (2020)
9. Lerdahl, F., Jackendoff, R.: A Generative Theory of Tonal Music. MIT Press, Cambridge (1985)
10. Meredith, D.: Point-set algorithms for pattern discovery and pattern matching in music. In: Dagstuhl Seminar Proceedings. Schloss Dagstuhl-Leibniz-Zentrum für Informatik (2006)
11. Meredith, D.: COSIATEC and SIATECCompress: pattern discovery by geometric compression. In: 14th International Society for Music Information Retrieval Conference (2013)
12. Meredith, D., Lemström, K., Wiggins, G.A.: Algorithms for discovering repeated patterns in multidimensional representations of polyphonic music. J. New Music Res. **31**(4), 321–345 (2002)
13. Meredith, D., Wiggins, G.A., Lemström, K.: Pattern induction and matching in polyphonic music and other multidimensional datasets. In: 5th World Multiconference on Systemics, Cybernetics and Informatics, vol. 10, pp. 61–66 (2001)
14. Minkowski, H.: Volumen und Oberfläche. Math. Ann. **57**, 447–495 (1903)
15. Serra, J.: Image Analysis and Mathematical Morphology. Academic Press, London (1982)

The Lattice Overparametrization Paradigm for the Machine Learning of Lattice Operators

Diego Marcondes[1,2]($\boxtimes$) and Junior Barrera[1]

[1] Department of Computer Science, Institute of Mathematics and Statistics, University of São Paulo, São Paulo, Brazil
dmarcondes@ime.usp.br
[2] Department of Electrical and Computer Engineering, Texas A&M University, College Station, USA

Abstract. The machine learning of lattice operators has three possible bottlenecks. From a statistical standpoint, it is necessary to design a constrained class of operators based on prior information with low bias, and low complexity relative to the sample size. From a computational perspective, there should be an efficient algorithm to minimize an empirical error over the class. From an understanding point of view, the properties of the learned operator need to be derived, so its behavior can be theoretically understood. The statistical bottleneck can be overcome due to the rich literature about the representation of lattice operators, but there is no general learning algorithm for them. In this paper, we discuss a learning paradigm in which, by overparametrizing a class via elements in a lattice, an algorithm for minimizing functions in a lattice is applied to learn. We present the stochastic lattice descent algorithm as a general algorithm to learn on constrained classes of operators as long as a lattice overparametrization of it is fixed, and we discuss previous works which are proves of concept. Moreover, if there are algorithms to compute the basis of an operator from its overparametrization, then its properties can be deduced and the understanding bottleneck is also overcome. This learning paradigm has three properties that modern methods based on neural networks lack: control, transparency and interpretability. Nowadays, there is an increasing demand for methods with these characteristics, and we believe that mathematical morphology is in a unique position to supply them. The lattice overparametrization paradigm could be a missing piece for it to achieve its full potential within modern machine learning.

Keywords: lattice overparametrization · discrete morphological neural networks · image processing · mathematical morphology · U-curve algorithms · stochastic lattice descent

D. Marcondes was funded by grants #22/06211-2 and #23/00256-7, São Paulo Research Foundation (FAPESP), and J. Barrera was funded by grants #14/50937-1 and #2020/06950-4, São Paulo Research Foundation (FAPESP).

S. Brunetti et al. (Eds.): DGMM 2024, LNCS 14605, pp. 204–216, 2024.
https://doi.org/10.1007/978-3-031-57793-2_16

1   Algebraic Representations of Operators

Let $(\mathcal{L}, \leq)$ be a complete lattice. A lattice operator $\psi : \mathcal{L} \to \mathcal{L}$ is a mapping from $\mathcal{L}$ into itself, and we denote by $\Psi = \mathcal{L}^{\mathcal{L}}$ the set of all lattice operators in $\mathcal{L}$. The collection Ψ inherits the complete lattice structure of $\mathcal{L}$ by considering the pointwise partial order. Let $\Omega \subset \Psi$ be a complete sublattice of Ψ.

An algebraic representation of $(\Omega, \leq)$ is any complete lattice $(\Theta, \leq)$ such that there exists a lattice isomorphism $R : \Omega \to \Theta$. The element $\theta \in \Theta$ is the parameter that represents the operator $\psi_\theta = R^{-1}(\theta)$ and (R, Θ) is a parametrization of Ω. The algebraic representations are not unique and, although they are all equivalent, some have advantages over others.

A general algebraic representation of a lattice operator ψ is through its kernel, as proposed in[1] [5]. Let $\Theta_{\mathcal{K}} = \mathcal{P}(\mathcal{L})^{\mathcal{L}}$ be the collection of all maps $\mathcal{F}$ from $\mathcal{L}$ to $\mathcal{P}(\mathcal{L})$ equipped with the pointwise partial order

$$\mathcal{F}_1 \leq \mathcal{F}_2 \iff \mathcal{F}_1(Y) \subset \mathcal{F}_2(Y) \ \forall Y \in \mathcal{L}$$

for $\mathcal{F}_1, \mathcal{F}_2 \in \Theta_{\mathcal{K}}$, and consider the lattice isomorphism $R_{\mathcal{K}} : \Omega \to \Theta_{\mathcal{K}}$ given by

$$R_{\mathcal{K}}(\psi)(Y) = \mathcal{K}(\psi)(Y) = \{X \in \mathcal{L} : Y \leq \psi(X)\} \qquad (Y \in \mathcal{L}).$$

See [5, Proposition 6.1] for a proof that $R_{\mathcal{K}}$ is a lattice isomorphism.

The operators in specific lattices, such as finite lattices, and subclasses of operators in general lattices, such as upper semi-continuous operators [6], have a minimal algebraic representation by the maximal intervals lesser or equal to the kernel. Formally, for $\psi \in \Psi$ let

$$A(\mathcal{K}(\psi)) = \{[\alpha, \beta] : [\alpha, \beta] \leq \mathcal{K}(\psi)\}$$

be the intervals[2] which are lesser or equal to the kernel of ψ. The basis of ψ is defined as the maximal intervals in $A(\psi)$, that is

$$
\begin{aligned}
B(\psi) &= \mathrm{Max}\,(A(\mathcal{K}(\psi))) \\
&= \{[\alpha, \beta] \in A(\psi) : [\alpha', \beta'] \in A(\psi), [\alpha, \beta] \leq [\alpha', \beta'] \implies [\alpha, \beta] = [\alpha', \beta']\}
\end{aligned}
$$

in which $\leq$ above is the partial order in $(\Theta_{\mathcal{K}}, \leq)$. Under certain conditions, of which more details may be found in [4–6], it follows that

$$\psi = \vee \{\lambda_{[\alpha, \beta]} = \overline{\alpha} \wedge \overline{\beta} : [\alpha, \beta] \in B(\psi)\} \tag{1}$$

in which

$$\overline{\alpha}(X) = \vee \{Y \in \mathcal{L} : \alpha(Y) \leq X\} \quad \text{and} \quad \overline{\beta}(X) = \vee \{Y \in \mathcal{L} : X \leq \beta(Y)\}$$

for $X \in \mathcal{L}$. Decomposition (1) is called sup-generating decomposition of ψ and it has a dual inf-generating decomposition.

---

[1] We are calling kernel what [5] defined as left-kernel.

[2] See [5] for the formal definition of interval in this context.

Assuming that (1) holds for all $\psi \in \Omega$, denote by

$$\Theta_B = \{\mathrm{Max}\,(\boldsymbol{A}(\mathcal{F})) : \mathcal{F} \in \Theta_{\mathcal{K}}\}$$

the maximal intervals associated to each $\mathcal{F} \in \Theta_{\mathcal{K}}$ and consider the map $R_B : \Theta_{\mathcal{K}} \to \Theta_B$ given by

$$R_B(\mathcal{F}) = \mathrm{Max}\,(\boldsymbol{A}(\mathcal{F})).$$

It follows that $(\Theta_B, \leq)$ is a complete lattice isomorphic to $(\Theta_{\mathcal{K}}, \leq)$ with partial order

$$\boldsymbol{B}_1 \leq \boldsymbol{B}_2 \iff \forall [\alpha, \beta] \in \boldsymbol{B}_1, \exists [\alpha', \beta'] \in \boldsymbol{B}_2 : [\alpha, \beta] \leq [\alpha', \beta']$$

in which the partial order on the right-hand side is that of $(\Theta_{\mathcal{K}}, \leq)$. From now on, we assume that Ω is a subclass of operators on $\mathcal{L}$ with a basis representation.

Specific classes of operators may have other algebraic representations. For instance, when $\mathcal{L} = \mathcal{P}(E)$ and $(E, +)$ is an Abelian group, then the class of translation invariant (t.i.) and locally defined lattice operators (i.e., W-operators), which in this case are set operators, can also be represented by a characteristic Boolean function. Denoting by Ψ_W the class of t.i. set operators locally defined within a window $W \in \mathcal{P}(E)$ and by $\mathfrak{B} = \{0, 1\}^{\mathcal{P}(W)}$ the Boolean functions in $\mathcal{P}(W)$, we consider the lattice isomorphism $R_{\mathfrak{B}} : \Psi_W \to \mathfrak{B}$ given by

$$R_{\mathfrak{B}}(\psi)(X) = \begin{cases} 1, & \text{if } o \in X \\ 0, & \text{otherwise} \end{cases} \qquad (X \in \mathcal{P}(W))$$

in which o is the zero element E. See [8] for more details.

Clearly, the isomorphisms may be composed to obtain isomorphisms between distinct algebraic representations and all algebraic representations are equivalent. For example, $R_{\mathfrak{B},B} : \mathfrak{B} \to \boldsymbol{B}$ given by $R_{\mathfrak{B},B} = R_B \circ R_{\mathcal{K}} \circ R_{\mathfrak{B}}^{-1}$ is an isomorphism between $(\mathfrak{B}, \leq)$ and $(\boldsymbol{B}, \leq)$. The isomorphisms defined so far are illustrated in Fig. 1.

From an algebraic perspective, the basis representation has some advantages over other representations, since algebraic properties of an operator may be deduced from its basis. For example, in the case of W-operators, the intervals in the basis of increasing operators are of form $[A, W]$ for $A \in \mathcal{P}(W)$; the basis of extensive increasing operators contains the interval $[o, W]$; and the basis of an increasing anti-extensive operator is such that $o \in A$ for all lower limits A of the intervals in its basis (see [21] for more details). Hence, reducing an operator to its basis representation is enough to verify its mathematical properties.

2 Lattice Overparametrization

An algebraic representation R is an isomorphism between a class Ω and a parametric set Θ. Such a representation is obtained by departing from a fixed Ω and defining an isomorphism $R : \Omega \to \Theta$. This is done in [4–6]. Another family of representations may be obtained by departing from a Θ and defining an onto map $\tilde{R} : \Theta \to \Omega$, so each parameter $\theta \in \Theta$ represents an operator $\psi_\theta = \tilde{R}(\theta) \in \Omega$ and for each $\psi \in \Omega$ there exists *at least one* $\theta \in \Theta$ such that $\psi = \psi_\theta$.

When $\tilde{R}$ is not injective, $(\tilde{R}, \Theta)$ is an overparametrization of Ω by the parameters in Θ since a same operator can be represented by more than one parameter. If $(\Theta, \leq)$ is a lattice, we say that $(\tilde{R}, \Theta)$ is a lattice overparametrization of Ω. Since $\tilde{R}$ is not an isomorphism, the partial relation in Θ is not equivalent to that in Ω. The basis of the operator represented by θ is given by $\tilde{R}_B(\theta) = (R_B \circ R_{\mathcal{K}} \circ \tilde{R})(\theta)$.

As an example, assume that $\mathcal{L} = \mathcal{P}(E)$ and $(E, +)$ is an Abelian group. For a finite subset $W \in \mathcal{P}(E)$ let

$$\Omega = \{\epsilon_A \vee \epsilon_B \vee \epsilon_C : A, B, C \in \mathcal{P}(W); \{[A, W], [B, W], [C, W]\} \text{ is maximal}\} \tag{2}$$

be the class of t.i. operators locally defined within W that can be written as the supremum of three erosions. We note that $B(\epsilon_A \vee \epsilon_B \vee \epsilon_C) = \{[A, W], [B, W], [C, W]\}$ and Ω is actually the class of the increasing W-operators with at most three elements in their basis[3]. By making $\Theta = \mathcal{P}(W)^3$ and $\tilde{R}((A, B, C)) = \epsilon_A \vee \epsilon_B \vee \epsilon_C$ we have a lattice overparametrization of Ω since $\tilde{R}((A, B, C)) = \psi$ for all $(A, B, C) \in \mathcal{P}(W)^3$ satisfying $\mathcal{K}(\psi) = [A, W] \cup [B, W] \cup [C, W]$. By lifting the restriction that the intervals $\{[A, W], [B, W], [C, W]\}$ are maximal, we depart from an algebraic representation of Ω to a lattice overparametrization by a Boolean lattice.

A lattice overparametrization may be useful for representing a constrained class of operators defined via the composition, supremum, and infimum of operators that can be parametrized by elements in a lattice. A special case is when the operators can be written as combinations of erosions and dilations with structural elements in a lattice. In [22] we proposed the discrete morphological neural networks (DMNN) to represent constrained classes of W-operators via the composition, supremum and infimum of W-operators, which are an example of overparametrizations of a class of operators. In special, the canonical DMNN are those in which the W-operators computed in the network can be written as the supremum, infimum, complement, or composition of erosions and dilations with a same structuring element, an example of which is the class in (2) (see Example 5.8 in [22]). The canonical DMNN are a specific example of lattice overparametrization.

The main advantage of considering a lattice overparametrization is the possibility of applying general, efficient algorithms to learn operators in a constrained class. This is the case since the lattice $(\Theta, \leq)$ is known and can be chosen with desired computational properties, so minimizing a function in it may be more efficient than doing so in $(\Omega, \leq)$, specially when Ω is not a lattice. We further discuss the advantages of considering a lattice overparametrization to learn lattice operators in Sect. 4.

---

[3] Observe that if some of the elements A, B, C are equal, then $|B(\epsilon_A \vee \epsilon_B \vee \epsilon_C)| < 3$.

3 The Machine Learning of Lattice Operators

The general framework for learning lattice operators consists of a class Ω, a sample $\mathcal{D}_N = \{(X_1, Y_1), \ldots, (X_N, Y_N)\}$ of N pairs of input and output elements X and Y in $\mathcal{L}$, in which Y is obtained by a possibly random transformation of X, and a loss function $\ell : \mathcal{L}^2 \times \Psi \to \mathbb{R}^2$ which evaluates the *error* $\ell((X, Y), \psi)$ incurred when $\psi(X)$ is applied to approximate Y, for each pair $(X, Y) \in \mathcal{L}^2$ and operator $\psi \in \Psi$.

It is assumed that the pairs in $\mathcal{D}_N$ are sampled from an unknown, but fixed, statistical distribution P over $\mathcal{L}^2$. Each $\psi \in \Psi$ has a mean expected error under distribution P defined as $L(\psi) = \mathbb{E}_P[\ell((X, Y), \psi)]$, in which the expectation is over a random vector (X, Y) with distribution P. A target operator of Ψ is a minimizer of L in Ψ and a target operator of Ω is a minimizer of L in Ω. We denote the target operators by $\psi^\star$ and $\psi_\Omega^\star$, respectively, and they satisfy $L(\psi^\star) \leq L(\psi), \forall \psi \in \Psi$, and $L(\psi_\Omega^\star) \leq L(\psi), \forall \psi \in \Omega$. For the sake of the argument, we assume that both target operators exist and are unique.

Defining

$$L_{\mathcal{D}_N}(\psi) = \frac{1}{N} \sum_{i=1}^N \ell((X_i, Y_i), \psi)$$

as the mean empirical error of $\psi \in \Psi$ in sample $\mathcal{D}_N$, the empirical risk minimization paradigm propose as an estimator for $\psi_\Omega^\star$ the operator that minimizes $L_{\mathcal{D}_N}$ in Ω:

$$\hat{\psi} = \arg \min_{\psi \in \Omega} L_{\mathcal{D}_N}(\psi) = \arg \min_{\theta \in \Theta} L_{\mathcal{D}_N}(\psi_\theta) \tag{3}$$

in which Θ is any representation, algebraic or otherwise, of Ω. The quality of the estimator $\hat{\psi}$ is measured by $L(\hat{\psi})$, which is called its generalization error, and assesses how it is expected to perform on data not in the sample, but generated by the same unknown distribution P.

The goal of learning is to obtain an estimator such that $L(\hat{\psi}) \approx L(\psi^\star)$ so its generalization quality is close to the best possible. On the one hand, it is necessary to have $L(\psi_\Omega^\star) \approx L(\psi^\star)$ for otherwise there is a systematic bias in the learning process since $\hat{\psi}$ cannot generalize better than $\psi_\Omega^\star$. On the other hand, if Ω is chosen as a class of complex operators, or as $\Omega = \Psi$, then, even if $\psi_\Omega^\star$ is as good as or equal to $\psi^\star$, if the sample size is not great enough, there may be a complex operator $\hat{\psi}$ in Ω that completely fits the data, so it has zero empirical error, but that does not generalize very well. When this happens, we say overfitting occurred. Actually, if $\Omega = \Psi$ and Ψ has infinite VC dimension, which is a measure of the complexity of a class of operators [29], not even an infinite sample suffices to guarantee that $L(\hat{\psi}) \approx L(\psi^\star)$. This is the usual bias-variance trade-off in machine learning [1].

Hence, we have the following statistical bottleneck for learning lattice operators:

(B1) *To fix a class of operators with low bias and relative low complexity*

The recipe to circumvent **(B1)** is the core of mathematical morphology: to design a class of operators based on prior information about the practical problem and on the mathematical properties of lattice operators. Geometrical and topological properties of the transformation applied to X_i to obtain Y_i in sample $\mathcal{D}_N$ are identified, and based on them a class Ω of lattice operators is designed via the mathematical morphology toolbox. If prior information is right, so the best operator in Ω well generalizes, and Ω is not too complex, then learning is feasible and $\hat{\psi}$ is expected to well generalize. As an example, the class in (2) can be applied to a problem in which it is known that an increasing transformation was applied to X_i to obtain Y_i, and the maximum number of elements in the basis controls the complexity of Ω.

There are almost 60 years of rich literature in mathematical morphology, that we could not possibly cite here without committing huge injustices, which can be directly applied to solving **(B1)**, so it is not really a bottleneck for learning lattice operators. However, there is a second, computational, bottleneck that has not yet been overcome in general:

(B2) *To compute $\hat{\psi}$ by solving* (3)

Despite their practical success, many proposed methods for the machine learning of lattice operators in the literature are heuristics that seek to control the complexity of the class of operators relative to the sample size, but do not strongly restrict the operator class based on prior information. The ISI algorithm [17], iterative designs [18] and multiresolution designs [13,19] offer methods to control the complexity of the class based on data, however are not flexible to represent specific classes of operators, but only general classes such as filters.

Furthermore, methods such as the those based on envelope constraints [9,10] can insert sharp prior information into the learning process by projecting the operator learned by a heuristic method into a constrained class, but do not guarantee that the projected operator well approximates the target of the class. Finally, we note that methods to solve (3) for specific classes, such as stack filters [16], have been proposed, but are not general methods that can be easily extended to other classes of operators. See [7] for more details on methods for the machine learning of operators.

We propose as a general paradigm to overcoming **(B2)** the development of algorithms to efficiently minimize, or approximately minimize, a function in a lattice so (3) can be at least approximately computed whenever Ω has a lattice overparametrization $(\Theta, \leq)$. Such an algorithm would be a general optimizer for learning operators once a subclass Ω and a lattice overparametrization for it is fixed. This abstract idea, which is behind the DMNN proposed in [22], can be a paradigm for the machine learning of lattice operators based on the stochastic lattice descent algorithm (SLDA). The general framework for the machine learning of lattice operators is depicted in Fig. 1.

4   Stochastic Lattice Descent as a General Learning Algorithm

The U-curve algorithm was first proposed by [27] for minimizing U-shaped functions in Boolean lattices, and was then improved by [3,14,26]. It has also been applied to solve other problems in mathematical morphology [25]. Inspired by this algorithm and by the success of stochastic gradient descent algorithms for minimizing overparametrized functions in $\mathbb{R}^d$, such as the regularized empirical error of a neural network, we propose the SLDA to learn operators in a class with lattice overparametrization $(\Theta, \leq)$.

Informally, the SLDA performs a greedy search of a lattice to minimize an empirical error. At each step, n neighbors of an element are sampled and the empirical error on a fixed sample batch of the operator represented by each sampled neighbor is calculated. The algorithm jumps to the sampled neighbor with the least empirical error on the sample batch. The algorithm starts again from this new element, by sampling n neighbors and calculating their empirical error on a new sample batch. This process goes on for a predetermined number of epochs. An epoch ends when all sample batches have been considered, and the algorithm returns the element visited at the end of an epoch with the least empirical error on the whole sample. We now formally define the SLDA.

For each $\theta \in \Theta$, let $N(\theta)$ be a *neighborhood* of θ in $(\Theta, \leq)$. If Θ is countable, then $N(\theta)$ may be composed by the elements of Θ at distance one from θ. When Θ is uncountable and $d(\theta, \theta')$ is a distance measure, with $d(\theta, \theta') = \infty$ whenever $\theta \not\leq \theta'$ and $\theta' \not\leq \theta$, then one could consider $N(\theta) = \{\theta' : d(\theta, \theta') < \delta\}$ for a fixed $\delta > 0$. Assume that, given θ and a constant n, there exists an algorithm which samples n elements from $N(\theta)$. If $N(\theta)$ is a finite set, then the elements may be sampled uniformly, while if it is countable or uncountable then other statistical distributions should be considered.

The SLDA is formalized in Algorithm 1. The initial point $\theta \in \Theta$, a batch size[4] b, the number n of neighbors to be sampled at each step, and the number of training epochs is fixed. The initial point is stored as the point with minimum empirical error visited so far. For each epoch, the sample $\mathcal{D}_N$ is randomly partitioned in N/b batches $\{\tilde{\mathcal{D}}_b^{(1)}, \ldots, \tilde{\mathcal{D}}_b^{(N/b)}\}$. For each batch $\tilde{\mathcal{D}}_b^{(j)}$, n neighbors of θ are sampled and θ is updated to a sampled neighbor with the least empirical error $L_{\tilde{\mathcal{D}}_b^{(j)}}$, that is calculated on the sample batch $\tilde{\mathcal{D}}_b^{(j)}$. Observe that θ is updated at each batch, so during an epoch, it is updated N/b times.

At the end of each epoch, the empirical error $L_{\mathcal{D}_N}(\theta)$ of θ on the whole sample $\mathcal{D}_N$ is compared with the error of the point with the least empirical error visited so far at the end of an epoch, and it is stored as this point if its empirical error is lesser. After the predetermined number of epochs, the algorithm returns the point with the least empirical error on the whole sample $\mathcal{D}_N$ visited at the end of an epoch. For finite lattices, if $b = N$ and n is equal to the number of neighbors of θ, i.e., $n = n(\theta) = |N(\theta)|$, then Algorithm 1 reduces to the (deterministic) lattice descent algorithm.

---

[4] We assume that N/b is an integer to easy notation. If this is not the case, the last batch will contain less than b points.

Algorithm 1. Stochastic lattice descent algorithm for learning lattice operators.

Ensure: $\theta \in \Theta, n, b, Epochs$

1: $L_{min} \leftarrow L_{\mathcal{D}_N}(\psi_\theta)$
2: $\widehat{\theta} \leftarrow \theta$
3: **for** run $\in \{1, \ldots, \text{Epochs}\}$ **do**
4:     $\{\tilde{\mathcal{D}}_b^{(1)}, \ldots, \tilde{\mathcal{D}}_b^{(N/b)}\} \leftarrow \text{SampleBatch}(\mathcal{D}_N, b)$
5:     **for** $j \in \{1, \ldots, N/b\}$ **do**
6:         $\tilde{N}(\theta) \leftarrow \text{SampleNeighbors}(\theta, n)$
7:         $\theta \leftarrow \theta'$ s.t. $\theta' \in \tilde{N}(\theta)$ and $L_{\tilde{\mathcal{D}}_b^{(j)}}(\psi_{\theta'}) = \min\{L_{\tilde{\mathcal{D}}_b^{(j)}}(\psi_{\theta''}) : \theta'' \in \tilde{N}(\theta)\}$
8:     **if** $L_{\mathcal{D}_N}(\psi_\theta) < L_{min}$ **then**
9:         $L_{min} \leftarrow L_{\mathcal{D}_N}(\psi_\theta)$
10:      $\widehat{\theta} \leftarrow \theta$
11: **return** $\widehat{\theta}$

An implementation of Algorithm 1 for a finite lattice has been done in [22] and good results were obtained in a simple binary image transformation problem. We note that in order for the algorithm to work for uncountable lattices, the statistical distribution applied to sample the neighbors should be chosen in a way to give a meaningful probability to chains in which the error decreases. The challenge of doing so is defining such a distribution without computing the error on the chains, what is computationally unfeasible. An implementation of the SLDA, or a modification of it, for uncountable lattices is currently an open problem.

We argue that, in general, it is not computationally feasible to apply the SLDA directly on lattice $(\Omega, \leq)$. On the one hand, since $(\Theta, \leq)$ is known a priori, for any $\theta \in \Theta$ the set $N(\theta)$ is known, so the complexity of sampling n neighbors should be that of sampling from a known statistical distribution, which is usually very low. On the other hand, if the SLDA was applied directly on $(\Omega, \leq)$, fixed a $\psi \in \Omega$, the computation of its neighborhood in $(\Omega, \leq)$ would be problem-specific and could have a great complexity. Therefore, suboptimally minimizing the empirical error in Θ via the SLDA should be less computationally complex than doing so in Ω. Furthermore, it is possible to learn on a poset $(\Omega, \leq)$ as long as it has a lattice overparametrization. In this case, minimizing the empirical error in $(\Omega, \leq)$ is a constrained optimization problem, while minimizing it in the lattice $(\Theta, \leq)$ is an unconstrained one which ought to be more efficiently solved.

We also note that the SLDA could be applied to the case in which Θ is a poset possibly contained in a lattice. In this case, the complexity of the algorithm could increase significantly due to the restrictions on $N(\theta)$. For example, sampling n neighbors of an element in a Boolean lattice is trivial, while sampling n neighbors which are also in a set of elements (the poset Θ) may be quite complex, specially when $\Theta \cap N(\theta)$ needs to be computed. In other cases, Θ being a poset may not meaningfully increase the complexity (see the application in [22]).

5   Degrees of Prior Information and Hierarchical SLDA

When one has strong prior information about the properties that $\psi^\star$ satisfies, then he can properly fix a constrained Ω and, having a lattice overparametrization of Ω, he can in principle approximately compute (3). However, when strong prior information is not available, Ω may be too complex, so overfitting occurs, or the lattice Θ may be too complex, so high computational resources are needed. Either way, if one can decompose Θ into a lattice $(\mathbb{L}(\Theta), \subset)$ of subsets of Θ then he can apply an algorithm analogous to the SLDA to minimize a validation error in $(\mathbb{L}(\Theta), \subset)$ to select a subset $\hat{\Theta} \subset \Theta$, which represents a constrained class $\{\psi_\theta : \theta \in \hat{\Theta}\} \subset \Omega$, and then learn an operator in it. This is a specific instance of learning via model selection and is also represented in Fig. 1 (see [24] for a formal definition of learning via model selection).

We proposed in [23] a hierarchical SLDA in the context of the unrestricted sequential DMNN proposed in [22] to represent W-operators. The class represented by these DMNN is composed of all operators that can be represented via the composition of d W-operators locally defined in $W_1, \ldots, W_d$, which is overparametrized by the Boolean characteristic functions of the W-operators. The set of possible sequences of Boolean functions is a Boolean lattice, and hence this is a lattice overparametrization. Since this class is quite complex, it is prone to overfit the data, so we propose a SLDA to select the windows of the W-operators, what is equivalent to creating equivalence classes on the characteristic functions' domain. Each possible sequence of windows defines a subset of Θ and varying all possible windows generates a lattice $(\mathbb{L}(\Theta), \subset)$ of subsets of Θ. This is an example where it is possible to learn lattice operators without strong prior information, and we refer to [23] for more details.

We are currently working on more general methods to learn lattice operators via a hierarchical SLDA in contexts where prior information is not available.

6   Control, Transparency and Interpretability

The lattice overparametrization paradigm for the machine learning of lattice operators has by design three important properties that modern learning methods lack in general: control, transparency and interpretability. Due to the extensive knowledge about lattice operators and the mathematical morphology toolbox, the practitioner has all the resources necessary to design Ω to fulfill its needs, so he has complete control over the class of operators. This is clear in the case of canonical DMNN in the context of set operators [22], which can represent any class of operators that can be decomposed via supremum, infimum, complement, and composition of erosions and dilations.

All the steps of the machine learning are transparent: the practitioner knows the properties of the operators in Ω since he can compute the basis of each one via $\tilde{R}_B$; he knows the lattice overparametrization, which he chose; and he can trace the path of the SLDA and inspect the choices of the algorithm at each step. By monitoring the properties of θ each time it is updated, one can make

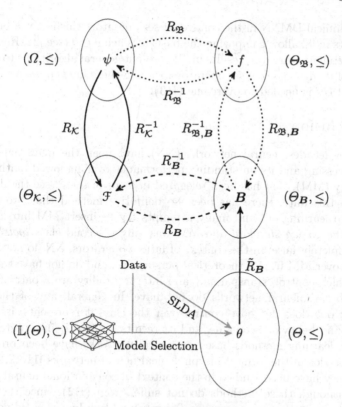

Fig. 1. The lattice isomorphisms between representations of $(\Omega, \leq)$. The dashed lines represent an isomorphism that holds when the operators in Ω have a basis representation. The dotted lines represent isomorphisms that hold for t.i. and locally defined set operators when $\mathcal{L} = \mathcal{P}(E)$ and $(E, +)$ is an Abelian group. The orange arrows represent frameworks for the learning of lattice operators via the SLDA and via model selection based on data.

sense of a possible logic that the algorithm is following. This monitoring may be that of the basis $\tilde{R}_B(\theta)$ or of the values of the parameters θ in case they have semantic information.

Finally, the mathematical properties of the learned operator $\psi_{\hat{\theta}}$ are completely known, since it suffices to compute its basis $\tilde{R}_B(\hat{\theta})$ from which its properties can be deduced. From these properties, it is possible to explain what the operator is doing, foresee cases in which it might not properly work and obtain insights about the relation between X and Y.

We note that these three properties are present in a learning framework only if $\tilde{R}_B$ can be computed, for otherwise, if one cannot reduce an operator to its basis, then he may not be able to deduce its properties. This is a possible bottleneck to this learning paradigm:

(B3) *To compute $\tilde{R}_B(\theta)$ for $\theta \in \Theta$*

For canonical DMNN in the context of set operators this is not a bottleneck since results in [8] allow computing the basis for each $\theta \in \theta$ (see [22, Remark 5.2] for more details). Moreover, results in [21] present general algorithms to compute the basis of many classes of set operators. Having these kinds of results for more general lattices is needed to overcome **(B3)**.

7   Conclusion

In the last decades, neural networks (NN) have been the main paradigm in image processing and its outstanding performance overshadowed mathematical morphology (MM), that has been relegated in favor of them (see the discussion in [2]). To this date, there has been no definitive method that brought MM to the deep learning era and many attempts try to insert MM into NN, as if NN were the *golden standard* and MM was only a second class *tool*. This last fact is completely false: in the context of lattice operators, NN do not have any advantage over MM from a theoretical perspective, and do not have the essence of MM, which control, transparency and interpretability are a part of. Indeed, needless to say, neural networks do not have, in general, any of these three properties: one does not have control over the class it represents, its learning algorithm and behavior is opaque, and its results are hardly interpretable.

Indeed, learning methods based on neural networks have been proposed in the last decades in the form of morphological neural networks [11,12,28]. More recently, they have been studied in the context of convolutional neural networks [15,20]. Although these methods do not suffer from **(B2)**, since they can be efficiently trained, they are opaque and as much a black-box as usual neural networks. In special, it is not trivial to insert constraint into them to achieve **(B1)** and once they are trained it is not possible to solve **(B3)** efficiently. Therefore, MNN do not address all the bottlenecks, but could maybe be adapted to fit the paradigm proposed in this paper. This is a current line of research.

However, NN do have great advantages from a practical standpoint, since they obtain good results and can be efficiently trained, and a great part of this success appears to be due to the possibility of proper learn in an overparametrized context. The paradigm proposed in this paper asks the following question: what if we could consider overparametrization to learn, but do not lose control and semantic understanding? To this day, there is no way of doing so with neural networks, but we argued in this paper that it is possible with MM as long as bottlenecks **(B1)**, **(B2)** and **(B3)** are overcome.

Since solving **(B1)** has been the purpose of MM for decades, it is necessary to overcome only **(B2)** and **(B3)**. The latter can be done by extending results such as those in [8,21] to general lattices, and the former can be overcome via implementations of algorithms analogous to the SLDA. We believe one should value MM and embrace aspects of NN, such as overparametrization, that can enhance MM without losing its spirit, instead of embracing NN and trying to insert MM into it.

The works in [22] and [23] are proofs of concept of the paradigm we discussed in this paper that, we believe, could be the guide for research in the machine learning of mathematical morphology in the deep learning era. The potential of such a line of research is enormous, since even if the performance of these methods come only close to that of neural networks, they may be preferred since they are controllable, transparent, and interpretable by design. Nowadays, there is an increasing demand for methods with these characteristics, and we believe that MM is in a unique position to supply them. The lattice overparametrization paradigm could be a missing piece for MM to achieve its full potential within modern machine learning.

References

1. Abu-Mostafa, Y.S., Magdon-Ismail, M., Lin, H.T.: Learning from Data, vol. 4. AMLBook, New York (2012)
2. Angulo, J.: Some open questions on morphological operators and representations in the deep learning era: a personal vision. In: Lindblad, J., Malmberg, F., Sladoje, N. (eds.) DGMM 2021. LNCS, vol. 12708, pp. 3–19. Springer, Cham (2021). https://doi.org/10.1007/978-3-030-76657-3_1
3. Atashpaz-Gargari, E., Reis, M.S., Braga-Neto, U.M., Barrera, J., Dougherty, E.R.: A fast branch-and-bound algorithm for u-curve feature selection. Pattern Recogn. **73**, 172–188 (2018)
4. Banon, G.J.F., Barrera, J.: Minimal representations for translation-invariant set mappings by mathematical morphology. SIAM J. Appl. Math. **51**(6), 1782–1798 (1991)
5. Banon, G.J.F., Barrera, J.: Decomposition of mappings between complete lattices by mathematical morphology, part i. general lattices. Sig. Process. **30**(3), 299–327 (1993)
6. Barrera, J., Banon, G.J.F.: Expressiveness of the morphological language. In: Image Algebra and Morphological Image Processing III, vol. 1769, pp. 264–275. SPIE (1992)
7. Barrera, J., Hashimoto, R.F., Hirata, N.S., Hirata, R., Jr., Reis, M.S.: From mathematical morphology to machine learning of image operators. São Paulo J. Math. Sci. **16**(1), 616–657 (2022)
8. Barrera, J., Salas, G.P.: Set operations on closed intervals and their applications to the automatic programming of morphological machines. J. Electron. Imaging **5**(3), 335–352 (1996)
9. Brun, M., Dougherty, E.R., Hirata, R., Jr., Barrera, J.: Design of optimal binary filters under joint multiresolution-envelope constraint. Pattern Recogn. Lett. **24**(7), 937–945 (2003)
10. Brun, M., Hirata, R., Barrera, J., Dougherty, E.R.: Nonlinear filter design using envelopes. J. Math. Imaging Vis. **21**, 81–97 (2004)
11. Davidson, J.L., Ritter, G.X.: Theory of morphological neural networks. In: Digital Optical Computing II, vol. 1215, pp. 378–388. SPIE (1990)
12. Dimitriadis, N., Maragos, P.: Advances in morphological neural networks: training, pruning and enforcing shape constraints. In: ICASSP 2021–2021 IEEE International Conference on Acoustics, Speech and Signal Processing (ICASSP), pp. 3825–3829. IEEE (2021)
13. Dougherty, E.R., Barrera, J., Mozelle, G., Kim, S., Brun, M.: Multiresolution analysis for optimal binary filters. J. Math. Imaging Vis. **14**, 53–72 (2001)

14. Estrela, G., Gubitoso, M.D., Ferreira, C.E., Barrera, J., Reis, M.S.: An efficient, parallelized algorithm for optimal conditional entropy-based feature selection. Entropy **22**(4), 492 (2020)
15. Franchi, G., Fehri, A., Yao, A.: Deep morphological networks. Pattern Recogn. **102**, 107246 (2020)
16. Hirata, N.S., Barrera, J., Dougherty, E.R.: Design of statistically optimal stack filters. In: XII Brazilian Symposium on Computer Graphics and Image Processing (Cat. No. PR00481), pp. 265–274. IEEE (1999)
17. Hirata, N.S.T., Barrera, J., Terada, R., Dougherty, E.R., Talbot, H., Beare, R.: The incremental splitting of intervals algorithm for the design of binary image operators. In: Proceedings of the 6th ISMM, pp. 219–228 (2002)
18. Hirata, N.S.T., Dougherty, E.R., Barrera, J.: Iterative design of morphological binary image operators. Opt. Eng. **39**(12), 3106–3123 (2000)
19. Hirata Junior, R., Brun, M., Barrera, J., Dougherty, E.R.: Multiresolution design of aperture operators. J. Math. Imaging Vis. **16**, 199–222 (2002)
20. Hu, Y., Belkhir, N., Angulo, J., Yao, A., Franchi, G.: Learning deep morphological networks with neural architecture search. Pattern Recogn. **131**, 108893 (2022)
21. Jones, R., Svalbe, I.D.: Basis algorithms in mathematical morphology. In: Advances in electronics and electron physics, vol. 89, pp. 325–390. Elsevier (1994)
22. Marcondes, D., Barrera, J.: Discrete morphological neural networks. arXiv preprint arXiv:2309.00588 (2023)
23. Marcondes, D., Feldman, M., Barrera, J.: An algorithm to train unconstrained sequential discrete morphological neural networks. arXiv preprint arXiv:2310.04584 (2023)
24. Marcondes, D., Peixoto, C.: Distribution-free deviation bounds of learning via model selection with cross-validation risk estimation. arXiv preprint arXiv:2303.08777 (2023)
25. Reis, M.S., Barrera, J.: Solving problems in mathematical morphology through reductions to the u-curve problem. In: Hendriks, C.L.L., Borgefors, G., Strand, R. (eds.) ISMM 2013. LNCS, vol. 7833, pp. 49–60. Springer, Heidelberg (2013). https://doi.org/10.1007/978-3-642-38294-9_5
26. Reis, M.S., Estrela, G., Ferreira, C.E., Barrera, J.: Optimal Boolean lattice-based algorithms for the u-curve optimization problem. Inf. Sci. **470**, 97–114 (2018)
27. Ris, M., Barrera, J., Martins, D.C.: U-curve: a branch-and-bound optimization algorithm for u-shaped cost functions on Boolean lattices applied to the feature selection problem. Pattern Recogn. **43**(3), 557–568 (2010)
28. Ritter, G.X., Sussner, P.: An introduction to morphological neural networks. In: Proceedings of 13th International Conference on Pattern Recognition, vol. 4, pp. 709–717. IEEE (1996)
29. Vapnik, V.: The Nature of Statistical Learning Theory. Springer, New York (1999). https://doi.org/10.1007/978-1-4757-3264-1

Convex Optimization for Binary Tree-Based Transport Networks

Raoul Sallé de Chou[1,2(✉)] , Mohamed Ali Srir[1], Laurent Najman[3] ,
Nicolas Passat[4] , Hugues Talbot[2] , and Irene Vignon-Clementel[1]

[1] Inria, Palaiseau, France
[2] CentraleSupelec, Inria, Université Paris-Saclay, Gif-sur-Yvette, France
raoul.salle-de-chou@inria.fr
[3] Université Gustave Eiffel, CNRS, LIGM, Champs-sur-Marne, France
[4] Université de Reims Champagne Ardenne, CReSTIC, Reims, France

Abstract. Optimizing transport networks is a well-known class of problems that have been extensively studied, with application in many domains. Here we are interested in a generalization of the Steiner problem, which entails finding a graph minimizing a cost function associated with connecting a given set of points. In this paper, we concentrate on a specific formulation of this problem which is applied to the generation of synthetic vascular trees. More precisely, we focus on the Constrained Constructive Optimization (CCO) tree algorithm, which constructs a vascular network iteratively, optimizing a blood transport energy efficiency. We show that the classical incremental construction method often leads to sub-optimal results, and that a better global solution can be reached.

Keywords: Constrained Constructive Optimization · Discrete optimal transport · Binary trees · Convex optimization · Numerical twin

1 Introduction

Identifying an optimal transportation network is a topic of significant interest across a wide range of fields, including urban modeling, water supply, telecommunication, and the study of vascular system [1,10]. The problem consists of finding a path that minimizes a cost function while interconnecting points of interest and utilizing possible intermediate branching points. This paper specifically explores a formulation of the problem where the points of interest are the terminal leaves of a rooted binary tree. More specifically, we study the application of this approach to the generation of synthetic vascular trees.

With the emergence of patient-specific models as non-invasive diagnostic tools in medicine, it has become important to generate realistic vessel networks efficiently and effectively. An example is to construct numerical twins of various organs, for example to study their associated vascular diseases [13,15]. Most

---

Supported by an industrial grant from Heartflow, Inc.

published methods construct these trees incrementally, branch by branch, with the underlying principle that blood vessels supply tissues by filling the tissue volume while minimizing transport costs.

A widely adopted method is the Constrained Constructive Optimization (CCO) tree algorithm, which locally optimizes the tree geometry after the addition of a new segment. This method aims to minimize total vascular volume and was initially developed by Schreiner et al. [16]. Several extensions of this method have been proposed for the generation of organ-specific vascular trees in 3D non-convex volumes [6,9,17,18] or the concurrent construction of multiple concurrent trees [3]. As stated by Keelman et al. [11], one of the main drawbacks of CCO is that it only performs local optimizations, leading to suboptimal results. In order to search for optimal global solutions, they devised a method based on simulated annealing (SA). This approach enabled them to explore multiple tree structures while trying to converge towards a global minimum with respect to a cost function. In a recent publication, Jessen et al. [7] combined the CCO algorithm with the SA method to address the local optimality issue. In order to reduce the search space of SA, and thus the high computational cost, they restrained the tree structure exploration to changes in the tree connectivity (or tree topology). The optimization of the branching node positions was computed thanks to a global non-linear solver inspired by CCO's local optimization. While they successfully reduced the total vascular volume, there is currently no proof of convergence to a global minimum.

Another tree generation approach, known as Global Constructive Optimization (GCO) [4], also aims at building optimal tree structures. It starts from uniformly distributed random points in the volume, which remain fixed during the tree generation. The method is initialized by connecting all points to a root node. Subsequently, through a series of iterative split and merge operations, it reshapes the tree topology. During each iteration, the method optimizes the bifurcation positions with the objective of volume minimization. Notably, this optimization problem has been proven to be convex, guaranteeing the achievement of optimal node positions. In contrast to CCO, it is essential to note that a key limitation of this method lies in its inability to generate multiple trees within the same volume.

In our present work, we introduce a novel approach for the CCO algorithm. Instead of minimizing the total volume, our approach seeks to minimize an energy cost function. Notably, we demonstrate that our optimization framework exhibits convexity, leading to improved global results. Moreover, our approach significantly reduces computational costs in comparison to the volume-based CCO method.

2   Method

In this section, we first describe the general formulation of the optimal transport problem considered in this paper. Subsequently, we delve into the original CCO method [16], and its model assumptions. Finally, we introduce our novel vision of the CCO framework. we emphasize that all the CCO implementations detailed in this paper are used for the generation of two-dimensional vascular trees, but that the construction generalizes to 3D trees without difficulty.

2.1 Optimal Transport Problem

The optimal transport problem explored in this paper can be formulated as follows: starting with a binary tree that originates from a root point, and connecting N terminal points through $N-1$ branching points (bifurcation points), each defined by two-dimensional Euclidean coordinates, the objective is to determine the optimal positions for these bifurcation points while keeping the terminal and root nodes fixed. The cost function is expressed as follows:

$$S = \sum_{i=1}^{2N-1} w_i l_i \tag{1}$$

Here, the l_i values ($1 \leq i \leq 2N - 1$) denote the lengths of the segments between each pair of connected points, and the w_i values are positive weights assigned to each segment. Notably, since the Euclidean distance of a segment is convex with respect to its endpoints, the sum S itself is convex [4]. Consequently, optimal bifurcation positions, that minimize S, can be determined. However, it is essential to note that the solution may not be unique, as the Euclidean distance is not strictly convex [2].

2.2 Constrained Constructive Optimization

Model Assumptions. Vascular trees are represented as rooted binary trees composed of one root node and N terminal nodes. The edges constituting the tree are represented as cylindrical segments, interconnected through $N - 1$ branching—or bifurcation—nodes.

Blood flow is modeled in the vascular network as an incompressible, homogeneous Newtonian fluid, and is assumed unidirectional, axisymmetric, developed and stationary. Thus for each segment j , the resistance to flow R_j and subsequent pressure drop Δp_j are related to its flow rate q_j through Poiseuille's law:

$$\Delta p_j = R_j q_j \tag{2}$$

$$R_j = \frac{8\mu l_j}{\pi r_j^4} \tag{3}$$

with μ the blood viscosity, and l_j and r_j, q the segment length and radius respectively. At all terminal nodes, the flow rates q_{term} are set (here to be equal). Thanks to mass conservation, the sum of the q_{term} corresponds to the perfusion flow rate at the tree root q_{tot}.

At a bifurcation level, the parent and children segments radii obey the following power law:

$$r_{parent}^{\gamma} = r_{child1}^{\gamma} + r_{child2}^{\gamma} \tag{4}$$

with γ a parameter which has been shown empirically to vary between 2 and 3, depending on the size of the arteries or veins. In the next sections, this parameter is set to 3 to simulate small arteries. This relationship has been defined by Murray [14], and is derived from the assumptions that vascular networks should be constructed so that the cost to transport and maintain/generate blood is minimal. Indeed, for a vessel modeled as a tube with Poiseuille's law, Murray defines this cost as the sum of a term due to viscous work, and a second one due to the generation of the blood volume in the tube. This sum is homogeneous to an energy. Based on Poiseuille law, (3), this energy (or rather power) is defined as:

$$E_j = \frac{8\eta q_j^2 l}{\pi r_j^4} + b\pi l_j r_j^2 \tag{5}$$

with b the volumetric metabolic cost. Searching the minimum of this equation, with respect to the radius, leads to Murray's law:

$$q_j = kr_j^3 \tag{6}$$

with k a constant independent of j. At a bifurcation, with one parent and two children segments, mass conservation reads:

$$q_{parent} = q_{child_1} + q_{child_2} \tag{7}$$

which, combined with (6) leads to (4).

Current CCO Algorithm. The CCO method aims at generating a synthetic vascular tree incrementally, by adding segments one by one, until a desired number of terminal segments is reached. At each growth step, the new local structure is optimized: the total vascular volume is minimized, by considering the cylindrical segment:

$$V = \pi \sum_{i=1}^{N_{segments}} l_i r_i^2 \tag{8}$$

The addition of a new terminal segment can be described as follows: first, a new terminal node location is randomly sampled in the perfusion territory. Then, the connection between two points defining a neighboring segment and the new node, is tested. For that purpose, the latter segment is replaced by 3 new segments and a bifurcation node connecting the 3 points. Local optimization of the bifurcation structure is then performed. The bifurcation point position and the 3 newly added segments radii are optimized with respect to total volume, while ensuring Poiseuille's law (3) and Murray's bifurcation law (4). Since a new terminal node is added, the total flow in the tree is incremented by q_{term}. Segment resistances upstream to the new bifurcation, need to be adjusted by rescaling recursively the segments radii. Finally, a fixed number of connections within the closest neighboring segments are tested and only the one inducing the lowest total volume is kept.

The volume-based minimization method comes from Kamya [8]. The method seems similar to the energy-based approach, as it had the following objective: *"the lower the blood volume in the vascular tree becomes, the smaller the energy for blood maintenance is spent"*. Regarding (5), it seems that minimizing volume shares similarities with minimizing energy, but does not take into account the effect of viscous forces. One can note that this original CCO bifurcation optimization is not completely volume-based, as it integrates Murray's bifurcation law (4). However, the use of the bifurcation relationship instead of (6) induces a loss of information, and the model is not mathematically well-posed. Consequently, the use of other equations such as Poiseuille's law becomes necessary to solve the local optimization problem, which becomes more complex. Solving this problem is often done by iterative solvers, and no convergence toward a global minimum has been demonstrated.

2.3 New CCO Approach

Local Optimization Algorithm. We now describe a new method for optimizing a bifurcation in CCO that relies solely on energy considerations, and is mathematically well-posed. This approach should provide a more consistent and robust solution.

Because terminal flows q_{term} are fixed in CCO and mass is conserved at each bifurcation, one can compute from the terminal nodes to the tree root, the flow q_{parent}, q_{child1} and q_{child2} for every segment of a bifurcation. Additionally, the best radii r_0, r_1, r_2 and bifurcation coordinates (x, y) can be found by minimizing the energy defined in (5):

$$\begin{cases} \dfrac{\partial E}{\partial x} = 0 \\[2mm] \dfrac{\partial E}{\partial y} = 0 \\[2mm] \dfrac{\partial E}{\partial r_i} = 0 \text{ for } i \in \{parent, child1, child2\} \end{cases} \tag{9}$$

As in Murray [14] for the case of a single bifurcation, with k a constant, the partial derivatives of E with regard to the radii, lead to:

$$q_i = k r_i^3$$

Since the flow rates q_i are known, this sets all the radii for a given k as:

$$r_i = \sqrt[3]{\frac{q_i}{k}} \tag{10}$$

Employing (10) in order to compute the total energy of a bifurcation, leads to:

$$E = \sum_i E_i = \left(\frac{8k^2\eta}{\pi} + b\pi \right) \sum_i l_i r_i^2 \tag{11}$$

This new form of the energy, when considering r_i as fixed values, can be expressed in a similar form as (1). As stated in Sect. 2.1, this objective function has been demonstrated to exhibit convexity with respect to the point coordinates defining the segment lengths l_i. Consequently, it becomes feasible to identify a global minimum solution for the new bifurcation point positions, ultimately leading to an optimal bifurcation geometry. This optimum can be reached under the condition that the constant k can be computed. In fact, it is easy to see by recursivity, that this constant is the same for the whole tree. In CCO, medical image data usually give the root radius. When such data is not available, an additional condition such as specifying a target pressure drop between the root and the leaves, needs to be given. This target pressure drop may, for instance, represent the average pressure drop occurring between arteries and arterioles. Based on this target value, and considering the overall tree resistance and total flow at growth step t, the root radius can be determined. Consequently, with the knowledge of the root radius, we can define at each growth step the constant k:

$$k^t = \frac{q_{tot}^t}{(r_{root})^3} \tag{12}$$

Global Optimization Algorithm. When adding a new terminal node in the current CCO implementation, the entire tree is adjusted to be able to sustain the incremented total flow. These modifications specifically pertain to the radii, while the previous bifurcation points remain untouched. Nevertheless, the creation of a new bifurcation modifies the geometry of the tree, which leads to uncertainties regarding the optimality of the bifurcation positions previously found in the existing tree.

In this section, we describe a method to optimize all bifurcation positions and all segment radii at once in a tree. This method starts from given terminal nodes positions and connectivity of a tree, for example here given by a CCO tree. It is based on an energy-based system of equations. This system has been shown to deliver an optimal solution [5].

Similarly, as in (9), one can define, for a vascular tree with N terminal nodes, a system of $3N - 1$ equations with $3N - 1$ unknowns:

$$\begin{cases} k = \frac{Q_{tot}}{(r_{root})^3} \\\\ \frac{\partial E}{\partial x_i} = 0 \\\\ \frac{\partial E}{\partial y_i} = 0 \\\\ \frac{\partial E}{\partial r_j} = 0 \end{cases} \tag{13}$$

with i iterating over the $N-1$ bifurcation nodes, and j over the $2N-1$ segments. Solving the partial derivatives of E with respect to the radii leads to:

$$r_j = \sqrt[3]{\frac{f_j}{k}} \tag{14}$$

After injecting (14) in the energy equation, we obtain:

$$E = \sum_{j=0}^{N_s-1} E_j = \sum_{j=0}^{N_s-1} \left(\frac{8k^2\eta}{\pi} + b\pi\right) l_j r_j^2 \tag{15}$$

We note that (15) has the same form as (8), thus a global solution minimizing this energy function with regard to the bifurcation nodes can be found. This time however, it is a minimum for all the bifurcations at once, so global to the whole tree.

2.4   Tree Comparison Metric

To be able to compare the trees in terms of morphometry, and more precisely to measure the asymmetry between branches, an asymmetry ratio is defined for every branches starting from a segment u as:

$$\mathcal{A}_u = \frac{|l_u - r_u|}{l_u + r_u - 2} \tag{16}$$

where l_u (resp. r_u) is the number of leaves node downstream to the left (resp. right) children segment of the parent segment u. An asymmetry ratio of 0 means the branch is completely symmetric, while a value of 1 exhibits an asymmetric branch.

3   Results

3.1   Comparison of Volume vs. Energy Approaches

To facilitate a comparative analysis of the two distinct approaches, one based on volume optimization (Sect. 2.2) and the other on energy optimization (local optimization algorithm, Sect. 2.3), we generated synthetic trees with varying numbers of terminal nodes: 100, 200, 300, and 400. For each specific number of terminal nodes, we constructed a total of 10 different trees, with different random seeds for the generation of new random node locations within the tree generation. All trees were grown within the same circular domain, as depicted in Fig. 1.

Figure 2 shows that the energy-based method yields significantly lower computational time than the volume-based approach. The time saved seems more or less constant as the relative time difference exhibits an average gain of 31.7%, 36.2%, 33.9% and 36.3% for trees with 100, 200, 300 and 400 terminal segments (or nodes), respectively.

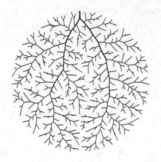 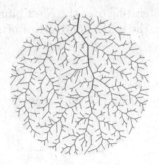

Fig. 1. Example of trees with 400 terminal segments simulated with the volume-based (left) and energy-based (right) methods.

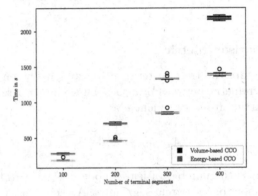

Fig. 2. Boxplot of the computational time vs the number of terminal segments for the volume and energy-based CCO (without global optimization) methods.

A comparison of morphometric features is conducted between the two methods, as illustrated in Fig. 3. These features encompass the mean diameters and mean lengths at each bifurcation level (the number of segments in the path from the current segment to the root segment), along with the radius-weighted average asymmetry ratio (see (16)). On average, it appears that both methods yield similar tree structures in terms of radii and length distributions. Notably, the energy-based method tends to generate deeper trees, evidenced by a maximum of 46 bifurcation levels, in contrast to 43 levels achieved through volume optimization. Additionally, the distribution of branch lengths exhibits a slightly broader variation. However, the asymmetry ratio reveals no significant differences between the two methods. It is worth noting that both approaches produce somewhat asymmetric trees, with smaller branches exhibiting greater symmetry.

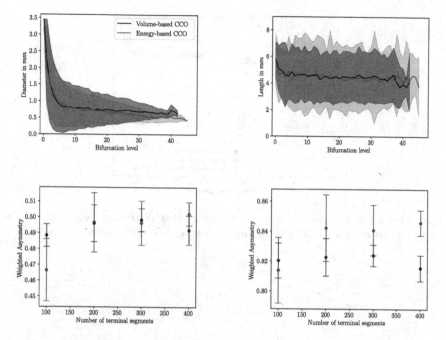

Fig. 3. Morphometric analysis along bifurcation levels: average diameter (top left), and average length (top right) along standard deviation envelope at each bifurcation level, on 10 simulated trees with 400 terminal segments. Weighted average asymmetry ratio, for both methods, for all segments (bottom left), and by taking into account only segments with at least 3 downstream children (bottom right). The average is weighted by the segment radius in both cases

3.2 Global Optimization Results

We applied our global optimization framework (Sect. 2.3) to the previously generated synthetic trees that were initially constructed with the energy-based approach. Consistently with the energy formulation outlined in (15), an energy reduction achieved through our global framework should be approximately equivalent to a volume reduction, up to a constant factor. Consequently, for the purpose of comparing the energy-based approaches (with and without global optimization) to the volume-based method, we have plotted the total vascular volume in Fig. 5. The results show that even prior to global optimization, the energy-based method already yields lower volumes compared to the volume-based method. The mean relative volume differences are 21.3%, 36.7%, 36.1%, and 37.2% for trees with 100, 200, 300, and 400 terminal segments, respectively. In addition, our global optimization method successfully leads to a reduction in the total vascular volume (and energy) across all cases when compared to the CCO energy-based trees. On average, the volumes are decreased by 3.0%, 7.9%, 7.8% and 7.7% for trees with 100, 200, 300 and 400 terminal segments, respectively (Fig. 4).

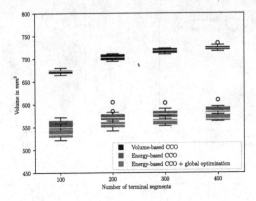

Fig. 4. Boxplot of the vascular volume vs the number of terminal segments for the volume and energy-based CCO (with and without global optimization) methods.

Fig. 5. Example of a tree generated with the energy-based CCO method (in blue), which is then optimized globally (in red). In the right image, two trees starting from two different root positions are constructed with the forest growth algorithm developed in [6] (Color figure online).

4   Discussion

Most synthetic vascular tree generative models are built on the foundational principle, first proposed by Murray [14], that vascular structures aim for optimally minimizing transport costs. The CCO method aims to minimize vascular volume. However, relying solely on a volume-based objective could lead to an incomplete transport cost. We introduce a term related to viscous work, consistently with the original Murray's derivation for a single bifurcation. Incorporating this new objective function leads to a well-posed and convex problem. Related objective functions have been mentioned in previous publications on tree generation methods, but never in the context of CCO. For instance, Kim *et al.* [12] developed a tissue-growth-based tree generation method, which shares similarities with CCO in its incremental tree construction approach, involving local optimization at each new bifurcation. Although they aim at minimizing the same energy function, it remains unclear if they also utilize this energy function to scale the radii. In contrast to our findings, they report achieving similar volumes between their method and CCO, with differences below 5%. Our approach results in a significant reduction in tree energy (or volume, see Sect. 2.3), approximately by 36% compared to the volume-based CCO. This reduction can be attributed to the optimality of our solution and by our more consistent way to compute the radii.

Concerning tree morphometry, our analysis tends to indicate that the energy-based method produces similar trees as the volume-based method. In order to fully capture the impact of the new method, we plan to extend our local and global optimization frameworks to generate 3D organ-specific trees in the future. This will enable comparisons with real data.

A major consideration in CCO is the reduction of computational costs: our approach notably achieves an average cost reduction of approximately 35%. This improvement is facilitated by the convex nature of our problem, which enhances convergence efficiency.

As highlighted by previous publications [7,11], one limitation of CCO lies in its focus on optimizations of one bifurcation at a time only. In that sense it provides a local tree optimization. Based on our energy-based CCO method, we developed a global optimization framework. We showed that a global optimum solution can be reached, and we succeeded to obtain an energy reduction up to 8%. In comparison, the branching node positions optimization developed by Jessen *et al.* achieved a volume reduction of 4% from a volume-based CCO tree. Unlike our approach, their system of equations relies on a CCO volume-based local optimization, yielding non-optimal solutions.

In contrast to the aforementioned works [7,11], our approach does not attempt to modify the tree topology: we maintain the connectivity from CCO. However, we have seen that our tree generation problem can be linked to the well-studied field of optimal transport (Sect. 2.1). Specifically, the task of finding optimal trees could be related to the Steiner tree problem [2]. Further research, potentially incorporating methods for solving this problem, could be addressed to identify an optimal tree topology.

To conclude, we have proposed a novel optimization framework for CCO, relying on an energy-based transport cost, which leads to a well-posed system of equations. While simulating similar 2D vascular trees as CCO in terms of morphometry, thus new method succeeded to produce better optimal solutions, and more efficiently.

References

1. Bohn, S., Magnasco, M.O.: Structure, scaling, and phase transition in the optimal transport network. Phys. Rev. Lett. **98**(8), 088702 (2007)
2. Brazil, M., Zachariasen, M.: Optimal Interconnection Trees in the Plane, vol. 29. Springer, Switzerland (2015)
3. Cury, L., Maso Talou, G., Younes-Ibrahim, M., Blanco, P.: Parallel generation of extensive vascular networks with application to an archetypal human kidney model. Roy. Soc. Open Sci. **8**(12), 210973 (2021)

4. Georg, M., Preusser, T., Hahn, H.K.: Global constructive optimization of vascular systems. Technical report. WUCSE-2010-11, Washington University in St. Louis (2010)
5. Hahn, H.K., Georg, M., Peitgen, H.O.: Fractal aspects of three-dimensional vascular constructive optimization. In: Losa, G.A., Merlini, D., Nonnenmacher, T.F., Weibel, E.R. (eds.) Fractals in Biology and Medicine, pp. 55–66. Springer, Basel (2005). https://doi.org/10.1007/3-7643-7412-8_5
6. Jaquet, C., et al.: Generation of patient-specific cardiac vascular networks: a hybrid image-based and synthetic geometric model. IEEE Trans. Biomed. Eng. **66**(4), 946–955 (2018)
7. Jessen, E., Steinbach, M.C., Debbaut, C., Schillinger, D.: Rigorous mathematical optimization of synthetic hepatic vascular trees. J. R. Soc. Interface **19**(191), 20220087 (2022)
8. Kamiya, A., Togawa, T.: Optimal branching structure of the vascular tree. Bull. Math. Biophys. **34**, 431–438 (1972)
9. Karch, R., Neumann, F., Neumann, M., Schreiner, W.: A three-dimensional model for arterial tree representation, generated by constrained constructive optimization. Comput. Biol. Med. **29**(1), 19–38 (1999)
10. Katifori, E., Szöllősi, G.J., Magnasco, M.O.: Damage and fluctuations induce loops in optimal transport networks. Phys. Rev. Lett. **104**(4), 048704 (2010)
11. Keelan, J., Chung, E.M., Hague, J.P.: Simulated annealing approach to vascular structure with application to the coronary arteries. Roy. Soc. Open Sci. **3**(2), 150431 (2016)
12. Kim, H.J., Rundfeldt, H.C., Lee, I., Lee, S.: Tissue-growth-based synthetic tree generation and perfusion simulation. Biomech. Model. Mechanobiol. **22**(3), 1095–1112 (2023)
13. Menon, K., Khan, M.O., Sexton, Z.A., Richter, J., Nieman, K., Marsden, A.L.: Personalized coronary and myocardial blood flow models incorporating CT perfusion imaging and synthetic vascular trees. medRxiv, 2023–08 (2023)
14. Murray, C.D.: The physiological principle of minimum work: I. The vascular system and the cost of blood volume. Proc. Nat. Acad. Sci. **12**(3), 207–214 (1926)
15. Papamanolis, L., et al.: Myocardial perfusion simulation for coronary artery disease: a coupled patient-specific multiscale model. Ann. Biomed. Eng. **49**, 1432–1447 (2021)
16. Schreiner, W., Buxbaum, P.F.: Computer-optimization of vascular trees. IEEE Trans. Biomed. Eng. **40**(5), 482–491 (1993)
17. Schwen, L.O., Preusser, T.: Analysis and algorithmic generation of hepatic vascular systems. Int. J. Hepatol. **2012**, 357687 (2012)
18. Talou, G.D.M., Safaei, S., Hunter, P.J., Blanco, P.J.: Adaptive constrained constructive optimisation for complex vascularisation processes. Sci. Rep. **11**(1), 6180 (2021)

Training Morphological Neural Networks with Gradient Descent: Some Theoretical Insights

Samy Blusseau[✉][iD]

Mines Paris, PSL University, Centre for mathematical morphology (CMM), Fontainebleau, France
`samy.blusseau@minesparis.psl.eu`

Abstract. Morphological neural networks, or layers, can be a powerful tool to boost the progress in mathematical morphology, either on theoretical aspects such as the representation of complete lattice operators, or in the development of image processing pipelines. However, these architectures turn out to be difficult to train when they count more than a few morphological layers, at least within popular machine learning frameworks which use gradient descent based optimization algorithms. In this paper we investigate the potential and limitations of differentiation based approaches and back-propagation applied to morphological networks, in light of the non-smooth optimization concept of Bouligand derivative. We provide insights and first theoretical guidelines, in particular regarding initialization and learning rates.

Keywords: Morphological neural networks · Nonsmooth optimization · Lattice operators

1 Introduction

Morphological neural networks were introduced in the late 1980s [5,17], and have been revisited in recent years [4,6,8,12,18]. With the growing maturity of deep learning science, new exciting perspectives seem to open and give hope for significant breakthroughs.

In image processing, with the development of libraries specialized in morphological architectures [15], where basic as well as advanced operators are implemented, such as geodesical reconstruction layers [16], it is now within reach to train end to end pipelines which include morphological preprocessing and postprocessing, and to use the know how of the morphological community to impose topological and geometrical constraints inside deep networks.

Furthermore, morphological networks can help investigate in practice the representation theory of lattice operators initiated by Georges Matheron [2,9,11]. Just as the universal approximation theorem for the multi-layer perceptron, the representation theorem of lattice operators with families of erosions and antidilations, is an existence one and is asymptotic, but does not provide any algorithm to actually exhibit such representations. Since these decompositions can

S. Brunetti et al. (Eds.): DGMM 2024, LNCS 14605, pp. 229–241, 2024.
https://doi.org/10.1007/978-3-031-57793-2_18

be implemented as morphological layers, we may hope to learn these representations *from data*.

Yet, the optimization of morphological architectures is still slow and difficult.
Despite the several contributions in this area, [1,6,12], architectures including
morphological layers are often quite shallow and do not compete with the state
of the art networks for image analysis. On the one hand, it may be due to
the Fréchet non differentiability of the morphological layers, reason for several
attempts to replace them by smooth approximations [7,8]. On the other hand,
non smooth operations such as the Rectifier Linear Unit (ReLU) or the max-
pooling, are commonly used in successfully trained architectures, while smooth
morphological ones do not seem to solve all the optimization issues.

In this paper we investigate the potential and limitations of training morphological neural networks with differentiation-based algorithms relying on back-
propagation and the chain rule. In Sect. 2 we introduce morphological networks,
and recall in Sect. 3 the principles of gradient descent, back-propagation and
chain rule. Section 4 presents the concept of Bouligand derivative, which is suited
to morphological layers. In Sect. 5 we expose the possibilities and issues of this
framework within the chain-rule paradigm, before concluding in Sect. 6.

2   Morphological Networks

There is no universal definition of morphological neural networks, but most architectures that are called so, are neural networks including at least a morphological
layer. In turn, a morphological layer is one computing a morphological operation
such as a dilation or an erosion, or sometimes a (weighted) rank filter. In this
paper we will focus on dilation and erosion layers, composed with each other or
with other classical (dense or convolutional) layers.

Dilation Layers. We will call dilation layer a function $\delta_W : \mathbb{R}^n \to \mathbb{R}^m$, $n, m \in$
$\mathbb{N}^*$, defined by

$$\delta_W : \mathbf{x} = (x_1, \ldots, x_n) \mapsto \left(\max_{1 \leq k \leq n} x_k + w_{i,k} \right)_{1 \leq i \leq m} \tag{1}$$

where the $w_{i,k}$ are the real valued coefficients of a matrix $W \in \mathbb{R}^{m \times n}$, and
the parameters (or *weights*) of the layer. Extended to the complete lattices $\bar{\mathbb{R}}^n$
and $\bar{\mathbb{R}}^m$, where $\bar{\mathbb{R}} := \mathbb{R} \cup \{-\infty, +\infty\}$, δ_W is a shift invariant morphological
dilation [3,10]. In practical neural architectures the input and output of a layer
are usually represented as sets of vectors, called feature maps. In such a setting,
each output feature map would be the supremum of dilations like δ_W, of the
input feature maps. By reshaping the set of input feature maps into one input
vector, and the set of output ones into one output vector, we get the equivalent
formulation (1), simpler to analyze.

Erosion Layers. Similarly, we will call erosion layer a function $\varepsilon_W : \mathbb{R}^m \to \mathbb{R}^n$,
$n, m \in \mathbb{N}^*$, defined by

$$\varepsilon_W : \mathbf{x} = (x_1, \ldots, x_m) \mapsto \left(\min_{1 \leq k \leq m} x_k - w_{k,j} \right)_{1 \leq j \leq n}. \tag{2}$$

Note that the sign "−" and the transposition ($w_{k,i}$ instead of $w_{i,k}$) in the definition are meant to make (ε_W, δ_W) a morphological adjunction.

Morphological Networks. As said earlier, in this paper any neural network including at least a morphological layer is considered a morphological network. This includes sequential compositions of dilations and erosions layers, supremum of erosion layers, infimum of dilation layers, and composition with classical layers (linear operators followed by a non-linear activation function). This also includes anti-dilations and anti-erosions, which are of the kind $\mathbf{x} \mapsto \delta_W(-\mathbf{x})$ and $\mathbf{x} \mapsto \varepsilon_W(-\mathbf{x})$. Note however that the composition $\delta_A \circ \delta_B$ of two dilation layers can be considered as one dilation layer δ_C where $C \in \mathbb{R}^{m \times n}$ is the max-plus matrix product of $A \in \mathbb{R}^{m \times p}$ by $B \in \mathbb{R}^{p \times n}$,

$$C_{ij} = \max_{1 \leq k \leq p} A_{ik} + B_{kj}, \quad 1 \leq i \leq m, \ 1 \leq j \leq n. \tag{3}$$

Furthermore, the pointwise maximum of l dilation layers $\delta_{W_1}, \ldots, \delta_{W_l}$ (where all W_is are the same size), is also equivalent to one dilation layer δ_{W^*} where W^* is the pointwise maximum the matrices W_is.

Similarly, in theory it is pointless to compose or take the minimum of erosion layers, since such operators can be represented (and learned) as one erosion layer.

3 Optimization with Gradient Descent

Let us consider a classic neural network setting where a function $f_\theta : \mathbb{R}^n \to \mathbb{R}^+$ depending on a parameter $\theta = [\theta_1, \ldots, \theta_L]$ is a composition of L functions

$$f_\theta := f_{L,\theta_L} \circ f_{L-1,\theta_{L-1}} \circ \cdots \circ f_{1,\theta_1}, \tag{4}$$

each f_{k,θ_k} depending on its own parameter $\theta_k \in \mathbb{R}^{p_k}$ and mapping $\mathbb{R}^{n_k}$ to $\mathbb{R}^{n_{k+1}}$, with $n_1 = n$ and $n_L = 1$ (we include the loss function as part of the last layer). Typically, we would like to find a parameter θ which minimizes the expectation $\mathbb{E}(f_\theta(X))$ where X is a random variable that models the distribution of the data we want to process[1]. In practice this can be done by applying f_θ to samples $x_1, \ldots, x_N$ of X and iteratively update $\theta \leftarrow \theta + \Delta\theta$ in a way that decreases the function at the current sample, $f_{\theta+\Delta\theta}(x_i) \leq f_\theta(x_i)$. Hence the change $\Delta\theta$ that is looked for is a *descent direction*.

3.1 Gradient Descent

Where it exists, the gradient of a function $g : \mathbb{R}^n \to \mathbb{R}$ precisely provides a descent direction. Indeed if g is Fréchet-differentiable[2] at $x \in \mathbb{R}^n$, then

$$\forall h \in \mathbb{R}^n, \ \forall \eta \geq 0, \quad g(x + \eta h) = g(x) + \eta(\langle \nabla g(x), h \rangle + \epsilon(\eta)) \tag{5}$$

---

[1] Recall that f_θ is real valued since we include the loss in the last layer f_{L,θ_L}.
[2] The Fréchet derivative is just the usual derivative, which is a linear function, like $h \mapsto \langle \nabla g(x), h \rangle$ in (5).

where $\langle \cdot, \cdot \rangle$ is the inner product in $\mathbb{R}^n$ and ϵ is a function that goes to zero when η goes to zero. Hence, if $\nabla g(x) \neq 0$, for η sufficiently small $|\epsilon(\eta)| < \|\nabla g(x)\|^2$ and therefore $g(x - \eta \nabla g(x)) < g(x)$, for which $-\nabla g(x)$ is called a descent direction of g at x. Eq. (5) also implies that any $h \in \mathbb{R}^n$ such that $\langle \nabla g(x), h \rangle \leq 0$ is a descent direction. Furthermore, it shows $-\nabla g(x)$ is the *steepest* descent direction: given $\eta > 0$ sufficiently small, any unit vector v verifies $g\left(x - \eta \frac{\nabla g(x)}{\|\nabla g(x)\|}\right) \leq g(x + \eta v)$.

These results can be applied to the function $g : \theta \mapsto f_\theta(x)$ for a fixed sample x, provided g is a Fréchet-differentiable (also called F-differentiable) function of θ. In that case we will note $\nabla_\theta f_\theta(x) := \nabla g$.

3.2   Back Propagation and the Chain Rule

To compute $\nabla_\theta f_\theta(x)$, it is sufficient to compute each $\nabla_{\theta_i} f_\theta(x)$, which can also be noted $\frac{\partial f_\theta(x)}{\partial \theta_i}$, and is the gradient of the function $g_i : \theta_i \mapsto f_\theta(x)$, x and the other parameters $\theta_j, j \neq i$, being fixed. Indeed, the gradient with respect to θ is the concatenation of the gradients with respect to the θ_is, $\nabla_\theta f_\theta(x) = [\nabla_{\theta_1} f_\theta(x), \ldots, \nabla_{\theta_L} f_\theta(x)]$.

To obtain these, the so called "chain rule" is applied, involving the (Fréchet) derivative of each layer with respect to its input variable and its derivative with respect to its parameter. The derivatives with respect to the input variables tell earlier layers (i.e. the layers that are closer to the input) how they should change their output values to eventually decrease the whole function $f_\theta(x)$. They play a role of message passing to earlier layers. The derivative with respect to a layer's parameter tells how to update this parameter in order to comply with the instruction received from later layers (that is, layers closer to the output).

More formally, we can see this in the case of f_θ as defined in (4). We note $\mathbf{x}_1 := x$ the input variable of f_{1,θ_1} (and therefore f_θ), and $\mathbf{x}_{k+1} := f_{k,\theta_k}(\mathbf{x}_k)$, $1 \leq k \leq L - 1$. For fixed $\theta_k, \mathbf{x}_k$, we denote by $f'_{k,\theta_k}(\mathbf{x}_k; \cdot)$ and $f'_{k,\mathbf{x}_k}(\theta_k; \cdot)$ the Fréchet derivatives of the k-th layer with respect to its input variable and parameter respectively. Then the chain rule algorithm can be summarized as follows (see Fig. 1).

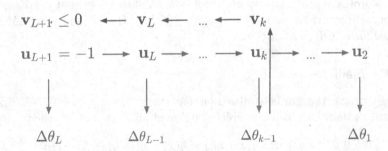

Fig. 1. illustration of the chain rule algorithm.

Initialize the Message $\mathbf{u}_{L+1}$: Since we want to decrease f_θ, the first target direction to be passed on to layer L is $\mathbf{u}_{L+1} = -1$.

Update θ_k, Given $\mathbf{u}_{k+1}$: Move θ_k in the direction

$$\Delta\theta_k := \arg\max_{\|\mathbf{h}\|=1} \langle f'_{k,\mathbf{x}_k}(\theta_k;\mathbf{h}), \mathbf{u}_{k+1} \rangle. \tag{6}$$

Pass on Message $\mathbf{u}_k$, Given $\mathbf{u}_{k+1}$: If $k \geq 2$ pass to layer $k-1$ the target direction

$$\mathbf{u}_k := \arg\max_{\|\mathbf{h}\|=1} \langle f'_{k,\theta_k}(\mathbf{x}_k;\mathbf{h}), \mathbf{u}_{k+1} \rangle. \tag{7}$$

Both problems (6) and (7) are easily solved using $f^*_{k,\mathbf{x}_k}(\theta_k; \ \cdot \ )$ and $f^*_{k,\theta_k}(\mathbf{x}_k; \ \cdot \ )$, the adjoint operators to the derivatives $f'_{k,\mathbf{x}_k}(\theta_k; \ \cdot \ )$ and $f'_{k,\theta_k}(\mathbf{x}_k; \ \cdot \ )$ respectively:

$$\Delta\theta_k = \frac{f^*_{k,\mathbf{x}_k}(\theta_k; \ \mathbf{u}_{k+1})}{\|f^*_{k,\mathbf{x}_k}(\theta_k; \ \mathbf{u}_{k+1})\|} \quad \text{and} \quad \mathbf{u}_k = \frac{f^*_{k,\theta_k}(\mathbf{x}_k; \ \mathbf{u}_{k+1})}{\|f^*_{k,\theta_k}(\mathbf{x}_k; \ \mathbf{u}_{k+1})\|}. \tag{8}$$

These solutions do not ensure a change of the output value of layer k in the direction $\mathbf{u}_{k+1}$, but they do guarantee

$$\langle f'_{k,\mathbf{x}_k}(\theta_k;\Delta\theta_k), \mathbf{u}_{k+1} \rangle \geq 0 \quad \text{and} \quad \langle f'_{k,\theta_k}(\mathbf{x}_k;\mathbf{u}_k), \mathbf{u}_{k+1} \rangle \geq 0. \tag{9}$$

Answer $\mathbf{v}_k$ to Message $\mathbf{u}_k$: Then, when layer $k-1$ ($k \geq 2$) updates its parameter in the direction $\Delta\theta_{k-1}$, its output does not move in the target direction $\mathbf{u}_k$, but in the direction $\mathbf{v}_k := f'_{k-1,\mathbf{x}_{k-1}}(\theta_{k-1};\Delta\theta_{k-1})$, which "only" verifies $\langle \mathbf{v}_k, \mathbf{u}_k \rangle \geq 0$, according to (9). Therefore, the output of layer k moves in the direction $\mathbf{v}_{k+1} := f'_{k,\theta_k}(\mathbf{x}_k;\mathbf{v}_k)$ instead of $f'_{k,\theta_k}(\mathbf{x}_k;\mathbf{u}_k)$, and so on. The linearity of $f'_{k,\theta_k}(\mathbf{x}_k; \ \cdot \ )$ ensures the crucial following property

$$\begin{cases} \langle f'_{k,\theta_k}(\mathbf{x}_k;\mathbf{u}_k), \mathbf{u}_{k+1} \rangle \geq 0 \\ \langle \mathbf{v}_k, \mathbf{u}_k \rangle \geq 0 \end{cases} \Rightarrow \langle \mathbf{v}_{k+1}, \mathbf{u}_{k+1} \rangle := \langle f'_{k,\theta_k}(\mathbf{x}_k;\mathbf{v}_k), \mathbf{u}_{k+1} \rangle \geq 0. \tag{10}$$

Hence, as soon as (9) and (10) hold for layer $k-1$ and later layers, the property $\langle \mathbf{v}_k, \mathbf{u}_k \rangle \geq 0$, triggered by the update $\Delta\theta_{k-1}$, propagates and eventually yields $\langle \mathbf{v}_{L+1}, \mathbf{u}_{L+1} \rangle \geq 0$, i.e. $\mathbf{v}_{L+1} \leq 0$, meaning the output of f_θ is decreased.

This quick reminder of the chain rule mechanism highlights that the layer derivatives have two goals: optimal message passing and optimal parameter update based on the message passed by later layers. Therefore, in the case of non Fréchet-differentiable layers, like dilation and erosion layers, we may investigate if these two targets, namely properties (9) and (10), can still be met somehow. In the next sections we will see that morphological layers are differentiable in the more general sense of the Bouligand differentiability, which makes this notion worth analyzing in the perspective of optimization with gradient-descent-like algorithms.

4 The Bouligand Derivative

The Bouligand derivative has been introduced in the nonsmooth analysis literature [13,14]. It is a directional derivative that provides a first order approximation of its function in all directions. Formally, given a function $g : \mathbb{R}^n \to \mathbb{R}^m$ and
$x \in \mathbb{R}^n$, if for every $y \in \mathbb{R}^n$ the limit

$$g'(x; y) := \lim_{\alpha \to 0, \alpha > 0} \frac{g(x + \alpha y) - g(x)}{\alpha} \tag{11}$$

exists, then g is directionally differentiable at x and $g'(x; .)$ is called its directional derivative at x. If additionally for any $h \in \mathbb{R}^n$

$$g(x + h) = g(x) + g'(x; h) + o_0(h) \tag{12}$$

then g is said to be Bouligand differentiable (or B-differentiable) at x, and $g'(x; .)$
is its Bouligand derivative, also called B-derivative[3]. If g is B-differentiable at
every $x \in \mathbb{R}^n$, then it simply said B-differentiable.

Fréchet differentiablilty implies B-differentiability, but what makes the latter
more general than the former is that the B-derivative does not need to be a
linear function. If $g'(x; .)$ is a linear function, then g is Fréchet differentiable at
x, and $g'(x; .)$ is its Fréchet derivative at that point.

The B-derivative has nice properties similar to the Fréchet derivative, in
particular [14]:

- **Positive homogeneity:** $g'(x; \lambda y) = \lambda g'(x; y)$ for any $\lambda \geq 0$.
- **Chain rule:** if $f : \mathbb{R}^n \to \mathbb{R}^m$ and $g : \mathbb{R}^p \to \mathbb{R}^n$ are continuous and B-
  differentiable at $x \in \mathbb{R}^p$ and $g(x)$ respectively, then $f \circ g$ is B-differentiable
  at x and
  $$(f \circ g)'(x; y) = f'(g(x); g'(x; y)) \tag{13}$$
- **Linearity of $f \mapsto f'(x; .)$:** if $f : \mathbb{R}^n \to \mathbb{R}^m$ and $g : \mathbb{R}^n \to \mathbb{R}^m$ are continuous
  and B-differentiable at $x \in \mathbb{R}^n$, then so is $\alpha f + \beta g$ for any $\alpha, \beta \in \mathbb{R}$ and
  $$(\alpha f + \beta g)'(x; y) = \alpha f'(x; y) + \beta g'(x; y). \tag{14}$$

- **Derivative of components:** $g : \mathbb{R}^n \to \mathbb{R}^m$ is B-differentiable at x if and
  only if each of its components $g_i : \mathbb{R}^n \to \mathbb{R}$ is, and in this case
  $$g'(x; y) = (g'_1(x; y), \ldots, g'_m(x; y)). \tag{15}$$

As we will see, the dilation and erosion layers are continuous and B-
differentiable functions of both their input variables and parameters, as well
as all the usual neural layers. Therefore, a neural network $f_\theta(x)$ is a continuous and B-differentiable function of its parameter $\theta \in \mathbb{R}^p$ for a fixed x. Noting
$g : \theta \mapsto f_\theta(x)$ we have for $h \in \mathbb{R}^p$ and any $\eta > 0$,

$$g(\theta + \eta h) = g(\theta) + \eta(g'(\theta; h) + \epsilon(\eta)) \tag{16}$$

---

[3] Recall that $o_0(h)$ denotes $h \cdot \epsilon(h)$ where ϵ is any function that goes to zero when h
goes to zero.

where ϵ is a function that goes to zero when η goes to zero. Hence, we are left with finding in which direction h we need to move the parameter θ in order to ensure $g(\theta + \eta h) < g(\theta)$ for η sufficiently small. Whereas this was straightforward when $g'(\theta; h) = \langle \nabla g(\theta), h \rangle$ in Eq. (5), the problem is open when $g'(\theta; \cdot)$ is not linear. The purpose of the next section is to focus on this problem in the case of morphological neural networks.

5   Optimization with the Bouligand Derivative

5.1   Derivatives of the Morphological Layers

The Bouligand derivatives of the dilation and erosion layers with respect to their input values and parameters, are well known in the nonsmooth optimization literature [14], since they are easy examples of *piecewise affine functions* for which formulas exist. Here we provide some details of their computation, that will matter in addressing the problem stated in the previous section. We focus on the dilation layers, the case of erosions being analogous.

With the same notations as in Sect. 2, we denote by $\mathbf{x} \in \mathbb{R}^n$ and $W \in \mathbb{R}^{m \times n}$ the input vector and parameter matrix of a dilation layer. We will note $\delta_W(\mathbf{x})$ to clarify that we are considering a function of $\mathbf{x}$ with fixed parameter W, and $\delta_{\mathbf{x}}(W)$ for a function of W with fixed $\mathbf{x}$.

Derivative with Respect to W. An interesting property of $\delta_{\mathbf{x}}$ is that, if we move away from W in the direction $H \in \mathbb{R}^{m \times n}$, with a sufficiently small step $\eta \geq 0$, $\delta_{\mathbf{x}}(W + \eta H)$ shows an exact affine behaviour. Proposition 1 below provides a sufficient and necessary condition on the step η for this to hold. It will also provide the Bouligand derivative of $\delta_{\mathbf{x}}$.

Given a fixed $\mathbf{x} \in \mathbb{R}^n$ and a variable $W \in \mathbb{R}^{m \times n}$ we note $\delta_{\mathbf{x}}(W) = \left(\varphi_{\mathbf{x},i}(W)\right)_{1 \leq i \leq m}$ with

$$\varphi_{\mathbf{x},i}(W) := \max_{1 \leq j \leq n} w_{ij} + x_j. \tag{17}$$

Additionally, for each index $1 \leq i \leq m$, let us note

$$J_{W_i, \mathbf{x}} := \{ j \in \{1, \dots, n\}, \ \varphi_{\mathbf{x},i}(W) = w_{ij} + x_j \} \tag{18}$$

the set of indices where the maximum is achieved in $\varphi_{\mathbf{x},i}(W)$. When W and $\mathbf{x}$ will be clear from the context, we shall just denote it by J_i.

Let $H \in \mathbb{R}^{m \times n}$. Then for each $1 \leq i \leq m$, we also introduce the set

$$K_i := \left\{ k \in \{1, \dots, n\}, \ h_{ik} > \max_{j \in J_i} h_{ij} \right\}. \tag{19}$$

Then we have the following result:

Proposition 1. *For fixed* $W, H \in \mathbb{R}^{m \times n}$ *and* $\mathbf{x} \in \mathbb{R}^n$, *let* $\varphi_{\mathbf{x},i}$, J_i *and* K_i *as defined by* (17), (18) *and* (19) *respectively for* $1 \leq i \leq m$. *Let*

$$\epsilon_i = \min_{k \in K_i} \frac{\varphi_{\mathbf{x},i}(W) - (w_{ik} + x_k)}{h_{ik} - \max_{j \in J_i} h_{ij}}, \quad 1 \leq i \leq m, \tag{20}$$

and $\epsilon = \min_{1 \le i \le m} \epsilon_i$. Then, for any $\eta \in \mathbb{R}^+$ we have

$$\eta \in [0, \epsilon] \iff \delta_{\mathbf{x}}(W + \eta H) = \delta_{\mathbf{x}}(W) + \eta \left(\max_{j \in J_i} h_{ij} \right)_{1 \le i \le m}. \tag{21}$$

Proof (Proposition 1). Let $\eta \in \mathbb{R}^+$, and let us note $\mathbf{b} := (\max_{j \in J_i} h_{ij})_{1 \le i \le m}$. Then $\delta_{\mathbf{x}}(W + \eta H) = \delta_{\mathbf{x}}(W) + \eta \mathbf{b}$ if and only if $\varphi_{\mathbf{x},i}(W + \eta H) = \varphi_{\mathbf{x},i}(W) + \eta b_i$ for all $1 \le i \le m$. Now, the left-hand term writes

$$\varphi_{\mathbf{x},i}(W + \eta H) = \max \left(\max_{j \in J_i} w_{ij} + x_j + \eta h_{ij}, \max_{k \notin J_i} w_{ik} + x_k + \eta h_{ik} \right). \tag{22}$$

Since by definition $\varphi_{\mathbf{x},i}(W) = w_{ij} + x_j$ for any $j \in J_i$, we get

$$\varphi_{\mathbf{x},i}(W + \eta H) = \max \left(\varphi_{\mathbf{x},i}(W) + \eta b_i, \max_{k \notin J_i} w_{ik} + x_k + \eta h_{ik} \right). \tag{23}$$

Therefore $\varphi_{\mathbf{x},i}(W + \eta H) = \varphi_{\mathbf{x},i}(W) + \eta b_i$ if and only if for any $k \notin J_i$, $\varphi_{\mathbf{x},i}(W) + \eta b_i \ge w_{ik} + x_k + \eta h_{ik}$ which is equivalent to $\eta \le \epsilon_i$, and the result follows. $\square$

Given the definitions of Sect. 4, Proposition 1 readily shows that $\delta_{\mathbf{x}}$ has a directional derivative in any direction. By a similar reasoning, one can show that $\delta_{\mathbf{x}}(W + H) = \delta_{\mathbf{x}}(W) + (\max_{j \in J_i} h_{ij})_{1 \le i \le m}$ as soon as all $|h_{ij}|$ are small enough[4]. Therefore $\delta_{\mathbf{x}}$ is Bouligand differentiable everywhere and its B-derivative is

$$\delta'_{\mathbf{x}}(W; H) = \left(\max_{j \in J_i} h_{ij} \right)_{1 \le i \le m}. \tag{24}$$

Furthermore, with the notations of Proposition 1,

$$\eta \in [0, \epsilon] \iff \delta_{\mathbf{x}}(W + \eta H) = \delta_{\mathbf{x}}(W) + \eta \delta'_{\mathbf{x}}(W; H). \tag{25}$$

It appears that for any W such that $J_i = \{j_i\}$ is a singleton for each $1 \le i \le m$ (the maximum is achieved only once for each $\varphi_{\mathbf{x},i}$), $\delta'_{\mathbf{x}}(W; H)$ is a linear function since $\max_{j \in J_i} h_{ij} = h_{ij_i} = \langle H_{i,:}, e_{j_i} \rangle$, where e_{j_i} is the vector with a one at index j_i and zeros elsewhere. Hence in that case $\delta_{\mathbf{x}}$ is Fréchet differentiable. One can check[5] that this happens for almost every W.

Derivative with Respect to x. With the same approach as previously, one can show that δ_W is B-differentiable with respect to $\mathbf{x}$, and its B-derivative is, for all $\mathbf{h} \in \mathbb{R}^n$,

$$\delta'_W(\mathbf{x}; \mathbf{h}) = \left(\max_{j \in J_i} h_j \right)_{1 \le i \le m}, \tag{26}$$

---

[4] Take for example $\|H\|_\infty < \frac{1}{2} \min_{1 \le i \le m} \min_{k \notin J_i} \varphi_{\mathbf{x},i}(W) - (w_{ik} + x_k)$.
[5] Indeed the set of matrices for which the maximum in $\varphi_{\mathbf{x},i}(W)$ is achieved more than once, for a given i, is of zero Lebesgue measure.

Furthermore, for a fixed $\mathbf{h} \in \mathbb{R}^n$, changing only h_{ij} and h_{ik} for h_j and h_k in (19) and (20), it comes that for any $\eta \in \mathbb{R}^+$

$$\eta \in [0, \epsilon] \iff \delta_W(\mathbf{x} + \eta\mathbf{h}) = \delta_W(\mathbf{x}) + \eta\delta'_W(\mathbf{x}; \mathbf{h}). \tag{27}$$

Again, $\delta'_W(\mathbf{x}; \mathbf{h})$ is a linear function of $\mathbf{h}$ as soon as the maximum is achieved only once in $\{w_{ij} + x_j, 1 \leq j \leq n\}$, i.e. $J_i = \{j_i\}$, for each $1 \leq i \leq m$. In that case $\max_{j \in J_i} h_j = h_{j_i} = \langle \mathbf{h}, e_{j_i} \rangle$, hence $\delta'_W(\mathbf{x}; \mathbf{h}) = E\mathbf{h}$, where E is the matrix whose rows are the e_{j_i}s. Again, this case holds for almost every $\mathbf{x}$.

5.2   Updating the Parameters

Problem Setting. Let us focus on the update of the parameter W of the dilation layer $\delta_W : \mathbb{R}^n \to \mathbb{R}^m$. In the context of the chain rule, we assume that later layers (those closer to the output) have transmitted an instruction direction $\mathbf{u} \in \mathbb{R}^m$, and δ_W is supposed to modify its parameter $W \leftarrow W + \Delta W$ so that $\delta_{\mathbf{x}}(W + \Delta W) - \delta_{\mathbf{x}}(W)$ maximizes the inner product with $\mathbf{u}$. More formally, just as in Eq. (6), we want to solve

$$\Delta W = \arg \max_{\|H\|=1} \langle \delta'_{\mathbf{x}}(W; H), \mathbf{u} \rangle, \tag{28}$$

where $\|\cdot\|$ denotes the Frobenius norm and, this time, we consider the Bouligand derivative (24) computed earlier. The reason for which we are faced with the same problem as with Fréchet derivative, is that the Bouligand one also provides the first order approximation (12). Without loss of generality, we assume $\|\mathbf{u}\| = 1$.

Solving (28) does not seem straightforward but an attempt could start by noticing that

$$\|\delta'_{\mathbf{x}}(W; H)\|^2 = \sum_{i=1}^{m} \left(\max_{j \in J_i} h_{ij} \right)^2 \leq \sum_{i=1}^{m} \sum_{j=1}^{n} h_{ij}^2 = \|H\|^2 \tag{29}$$

hence $\|\delta'_{\mathbf{x}}(W; H)\| \leq 1$ for $\|H\| = 1$, therefore $\langle \delta'_{\mathbf{x}}(W; H), \mathbf{u} \rangle \leq \|u\| = 1$. This upper-bound is obviously achieved when $\delta'_{\mathbf{x}}(W; \cdot)$ is linear (i.e. $\delta_{\mathbf{x}}$ is F-differentiable at W), and for H such that $h_{ij_i} = u_i$, $1 \leq i \leq m$, and zero elsewhere, where we recall that in the F-differentiable case, j_i is the only index achieving the maximum in $\varphi_{\mathbf{x},i}(W)$, i.e. $J_i = \{j_i\}$. Indeed in that case, $\|H\| = 1$ and $\delta'_{\mathbf{x}}(W; H) = \mathbf{u}$.

Proposition of Candidates ΔW. In the non-F-differentiable case (i.e. when at least one J_i has more than one element), without analytically solving (28), we can at least propose decent candidates, inspired by the F-differentiable case. Let $I^+ := \{1 \leq i \leq m, u_i \geq 0\}$, $I^- := \{1 \leq i \leq m, u_i < 0\}$ and for $1 \leq i \leq m$ let us note $p_i := |J_i|$ the number of indices achieving the maximum in $\varphi_{\mathbf{x},i}(W)$. To make $\delta'_{\mathbf{x}}(W; H)$ similar to $\mathbf{u}$ while keeping $\|H\| = 1$, we propose:

- For $i \in I^+$, set $h_{ij_0} = u_i$ for any *one* $j_0 \in J_i$, and zero for $j \neq j_0$
- For $i \in I^-$, set $h_{ij} = \frac{u_i}{\sqrt{p_i}}$ for all $j \in J_i$, and zero for $j \notin J_i$.

Any such H verifies $\|H\| = 1$ and

$$\langle \delta'_{\mathbf{x}}(W; H), \mathbf{u} \rangle = \sum_{i \in I^+} u_i^2 + \sum_{i \in I^-} \frac{u_i^2}{\sqrt{p_i}} = 1 - \sum_{i \in I^-} \left(1 - \frac{1}{\sqrt{p_i}}\right) u_i^2. \qquad (30)$$

We see that this quantity gets closer to one as the p_i get closer to one, and we recover the optimal bound in the F-differentiable case, which corresponds to $p_i = 1$ for all $1 \le i \le m$, or when all $u_i \ge 0$. Furthermore, we have the lower bound $\langle \delta'_{\mathbf{x}}(W; H), \mathbf{u} \rangle \ge \frac{1}{\sqrt{n}} > 0$, which is the left hand part of property (9). Note that numerical experiments show that better H can be found (for example in the neighbourhood of the proposed ones).

Choosing the Learning Rate. Recall that solving problem (28) is relevant as long as a good first order approximation $\delta_{\mathbf{x}}(W + \eta H) \approx \delta_{\mathbf{x}}(W) + \eta \delta'_{\mathbf{x}}(W; H)$ holds, since only in this case does the parameter update ensure a change in the output value towards a descent direction. Proposition 1 provides the exact range of learning rates for which this approximation is an equality. For our proposed H, it holds if and only if $\eta \in [0, \epsilon] \cap \mathbb{R}^+$, with

$$\epsilon = \min_{i \in I^-} \frac{\eta_i \sqrt{p_i}}{|u_i|}, \qquad (31)$$

where $\eta_i := \min_{k \notin J_i} \varphi_{\mathbf{x}, i}(W) - (w_{ik} + x_k) = \varphi_{\mathbf{x}, i}(W) - \max_{k \notin J_i}(w_{ik} + x_k)$.

5.3 Message Passing

Problem Setting. For the message passing, we are first faced with the same problem as (7) for F-differentiable functions, but with the B-derivative. Namely, given the received target direction $\mathbf{u}$, we want to find the best update direction $\Delta \mathbf{x}$ for $\mathbf{x}$,

$$\Delta \mathbf{x} = \arg \max_{\|\mathbf{h}\|=1} \langle \delta'_W(\mathbf{x}; \mathbf{h}), \mathbf{u} \rangle. \qquad (32)$$

Assuming we can find a good enough $\mathbf{h}$, which would ensure $\langle \delta'_W(\mathbf{x}; \mathbf{h}), \mathbf{u} \rangle \ge 0$, i.e. the right hand part of property (9), then we have another problem, which is to guarantee property (10): that if $\langle \mathbf{v}, \mathbf{h} \rangle \ge 0$ for some $\mathbf{v}$, then $\langle \delta'_W(\mathbf{x}; \mathbf{v}), \mathbf{u} \rangle \ge 0$. To make sure the chain rule works, we could therefore focus on the problem

$$\text{Find } \mathbf{h} \in \mathbb{R}^n \text{ such that } \begin{cases} \|\mathbf{h}\| = 1 \\ \langle \delta'_W(\mathbf{x}; \mathbf{h}), \mathbf{u} \rangle \ge 0 \\ \forall \mathbf{v} \in \mathbb{R}^n, \ \langle \mathbf{v}, \mathbf{h} \rangle \ge 0 \Rightarrow \langle \delta'_W(\mathbf{x}; \mathbf{v}), \mathbf{u} \rangle \ge 0. \end{cases} \qquad (33)$$

Proposition of Candidates $\Delta \mathbf{x}$. Recall that $\delta'_W(\mathbf{x}; \mathbf{h}) = (\max_{j \in J_i} h_j)_{1 \le i \le m}$, hence contrary to the case of parameter update (Sect. 5.2), the same h_j can contribute to different J_i, which makes a heuristic construction of $\mathbf{h}$ much more complicated. One exception is the case where the sets $J_{W_i,\mathbf{x}}$ are pairwise disjoint, as with the max-pooling layer with strides, for which the same kind of construction as in Sect. 5.2 can be done. However, this guarantees only the first two conditions of (33), but we cannot say much about the last one.

At this stage we have no provable solution for (33) except, obviously, in the F-differentiable case, where each J_i is a singleton $\{j_i\}$. In that case, as presented in Sect. 5.1, $\delta'_W(\mathbf{x}; \mathbf{h}) = E\mathbf{h}$, where E is the $m \times n$ matrix whose rows are the e_{j_i}s, each e_{j_i} being the vector with a one at index j_i and zeros elsewhere. Hence the solution of (32), and a solution of (33), is $\mathbf{h} = \frac{E^T\mathbf{u}}{\|E^T\mathbf{u}\|}$ if $E^T\mathbf{u} \neq 0$, and any unit vector $\mathbf{h}$ otherwise.

Therefore we propose as update candidate, one that generalizes the F-differentiable case, namely $\mathbf{h} = \frac{E^T\mathbf{u}}{\|E^T\mathbf{u}\|}$ but with E the matrix whose i-th row is $E_{i,:} = \sum_{j \in J_i} e_j$. Numerical experiments show that this proposition can sometimes violate the last two conditions of (33), but often behaves well.

Choosing the Learning Rate Hoping that the chosen $\mathbf{h}$ fulfills (33), we make the best of it by choosing a learning rate ensuring the first order equality (27). Hence once again we follow the construction inspired by Proposition 1. The choice of $\mathbf{h} = \frac{E^T\mathbf{u}}{\|E^T\mathbf{u}\|}$ yields no simplification of the expression of ϵ.

5.4   The Convolutional Case

The definitions (1) and (2) cover translation invariant dilations and erosions, as soon as $W \in \mathbb{R}^{n \times n}$ is a Toeplitz matrix. However, in Sect. 5.2, we assumed no "shared weights", i.e. each row of W was considered independent from the others, which allowed an easy choice for the parameter update.

To model the constraint on W due to translation invariance, we assume δ_W is represented by a vector $\mathbf{w} \in \mathbb{R}^p$, $p \leq n$, and the input variable $\mathbf{x} \in \mathbb{R}^n$ is now seen as a matrix $X \in \mathbb{R}^{n \times p}$ containing n blocks of length p. The dilation now writes

$$\delta_\mathbf{w}(X) = \delta_X(\mathbf{w}) := \left(\max_{1 \leq j \leq p} x_{ij} + w_j \right)_{1 \leq i \leq n}. \tag{34}$$

Unfortunately, we see that even for the parameter update, which was rather favorable in the "dense" layer case of Sect. 5.2, we are in the same situation as in the message passing of Sect. 5.3, in the sense that finding good candidate for $\Delta\mathbf{w}$ is as difficult as solving (32). We would therefore apply the same heuristics, i.e. $\mathbf{h} = \frac{E^T\mathbf{u}}{\|E^T\mathbf{u}\|}$, where $E_{ij} = 1$ if j achieves the maximum in $\max_{1 \leq j \leq p} x_{ij} + w_j$ and zero elsewhere. Concerning the learning rate, (27) holds.

5.5   Practical Consequences

Position in the Network, Dense or Convolutional Layer. We saw that the chain rule mechanism is not guaranteed with morphological layers because of uncertainties in the message passing in general, and even in the parameter update for convolutional operators. Therefore, we expect better performance as a morphological layer is closer to the input of the network, and even more so if it is a dense layer. Typically, starting a neural pipeline with a dense dilation or erosion is the most favorable case with the update and learning rate propositions of Sect. 5.2. Furthermore, if each morphological layer is seen as a noisy message transmitter, then it is expected that many such layers in the same network may be hard to train with the chain rule paradigm.

Initialization. In both the dense and convolutional cases, according to our propositions or even in the F-differentiable case, a parameter coefficient is not updated if it does not contribute to a maximum. In the dense case, w_{ij} is not modified if $j \notin J_i$, and in the convolutional one, w_j remains unchanged if $j \notin J_i$ for all i. In particular, if such coefficient is moved to $-\infty$, it will never be updated anymore. Now, consider for example that if the input variable $\mathbf{x}$ has values in $[0, 1]$ and at least one weight $w_{ij_1} \geq 0$, then the closer another weight w_{ij_2}, on the same line, will be to -1 the less likely it will be to achieve the maximum, and $w_{ij_2} \leq -1$ is equivalent to $w_{ij_2} = -\infty$. Therefore it seems preferable to initialize the parameters with non-negative values (typically, zero if input values in $[0, 1]$). Then, the proposed adaptive learning rates should avoid a divergence of weights to values from where they cannot come back.

6   Conclusion

In this paper we investigated the optimization of morphological layers based on the Bouligand derivative and the chain rule. We showed that despite the first order approximation of the B-derivative, its non-linearity makes morphological layers noisy message transmitter in the chain rule, where they are not F-differentiable. We clearly stated the problems to overcome in order to make this framework compatible with the chain rule. We also provided insights regarding the choice of the learning-rate for these layers, which seems much clearer than with classic layers. Future work will deal with addressing the stated problems and show the experimental consequences of the theoretical results presented here.

Acknowledgments. I would like to thank François Pacaud and Santiago Velasco-Forero for fruitful discussions on this topic.

References

1. Aouad, T., Talbot, H.: Binary morphological neural network. In: 2022 IEEE International Conference on Image Processing (ICIP), pp. 3276–3280 (2022)
2. Banon, G.J.F., Barrera, J.: Decomposition of mappings between complete lattices by mathematical morphology, Part I. general lattices. Sig. Process. **30**(3), 299–327 (1993)

3. Blusseau, S., Velasco-Forero, S., Angulo, J., Bloch, I.: Morphological adjunctions represented by matrices in max-plus algebra for signal and image processing. In: Baudrier, É., Naegel, B., Krähenbühl, A., Tajine, M. (eds.) DGMM 2022. LNCS, vol. 13493, pp. 206–218. Springer, Cham (2022). https://doi.org/10.1007/978-3-031-19897-7_17

4. Charisopoulos, V., Maragos, P.: Morphological perceptrons: geometry and training algorithms. In: Angulo, J., Velasco-Forero, S., Meyer, F. (eds.) ISMM 2017. LNCS, vol. 10225, pp. 3–15. Springer, Cham (2017). https://doi.org/10.1007/978-3-319-57240-6_1

5. Davidson, J.L., Ritter, G.X.: Theory of morphological neural networks. In: Digital Optical Computing II, vol. 1215, pp. 378–389 (1990)

6. Franchi, G., Fehri, A., Yao, A.: Deep morphological networks. Pattern Recogn. **102**, 107246 (2020)

7. Hermary, R., Tochon, G., Puybareau, É., Kirszenberg, A., Angulo, J.: Learning grayscale mathematical morphology with smooth morphological layers. J. Math. Imaging Vis. **64**(7), 736–753 (2022)

8. Kirszenberg, A., Tochon, G., Puybareau, É., Angulo, J.: Going beyond p-convolutions to learn grayscale morphological operators. In: Lindblad, J., Malmberg, F., Sladoje, N. (eds.) DGMM 2021. LNCS, vol. 12708, pp. 470–482. Springer, Cham (2021). https://doi.org/10.1007/978-3-030-76657-3_34

9. Maragos, P.: A representation theory for morphological image and signal processing. IEEE Trans. Pattern Anal. Mach. Intell. **11**(6), 586–599 (1989)

10. Maragos, P.: Chapter two - representations for morphological image operators and analogies with linear operators. Adv. Imaging Electron Phys. **177**, 45–187 (2013)

11. Matheron, G.: Random Sets and Integral Geometry. Wiley, New York (1975)

12. Mondal, R., Dey, M.S., Chanda, B.: Image restoration by learning morphological opening-closing network. Mathem. Morphol. Theory Appl. **4**(1), 87–107 (2020)

13. Robinson, S.M.: Local structure of feasible sets in nonlinear programming, stability and sensitivity. In: Cornet, B., Nguyen, V.H., Vial, J.P. (eds.) Nonlinear Analysis and Optimization. Mathematical Programming Studies Part III, vol. 30, pp. 45–66. Springer, Heidelberg (1987). https://doi.org/10.1007/bfb0121154

14. Scholtes, S.: Introduction to Piecewise Differentiable Equations. Springer, New York (2012). https://doi.org/10.1007/978-1-4614-4340-7

15. Velasco-Forero, S.: Morpholayers (2020)

16. Velasco-Forero, S., Rhim, A., Angulo, J.: Fixed point layers for geodesic morphological operations. In: BMVC. London, United Kingdom (2022)

17. Wilson, S.S.: Morphological networks. In: Visual Communications and Image Processing IV, vol. 1199, pp. 483–496 (1989)

18. Zhang, Y., Blusseau, S., Velasco-Forero, S., Bloch, I., Angulo, J.: Max-plus operators applied to filter selection and model pruning in neural networks. In: Burgeth, B., Kleefeld, A., Naegel, B., Passat, N., Perret, B. (eds.) ISMM 2019. LNCS, vol. 11564, pp. 310–322. Springer, Cham (2019). https://doi.org/10.1007/978-3-030-20867-7_24

HaarNet: Large-Scale Linear-Morphological Hybrid Network for RGB-D Semantic Segmentation

Rick Groenendijk$^{(\boxtimes)}$ ⓘ, Leo Dorst ⓘ, and Theo Gevers ⓘ

University of Amsterdam, Science Park 900, 1098XH Amsterdam, The Netherlands
{r.w.groenendijk,l.dorst,th.gevers}@uva.nl

Abstract. Signals from different modalities each have their own combination algebra which affects their sampling processing. RGB is mostly linear; depth is a geometric signal following the operations of mathematical morphology. If a network obtaining RGB-D input has both kinds of operators available in its layers, it should be able to give effective output with fewer parameters. In this paper, morphological elements in conjunction with more familiar linear modules are used to construct a mixed linear-morphological network called **HaarNet**. This is the first large-scale linear-morphological hybrid, evaluated on a set of sizeable real-world datasets. In the network, morphological Haar sampling is applied to both feature channels in several layers, which splits extreme values and high-frequency information such that both can be processed to improve both modalities. Moreover, morphologically parameterised ReLU is used, and morphologically-sound up-sampling is applied to obtain a full-resolution output. Experiments show that HaarNet is competitive with a state-of-the-art CNN, implying that morphological networks are a promising research direction for geometry-based learning tasks.

Keywords: Machine learning · Morphological neural networks · Multi-modal learning

1 Introduction

Semantic segmentation is a challenging pixel-level prediction task in computer vision. It has become common practice [15,31] to use multi-modal *colour* and *depth* (RGD-D) data since failure modes arising from one modality (*e.g.* homogeneity, noise, sparsity, etc.) can often be avoided when relying on multiple modalities. Datasets such as [1,8,31] facilitate research that delves into combining features from different signals by providing data from different sensors. However, optimal fusion strategies for the colour and depth data modalities are still an open area of research.

Through the works of Serra [28], the computer vision community is familiarised with the underlying algebraic structure of data that is acquired using

S. Brunetti et al. (Eds.): DGMM 2024, LNCS 14605, pp. 242–254, 2024.
https://doi.org/10.1007/978-3-031-57793-2_19

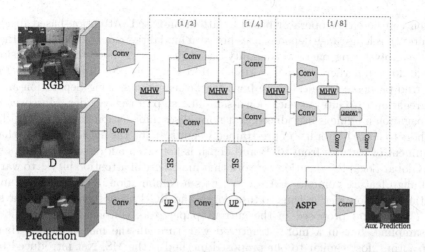

Fig. 1. The Morphological HaarNet. Depth and RGB modalities are separately encoded, but during down-sampling the Morphological Haar Wavelet (MHW, see Fig. 2 for more details) blocks ensure that the down-sampled signal is improved using high-frequency details from both modality feature signals. Up-sampling happens through morphological up-sampling and skip connections (UP). Both morphological types are underlined in this overview. The skip connections receive channel-wise attention from (multi-modal) SE [20] blocks. Multi-resolution features are combined using ASPP [5]. All non-linearities are replaced by the morphological dilation.

probing contact (*e.g.* LiDAR and radar). [26, 29, 32]. This field is called *mathematical morphology*, and it differs from the algebra of linear diffusion that is used to build convolutional neural networks (CNNs). Recently, it was shown that morphological operations are suitable for processing specific types of data, such as depth data [13]. Besides the benefits for encoding input signals, morphological sampling and interpolation may help to predict sharply delineated semantic boundaries. While there exist examples of methods that explore morphological networks [23], or embed morphological operations in few-layer linear networks [10, 11, 19, 27, 30, 35], large-scale linear-morphological hybrids have not received much research attention.

This paper proposes the construction of a mixed linear-morphological network, coined HaarNet, to exploit linear and morphological operations jointly. The proposed morphological modules are: a down-sampling operation based on the morphological Haar transform; non-linearities formulated as parameterised dilations; and an up-sampling procedure that can innately delineate semantic boundaries.

2 Related Work

Fusion of RGB and depth images in neural networks may be done at the input or output stage, although the most common method is "middle" fusion. Middle fusion can be either operator-based or attention-based. Attention-based fusion ensures that long-range dependencies between modalities can be modelled, rather than simply fusing on local modality-based features as operator-based fusion would do. While there are many articles addressing this type of fusion, of note is [7], who proposes to re-calibrate both feature encodings using Separation-and-Aggregation (SAGate) at each down-sampling step in the encoder. Moreover, a propagation module is introduced that normalises information flow, allowing the authors to make better use of pre-trained weights. In [6], the authors observe that alignment of features from RGB and depth is not often addressed, and propose the Global-Local Propagation network that makes use of attention blocks to warp and align feature volumes. CANet [40] uses a combination of channel-wise and spatial attention mechanisms to fuse depth and RGB using separate pathways.

The method proposed in the current paper, too, considers the attention-by-sampling, but in a more structured way through the morphological Haar transform. Most similar to our proposed method is the MSFNet introduced by [22]. In MSFNet, a *linear* Haar Wavelet transform is used to decompose high resolution signals into constituents, ensuring high frequency information is preserved throughout down-sampling. However, instead of using the wavelets as a means of down-sampling, the inverse transform is used to reconstruct the feature signal at the original resolution. In addition, the depth features are only used to attend the RGB features, so no joint optimisation of feature representations is being performed. HaarNet, on the other hand, allows to freely optimise the set of features from both modalities.

3   Method

To specify the composition of a Morphological Haar Wavelet Network, a brief overview of the fundamental morphological operations is required: the morphological *dilation* is defined on the semi-ring $\{\mathbb{R}_{-\infty}, \bigvee, +\}$ where $\bigvee$ denotes the supremum operation and $+$ is addition. This algebraic system extends the set of real numbers $\mathbb{R}$ with minus infinity: $\mathbb{R}_{-\infty} \equiv \mathbb{R} \cup -\infty$ [26]. The morphological *erosion* is the dual of the morphological dilation, and is constructed with infimum $\bigwedge$ over the extended set of reals $\{\mathbb{R} \cup \infty\}$. The dilation and erosion can be parameterised by means of a structuring element h. The dilation is defined as

$$g(\mathbf{x}) = [f \boxplus h](\mathbf{x}) = \bigvee_{\mathbf{z}} f(\mathbf{x} - \mathbf{z}) + h(\mathbf{z}) , \qquad (1)$$

where f, g are the inputs and outputs of a layer in a neural network respectively, and $\mathbf{x}, \mathbf{z}$ are taken from indicator sets. Unlike convolutions in CNNs, morphological operations are often only applied in a channel-wise manner [13]; back-propagation over the maximum element results in slow and complicated learning [14]. Note that dilation is algebraically similar to convolution, though in the *logarithmic* sense [2]. As a consequence, it can serve as a signal processing framework with an essentially analogous structure to linear signal processing.

There are four novel elements on which the morphological HaarNet is built:

– The morphological Haar Wavelet module, which improves down-sampling compared to common strategies like strided convolutions or pooling.
– The ReLU non-linearity that is shown to be a special case of the morphological dilation; it can be parameterised to be a learnable threshold-filter.
– An improved sampling procedure relative to [13], following seminal papers on sampling theory in morphology [17,18] rather than being provenance-based.
– The use of additional pre-training to compensate for the reduced effectiveness of transfer learning when compared to baselines similar to a pre-trained ResNet [16].

The full network overview is shown in Fig. 1, which details at which stages multi-modal features are fused. Note that the network uses Squeeze-and-Excitation blocks [20] for attending channel-wise information during skip connections, although they are modified to attend separately to each modality feature volume before feature fusion. Another module that is used, and that is prevalent in pixel-level semantic prediction networks, is the multi-scale feature pyramid ASPP [5] to improve predictive performance for long-range semantic dependencies. These blocks are naturally augmented by morphological non-linearities (see Subsect. 3.2), to truly make this network a linear-morphological hybrid.

3.1   Morphological Haar Wavelets

The morphological Haar wavelet transform decomposes signal $f^\sigma : \mathbb{R}^{H \times W} \to \mathbb{R}$ at a particular scale σ into four subband signals: Down-sampled $f^{\sigma/2} : \mathbb{R}^{H/2 \times W/2} \to \mathbb{R}$ and vertical, horizontal, and diagonal details $\phi_f^v, \phi_f^h, \phi_f^d : \mathbb{R}^{H/2 \times W/2} \to \mathbb{R}$ respectively. A particularly helpful property of morphological wavelets over linear wavelets is that they retain local maxima of the signal over multiple scales; similar to pooling, this implies the network achieves invariance to particular kinds of noise [21]. The morphological Haar wavelet transform [18] is defined

$$f^{\sigma/2} = f^\sigma \boxplus \begin{bmatrix} 0 & 0 \\ 0 & 0 \end{bmatrix}, \tag{2a}$$

$$\phi_f \equiv \{\phi_f^v, \phi_f^h, \phi_f^d\} = f^\sigma * \left\{ \begin{bmatrix} -1 & -1 \\ 1 & 1 \end{bmatrix}, \begin{bmatrix} -1 & 1 \\ -1 & 1 \end{bmatrix}, \begin{bmatrix} 1 & -1 \\ -1 & 1 \end{bmatrix} \right\}, \tag{2b}$$

where both morphological dilation $\boxplus$ and convolution $*$ are applied at a stride of 2 to get non-overlapping windows, and to ensure that the spatial dimensions of the signal f^σ are reduced by a factor 2. Clearly, $f^{\sigma/2}$ is similar to the application of max pooling which computes the local maximum value in a 2-by-2 neighbourhood. Max pool and $f^{\sigma/2}$ both reduce spatial feature size in addition to obtaining maximum-amplitude coefficients [4,36]. On top of this, detail signals ϕ encode all directional high-frequency information that was lost during down-sampling. A property of the wavelet transform is that it is possible to fully restore f by the inverse transform using subbands $f^{\sigma/2}$ and ϕ_f.

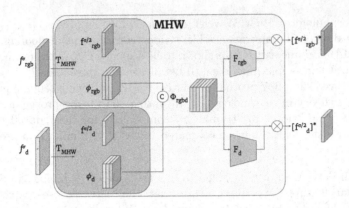

Fig. 2. The Morphological Haar Wavelet (MHW) block, which down-samples the signal f^σ by decomposing into subbands $f^{\sigma/2}$ and ϕ. A 2-layer MLP F is used to attend to channel-wise information based on combined high-frequency details Φ_{rgbd} for both modalities. This yields updated feature representations $\left[f_{rgb}^{\sigma/2}\right]^*$ and $\left[f_d^{\sigma/2}\right]^*$.

The idea of the Morphological Haar Wavelet (MHW) block is to use combined directional details of RGB and depth modalities $\Phi_{rgbd} \equiv \{\phi_{rgb}, \phi_d\}$ to simultaneously improve both the down-sampled RGB feature signal $f_{rgb}^{\sigma/2}$ and the depth feature signal $f_d^{\sigma/2}$. To this end, MHW uses two distinct 2-layer CNNs F_{rgb}, F_d. Each F takes as input the combined high-frequency signal Φ_{rgbd} to obtain the updated and improved down-sampled signals $[f^{\sigma/2}]^*$. For RGB features this can be expressed as

$$\left[f_{rgb}^{\sigma/2}\right]^* (\mathbf{x}) = f_{rgb}^{\sigma/2}(\mathbf{x}) \otimes \sigma\left(F_{rgb}\left(\Phi_{rgbd}\left(\mathbf{x}\right)\right)\right), \tag{3}$$

where $\sigma(\cdot)$ is a sigmoid, $\otimes$ is element-wise multiplication, and $\mathbf{x}$ taken from the indicator set over the signal. Similarly, for the improved depth features $[f_d^{\sigma/2}]^*$. The combined operation ensures not only that high frequency information is used explicitly during down-sampling, but also forces the resulting information to be dependent on both modalities through their common Φ_{rgbd}. The full procedure is depicted schematically in Fig. 2.

3.2 Formulating ReLU as Dilation

The ReLU activation layer [12] is a special case of the morphological dilation with a flat structuring element that is the morphological delta-function, and a morphological bias term. The morphological delta-function $\delta(\cdot)$ is 0 only at location 0, and $-\infty$ elsewhere. To demonstrate the equivalence between ReLU and morphological dilation, first consider the definition of the ReLU:

$$\xi(\mathbf{y}) = \begin{cases} \mathbf{y} & \text{if } \mathbf{y} \geq 0 \\ 0 & \text{otherwise.} \end{cases} \tag{4}$$

Similar to the convolutional bias, the dilation can be written to include a morphological bias term [3], denoted h_0. First, consider the a specific instance of the morphological dilation:

$$\xi\left(f\left(\mathbf{x}\right)\right) = 0 \vee f\left(\mathbf{x}\right) = 0 \vee \left[f \boxplus \delta\right]\left(\mathbf{x}\right), \tag{5}$$

using the morphological delta-function $\delta(\cdot)$. The above can be generalised by setting $h_0 = 0$ and allowing an arbitrary structuring element h:

$$\xi\left(f\left(\mathbf{x}\right)\right) = h_0 \vee \left[f \boxplus h\right]\left(\mathbf{x}\right). \tag{6}$$

Thus, a form is reached that suggests a natural extension of ReLU by morphological kernel h and bias h_0. In HaarNet, all activation layers are replaced by the morphological dilation in which the morphological bias term h_0 is learnable. Such a learnable ReLU was coined Flexible ReLU (FReLU) in [25], but the connection to the morphological dilation was not recognised in this work. A strong belief is held by the authors that understanding of networks improves by recognising that nearly all modern CNNs exploit inherently morphological layers through ReLU [34] and pooling [13].

3.3   Morphological Up-Sampling

In encoder-decoder architectures, up-sampling from low-resolution features is vital to obtain high-quality predictions. In [13], it was shown that non-linear up-sampling can aid semantic networks, since it delineate boundaries better than linear up-sampling. The procedure also did not introduce sparsity the way standard unpooling does [38,39]. However, morphological up-sampling can be improved using insights from [18], where it is proven by the theory of adjunctions that the down-sampling dilation has an up-sampling dual that is an erosion with the same structuring element. In addition, up-sampling is of equidistant nature: intervals at which to place back values are kept constant. As a consequence, the total reconstruction operator ρ is a closing: $\rho[f] = [f \boxplus h] \boxminus h$. Note, however, that in prediction tasks *reconstruction* of the same signal is not the goal, rather the input signal must be *processed* for analysis. Unlike [18], this paper is concerned with obtaining the most expressive sampling processing operator.

Based on experimentation, the proposed morphological up-sampling operator for prediction networks is as follows: First, equidistant sampling is used, and not provenance mapping as in [13]. This means low-resolution features are placed back at a regular interval. Second, all sampling operators are dilations. Third, structuring elements are not constrained to be the same across down- and up-sampling operators. That is, they are all parameterised individually. Lastly, in HaarNet, the up-sampling procedure is combined with a skip connection at the same resolution.

4   Experiments

In this section, HaarNet is applied to an RGB-D semantic segmentation task on three different datasets, both indoor and outdoor. HaarNet is compared to most notably SAGate [7] as it is encoder-decoder-based, reaches state-of-the-art consistently, and public code is available for replication of the results and for a fair comparison. Large-scale linear-morphological hybrids have not received research attention; the goal of the experiments is to show that even when network design principles from linear networks are used, linear-morphological hybrids still perform at least on par with linear counterparts.

Implementation. All modules are implemented in PyTorch [24], except for morphological operations, which are absent in PyTorch and potentially intractable when naively done. These operations are implemented directly in CUDA/C++. Source code is made available at https://github.com/rickgroen/haarnet.

Dataset and Training. The experiments are performed on three datasets: NYUv2 ($N_{train} = 795$, $N_{test} = 654$, $C = 40$), 2D-3D-S ($N_{train} = 52903$, $N_{test} = 17593$, $C = 13$), and CityScapes ($N_{train} = 2975$, $N_{test} = 500$, $C_{train} = 30$, $C_{test} = 19$). The first two datasets are indoor datasets, the last is an outdoor autonomous driving dataset. RGB images are normalised with training set statistics to follow a zero-mean Gaussian. According to convention, depth images are converted to the HHA encoding [15] and normalised in the range $[0, 1]$. Based on initial experimentation, all networks are trained using SGD with Nesterov Momentum [33] with a learning rate λ_0 of $5e-3$. The learning rate is updated at each epoch by a polynomial scheduler according to $\lambda_{epoch} = \lambda_0 \left(1 - \frac{epoch}{500}\right)^{0.9}$. The networks are trained for 500, 12, and 200 epochs for the NYU, 2D-3D-S, and CityScapes datasets respectively. Models are evaluated using mean Intersection over Union (mIoU), pixel accuracy, and a boundary F1-score [9] to measure performance at semantic edges.

4.1   The Necessity of Pre-training

It is hypothesised that many contemporary networks are architecturally biased to the pre-trained ResNet weights on ImageNet, since pre-training is such a powerful means to obtain state-of-the-art performance [37]. Specifically, when the architecture starts to significantly differ from that of the ResNet, exploiting the pre-trained weights becomes less valuable. When a fundamentally different network, like HaarNet, is introduced, it should undergo pre-training for fair assessment of its novel capabilities.

Pre-training new encoders on ImageNet, however, is not computationally feasible in most scenarios because of the scale of this dataset. To adapt, this paper joins the three datasets (*i.e.* NYUv2, 2D-3D-S, and CityScapes) and maps all semantic labels to a joint set of 67 labels, including semantic classes from both indoor and outdoor scenes. Pre-training is performed in two phases: first, the decoder is frozen and the encoder is trained using the joint dataset. Second, the decoder is trained on a specific dataset with a frozen encoder.

Consider Table 1, in which the pre-training hypothesis is tested. All networks use a ResNet50 encoder, although both the SAGate baseline and HaarNet

Table 1. Effects of Pre-trained Weights. ResNet50 encoders are initialised either randomly or use weights pre-trained from ImageNet. HaarNet performs better on two out of three metrics when randomly initialised. When ImageNet weights are used, a basic ResNet benefits most.

	mIoU	pixel accuracy	Boundary F1
Random Initialisation			
Middle Fusion Dual ResNet	0.193	0.582	0.220
SA-Gate [7]	**0.266**	0.603	0.213
HaarNet	0.252	**0.624**	**0.231**
Naive Transfer of ImageNet Weights			
Middle Fusion Dual ResNet	0.382	0.687	**0.314**
SA-Gate [7]	**0.408**	**0.697**	0.267
HaarNet	0.360	0.687	0.292

Table 2. Indoor Segmentation Results. When trained using the same set-up, HaarNet performs at least on par with the baselines using the least amount of additional parameters, as reported in the second column.

Network		NYUv2			2D-3D-S		
	+#param	mIoU	pixel accuracy	Boundary F1	mIoU	pixel accuracy	Boundary F1
Dual ResNet Baseline	+25.8M	0.474	0.749	0.373	0.565	0.787	0.564
SAGate [7]	+21.8M	0.504	0.768	0.381	0.561	**0.806**	**0.577**
HaarNet	+13.1M	**0.507**	**0.770**	**0.392**	**0.568**	0.794	0.567

make significant changes to their encoder. SAGate compensates for architectural changes by the normalising the outputs of their fusion blocks; HaarNet has no such inherent rectification. It is shown that without any pre-training, HaarNet performs best on two out of three metrics. However, when ImageNet weights for ResNet are used as a pre-training procedure, HaarNet benefits least; a basic ResNet benefits most.

4.2   Segmentation Performance

Indoor Semantic Segmentation. The predictive performance of HaarNet on indoor segmentation datasets is shown in Table 2. For NYUv2, HaarNet outperforms the general baseline and state-of-the-art SAGate on all metrics, provided the networks are trained on the same system, using the same pre-training, and with the same hyper-parameters. HaarNet is also more efficient, given that it only uses 13.1M additional parameters on top of the encoder, versus 20-25M for the other baselines. For 2D-3D-S, performance is on par with SAGate. The main difference between NYUv2 and 2D-3D-S, apart from size, is the quality of the depth signals. To see so, review the qualitative results, as depicted in Fig. 3, in which the relative best and worst predictions of HaarNet versus SAGate are shown. From the centre rows, it can be seen that the depth signals of 2D-3D-S contain many reconstruction artefacts. Since morphological operations are more

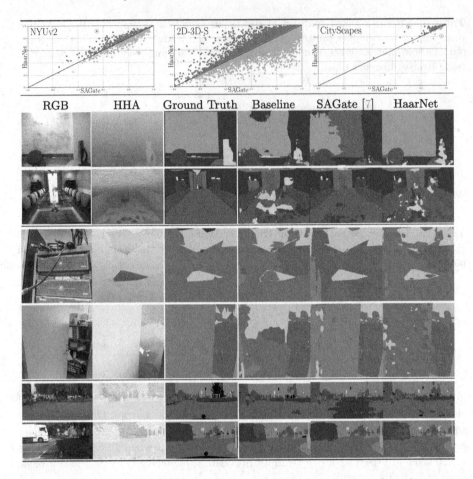

Fig. 3. Qualitative Results of HaarNet. The top three scatter plots show accuracy per image for HaarNet versus SAGate, and select the relative best and worst sample. These samples from NYUv2, 2D-3D-S and Cityscapes are shown in the rows with images, interweaving relative best (top, HaarNet outperforms SAGate) and worst (bottom, vice versa) samples. In the second row (NYUv2, worst), a reflective surface results in a low-quality depth signal. The morphological HaarNet misclassifies at those depth artefacts, although it does recover the lamp situated on the table, which is not labelled as lamp. The third and fourth row show that 2D-3D-S has relatively coarse depth and semantics, complicating learning for all networks. Here, the third best and tenth worst images are shown, since the other failure modes are completely homogeneous images. Finally, CityScapes images show that HaarNet is able to classify the road mostly correctly. However, the monochromatic truck poses a challenge. Best viewed in colour.

sensitive to geometric modalities, it is possible the depth artefacts complicate learning for HaarNet.

Table 3. Outdoor Segmentation Results. Unlike results from indoor datasets, HaarNet significantly outperforms the baselines. It could be this is due to the high amount of structure present in the scenes, which allows absolute operators like morphological ones to make more rigid predictions.

Network	mIoU	pixel accuracy	Boundary F1
Dual ResNet Baseline	0.709	0.952	0.714
SAGate [7]	0.713	0.953	0.728
HaarNet	**0.762**	**0.961**	**0.780**

Table 4. Ablation Study. Rows show performance due to the morphological up-sampling layers (M-UP), morphological non-linearities (M-ReLU), and the Haar module (MHW). Modules are replaced with simple (linear) counterparts when possible. Results indicate that, while networks adapt to the available modules, a network that is a full linear-morphological hybrid performs best.

M-UP	M-ReLU	MHW	mIoU	pixel accuracy	Boundary F1
			0.469	0.751	0.373
✓			0.475	0.752	0.367
✓	✓		0.469	0.753	0.371
✓		✓	0.481	0.755	0.363
✓	✓	✓	**0.510**	**0.771**	**0.385**

Outdoor Semantic Segmentation. HaarNet is also evaluated on the outdoor CityScapes dataset; results are shown in Table 3. Here, HaarNet significantly outperforms the baselines. This may be because autonomous driving datasets are much less varied than indoor datasets since traffic is structured to be predictable. Morphological operators are more calibrated than linear operators, since they act absolutely rather than relatively.

The performance due to three core modules of HaarNet are evaluated: morphological up-sampling layers, morphological non-linearities (Section 3.2), and the morphological Haar Wavelet (MHW) module (Section 3.1). Results are shown in Table 4. In the ablation study, the base class without any of the introduced modules is architecturally equivalent to a DeeplabV3 [5]. All networks exploit ResNet-101 backbones, and are trained for 100 epochs on NYUv2 40-class setting. Results indicate that the Haar module is important for achieving best performance, since it enables modality fusion at the same time as informed down-sampling. The ablation also shows that networks are flexible: when modules are replaced by simple (linear) counterparts, the network can come to depend on the reliable DeepLab architecture.

5   Conclusions

It is the authors' strong belief that the set of operators made available to a network should reflect the admissible transformations on the input modalities. Morphological operations are more suitable to analysis of geometry-based modalities than convolutions are. In this paper, all non-linearities and pooling are replaced by their parameterised morphological equivalents, which entails natural and structural generalisation of core network modules. Leveraging these generalised modules, resulting from the field of mathematical morphology, yields promising results, even when core network design principles are kept similar to those of CNNs. This was achieved by utilising three core MM concepts and introducing a new pre-training regime, resulting in the large-scale linear-morphological hybrid network called **HaarNet**. Further research should address attention mechanisms that enable networks to focus on specific features using modality-specific operators. This should be explored both across scales and along feature depth; the latter has yet to be investigated in the literature.

References

1. Armeni, I., Sax, S., Zamir, A.R., Savarese, S.: Joint 2d-3d-semantic data for indoor scene understanding. arXiv preprint arXiv:1702.01105 (2017)
2. Burgeth, B., Weickert, J.: An explanation for the logarithmic connection between linear and morphological system theory. IJCV **64**, 157–169 (09 2005)
3. Charisopoulos, V., Maragos, P.: Morphological perceptrons: geometry and training algorithms. In: ISMM. pp. 3–15. Springer (2017)
4. Chen, B., Polatkan, G., Sapiro, G., Blei, D., Dunson, D., Carin, L.: Deep learning with hierarchical convolutional factor analysis. TPAMI **35**(8), 1887–1901 (2013)
5. Chen, L.C., Papandreou, G., Kokkinos, I., Murphy, K., Yuille, A.L.: Deeplab: Semantic image segmentation with deep convolutional nets, atrous convolution, and fully connected CRFs. TPAMI **40**(4), 834–848 (2017)
6. Chen, S., Zhu, X., Liu, W., He, X., Liu, J.: Global-local propagation network for rgb-d semantic segmentation. arXiv preprint arXiv:2101.10801 (2021)
7. Chen, X., Lin, K.Y., Wang, J., Wu, W., Qian, C., Li, H., Zeng, G.: Bi-directional cross-modality feature propagation with separation-and-aggregation gate for rgb-d semantic segmentation. In: ECCV. pp. 561–577. Springer (2020)
8. Cordts, M., Omran, M., Ramos, S., Rehfeld, T., Enzweiler, M., Benenson, R., Franke, U., Roth, S., Schiele, B.: The cityscapes dataset for semantic urban scene understanding. In: CVPR (2016)
9. Csurka, G., Larlus, D., Perronnin, F., Meylan, F.: What is a good evaluation measure for semantic segmentation? TPAMI **26**(1) (2004)
10. Dimitriadis, N., Maragos, P.: Advances in the training, pruning and enforcement of shape constraints of morphological neural networks using tropical algebra. arXiv preprint arXiv:2011.07643 (2020)
11. Franchi, G., Fehri, A., Yao, A.: Deep morphological networks. Pattern Recogn. **102**, 107246 (2020)
12. Fukushima, K.: Cognitron: A self-organizing multilayered neural network. Biol. Cybern. **20**(3–4), 121–136 (1975)
13. Groenendijk, R., Dorst, L., Gevers, T.: Morphpool: Efficient non-linear pooling & unpooling in cnns. In: 33rd British Machine Vision Conference 2022, BMVC 2022, London, UK, November 21-24, 2022. BMVA Press (2022)
14. Groenendijk, R., Dorst, L., Gevers, T.: Geometric back-propagation in morphological neural networks. TPAMI pp. 1–8 (2023)

15. Gupta, S., Girshick, R., Arbeláez, P., Malik, J.: Learning rich features from rgb-d images for object detection and segmentation. In: ECCV. pp. 345–360. Springer (2014)
16. He, K., Zhang, X., Ren, S., Sun, J.: Deep residual learning for image recognition. In: CVPR. pp. 770–778 (2016)
17. Heijmans, H.J., Goutsias, J.: Nonlinear multiresolution signal decomposition schemes. ii. morphological wavelets. TIP 9(11), 1897–1913 (2000)
18. Heijmans, H.J., Toet, A.: Morphological sampling. CVGIP: Image understanding 54(3), 384–400 (1991)
19. Hernández, G., Zamora, E., Sossa, H., Téllez, G., Furlán, F.: Hybrid neural networks for big data classification. Neurocomputing 390, 327–340 (2020)
20. Hu, J., Shen, L., Sun, G.: Squeeze-and-excitation networks. In: CVPR. pp. 7132–7141 (2018)
21. Jarrett, K., Kavukcuoglu, K., Ranzato, M., LeCun, Y.: What is the best multistage architecture for object recognition? In: ICCV. pp. 2146–2153. IEEE (2009)
22. Jiang, S., Xu, Y., Li, D., Fan, R.: Multi-scale fusion for rgb-d indoor semantic segmentation. Sci. Rep. 12(1), 20305 (2022)
23. Mondal, R., Santra, S., Mukherjee, S.S., Chanda, B.: Morphological network: How far can we go with morphological neurons? In: BMVC. BMVA Press (2022)
24. Paszke, A., Gross, S., Massa, F., Lerer, A., Bradbury, J., Chanan, G., Killeen, T., Lin, Z., Gimelshein, N., Antiga, L., Desmaison, A., Kopf, A., Yang, E., DeVito, Z., Raison, M., Tejani, A., Chilamkurthy, S., Steiner, B., Fang, L., Bai, J., Chintala, S.: Pytorch: An imperative style, high-performance deep learning library. In: NeuRIPS, pp. 8024–8035. Curran Associates, Inc. (2019)
25. Qiu, S., Xu, X., Cai, B.: Frelu: flexible rectified linear units for improving convolutional neural networks. In: ICPR. pp. 1223–1228. IEEE (2018)
26. Ritter, G.X., Sussner, P.: An introduction to morphological neural networks. In: ICPR. vol. 4, pp. 709–717. IEEE (1996)
27. Roy, S.K., Mondal, R., Paoletti, M.E., Haut, J.M., Plaza, A.: Morphological convolutional neural networks for hyperspectral image classification. IEEE Journal of Selected Topics in Applied Earth Observations and Remote Sensing 14, 8689–8702 (2021)
28. Serra, J.: Image analysis and mathematical morphology (1983)
29. Serra, J., Vincent, L.: An overview of morphological filtering. Circuits Systems Signal Process. 11(1), 47–108 (1992)
30. Shen, Y., Shih, F.Y., Zhong, X., Chang, I.C.: Deep morphological neural networks. PRAI 36(12), 2252023 (2022)
31. Silberman, N., Hoiem, D., Kohli, P., Fergus, R.: Indoor segmentation and support inference from RGBD images. ECCV 7576, 746–760 (2012)
32. Sussner, P.: Morphological perceptron learning. In: Proceedings of the 1998 ISIC/CIRA. pp. 477–482. IEEE (1998)
33. Sutskever, I., Martens, J., Dahl, G., Hinton, G.: On the importance of initialization and momentum in deep learning. In: ICML. pp. 1139–1147. PMLR (2013)
34. Velasco-Forero, S., Angulo, J.: Morphoactivation: Generalizing relu activation function by mathematical morphology. In: DGMM. pp. 449–461. Springer (2022)
35. Velasco-Forero, S., Rhim, A., Angulo, J.: Fixed point layers for geodesic morphological operations. In: BMVC. BMVA Press (2022)
36. Xie, L., Tian, Q., Wang, M., Zhang, B.: Spatial pooling of heterogeneous features for image classification. TIP 23(5), 1994–2008 (2014)
37. Yosinski, J., Clune, J., Bengio, Y., Lipson, H.: How transferable are features in deep neural networks? NeuRIPS 27 (2014)

38. Zeiler, M.D., Fergus, R.: Visualizing and understanding convolutional networks. In: ECCV. pp. 818–833. Springer (2014)
39. Zeiler, M.D., Taylor, G.W., Fergus, R.: Adaptive deconvolutional networks for mid and high level feature learning. In: ICCV. pp. 2018–2025. IEEE (2011)
40. Zhou, H., Qi, L., Huang, H., Yang, X., Wan, Z., Wen, X.: Canet: Co-attention network for rgb-d semantic segmentation. Pattern Recogn. **124**, 108468 (2022)

Nonlinear Representation Theory of Equivariant CNNs on Homogeneous Spaces Using Group Morphology

Jesús Angulo-Lopez[(✉)]

Mines Paris, PSL University, CMM-Centre de Morphologie Mathématique,
Paris, France
jesus.angulo@mines-paristech.fr

Abstract. This paper deals with a nonlinear theory of equivariant convolutional neural networks (CNNs) on homogenous spaces under the action of a group. Many groups of image transforms fit this framework.

The purpose of our work is to have a universal equivariant representation of nonlinear maps between image features which is based on mathematical morphology operators for groups. In particular, we combine some powerful results of universal representation of nonlinear mappings with the equivariance properties of morphological group operators.

The approach considered here is significantly different from other theories of representation of equivariant CNNs. On the one hand, it is founded on results from lattice theory and other hand, it deals with the universal representation of nonlinear maps, which can involve in a unified framework (linear) convolutions, activation functions and other nonlinear layers.

Keywords: deep learning · mathematical morphology · group morphology · equivariant CNN · nonlinear representation theory

1 Introduction

Figure 1 depicts the typical operators (layers) in a convolutional neural network (CNN) block formalized by the nonlinear map Ψ. Let us consider a group $\mathbb{G}$, now $\mathbb{G}$-CNN means that Ψ is equivariant to the group action T_g, $g \in \mathbb{G}$; i.e., $\Psi(T_g f) = T'_g \Psi(f)$. We consider that feature maps in these networks represent functions (or image fields) $f(x)$ on a homogeneous space and the layers are equivariant maps between spaces of functions, i.e., $[T_g f](x) = [f \circ g^{-1}](x) = f(g^{-1}x)$.

The theoretical study and implementation of equivariant CNNs and more generaly of equivariance in deep learning is an active area with many contributions. General theory of equivariant CNNs on homogenous spaces [5] or on compact groups [12] considers mainly the generalization of convolution to groups and

J. Angulo—This work was done during my academic visiting period to NYU in 2023, partially funded by the Foundation MINES Paris.

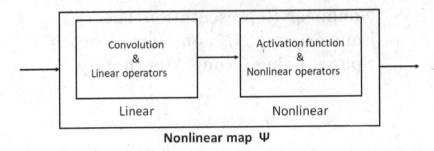

Fig. 1. Typical operators (layers) in a CNN block.

how any linear equivariant layer can be represented by combinations of equivariant group convolutions. The nonlinearity aspects of deep learning are secondary in this kind of approach and basically limited to point-wise nonlinearities. We propose to address in this paper the representation of nonlinear equivariant layers using group morphology and general morphological representation theory.

The interest of morphological scale-spaces in the context of equivariant deep learning has been explored previously from the perspective of Hamilton–Jacobi PDEs on groups [6,24]. This work proposes an initial contribution on the interest of group morphological operators for equivariant deep learning sketched in [1]. More precisely, our approach is a representation theory for nonlinear equivariant deep learning operators Ψ based on ordered structures (lattice algebra, tropical semirings) which will bring up the relevance of group morphological operators. However note that we are not dealing with the implementation of group morphological operators.

The main references of group morphology are the work done by Roerdink in [20,22], see also his papers on the commutative group case [19] or on the group of transformations for the camera projective model [21]. Another interesting theoretical contribution, in particular to the case of the affine group was proposed by Maragos [15]. The case of abelian groups was studied in detail in these references too [8,10]. Another related work which can be of interest for the reader is the definition of Minkowski product, instead of the sum, of two sets on the complex plane [7]. To the best of our knowledge, group morpholoy has been mainly used to solve problems on robotics, i.e., finding obstacles and free space, and on symmetry detection, both applications in [13].

Organisation of the Paper. The rest of the paper is organized as follows. Section 2 provides a review of the main theoretical results of universal representation of nonlinear (increasing) operators using mathematical morphology. An introduction to group morphology on homogeneous spaces is given in Sect. 3. The main contribution of our work is considered on Sect. 4 with a first general result on the morphological representation of equivariant CNNs. Some perspectives in Sect. 5 close the paper.

2   Universal Representation of Nonlinear Operators

In this section, we review the fundamental results for the morphological representation of nonlinear operators based on lattice algebra. We focus in particular on the case of increasing operators and provide just the references for the extension to non-increasing mappings.

2.1   Characterization of Increasing Operators on a Complete Lattice

Let us start by the most general result for the representation of any *increasing* nonlinear operator Ψ in a *lattice $\mathcal{L}$ with partial order* $\leq$, i.e., $X, Y \in \mathcal{L}$, $X \leq Y$ $\iff \Psi(X) \leq \Psi(Y)$.

Theorem 1 (Serra (1988) [23], Heijmans & Ronse (1990) [8]). *Let us consider a complete lattice $\mathcal{L}$ and an increasing operator $\Psi : \mathcal{L} \to \mathcal{L}$, which preserves the greatest element $\top$; i.e., satisfies $\Psi(\top) = \top$.*
  Then Ψ is the supremum of a non-empty set of erosions $\mathcal{E}$:

$$\Psi = \bigvee_{\varepsilon \in \mathcal{E}} \varepsilon.$$

If the operator preserves the smallest element $\perp$; i.e., $\Psi(\perp) = \perp$, the operator Ψ can be written as an infimum of dilations from a set $\mathcal{D}$:

$$\Psi = \bigwedge_{\delta \in \mathcal{D}} \delta.$$

That abstract universal representation theorem in terms of erosions and dilations can be instantiated in a particular case useful as starting point for this paper. Let us consider a translation equivariant[1] (TE) increasing operator Ψ. The domain of the functions considered here is either $E = \mathbb{R}^n$ or $E = \mathbb{Z}^n$, with the additional condition that we consider only closed subsets of E. We focus first on the set operator case applied on $\mathcal{P}(E)$ and then that for functions $f : E \to \bar{\mathbb{R}}$. Here are the main results from Matheron [17] and Maragos [16].

Kernel and Basis Representation of TE Increasing Set Operators. The kernel of the TE operator Ψ is defined as the following collection of input sets [17]:

$$\mathrm{Ker}(\Psi) = \{A \subseteq E \ : \ \mathbf{0} \in \Psi(A)\},$$

where $\mathbf{0}$ denotes the origin of E. In the following, we use the classic definitions of Minkowski sum $X \oplus B$ and Minkowski difference $X \ominus B$ of sets $X, B \in \mathcal{P}(E)$.

Theorem 2 (Matheron (1975) [17]). *Consider set operators on $\mathcal{P}(E)$. Let $\Psi : \mathcal{P}(E) \to \mathcal{P}(E)$ be a TE increasing set operator. Then*

$$\Psi(X) = \bigcup_{A \in Ker(\Psi)} X \ominus A = \bigcap_{B \in Ker(\bar{\Psi})} X \oplus \check{B}.$$

---

[1] In the classic literature of image processing and mathematical morphology, it is used the term translation-invariant (TI) filters and operators.

where the dual set operator is $\bar{\Psi}(X) = [\Psi(X^c)]^c$ and $\check{B}$ is the transpose structuring element.

The kernel of Ψ is a partially ordered set under set inclusion which has an infinity number of elements. In practice, by the property of absorption of erosion, that means that the erosion by B contains the erosions by any other kernel set larger than B and it is the only one required when taking the supremum of erosions. The morphological basis of Ψ is defined as the minimal kernel sets [16]:

$$\mathrm{Bas}(\Psi) = \{M \in \mathrm{Ker}(\Psi) : [A \in \mathrm{Ker}(\Psi) \text{ and } A \subseteq M] \implies A = M\}.$$

A sufficient condition for the existence of $\mathrm{Bas}(\Psi)$ is for Ψ to be an upper semi-continuous operator. We also consider closed sets on $\mathcal{P}(E)$.

Theorem 3 (Maragos (1989) [16]). *Let $\Psi : \mathcal{P}(E) \to \mathcal{P}(E)$ be a TE, increasing and upper semi-continuous set operator. Then*

$$\Psi(X) = \bigcup_{M \in Bas(\Psi)} X \ominus M = \bigcap_{N \in Bas(\bar{\Psi})} X \oplus \check{N}.$$

Kernel and Basis Representation of TE Increasing Operators on Functions. Previous set theory was extended [16] to the case of mappings on functions $\Psi(f)$ and therefore useful for signal or gray-scale image operators. We focus on the case of closed functions f, i.e., its epigraph is a closed set. In that case, the dual operator is $\bar{\Psi}(f) = -\Psi(-f)$ and the transpose function is $\check{f}(x) = f(-x)$. Let

$$\mathrm{Ker}(\Psi) = \{f : \Psi(f)(0) \geq 0\}$$

be the kernel of operator Ψ. As for the TE set operators, a basis can be obtained from the kernel functions as its minimal elements with respect to the partial order $\leq$, i.e.,

$$\mathrm{Bas}(\Psi) = \{g \in \mathrm{Ker}(\Psi) : [f \in \mathrm{Ker}(\Psi) \text{ and } f \leq g] \implies f = g\}.$$

This collection of functions can uniquely represents the operator Ψ.

Theorem 4 (Maragos (1989) [16]). *Consider an upper semi-continuous operator Ψ acting on an upper semi-continuous function f. Let $Bas(\Psi) = \{g_i\}_{i \in I}$ be its basis and $Bas(\bar{\Psi}) = \{h_j\}_{j \in J}$ the basis of the dual operator. If Ψ is a TE and increasing operator then it can be represented as*

$$\Psi(f)(x) = \sup_{i \in I} [(f \ominus g_i)(x)] = \sup_{i \in I} \left[\inf_{y \in \mathbb{R}^n} \{f(x+y) - g_i(y)\} \right] \tag{1}$$

$$= \inf_{j \in J} [(f \oplus \check{h}_j)(x)] = \inf_{j \in J} \left[\sup_{y \in \mathbb{R}^n} \{f(x-y) + \check{h}_j(y)\} \right] \tag{2}$$

The converse is true: given a collection of functions $\mathcal{B} = \{g_i\}_{i \in I}$ such that all elements of it are minimal in $(\mathcal{B}, \leq)$, the operator $\Psi(f) = \sup_{i \in I} \{f \ominus g_i\}$ is a TE increasing operator whose basis is equal to $\mathcal{B}$.

For some operators, the basis can be very large (potentially infinity) and even if the above theorem represents exactly the operator by using a full expansion of all erosions, we can obtain an approximation based on smaller collections or truncated bases $\mathcal{B} \subset \mathrm{Bas}(\Psi)$ and $\bar{\mathcal{B}} \subset \mathrm{Bas}(\bar{\Psi})$. Then, from the operators $\Psi_l(f) = \sup_{g \in \mathcal{B}} \{f \ominus g\}$ and $\Psi_u(f) = \inf_{h \in \bar{\mathcal{B}}} \{f \oplus \bar{h}\}$ the original Ψ is bounded from below and above, i.e., $\Psi_l(f) \leq \Psi(f) \leq \Psi_u(f)$. Note also that in the case of a non minimal representation by a subset of the kernel functions larger than the basis, one just gets a redundant still satisfactory representation.

2.2   Extension to Non-increasing Mappings

The extension of this theory to TE *non necessarily increasing mappings* was introduced by Bannon and Barrera in [3]. It involves a supremum of a basis of operators combining an erosion and an anti-dilation. Other than that additional level of complexity of the underlying operators, the results are structurally similar to those that we discussed above.

3   Morphological Group Equivariant Operators on Homogeneous Spaces

We revisit in this section the main results on Roerdink group morphology [20, 22].

3.1   Group Morphological Operators for Boolean Lattices

Let us consider E is now a homogeneous space under a group $\mathbb{G}$ acting transitively on E. The object space of interest is the Boolean lattice $\mathcal{P}(E)$ of all subsets of E.

The strategy to introduce the group operators on $\mathcal{P}(E)$ will consist in

1. defining Minkowski operators on $\mathcal{P}(\mathbb{G})$, then
2. using a lifting of subsets of E to subsets of $\mathbb{G}$, apply these operators, and finally
3. projecting the corresponding result back to the original space E.

Dilation and Erosion on $\mathcal{P}(\mathbb{G})$. A mapping $\Psi : \mathcal{P}(\mathbb{G}) \to \mathcal{P}(\mathbb{G})$ is called $\mathbb{G}$-left-equivariant when, for all $g \in \mathbb{G}$, $\Psi(gG) = g\Psi(G)$, $\forall G \in \mathcal{P}(\mathbb{G})$. And similarly, a $\mathbb{G}$-right-equivariant implies for all $\forall G \in \mathcal{P}(\mathbb{G})$, $\Psi(Gg) = \Psi(G)g$.

The dilation and erosion on $\mathcal{P}(\mathbb{G})$ will be defined as the $\mathbb{G}$-equivariant mappings commuting with unions and intersections respectively. Let H be a fixed subset of $\mathbb{G}$, called the group structuring element, we define the $\mathbb{G}$-left-equivariant dilation and erosion of G by H as

$$\delta_H^l(G) = G \oplus_{\mathbb{G}}^l H = \bigcup_{h \in H} Gh = \bigcup_{g \in G} gH = \{k \in \mathbb{G} : (k\check{H}) \cap G \neq \emptyset\}, \quad (3)$$

$$\varepsilon_H^l(G) = G \ominus_{\mathbb{G}}^l H = \bigcap_{h \in H} Gh^{-1} = \{g \in \mathbb{G} : gH \subseteq G\}, \quad (4)$$

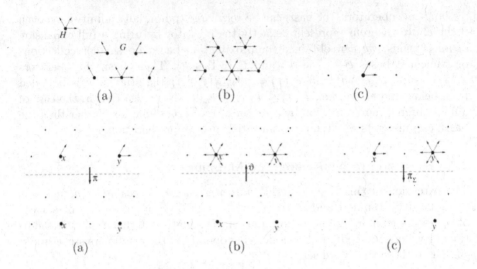

Fig. 2. Top, morphological operations on the motion group $SE(2)$: (a) group set G and group structuring element H, (b) dilation of G by H and (c) erosion of G by H. Bottom, illustration of the actions of operators ϑ (a), π (b) and (c) π_Σ on the rotation-translation group $SE(2)$. Extracted from [22].

where $gH = \{gh : h \in H\}$, $Hg = \{hg : h \in H\}$.

An example of dilation and erosion for the rotation-translation group $SE(2)$ is depicted in Fig. 2-top.

Proposition 1 (Roerdink (2000) [22]). *The pair $(\delta_H^l, \varepsilon_H^l)$ forms an adjunction and all $\mathbb{G}$-left-equivariant adjunctions on $\mathcal{P}(\mathbb{G})$ are of this form.*

The duality by complement is given by the fact that $\left(G \oplus_{\mathbb{G}}^l H\right)^c = G^{-1} \ominus_{\mathbb{G}}^l H^{-1}$. Because of the non-commutativity of the set product $G \oplus_{\mathbb{G}}^l H$, it is possible to introduce $\mathbb{G}$-right-equivariant dilation and erosion.

Lifting and Projections Operators. We remind that work on the case of $\mathbb{G}$ is acting transitively on E. Let the origin ω be an arbitrary point of E. The lifting operator $\vartheta : \mathcal{P}(E) \to \mathcal{P}(\mathbb{G})$ is the mapping defined for any subset $X \in \mathcal{P}(E)$ as

$$\vartheta(X) = \{g \in \mathbb{G} : g\omega \in X\}, \tag{5}$$

associates to X all group elements which map the origin ω to an element of X. The projection operator $\pi : \mathcal{P}(\mathbb{G}) \to \mathcal{P}(E)$ for any $G \in \mathcal{P}(\mathbb{G})$ as

$$\pi(G) = \{g\omega : g \in \mathbb{G}\}, \tag{6}$$

maps to each subset G of $\mathbb{G}$ the collection of points $g\omega \in E$, where g ranges over G. The main benefit of creating these maps it that they translate the group action on X into multiplication in $\mathbb{G}$.

The stabilizer–projection operator $\pi_\Sigma : \mathcal{P}(\mathbb{G}) \to \mathcal{P}(E)$ first extracts the cosets and then carries out the projection π:

$$\pi_\Sigma = \pi \varepsilon_\Sigma^l(G),$$

where the erosion by the stabilizer σ; i.e., $\varepsilon_\Sigma^l(G) = G \ominus_\mathbb{G}^l$ has the property $\varepsilon_\Sigma^l(G) = \varepsilon_\Sigma^l(\varepsilon_\Sigma^l(G)) = \delta_\Sigma^l(\varepsilon_\Sigma^l(G))$ which implies that $\varepsilon_\Sigma^l(G)$ is an idempotent operator in $\mathcal{P}(\mathbb{G}) \to \mathcal{P}(\mathbb{G})$ providing the invariant elements to Σ.

All the operators ϑ, π, π_Σ are increasing and $\mathbb{G}$-equivariant. It is obvious that $\pi\vartheta = \mathrm{Id}_{\mathcal{P}(E)}$ and $\pi_\Sigma\vartheta = \mathrm{Id}_{\mathcal{P}(E)}$.

The illustration of the actions of these operators for the example of rotation-translation group $SE(2)$ is given in Fig. 2-bottom.

$\mathbb{G}$-Equivariant Dilation and Erosion on $\mathcal{P}(E)$

A $\mathbb{G}$-equivariant operator Ψ on $\mathcal{P}(E)$ can be constructed by using the group operator $\tilde\Psi$ according to the following commuting diagram:

$$
\begin{array}{ccc}
\mathcal{P}(\mathbb{G}) & \xrightarrow{\tilde\Psi} & \mathcal{P}(\mathbb{G}) \\
\uparrow{\vartheta} & & \downarrow{\pi} \\
\mathcal{P}(E) & \xrightarrow{\Psi} & \mathcal{P}(E)
\end{array}
$$

Let us consider in particular the $\mathbb{G}$-equivariant dilation and erosion on $\mathcal{P}(E)$. For any set X and structuring element B, $X, B \in \mathcal{P}(E)$:

$$\delta_B^\mathbb{G}(X) = \pi\left[\vartheta(X) \oplus_\mathbb{G}^l \vartheta(B)\right] = \bigcup_{g \in \vartheta(X)} gB, \tag{7}$$

$$\varepsilon_B^\mathbb{G}(X) = \pi_\Sigma\left[\vartheta(X) \ominus_\mathbb{G}^l \vartheta(B)\right] = \bigcap_{g \in \vartheta(X^c)} g\hat{B}^*, \tag{8}$$

with $\hat{Y}^* = \left(\pi(\check\vartheta(Y))\right)^c$.

3.2 $\mathbb{G}$-Equivariant Dilation and Erosion on $\mathcal{L}$

The generalization to non-Boolean lattices and particular the case of numerical functions is based on the notion of sup-generating families of a lattice. A subset l of a complete lattice $\mathcal{L}$ is called sup-generating if every element of $\mathcal{L}$ can be written as a supremum of elements of l.

For every $X \in \mathcal{L}$, let $l(X) = \{x \in l : x \leq X\}$ and $X = \bigvee l(X)$. Let $\mathcal{L}$ be a complete lattice with an automorphism group $\mathbb{G}$ and a sup-generating subset l such that:

1. l is $\mathbb{G}$-equivariant; i.e., for every $g \in \mathbb{G}$ and $x \in l$, $gx \in l$;
2. $\mathbb{G}$ is transitive on l: for every $x, y \in l$ there exists at least one $g \in \mathbb{G}$ such that $gx = y$.

In that case, the construction of operators follows the commuting diagram:

$$
\begin{array}{ccc}
\mathcal{P}(\mathbb{G}) & \xrightarrow{\bar{\Psi}} & \mathcal{P}(\mathbb{G}) \\
\Big\uparrow\vartheta & & \Big\downarrow\pi \\
\mathcal{P}(l) & \xrightarrow{\bar{\bar{\Psi}}} & \mathcal{P}(l) \\
\Big\uparrow l & & \Big\downarrow\vee \\
\mathcal{L} & \xrightarrow{\Psi} & \mathcal{L}
\end{array}
$$

The lattice of numerical functions has a natural sup-generating family given by the impulse functions $f_{x,t}$, $x \in E$, $t \in \overline{\mathbb{R}}$ defined by

$$
f_{x,t}(y) = \begin{cases} t, & y = x \\ -\infty, & y \neq x \end{cases}
$$

For the complete results on that case, the reader is invited to [22]. In the framework of this paper, we propose to introduce numerical group operators as group convolutions in $(\max, +)$-algebra which could be represented by the impulse functions.

4   Morphological Representation of Equivariant CNNs on Homogeneous Spaces

Let us come back to the diagram of Fig. 1. The network Ψ as a whole is $\mathbb{G}$-equivariant if all its layers are $\mathbb{G}$-equivariant. One can consider group convolution for the linear components of Ψ and to characterize the nonlinear ones using a group morphological representation. Or alternatively, to represent the network Ψ as a nonlinear map of $\mathbb{G}$-equivariant erosions (or dilations). As we mentioned above, for the sake of simplicity of this paper we assume Ψ is increasing.

4.1   From Group Convolution to Group Dilations/Erosions for Functions

Given a compact group $\mathbb{G}$, the group convolution (or more precisely, "correlation") layer between $\mathbb{G}$-feature maps in $L_2(\mathbb{G})$ with kernel k is given by [4]

$$
(f \star_{\mathbb{G}} k)(g) = \int_{\mathbb{G}} f(h)k(g^{-1}h)dh, \tag{9}
$$

where dh is the left Haar measure on $\mathbb{G}$. Note that the feature map $f \in \mathcal{F}(\mathbb{G}, \mathbb{R})$ has been lifted to $\mathbb{G}$. A typical example is the group $SE(2)$:

$$\left(f \star_{SE(2)} k \right)(x, \theta) = \int_{\mathbb{R}^2} \int_{\mathbb{S}^1} f(x', \theta') k \left(R_\theta^{-1}(x' - x), \theta' - \theta \right) dx' d\theta'.$$

Let us consider the counterpart group convolution in tropical semirings. We work on the set of functions, or $\mathbb{G}$-feature maps, $f : \mathbb{G} \to \bar{\mathbb{R}} = \mathbb{R} \cup \{+\infty, +\infty\}$, where instead of square integrability we need upper (or lower) semi-continuity on $\mathbb{G}$. The $\mathbb{G}$-equivariant max-plus dilation and adjoint erosion of function f by the structuring function b, with $f, b \in \mathcal{F}(\mathbb{G}, \bar{\mathbb{R}})$, are defined as: $\forall g \in \mathbb{G}$

$$\left(f \oplus_{\mathbb{G}} b \right)(g) = \sup_{h \in \mathbb{G}} \left\{ f(h) + b(gh^{-1}) \right\}, \tag{10}$$

$$\left(f \ominus_{\mathbb{G}} b \right)(g) = \inf_{h \in \mathbb{G}} \left\{ f(h) - b(g^{-1}h) \right\}. \tag{11}$$

For readers interested on the relationship between morphological operators and PDE models, we note that similar operators appears in the context of Hamilton–Jacobi equations and its viscosity solutions as the Hopf–Lax formula for the Heisenberg group [14] or the Carnot group [2], relevant in control theory.

Because of the combination of weighting and nonlinearities which are used in deep learning, which can be viewed from a morphological perspective [25], we can propose more general equivariant morphological operators. The $\mathbb{G}$-equivariant max-times-plus dilation and and erosion of function f by the pair of structuring functions $\{a, b\}$, $a(g) > 0$, $\forall g \in \mathbb{G}$, $f, a, b \in \mathcal{F}(\mathbb{G}, \bar{\mathbb{R}})$, are defined as

$$\left(f \oplus_{\mathbb{G}} \{a, b\} \right)(g) = \sup_{h \in \mathbb{G}} \left\{ a(gh^{-1}) f(h) + b(gh^{-1}) \right\}, \tag{12}$$

$$\left(f \ominus_{\mathbb{G}} \{a, b\} \right)(g) = \inf_{h \in \mathbb{G}} \left\{ \frac{1}{a(g^{-1}h)} \left(f(h) - b(g^{-1}h) \right) \right\}. \tag{13}$$

These operators are the generalization to group morphology of the H-operators dilation and erosion studied by Heijmans in [9].

4.2   Representation of Increasing $\mathbb{G}$-Equivariant Operators

Consider an increasing $\mathbb{G}$-equivariant group operator

$$\tilde{\Psi} : \mathcal{F}(\mathbb{G}, \bar{\mathbb{R}}) \to \mathcal{F}(\mathbb{G}, \bar{\mathbb{R}})$$

The kernel of the operator $\tilde{\Psi}$ is given by:

$$\mathrm{Ker}(\tilde{\Psi}) = \left\{ b : \tilde{\Psi}(b)(\omega) \geq \perp \right\}, \quad b \in \mathcal{F}(\mathbb{G}, \bar{\mathbb{R}})$$

and the corresponding morphological minimal basis of $\tilde{\Psi}$ is obtained from the kernel functions as its minimal elements with respect to the partial order $\leq$, i.e.,

$$\mathrm{Bas}(\tilde{\Psi}) = \left\{ b' \in \mathrm{Ker}(\tilde{\Psi}) : [b \in \mathrm{Ker}(\tilde{\Psi}) \text{ and } b \leq b'] \implies b = b' \right\}$$

This collection of functions can uniquely represent the $\tilde{\Psi}$ operator as follows.

Theorem 5. *Consider a group operator $\tilde{\Psi}$ acting on an upper semi-continuous function f and satisfying $\tilde{\Psi}(\top) = \top$. Let $Bas(\tilde{\Psi}) = \{b_i\}_{i \in I}$ be its basis. If $\tilde{\Psi}$ is a $\mathbb{G}$-equivariant and increasing operator then it can be represented as a supremum of $\mathbb{G}$-equivariant erosions. Every $\mathbb{G}$-equivariant increasing operator $\tilde{\Psi}$ satisfying $\tilde{\Psi}(\bot) = \bot$ can be written as an infimum of $\mathbb{G}$-equivariant dilation. We have therefore the alternative representations:*

$$\tilde{\Psi}(f)(g) = \sup_{i \in I} \left[(f \ominus_{\mathbb{G}} b_i)(g) \right] = \sup_{i \in I} \inf_{h \in G} \left\{ f(h) - b_i(g^{-1}h) \right\} \tag{14}$$

$$= \inf_{j \in J} \left[(f \oplus_{\mathbb{G}} b_j)(g) \right] = \inf_{j \in J} \sup_{h \in G} \left\{ f(h) + b_i(gh^{-1}) \right\} \tag{15}$$

The converse is true.

Proof. This proof follows the same line as the proof in [9] (Proposition 7.1). Let us prove the expression of the infimum of dilations. Then the second follows by duality.

Let $\tilde{\Psi}(f)(g) = \bot$ if $f(g) = \bot$, and let $\mathcal{D}$ be the set of all dilations $f \oplus_{\mathbb{G}} b = \delta_b^{\mathbb{G}}(f)$ which dominate $\tilde{\Psi}$, that is $\delta_b^{\mathbb{G}} \geq \tilde{\Psi}$. It is clear that $\inf \mathcal{D} \geq \tilde{\Psi}$.

To prove the reverse inequality, it suffices to show that for every $f \in \mathcal{F}(\mathbb{G}, \mathbb{R})$ there is a $\delta_b^{\mathbb{G}} \in \mathcal{D}$ such that

$$\tilde{\Psi}(f)(\omega) \geq \delta_b^{\mathbb{G}}(\omega), \tag{16}$$

where w is the origin. That yields

$$\tilde{\Psi}(f)(\omega) \geq (\inf \mathcal{D})(f)(\omega),$$

and therefore using the transitivity and the notation $f_{g^{-1}}(\omega) = f(g)$, we have that

$$\tilde{\Psi}(f)(g) = \tilde{\Psi}(f_{g^{-1}})(\omega) \geq (\inf \mathcal{D}) f_{g^{-1}}(\omega) = (\inf \mathcal{D})(f)(g).$$

To prove (16), take $f \in \mathcal{F}(\mathbb{G}, \overline{\mathbb{R}})$. For $h \in \mathbb{G}$, we define the mapping $d_h : \overline{\mathbb{R}} \to \overline{\mathbb{R}}$ by

$$d_h(t) = \begin{cases} -\infty, & \text{if } t = -\infty \\ \tilde{\Psi}(f)(\omega), & \text{if } -\infty < t \leq f(h^{-1}) \\ +\infty, & \text{if } t = f(h^{-1}). \end{cases}$$

We can prove that d_h is dilation in t. Let $\delta_b^{\mathbb{G}}$ be the $\mathbb{G}$-equivariant group dilation given by

$$\delta_b^{\mathbb{G}}(f)(g) = (f \oplus_{\mathbb{G}})(g) = \sup_{h \in G} d_h \left(f(gh^{-1}) \right),$$

with $d_h(t) = t + b(h)$, $t \in \overline{\mathbb{R}}$ being a dilation [9] (Example 2.1).

It follows that for $g = \omega$ one has

$$\delta_b^{\mathbb{G}}(f)(\omega) = \sup_{h \in G} d_h \left(f(h^{-1}) \right) = \tilde{\Psi}(f)(\omega).$$

Next, it should be shown that $\delta_b^{\mathbb{G}} \in \mathcal{D}$, or in other words that $\tilde{\Psi} \leq \delta_b^{\mathbb{G}}$. Note that the later will be case only if

$$\tilde{\Psi}(f')(\omega) \leq \delta_b^{\mathbb{G}}(f')(0), \quad \forall f',$$

There are three possibilities for that

i) $f' = \perp$ (trivial);

ii) $\perp \neq f' \leq f$, then $\delta_b^{\mathbb{G}}(f')(0) = \sup_{h \in \mathbb{G}} d_h \left(f'(h^{-1}) \right) = \tilde{\Psi}(f)(\omega) \geq \tilde{\Psi}(f')(\omega)$;

iii) $f' > f$, then $\delta_b^{\mathbb{G}}(f')(0) = \infty \geq \tilde{\Psi}(f')(\omega)$.

We can now state the final result of the paper which provide the universal representation of $\mathbb{G}$-equivariant and increasing operators using group morphological dilations and erosions.

Theorem 6. *Let Ψ be an operator acting on upper semi-continuous functions $f : \mathcal{F}(E, \bar{\mathbb{R}}) \to \mathcal{F}(E, \bar{\mathbb{R}})$ on a homogeneous space E.*

If Ψ is a $\mathbb{G}$-equivariant and increasing operator then it can be represented as

$$\Psi(f)(x) = \pi \left[\sup_{i \in I} \left(\vartheta(f) \ominus_{\mathbb{G}} b_i \right) \right] = \pi \left[\sup_{i \in I} \inf_{h \in \mathbb{G}} \left\{ \vartheta(f)(h) - b_i(g^{-1}h) \right\} \right]. \quad (17)$$

The proof makes use of 14 and the construction of $\mathbb{G}$-equivariant operators using the paradigm of Sect. 3.2.

5 Perspectives

The first step forward in order to complete the scope of our program is to consider the case of non-increasing equivariant operators. Technically there is no major challenge and the representation combining both supremum of erosions and infimum of dilations will provide a sound framework to explore innovative deep learning architectures.

A second element to be studied is the particular case of the morphological representation of group convolution by morphological group operators. That makes sense in the particular case of finite discrete operators and functions, see [11].

Obviously the final perspective for us is the practical implementation of group morphology and its interest on equivariant deep learning. We do believe morphological representations yield the general structure of equivariant layers learning the nonlinear components of CNNs and other neural networks, as the companions to the equivariant convolution layers. For initial results, see [18].

References

1. Angulo, J.: Some open questions on morphological operators and representations in the deep learning era. In: Lindblad, J., Malmberg, F., Sladoje, N. (eds.) DGMM 2021. LNCS, vol. 12708, pp. 3–19. Springer, Cham (2021). https://doi.org/10.1007/978-3-030-76657-3_1

2. Balogh, Z.M., Calogero, A., Pini, R.: The Hopf-Lax formula in Carnot groups: a control theoretic approach. Calc. Var. **49**, 1379–1414 (2014)

3. Banon, G.J.F., Barrera, J.: Minimal representations for translation-invariant set mappings by mathematical morphology. SIAM J. Appl. Math. **51**(6), 1782–1798 (1991)

4. Cohen, T.S., Welling, M.: Group equivariant convolutional networks. In: International of Conference on Machine Learning, pp. 2990–2999 (2016)

5. Cohen, T.S., Geiger, M., Weiler, M.: A general theory of equivariant CNNs on homogeneous spaces. In: Advances in Neural Information Processing Systems, vol. 32 (2019)

6. Duits, R., Smets, B., Bekkers, E., Portegies, J.: Equivariant deep learning via morphological and linear scale space PDEs on the space of positions and orientations. In: Elmoataz, A., Fadili, J., Quéau, Y., Rabin, J., Simon, L. (eds.) SSVM 2021. LNCS, vol. 12679, pp. 27–39. Springer, Cham (2021). https://doi.org/10.1007/978-3-030-75549-2_3

7. Farouki, R.T., Moon, H.P., Ravani, B.: Minkowski geometric algebra of complex sets. Geom. Dedicata. **85**, 283–315 (2001)

8. Heijmans, H.J.A.M., Ronse, C.: The algebraic basis of mathematical morphology I. Dilations and erosions. Comput. Vision Graph. Image Process. **50**(3), 245–295 (1990)

9. Heijmans, H.J.A.M.: Theoretical aspects of gray-level morphology. IEEE Trans. PAMI **13**(6), 568–582 (1991)

10. Heijmans, H.J.A.M.: Mathematical morphology: a modern approach in image processing based on algebra and geometry. SIAM Rev. **37**(1), 1–36 (1995)

11. Khosravi, M., Schafer, R.W.: Implementation of linear digital filters based on morphological representation theory. IEEE Trans. Signal Process. **42**(9), 2264–2275 (1994)

12. Kondor, R., Trivedi, S.: On the generalization of equivariance and convolution in neural networks to the action of compact groups. Proc. Mach. Learn. Res. **80**, 2747–2755 (2018)

13. Lysenko, M., Nelaturi, S., Shapiro, V.: Group morphology with convolution algebras. In: Proceedings of the 14th ACM Symposium on Solid and Physical Modeling, pp. 11–22 (2010)

14. Manfredi, J., Stroffolini, B.: A Version of the Hopf-Lax Formula in the Heisenberg Group. Commun. Partial Differ. Eqn. **27**, 1139–1159 (2002)

15. Maragos, P.: Affine morphology and affine signal models. In: Proceedings of SPIE Vol. 1350 Image Algebra and Morphological Image Processing, pp. 31–44 (1990)

16. Maragos, P.: A representation theory for morphological image and signal processing. IEEE Tran. Pattern Anal. Mach. Intell. **11**(6), 586–599 (1989)

17. Matheron, G.: Random Sets and Integral Geometry. Wiley, Hoboken (1974)

18. Penaud-Polge, V., Velasco-Forero, S., Angulo, J.: Group equivariant networks using morphological operators. In: Rinaldi, S. (ed.) DGMM 2024. LNCS, vol. 14605, pp. 165–177. Springer, Cham (2024)

19. Roerdink, J.B.T.M., Heijmans, H.J.A.M.: Mathematical morphology for structures without translation symmetry. Signal Process. **15**(3), 271–277 (1988)

20. Roerdink, J.B.T.M.: Mathematical morphology with noncommutative symmetry groups. In: Mathematical Morphology in Image Processing, chap. 7, Marcel Dekker Press (1992)

21. Roerdink, J.B.T.M.: Computer vision and mathematical morphology. In: Kropatsch, W., Klette, R., Solina, F., Albrecht, R. (eds.) Theoretical Foundations of Computer Vision. Computing Supplement, vol. 11, pp. 131–148. Springer, Vienna (1996). https://doi.org/10.1007/978-3-7091-6586-7_8

22. Roerdink, J.B.T.M.: Group morphology. Pattern Recogn. **33**(6), 877–895 (2000)

23. Serra, J. (ed.): Image Analysis and Mathematical Morphology: Theoretical Advances. Academic Press (1988)

24. Smets, B., Portegies, J., Bekkers, E.J., Duits, R.: PDE-based group equivariant convolutional neural networks. J. Math. Imaging Vis. **65**, 209–239 (2023)
25. Velasco-Forero, S., Angulo, J.: MorphoActivation: generalizing ReLU activation function by mathematical morphology. In: Baudrier, É., Naegel, B., Krähenbühl, A., Tajine, M. (eds.) DGMM 2022. LNCS, vol. 13493, pp. 449–461. Springer, Cham (2022). https://doi.org/10.1007/978-3-031-19897-7_35

Hierarchical and Graph-Based Models, Analysis and Segmentation

Building the Topological Tree of Shapes from the Tree of Shapes

Julien Mendes Forte[1]([envelope]) [iD], Nicolas Passat[2] [iD], and Yukiko Kenmochi[1] [iD]

[1] Normandie Univ, UNICAEN, ENSICAEN, CNRS, GREYC, 14050 Caen, France
`julien.mendes-forte@unicaen.fr`
[2] Université de Reims Champagne-Ardenne, CReSTIC, 51100 Reims, France

Abstract. The topological tree of shapes was recently introduced as a new hierarchical structure within the family of morphological trees. Morphological trees are efficient models for image processing and analysis. For such applications, it is of paramount importance that these structures be built and manipulated with optimal complexity. In this article, we focus on the construction of the topological tree of shapes. We propose an algorithm for building the topological tree of shapes from the tree of shapes. In particular, a cornerstone of this algorithm is the construction of the complete tree of shapes, another recently introduced tree unifying both the tree of shapes and the topological tree of shapes. We also discuss the cost of the computation of these structures.

1 Introduction

Graphs are popular mathematical structures for the representation of objects and their relationships. In particular, they are widely used in mathematical morphology, especially in the field of connected operators [22]. In this context, they offer an efficient means for organizing information hierarchically, for modelling and manipulation purpose. The induced structures are often trees which encode hierarchies of partitions.

In mathematical morphology these trees often model the structure of images, with regard to their spatial and spectral information, via the topological organization of the connected components related to their flat zones. In the family of partial partitions, the main two subfamilies of hierarchical structures are component trees and trees of shapes. The component tree [21] models the inclusion relation between the connected components derived from the threshold sets of an image. The tree of shapes [13] models the nested inclusion between (the boundaries of) these connected components.

At the confluence of these two subfamilies, the topological tree of shapes was recently introduced [15]. It was initially designed for modelling the two orders—inclusion and nesting—associated to component trees and trees of shapes. In

This work was supported by Région Normandie (thesis grant RIN), Partenariats Hubert Curien (grant Sakura 49674RK) and ANR (Grants ANR-22-CE45-0034 and ANR-23-CE45-0015).

S. Brunetti et al. (Eds.): DGMM 2024, LNCS 14605, pp. 271–285, 2024.
https://doi.org/10.1007/978-3-031-57793-2_21

particular, the topological tree of shapes is structurally derived from the min- and max-trees of an image (two dual component trees) and the adjacency-tree (a binary variant of the tree of shapes) [20] of each threshold set.

The topological tree of shapes and the tree of shapes derive (by a decreasing homeomorphism) from the complete tree of shapes, which is also introduced in [15]. This led to a construction scheme of the topological tree of shapes from the tree of shapes, via the complete tree of shapes. A first non-optimal algorithm was proposed in [16].

In this article, we propose an algorithm for building the topological tree of shapes that improves the one of [16]. It aims to minimize the space cost of intermediate structures by avoiding the storage of redundant information. Doing so, it also reduces the time cost for their construction and handling, leading to an overall cost optimization.

This article is organized as follows. Section 2 recalls previous works on component trees and trees of shapes. Section 3 provides useful definitions. Section 4 describes the algorithmic scheme for building the topological tree of shapes from the tree of shapes. Section 5 provides a complexity analysis of this algorithm. Section 6 concludes this article.

2   Related Works

The trees developed in mathematical morphology can be divided into those that model total partitions (e.g. binary partition trees, watershed trees) or partial partitions. We focus on the second, which mainly include the component tree and the tree of shapes.

The component tree [2,21] is based on the inclusion of the threshold sets of an image. Initially designed for grey-level images, it also led to variants dedicated to images on partially ordered values (e.g. [11]), and alternative connectivity paradigms. It was involved in various applications, mostly for filtering and segmentation based on attribute selection or optimal cut computation. More recently, it was also investigated as a way to embed topological information in deep-learning frameworks [18]. Beyond its potential applications, many efforts were geared towards developing efficient construction algorithms for component trees [4]. More recently, parallel algorithms were proposed, including distributed paradigms [8,10] and GPU-based approaches [1].

The tree of shapes [13] is based on the nested relation between the level-lines of an image. It is often described as a self-dual version of the component tree, since these level-lines are defined by the hole-closing of the connected components of the min- and max-trees. The tree of shapes also led to variants dedicated to multivalued images [5], or compact versions focusing on the topological structure of the level-lines [23]. The adjacency tree [20], introduced as a topological descriptor for binary images, is also a variant of tree of shapes. The tree of shapes was involved in various image processing and analysis applications, mainly including segmentation. It is for instance related to the minimum barrier distance in images, and allows in particular to estimate it with various potential applications [14]. Many strategies were proposed for efficiently building the tree of shapes, e.g. based on a union-find structure [9] or via root-to-leave strategies [12]. Some parallel approaches were also investigated [7].

Within the literature dedicated to the construction of the component tree and the tree of shapes, some algorithms rely on the first to build the second [3,6] or vice versa [24]. This is motivated by the strong links that exist between them. In [16], we introduced the unifying notion of a complete tree of shapes, that provides a continuum between both structures. The complete tree of shapes contains the nodes of the min- and max-trees, and allows to derive the tree of shapes by a decreasing (reversible) homeomorphism. Since the topological tree of shapes can also be derived from the complete tree of shapes by a decreasing (yet non-reversible) homeomorphism, we propose to build the topological tree of shapes from the tree of shapes via the complete tree of shapes.

3   Background Notions

We consider images defined on a discrete support $\mathbb{U}$ where the Jordan-Brouwer property holds. In practice, we assume $\mathbb{U} = \mathbb{Z}^d$ $(d \geqslant 2)$, endowed with the digital topology framework. We also consider that images take their values in a finite, totally ordered set $(\mathbb{V}, \leqslant)$. Without loss of generality, we assume that $\mathbb{V} = [\![\perp, \top]\!] \subset \mathbb{Z}$. An image is defined as a function $\mathcal{F} : \mathbb{U} \to \mathbb{V}$. We assume that the number of points $\mathbf{x} \in \mathbb{U}$ such that $\mathcal{F}(\mathbf{x}) > \perp$ is finite. This number n is considered as the size of the image.

Let $v \in \mathbb{V}$. The upper- and lower-threshold sets of $\mathcal{F}$ (see Fig. 1(a)) at value v are the subsets of $\mathbb{U}$ defined as

$$\begin{aligned}
\Lambda_v^\circ(\mathcal{F}) &= \{\mathbf{x} \in \mathbb{U} \mid v \leqslant \mathcal{F}(\mathbf{x})\} \\
\Lambda_v^\bullet(\mathcal{F}) &= \{\mathbf{x} \in \mathbb{U} \mid v > \mathcal{F}(\mathbf{x})\}
\end{aligned} \tag{1}$$

Let $X \subseteq \mathbb{U}$. We note $\Pi[X] \subseteq 2^{\mathbb{U}}$ the set of the connected components of X. For all $v \in \mathbb{V}$, we define the following sets

$$\begin{aligned}
\Theta_v^\circ &= \Pi[\Lambda_v^\circ(\mathcal{F})] & & & \Theta^\circ &= \bigcup_{v \in \mathbb{V}} \Theta_v^\circ \\
\Theta_v^\bullet &= \Pi[\Lambda_v^\bullet(\mathcal{F})] & \text{and} & & \Theta^\bullet &= \bigcup_{v \in \mathbb{V}} \Theta_v^\bullet \\
\Theta_v &= \Theta_v^\circ \cup \Theta_v^\bullet & & & \Theta &= \Theta^\circ \cup \Theta^\bullet = \bigcup_{v \in \mathbb{V}} \Theta_v
\end{aligned} \tag{2}$$

3.1   Trees

We recall the definition of classic and more recently introduced trees. We consider the partial order relation $\subseteq$ on Θ° (resp. $\Theta^\bullet$) and we note $\lhd^\circ$ (resp. $\lhd^\bullet$) the reflexive-transitive reduction of $\subseteq$ on Θ° (resp. $\Theta^\bullet$). We define a tree by a set of elements linked by arcs. Here, these arcs are represented via a relation notation (and not via a set notation).

Definition 1 (Component tree(s) [21]). *The max-tree (resp. min-tree) of $\mathcal{F}$ is the tree $\mathfrak{T}_{\Theta^\circ} = (\Theta^\circ, \lhd^\circ)$ (resp. $\mathfrak{T}_{\Theta^\bullet} = (\Theta^\bullet, \lhd^\bullet)$). Both trees are also called component trees.*

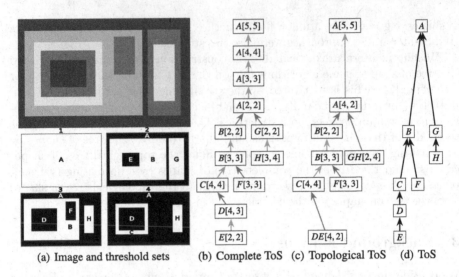

(a) Image and threshold sets   (b) Complete ToS   (c) Topological ToS   (d) ToS

Fig. 1. (a) An image $\mathcal{F}$ (top) and its upper (white) and lower (black) threshold sets (bottom). (b) The complete tree of shapes of $\mathcal{F}$ represents the inclusion (green arcs) or nesting (red arcs) of the connected components of the threshold sets of $\mathcal{F}$. (c) The topological tree of shapes of $\mathcal{F}$ "compresses" the nodes of the complete tree of shapes with regard to their topological relationship (strong deletability). (d) The tree of shapes of $\mathcal{F}$ represents the nesting of the outer boundaries of the threshold sets of $\mathcal{F}$. The topological tree of shapes and the tree of shapes are obtained by a decreasing homeomorphism from the complete tree of shapes. (Color figure online)

Let $X \subset \mathbb{U}$ be a (connected) set. We note $\tau(X) = X^\tau \supseteq X$ the set obtained by closing the holes of X. We note $\Theta^\tau = \{X^\tau \mid X \in \Theta\}$. We consider the partial order relation $\subseteq$ on Θ^τ and we note $\vartriangleleft^\tau$ the reflexive-transitive reduction of $\subseteq$ on Θ^τ.

Definition 2 (Tree of shapes [13]). *The tree of shapes of $\mathcal{F}$ is the tree $\mathfrak{T}_{\Theta^\tau} = (\Theta^\tau, \vartriangleleft^\tau)$. (See Fig. 1(d)).*

Let us suppose that $\mathbb{V} = \{\bot, \top\}$ with $\bot \neq \top$. Then $\mathcal{F} : \mathbb{U} \to \mathbb{V}$ is a binary image equivalent to the binary set $\Lambda_\top^\circ(\mathcal{F}) \subset \mathbb{U}$. Reversely, any (finite) set $X \subset \mathbb{U}$ is equivalent to a binary function $\mathbf{1}_X : \mathbb{U} \to \{\bot, \top\}$ defined such that $\mathbf{1}_X(\mathbf{x}) = \top \Leftrightarrow \mathbf{x} \in X$.

Definition 3 (Adjacency tree [20]). *The adjacency tree of a set $X \subset \mathbb{U}$ is the tree of shapes of the binary image $\mathbf{1}_X$.*

Remark 4. *Let $\mathcal{F} : \mathbb{U} \to \mathbb{V}$, and $v \in \mathbb{V}$. Each upper-threshold set $\Lambda_v^\circ(\mathcal{F})$ of $\mathcal{F}$ is composed by a set of connected components $\Theta_v^\circ = \Pi[\Lambda_v^\circ(\mathcal{F})]$. The adjacency tree of $\Lambda_v^\circ(\mathcal{F})$ is the tree $\mathfrak{T}_{\Theta_v} = (\Theta_v, \vartriangleleft^v)$ which is the tree of shapes of $\mathbf{1}_{\Lambda_v^\circ(\mathcal{F})}$. This allows us to link the component trees $\mathfrak{T}_{\Theta^\circ}$ and $\mathfrak{T}_{\Theta^\bullet}$, as $(\Theta_v, \vartriangleleft^v)$ models nested relations between elements of Θ_v° and $\Theta_v^\bullet$ for any $v \in \mathbb{V}$.*

Let $\sqsubseteq$ be the partial order relation defined on Θ by $X \sqsubseteq Y \Leftrightarrow (X \subseteq Y) \vee (X^\tau \subseteq Y^\tau)$. We note $\lhd$ the reflexive-transitive reduction of $\sqsubseteq$ on Θ.

Definition 5 (Complete tree of shapes [16]). *The complete tree of shapes of $\mathcal{F}$ is the tree $\mathfrak{T}_\Theta = (\Theta, \lhd)$. (See Fig. 1(b)).*

Let $D \subset \Lambda \subseteq \mathbb{U}$. Let $\iota : \Pi[\Lambda \setminus D] \to \Pi[\Lambda]$ and $\bar{\iota} : \Pi[\overline{\Lambda}] \to \Pi[\overline{\Lambda \setminus D}]$ be the two functions defined by $X \subseteq \iota(X)$ and $Y \subseteq \bar{\iota}(Y)$. We say that D is a strongly deletable set (of Λ) if ι and $\bar{\iota}$ are bijective [19]. Let $X, Y \in \Theta$ such that $X \lhd_\Theta Y$. If X is unique for this property and $Y \setminus X$ is a strongly deletable set of Y, then we note $Y \setminus\!\!\!\setminus X$. We note $\sim$ the equivalence relation on Θ derived from $\setminus\!\!\!\setminus$. We note $H = \Theta/\!\sim$. Let $\sqsubseteq_H$ be the order relation on H defined by $X \sqsubseteq_H Y \Leftrightarrow \bigwedge^{\sqsubseteq} X \sqsubseteq \bigwedge^{\sqsubseteq} Y$. We note $\lhd^H$ the reflexive-transitive reduction of $\sqsubseteq_H$ on H.

Definition 6 (Topological tree of shapes [16]). *The topological tree of shapes of $\mathcal{F}$ is the tree $\mathfrak{T}_H = (H, \lhd^H)$. (See Fig. 1(c)).*

Property 7 ([16]). *There exists a decreasing homeomorphism from the complete tree of shapes to the tree of shapes (resp. to the topological tree of shapes).*

3.2 Composition of Tree Nodes

The trees defined above allow to model images. This requires in particular to establish a correspondence between the nodes of these trees and their corresponding regions within the image. These regions are named proper parts.

Definition 8 (Proper part in the component tree(s)). *Let $X \in \Theta^\circ$ (the same definition holds for $\Theta^\bullet$). The proper part of X in the component tree $\mathfrak{T}_{\Theta^\circ} = (\Theta^\circ, \lhd^\circ)$ is defined by $\rho(X, \Theta^\circ) = X \setminus \bigcup_{Y \lhd^\circ X} Y$.*

Definition 9 (Proper part in the tree of shapes). *Let $X \in \Theta^\tau$. The proper part of X in the tree of shapes $\mathfrak{T}_{\Theta^\tau} = (\Theta^\tau, \lhd^\tau)$ is defined by $\rho(X, \Theta^\tau) = X \setminus \bigcup_{Y \lhd^\tau X} Y$.*

Definition 10 (Proper part in the complete tree of shapes). *Let $X \in \Theta$. The proper part of X in the complete tree of shapes $\mathfrak{T}_\Theta = (\Theta, \lhd)$ is defined by $\rho(X, \Theta) = X \setminus \bigcup_{Y \lhd X} Y^\tau$.*

Remark 11. *Each point $\mathbf{x} \in \mathbb{U}$ is contained in exactly one $\rho(X, \Theta^\circ)$ (resp. $\rho(X, \Theta^\bullet)$, $\rho(X, \Theta^\tau)$, $\rho(X, \Theta)$). More precisely, $\{\rho(X, \Theta^\circ) \mid X \in \Theta^\circ\}$, $\{\rho(X, \Theta^\bullet) \mid X \in \Theta^\bullet\}$ and $\{\rho(X, \Theta^\tau) \mid X \in \Theta^\tau\}$ are partitions of $\mathbb{U}$. The set $\{\rho(X, \Theta) \mid X \in \Theta\}$ may not be a partition since some proper parts may be empty.*

4   Building the Topological Tree of Shapes

We now describe an algorithmic scheme for building the topological tree of shapes of an image from a precomputed [9] tree of shapes. This scheme relies on four steps: enrichment of the tree of shapes (Sect. 4.1); construction of the graph of shapes from the enriched tree of shapes (Sect. 4.2); construction of the complete tree of shapes from the graph of shapes and the tree of shapes (Sect. 4.3); construction of the topological tree of shapes from the complete tree of shapes and the graph of shapes (Sect. 4.4).

4.1   Enriching the Tree of Shapes

The cornerstone of the construction of the topological tree of shapes is the construction of the complete tree of shapes. Indeed, the first is obtained from the second via a decreasing homeomorphism. The complete tree of shapes is made of the set $\Theta = \Theta^\circ \cup \Theta^\bullet$ of the nodes of the min- and max-trees, whereas the tree of shapes is made of the (smaller) set Θ^τ where each node is an equivalence class of nodes of Θ° or $\Theta^\bullet$.

Based on this fact, it is required to assign each node of Θ^τ to the class $\circ$ (resp. $\bullet$) if it is related to nodes of Θ° (resp. $\Theta^\bullet$). In the tree of shapes, it is usual to associate each node $Y \in \Theta^\tau$ to its "altitude" $Alt(Y) \in \mathbb{V}$, which is characterized by

$$Alt(Y) = \begin{cases} \bigvee\{v \in \mathbb{V} \mid X \in \Lambda_v^\circ(\mathcal{F}) \wedge X^\tau = Y\} & \text{if } Y \text{ is in the class } \circ \\ \bigwedge\{v \in \mathbb{V} \mid X \in \Lambda_v^\bullet(\mathcal{F}) \wedge X^\tau = Y\} & \text{if } Y \text{ is in the class } \bullet \end{cases} \quad (3)$$

where $\bigvee$ and $\bigwedge$ are the supremum and infimum respectively. We assume that the tree of shapes $\mathfrak{T}_{\Theta^\tau}$ is natively endowed with this function $Alt : \Theta^\tau \to \mathbb{V}$. Then, we can classify each node of Θ^τ into the class $\circ$ or $\bullet$ from Alt and the structural information of $\mathfrak{T}_{\Theta^\tau}$ by defining the function $Class : \Theta^\tau \to \{\circ, \bullet\}$ as

$$Class(Y) = \begin{cases} \circ & \text{if } Y = \mathbb{U} \\ \circ & \text{if } (Y \lhd^\tau X) \wedge (Class(X) = \circ) \wedge (Alt(Y) > Alt(X)) \\ \bullet & \text{if } (Y \lhd^\tau X) \wedge (Class(X) = \circ) \wedge (Alt(Y) \leqslant Alt(X)) \\ \bullet & \text{if } (Y \lhd^\tau X) \wedge (Class(X) = \bullet) \wedge (Alt(Y) < Alt(X)) \\ \circ & \text{if } (Y \lhd^\tau X) \wedge (Class(X) = \bullet) \wedge (Alt(Y) \geqslant Alt(X)) \end{cases} \quad (4)$$

From this classification, two useful information can be derived for each node $Y \in \Theta^\tau$: (1) the "origin" of each edge incident to Y, and (2) the interval $\mathbb{I}(Y)$ of values associated to Y. Regarding the origin of each edge $(Y \lhd^\tau X)$, if $Class(X) = Class(Y)$ (resp. $Class(X) \neq Class(Y)$) then this edge derives from the component trees (resp. the adjacency trees) and will be referred to the function φ (resp. ψ). The functions ψ and φ will be defined in the next section.

The interval $\mathbb{I}(Y) = [\![\alpha(Y), \omega(Y)]\!]$ of each node $Y \in \Theta^\tau$ can be defined by

$$\mathbb{I}(Y) = \begin{cases} [\![\bot, \bot]\!] & \text{if } Y = U \\ [\![Alt(X) + 1, Alt(Y)]\!] & \text{if } (Y \lhd^\tau X) \wedge (Class(Y) = Class(X) = \circ) \\ [\![Alt(X) - 1, Alt(Y)]\!] & \text{if } (Y \lhd^\tau X) \wedge (Class(Y) = Class(X) = \bullet) \\ [\![Alt(X), Alt(Y)]\!] & \text{if } (Y \lhd^\tau X) \wedge (Class(Y) \neq Class(X)) \end{cases} \tag{5}$$

Note that we may have $\alpha(Y) \leqslant \omega(Y)$ or $\alpha(Y) \geqslant \omega(Y)$, i.e. the intervals are "oriented".

4.2   Building the Graph of Shapes

Each $Y \in \Theta^\tau$ of the tree of shapes corresponds to an equivalence class $T(Y)$ of nodes of either Θ° or $\Theta^\bullet$. The nodes $X \in T(Y)$ are characterized by $\tau(X) = Y$. If we assume, without loss of correctness, that the nodes of the component trees are defined at each threshold set, $\mathbb{I}(Y)$ and $T(Y)$ are in bijection, since each node $X \in T(Y)$ is a connected component of the threshold set $\Lambda_v^\circ(\mathcal{F})$ (or $\Lambda_v^\bullet(\mathcal{F})$) for a specific value $v \in \mathbb{I}(Y)$. In other words, $\{Y\} \times \mathbb{I}(Y)$ models a subset of nodes of Θ° (or $\Theta^\bullet$). More generally, $\bigcup_{Y \in \Theta^\tau} \{Y\} \times \mathbb{I}(Y)$ models the set Θ. In particular, each node $X \in \Theta_v \subseteq \Theta$ is modeled by the couple (X^τ, v) $(v \in \mathbb{I}(X^\tau))$.

The set Θ^τ endowed with the set of intervals $\mathbb{I}_{\Theta^\tau} = \{\mathbb{I}(Y) \mid Y \in \Theta^\tau\}$ is a compact model of the nodes of the component trees and the adjacency trees (since $|\Theta^\tau| \leqslant |\Theta|$). Similarly, the edges of the tree of shapes, i.e. the elements of $\lhd_{\Theta^\tau}$, represent some edges of the component trees and the adjacency trees. However, this representation is partial. Our purpose is to build the graph of shapes, which enriches the tree of shapes with additional edges that will model all the edges of both the component trees and the adjacency trees. We first define the two functions φ and ψ that model these edges.

Definition 12. *The function $\varphi : \Theta \to \Theta$ associates each node of the component trees to its parent. It is defined, for any $X \in \Theta^\circ \setminus \{U\}$ (resp. $\Theta^\bullet \setminus \{U\}$) by $X \lhd^\circ \varphi(X)$ (resp. $X \lhd^\bullet \varphi(X)$). The function $\psi : \Theta \to \Theta$ associates each node of the adjacency trees to its parent. It is defined, for any $X \in \Theta_v \setminus \{\bigvee^\subseteq \Theta_v\}$ $(v \in \mathbb{V})$ by $X \lhd^v \psi(X)$.*

As Θ and $\bigcup_{Y \in \Theta^\tau} \{Y\} \times \mathbb{I}(Y)$ are in bijection, we can rewrite the functions φ and ψ as

$$\begin{vmatrix} \varphi : \Theta^\tau \times \mathbb{V} \to \Theta^\tau \\ (X^\tau, v) \mapsto Y^\tau \text{ s.t. } \begin{cases} X \lhd^\circ Y \in \Theta_{v-1}^\circ & \text{if } X \in \Theta_v^\circ \\ X \lhd^\bullet Y \in \Theta_{v+1}^\bullet & \text{if } X \in \Theta_v^\bullet \end{cases} \end{vmatrix} \tag{6}$$

$$\begin{vmatrix} \psi : \Theta^\tau \times \mathbb{V} \to \Theta^\tau \\ (X^\tau, v) \mapsto Y^\tau \text{ s.t. } X \lhd^v Y \in \Theta_v \end{vmatrix} \tag{7}$$

The graph of shapes is defined by $(\Theta^\tau, \blacktriangleleft)$, where the set of edges $\blacktriangleleft$ is partitioned into $\{\blacktriangleleft_\varphi, \blacktriangleleft_\psi\}$ such that $\blacktriangleleft_\varphi$ (resp. $\blacktriangleleft_\psi$) contains the edges related to φ (resp. ψ).

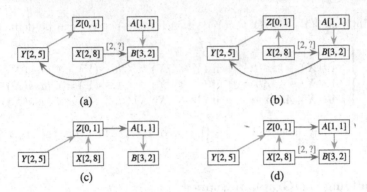

Fig. 2. Partial definition of the edges of $\blacktriangleleft_\varphi$ and $\blacktriangleleft_\psi$ in the graph of shapes from the information φ and ψ in the tree of shapes. The elements $A, \ldots, Z$ correspond to nodes of the tree of shapes (Θ^τ). The elements endowed with their interval ($A[1,1], \ldots, Z[0,1]$) correspond to nodes of the graph of shapes (Θ). The arrows in (a,c) correspond to edges of the tree of shapes; those in (b,d) to edges of the graph of shapes. The green (resp. red) edges relate to φ and $\blacktriangleleft_\varphi$ (resp. ψ and $\blacktriangleleft_\psi$). We focus on the node X. We recall that the intervals are oriented. As such, depending on the class of a node N, it is possible that $\alpha(N) \leqslant \omega(N)$ or $\alpha(N) \geqslant \omega(N)$. In the tree of shapes, X has exactly one outer arc $\mathcal{E}$ which is related to either ψ (a) or φ (c). In case (a), $\mathcal{E}$ is related to $\blacktriangleleft_\psi$ in the graph of shapes (b); the bound $\alpha(\mathcal{E})$ of $\mathbb{I}(\mathcal{E})$ is defined by the bound $\alpha(X)$ of $\mathbb{I}(X)$. In case (c), $\mathcal{E}$ is related to $\blacktriangleleft_\varphi$ in the graph of shapes (d). The transitive relations (Eqs. (8–9)) allow to determine in the graph of shapes the edge φ and $\blacktriangleleft_\varphi$ (b) and the edge ψ and $\blacktriangleleft_\psi$ (d). See Algorithm 1, lines 7–14. (Color figure online)

Let $X^\tau \in \Theta^\tau$. If $X^\tau \neq \mathbb{U}$ there exists exactly one edge (X^τ, Y^τ) in $\blacktriangleleft_\varphi$, i.e. exactly one $Y^\tau \in \Theta^\tau$ such that $X^\tau \blacktriangleleft_\varphi Y^\tau$. This edge (X^τ, Y^τ) defines the function φ for $(X^\tau, \alpha(X^\tau))$ by $\varphi(X^\tau, \alpha(X^\tau)) = Y^\tau$, while the function φ for the other (X^τ, v) with $v \in \mathbb{I}(X^\tau) \setminus \{\alpha(X^\tau)\}$ is defined by $\varphi(X^\tau, v) = X^\tau$ and then need not to be modelled by an edge of $\blacktriangleleft_\varphi$. Then, φ is fully defined by Θ^τ endowed with $\mathbb{I}_{\Theta^\tau}$ and by $\blacktriangleleft_\varphi$ with a space cost $\mathcal{O}(|\Theta^\tau|)$.

Let $X^\tau \in \Theta^\tau$ and $v \in \mathbb{I}(X^\tau)$. If $X^\tau \neq \mathbb{U}$ then there exist(s) $\sigma(X^\tau)$ edge(s) (X^τ, Y^τ) in $\blacktriangleleft_\psi$, i.e. there exist(s) $\sigma(X^\tau)$ node(s) $Y^\tau \in \Theta^\tau$ such that $X^\tau \blacktriangleleft_\psi Y^\tau$. Each such edge (X^τ, Y^τ) defines the function ψ for (X^τ, v) for a given $v \in \mathbb{I}(X^\tau)$ by $\psi(X^\tau, v) = Y^\tau$. The function ψ may be constant over successive values $v \in \mathbb{I}(X^\tau)$. In practice, the required number $\sigma(X^\tau)$ of edges of $\blacktriangleleft_\psi$ needed to model ψ for $\{X^\tau\} \times \mathbb{I}(X^\tau)$ satisfies $1 \leqslant \sigma(X^\tau) \leqslant |\alpha(X^\tau) - \omega(X^\tau)| + 1$. As a counterpart, each edge $\mathcal{E} = (X^\tau, Y^\tau)$ of $\blacktriangleleft_\psi$ has to be endowed with the interval $\mathbb{I}(\mathcal{E}) = [\![\alpha(\mathcal{E}), \omega(\mathcal{E})]\!]$ where it is defined. In other words, the function ψ is fully defined by $\blacktriangleleft_\psi$ endowed with the set of intervals $\mathbb{I}_{\blacktriangleleft_\psi} = \{\mathbb{I}(\mathcal{E}) \mid \mathcal{E} \in \blacktriangleleft_\psi\}$. Note that $\sum_{X^\tau \in \Theta^\tau} |\sigma(X^\tau)| = \mathcal{O}(|\Theta|)$. The space cost of $\blacktriangleleft_\psi$ is then $\mathcal{O}(|\Theta|)$.

Our purpose is to build the two sets of edges $\blacktriangleleft_\varphi$ and $\blacktriangleleft_\psi$, in order to obtain the graph of shapes. One important pre-processing step of the algorithm concerns the node $X^\tau = \mathbb{U}$. As initially computed by the tree of shapes, we have $Class(\mathbb{U}) = \circ$ and $\mathbb{I}(\mathbb{U}) = [\![\bot, \bot]\!]$. It is however important to note that X^τ

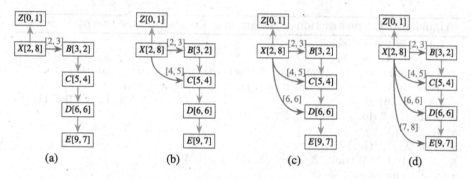

Fig. 3. Complete definition of the edges of ◀$_\psi$ in the graph of shapes. The process starts from a configuration illustrated in Fig. 2(b, d). The successive edges $\mathcal{E}$ of ◀$_\psi$ and their interval $\mathbb{I}(\mathcal{E})$ are obtained from the transitive property of Eq. (9)(a–d). See Algorithm 1, lines 18–25.

actually belongs to both the threshold set $\Lambda_\perp^\circ(\mathcal{F})$ and $\Lambda_\top^\bullet(\mathcal{F})$, and thus X^τ is part of both classes. Moreover, one node of the equivalence class of X^τ is present in each $\Lambda_v^\bullet(\mathcal{F})$, which is not explicitly encoded by the tree of shapes. Thus, we set $Class(X^\tau) = \bullet$ and $\mathbb{I}(X^\tau) = [\![\top, \perp + 1]\!]$. A part of ◀$_\varphi$ and ◀$_\psi$ is natively provided by the set of edges $\lhd^\tau$ of the tree of shapes. Let $X^\tau \in \Theta^\tau$. If $X^\tau \neq \mathbb{U}$ then there exists exactly one edge (X^τ, Y^τ) in $\lhd_{\Theta^\tau}$. If $Class(X^\tau) = Class(Y^\tau)$ then this edge defines φ for $(X^\tau, \alpha(X^\tau))$ by $\varphi(X^\tau, \alpha(X^\tau)) = Y^\tau$ and we have X^τ ◀$_\varphi$ Y^τ. If $Class(X^\tau) \neq Class(Y^\tau)$ then this edge defines ψ for $(X^\tau, \alpha(X^\tau))$ by $\psi(X^\tau, \alpha(X^\tau)) = Y^\tau$ and we have X^τ ◀$_\psi$ Y^τ. In other words, for each X^τ, either the (unique) edge of ◀$_\varphi$ is already defined while the $\sigma(X^\tau)$ edges of ◀$_\psi$ remain to be defined, or one edge of ◀$_\psi$ is already defined while the unique edge of ◀$_\varphi$ and the remaining $\sigma(X^\tau) - 1$ edges ◀$_\psi$ remain to be defined. Figure 2(a, c) illustrates how these properties allow to initialize the definition of ◀$_\varphi$ and ◀$_\psi$ in the graph of shapes from the information on φ and ψ carried by the tree of shapes.

These remaining edges to be defined can be computed by a transitive closure procedure that relies on tree properties that link φ and ψ [16]

$$\varphi(X) = [\varphi \circ \psi \circ \psi](X) \tag{8}$$

$$\psi(X) = [\varphi \circ \psi \circ \varphi](X) \tag{9}$$

$$\varphi(X) = [\varphi^{|\mathbb{V}|-2} \circ \psi](X) \tag{10}$$

This transitive closure, illustrated in Fig. 2(b, d) regarding its initialization and Fig. 3 for its iterative part, is performed on the tree of shapes, from its root to its leaves. Note that the third property effectively applies only for the nodes $X^\tau \in \Theta^\tau$ such that $X^\tau \lhd_{\Theta^\tau} \mathbb{U}$. This is summarized in Algorithm 1.

Algorithm 1: Construction of the graph of shapes from the tree of shapes.

Input: Θ^τ, $\lhd^\tau$, $Class$, $\mathbb{I}_{\Theta^\tau}$

Output: $\blacktriangleleft_\varphi$, $\blacktriangleleft_\psi$, $\mathbb{I}_{\blacktriangleleft_\psi}$

Notation: "$A := B$" means that A is set with B; "$A \leftarrow B$" means that B is added to A; "$A \rightarrow B$" means that B is set as an element removed from A

1   $\mathcal{L} := \{(\mathbb{U}, \emptyset)\}$ // $\mathcal{L}$ is a FIFO list

2   **while** $\mathcal{L} \neq \emptyset$ **do**

3      $\mathcal{L} \rightarrow (X, Y)$

4   **if** $X \neq \mathbb{U}$ **then**

5   **if** $Y \neq \mathbb{U}$ **then** $Z := \varphi(\psi(Y, \omega(Y)), \omega(Y))$

6   **else** $Z := Y$

7   **if** $Class(X) \neq Class(Y)$ **then**

8             $\blacktriangleleft_\varphi \leftarrow (X, Z)$

9             $\varphi(X, \alpha(X)) := Z$

10            $\psi(X, \alpha(X)) := Y$

11   **else**

12             $\blacktriangleleft_\varphi \leftarrow (X, Y)$

13             $\varphi(X, \alpha(X)) := Y$

14             $\psi(X, \alpha(X)) := Z$

15          $\widehat{Y} := \psi(X, \alpha(X))$

16          $\widehat{\alpha} := \alpha(X)$

17   **repeat**

18             $\mathcal{E} := (X, \widehat{Y})$

19             $\blacktriangleleft_\psi \leftarrow \mathcal{E}$

20             $\widehat{\omega} := \begin{cases} \bigwedge\{\omega(X), \alpha(\widehat{Y})\} & \text{if } Class(X) = \circ \\ \bigvee\{\omega(X), \alpha(\widehat{Y})\} & \text{if } Class(X) = \bullet \end{cases}$

21             $\mathbb{I}_{\blacktriangleleft_\psi}(\mathcal{E}) = [\![\alpha(\mathcal{E}), \omega(\mathcal{E})]\!] := [\![\widehat{\alpha}, \widehat{\omega}]\!]$

22             $\psi(X, \widehat{\omega}) := \widehat{Y}$

23             $\widehat{\alpha} := \begin{cases} \widehat{\omega} + 1 & \text{if } Class(X) = \circ \\ \widehat{\omega} - 1 & \text{if } Class(X) = \bullet \end{cases}$

24             $\widehat{Y} := \varphi(\widehat{Y}, \alpha(\widehat{Y}))$

25   **until** $\widehat{\omega} = \omega(X)$

26   **foreach** $Z \lhd_{\Theta^\tau} X$ **do** $\mathcal{L} \leftarrow (Z, X)$

4.3   Building the Complete Tree of Shapes

The tree of shapes can be obtained from the complete tree of shapes by a decreasing homeomorphism [16]. This homeomorphism consists of collapsing the branches of the complete tree of shapes with respect to the equivalence relation on Θ induced by the hole-closing τ (following the same way as for turning the nodes of the component trees into those of the tree of shapes). This homeomorphism is reversible. Each node of the tree of shapes can be duplicated to form a branch of nodes linked by edges of φ. The definition of these branches (by the nodes composing them) is given by the information carried by the relation $\blacktriangleleft_\psi$. This is exemplified by Fig. 4.

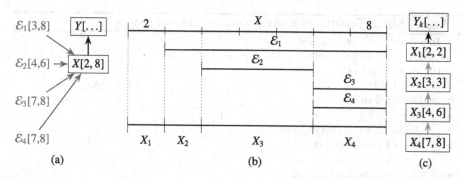

$\mathcal{E}_1[3,8]$ $Y[\ldots]$

$\mathcal{E}_2[4,6] \to X[2,8]$

$\mathcal{E}_3[7,8]$

$\mathcal{E}_4[7,8]$

(a)

2   X   8

$\mathcal{E}_1$

$\mathcal{E}_2$

$\mathcal{E}_3$

$\mathcal{E}_4$

X_1 X_2 X_3 X_4

(b)

$Y_k[\ldots]$

$X_1[2,2]$

$X_2[3,3]$

$X_3[4,6]$

$X_4[7,8]$

(c)

Fig. 4. Definition of the nodes of the complete tree of shapes (c) from the nodes of the graph of shapes (a) and the intervals of the edges of $\blacktriangleleft_\psi$ (b). (a) The node X in the graph of shapes is associated to 4 inner edges $\mathcal{E}_i$ of $\blacktriangleleft_\psi$ with intervals $\mathbb{I}(\mathcal{E}_i)$. (b) These intervals induce a partition of $\mathbb{I}(X)$. This partition allows to define the nodes X_j derived from X in the complete tree of shapes.

Let $X \in \Theta^\tau$, associated to the interval $\mathbb{I}(X) = [\![\alpha(X), \omega(X)]\!]$. Let us consider the k edges ($k \in \mathbb{N}$) $\mathcal{E}_i = (Y_i, X)$ ($1 \leqslant i \leqslant k$) of $\blacktriangleleft_\psi$ that link other nodes Y_i of Θ^τ to X. Each edge $\mathcal{E}_i$ is associated to an interval $\mathbb{I}(\mathcal{E}_i) = [\![\alpha(\mathcal{E}_i), \omega(\mathcal{E}_i)]\!]$ (Fig. 4(a)). These nodes Y_i define the holes of the nodes of the tree of shapes that derive from X over $\mathbb{I}(X)$. A node is in particular defined by a specific combination of holes Y_i. It follows that the different nodes of the complete tree of shapes derived from X are in bijection with the partition of intervals induced by the various $\mathbb{I}(\mathcal{E})$, that refines the interval $\mathbb{I}(X)$. We note $\Omega(X) = \{[\![\alpha(j), \omega(j)]\!]\}_{j=1}^{\ell}$ (with $1 \leqslant \ell \leqslant 2k$) this partition of intervals, with $\alpha(1) = \alpha(X)$, $\omega(\ell) = \omega(X)$, $\alpha(j) \leqslant \omega(j)$ for all $1 \leqslant j \leqslant \ell$ and $\omega(j) + 1 = \alpha(j+1)$ for all $1 \leqslant j \leqslant \ell - 1$ if $X \in \Theta^\circ$ (the same holds with $X \in \Theta^\bullet$ by substituting $\alpha(j) \geqslant \omega(j)$ to $\alpha(j) \leqslant \omega(j)$ and $\omega(j) - 1$ to $\omega(j) + 1$), see Fig. 4(b). The partition $\Omega(X)$ can be built in a time $\mathcal{O}(k \log k)$. Once $\Omega(X)$ is built for each node X, the structure of the complete tree of shapes can be obtained from the tree of shapes by substituting a branch composed by a number of nodes equal to the size of $\Omega(X)$ to the node X (Fig. 4(c)). This procedure is summarized in Algorithm 2.

The tree generated from Algorithm 2 is isomorphic to the complete tree of shapes. However, the nodes of this tree do not contain the information required to link that tree to the image. More precisely, their proper part (Definition 10) is not yet defined. It was proved in [16] that when splitting a node X of the tree of shapes into ℓ nodes X_i of the complete tree of shapes, the "first" node X_1, characterized by $\alpha(X_1) = \alpha(X)$ inherits the proper part of X, while the other $\ell - 1$ nodes X_i ($2 \leqslant i \leqslant \ell$) have an empty proper part. Algorithm 2 can be slightly modified (without extra time cost) to deal with this procedure by embedding the required proper part at the creation of each node (line 7).

4.4 Building the Topological Tree of Shapes

The construction of the topological tree of shapes from the complete tree of shapes is a procedure that consists of gathering the nodes of Θ into equivalence

---

Algorithm 2: Construction of the complete tree of shapes.

Input: Θ^τ, $\lhd^\tau$, $\blacktriangleleft_\psi$, $\mathbb{I}_{\blacktriangleleft_\psi}$, $\mathbb{I}_{\Theta^\tau}$
Output: Θ, $\lhd$, $\mathbb{I}_\Theta = \{\mathbb{I}(X) = [\![\alpha(X), \omega(X)]\!] \mid X \in \Theta\}$

1 $(\Theta, \lhd_\Theta) := (\emptyset, \emptyset)$
2 $\mathcal{L} := \{(\mathbb{U}, \emptyset)\}$ // $\mathcal{L}$ is a FIFO list
3 **while** $\mathcal{L} \neq \emptyset$ **do**
4     $\mathcal{L} \rightarrow (X, Y)$
5   Build $\Omega(X) = \{[\![\alpha(j), \omega(j)]\!]\}_{j=1}^\ell$
6   **for** j **from** 1 **to** ℓ **do**
7   Build X_j
8        $\mathbb{I}(X_j) := [\![\alpha(j), \omega(j)]\!]$
9        $\Theta \leftarrow X_j$
10   **if** $Y \neq \emptyset$ **then** $\lhd \leftarrow (X_j, Y)$
11       $Y := X_j$
12   **foreach** $Z \lhd^\tau X$ **do** $\mathcal{L} \leftarrow (Z, Y)$

---

classes with respect to the notion of strong deletability. In [16], it was observed that this equivalence between two nodes $X \blacktriangleleft_\varphi Y$ could be assessed by locally observing the structure of their children nodes with respect to $\blacktriangleleft_\psi$ in order to characterize a putative bijection between them. In [16], this checking was carried out in the so-called graph of valued shapes, a non-compact structure equivalent to the current graph of shapes. The characterization established in the graph of valued shapes can be rewritten in the graph of shapes.

Let $X, Y \in \Theta$ be such that $X \lhd Y$. If $Y = \psi(X)$ then X and Y cannot be equivalent. Let us assume that $Y = \varphi(X)$. Let X^τ and Y^τ be the nodes of Θ^τ associated to X and Y, respectively. We assume that X^τ is the only node such that $X^\tau \blacktriangleleft_\varphi Y^\tau$ (otherwise, X and Y are not equivalent). Let $\mathcal{E}_{X^\tau}$ be the set of edges of $\blacktriangleleft_\psi$ of the form (X_i, X^τ) such that $\alpha((X_i, X^\tau)) = \alpha(X^\tau)$. Let $\mathcal{E}_{Y^\tau}$ be the set of edges of $\blacktriangleleft_\psi$ of the form (Y_j, Y^τ) such that $\omega((Y_j, Y^\tau)) = \omega(Y^\tau)$. If these two sets of edges have distinct cardinals, X and Y are not equivalent. Otherwise, X and Y are equivalent iff there is a bijection between the elements X_i and the elements Y_i characterized by $X_i = Y_i$ or $Y_i \blacktriangleleft_\varphi X_i$.

Based on this characterization of the equivalence between the nodes of Θ, we can derive an algorithmic process that involves exactly twice (once for the value $\alpha(\mathcal{E})$ and once for the value $\omega(\mathcal{E})$) each edge of the set $\blacktriangleleft_\psi$ in the graph of shapes. This algorithmic process, which allows to gather the nodes of Θ into equivalence classes leading to the set of nodes H of the topological tree of shapes, then presents a time cost $\mathcal{O}(|\Theta|)$.

5   Complexity Analysis

The image $\mathcal{F}$ has a space cost n. The construction of the tree of shapes, which is the input of the proposed algorithm, has a time cost $\mathcal{O}(n \log n)$ [9]. The space cost of $\mathfrak{T}_{\Theta^\tau}$ is $\mathcal{O}(|\Theta^\tau|)$ with $|\Theta^\tau| \leqslant n$. The enrichment of the tree of shapes (Sect. 4.1),

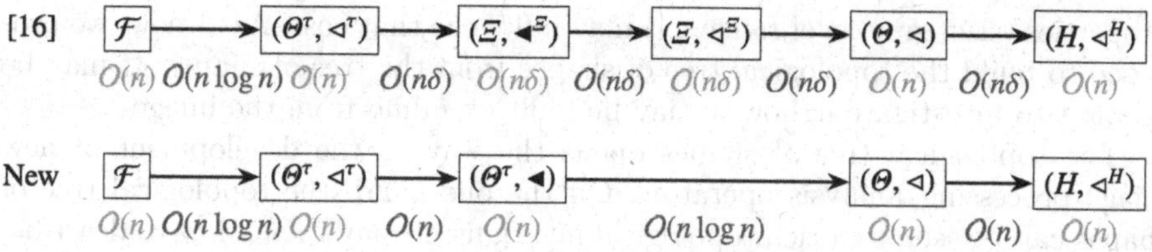

[16] $\mathcal{F} \longrightarrow (\Theta^\tau, \vartriangleleft^\tau) \longrightarrow (\Xi, \blacktriangleleft^\Xi) \longrightarrow (\Xi, \vartriangleleft^\Xi) \longrightarrow (\Theta, \vartriangleleft) \longrightarrow (H, \vartriangleleft^H)$

$O(n) \; O(n\log n) \; O(n) \quad O(n\delta) \quad O(n\delta) \; O(n\delta) \quad O(n\delta) \; O(n\delta) \quad O(n) \quad O(n\delta) \quad O(n)$

New $\mathcal{F} \longrightarrow (\Theta^\tau, \vartriangleleft^\tau) \longrightarrow (\Theta^\tau, \blacktriangleleft) \longrightarrow (\Theta, \vartriangleleft) \longrightarrow (H, \vartriangleleft^H)$

$O(n) \; O(n\log n) \; O(n) \quad O(n) \quad O(n) \qquad O(n\log n) \qquad O(n) \quad O(n) \quad O(n)$

Fig. 5. Synthetic comparison of the two algorithms for building the topological tree of shapes: the former algorithm [16] on the first line and the new algorithm on the second line. The intermediate structures are depicted by the boxes with their space cost in orange. The algorithmic steps for building each structure from the previous are depicted by arrows with their time cost in blue. Both space and time costs are expressed with respect to the size n of the image $\mathcal{F}$.

i.e. the construction of $Class(Y)$ and $\mathbb{I}(Y)$ for each $Y \in \Theta^\tau$, and the classification of each edge of $\vartriangleleft^\tau$ into either φ or ψ presents a time cost $\mathcal{O}(|\Theta^\tau|)$. The overall space cost of the enriched tree of shapes remains $\mathcal{O}(|\Theta^\tau|)$. The construction of the graph of shapes (Sect. 4.2) of size $\mathcal{O}(|\Theta|)$ from Algorithm 1 has a time cost $\mathcal{O}(|\Theta|)$. The construction of the complete tree of shapes (Sect. 4.3) of size $\mathcal{O}(|\Theta|)$ from Algorithm 2 has a time cost $\mathcal{O}(|\Theta|\log|\Theta^\tau|)$. The construction of the topological tree of shapes from the complete tree of shapes and the graph of shapes (Sect. 4.4) has a time cost $\mathcal{O}(|\Theta|)$. The overall process of building the topological tree of shapes from the tree of shapes then presents a space cost $\mathcal{O}(|\Theta|) = \mathcal{O}(n)$ related to the intermediate structures and a time cost $\mathcal{O}(|\Theta|\log|\Theta^\tau|) = \mathcal{O}(n\log n)$.

This algorithm presents a lower complexity, compared to the former approach in [16] (see Fig. 5). Indeed, we now use the graph of shapes, a compact structure where both redundant nodes and edges are modeled only once and endowed with an interval of definition, while in [16], they were modeled extensively in a structure called the graph of valued shapes $(\Xi, \vartriangleleft^\Xi)$ of size $\mathcal{O}(n\delta)$ where $\delta = \mathcal{O}(|\mathbb{V}|)$ is the average size of the interval of values where the nodes in the component trees of $\mathcal{F}$ exist. The optimization factor in terms of space (resp. time) cost between both algorithms is then δ (resp. $\frac{\delta}{\log n}$).

6 Conclusion

In this article, we continued our study of the topological tree of shapes initiated in [15,16]. Here, we focused on the design of an efficient algorithm for building the topological tree of shapes from the tree of shapes. We aim to release an implementation of this algorithm in libraries dedicated to mathematical morphology, such as Higra [17]. We presented a theoretical analysis of the cost of this algorithm, that focuses on asymptotic behaviour. Short term perspectives will be to refine this theoretical analysis, and to experimentally assess the cost of its implementation on various datasets of images, to understand how the image properties (size, grey-level range...) may influence that cost. As of now, we proposed to build the topological tree of shapes from the tree of shapes. It may be relevant to investigate if/how it may be built ex nihilo from the image.

The topological tree of shapes opens the way to the development of new image processing/analysis operators. On the one hand, the topological tree of shapes can be seen as a rich topological invariant; it may then allow to describe, analyze and/or compare images. On the other hand, as any other tree in the field of morphological hierarchies, it may be used to develop connected operators, e.g. for filtering, segmentation or simplification. Longer term perspectives will consist of investigating such applications.

References

1. Blin, N., Carlinet, E., Lemaitre, F., Lacassagne, L., Géraud, T.: Max-tree computation on GPUs. IEEE Trans. Parallel Distrib. Syst. **33**, 3520–3531 (2022)
2. Breen, E.J., Jones, R.: Attribute openings, thinnings, and granulometries. Comput. Vis. Image Underst. **64**(3), 377–389 (1996)
3. Carlinet, E., Crozet, S., Géraud, T.: The tree of shapes turned into a max-tree: a simple and efficient linear algorithm. In: ICIP, pp. 1488–1492 (2018)
4. Carlinet, E., Géraud, T.: A comparative review of component tree computation algorithms. IEEE Trans. Image Process. **23**, 3885–3895 (2014)
5. Carlinet, E., Géraud, T.: MToS: a tree of shapes for multivariate images. IEEE Trans. Image Process. **24**, 5330–5342 (2015)
6. Caselles, V., Meinhardt, E., Monasse, P.: Constructing the tree of shapes of an image by fusion of the trees of connected components of upper and lower level sets. Positivity **12**, 55–73 (2008)
7. Crozet, S., Géraud, T.: A first parallel algorithm to compute the morphological tree of shapes of nD images. In: ICIP, pp. 2933–2937 (2014)
8. Gazagnes, S., Wilkinson, M.H.F.: Distributed connected component filtering and analysis in 2D and 3D tera-scale data sets. IEEE Trans. Image Process. **30**, 3664–3675 (2021)
9. Géraud, T., Carlinet, E., Crozet, S., Najman, L.: A quasi-linear algorithm to compute the tree of shapes of nD images. In: Hendriks, C.L.L., Borgefors, G., Strand, R. (eds.) ISMM 2013. LNCS, vol. 7883, pp. 98–110. Springer, Berlin (2013). https://doi.org/10.1007/978-3-642-38294-9_9
10. Götz, M., Cavallaro, G., Géraud, T., Book, M., Riedel, M.: Parallel computation of component trees on distributed memory machines. IEEE Trans. Parallel Distrib. Syst. **29**, 2582–2598 (2018)
11. Kurtz, C., Naegel, B., Passat, N.: Connected filtering based on multivalued component trees. IEEE Trans. Image Process. **23**, 5152–5164 (2014)
12. Monasse, P.: A root-to-leaf algorithm computing the tree of shapes of an image. In: Kerautret, B., Colom, M., Lopresti, D., Monasse, P., Talbot, H. (eds.) RRPR 2018. LNCS, vol. 11455, pp. 43–54. Springer, Cham (2019). https://doi.org/10.1007/978-3-030-23987-9_3
13. Monasse, P., Guichard, F.: Scale-space from a level lines tree. J. Vis. Commun. Image Represent. **11**(2), 224–236 (2000)
14. Ngoc, M.O.V., Boutry, N., Fabrizio, J., Géraud, T.: A minimum barrier distance for multivariate images with applications. Comput. Vis. Image Underst. **197–198**, 102993 (2020)

15. Passat, N., Kenmochi, Y.: A Topological Tree of Shapes. In: Baudrier, É., Naegel, B., Krähenbühl, A., Tajine, M. (eds.) DGMM 2022. LNCS, vol. 13493, pp. 221–235. Springer, Cham (2022). https://doi.org/10.1007/978-3-031-19897-7_18
16. Passat, N., Mendes Forte, J., Kenmochi, Y.: Morphological hierarchies: a unifying framework with new trees. J. Math. Imaging Vis. 65(5), 718–753 (2023)
17. Perret, B., Chierchia, G., Cousty, J., Ferzoli Guimarães, S.J., Kenmochi, Y., Najman, L.: Higra: hierarchical graph analysis. SoftwareX 10, 100335 (2019)
18. Perret, B., Cousty, J.: Component tree loss function: definition and optimization. In: Baudrier, É., Naegel, B., Krähenbühl, A., Tajine, M. (eds.) DGMM 2022. LNCS, vol. 13493, pp. 248–260. Springer, Cham (2022). https://doi.org/10.1007/978-3-031-19897-7_20
19. Ronse, C.: A topological characterization of thinning. Theor. Comput. Sci. 43, 31–41 (1986)
20. Rosenfeld, A.: Adjacency in digital pictures. Inf. Control 26, 24–33 (1974)
21. Salembier, P., Oliveras, A., Garrido, L.: Anti-extensive connected operators for image and sequence processing. IEEE Trans. Image Process. 7, 555–570 (1998)
22. Salembier, P., Serra, J.: Flat zones filtering, connected operators, and filters by reconstruction. IEEE Trans. Image Process. 4, 1153–1160 (1995)
23. Song, Y., Zhang, A.: Monotonic tree. In: Braquelaire, A., Lachaud, J.-O., Vialard, A. (eds.) DGCI 2002. LNCS, vol. 2301, pp. 114–123. Springer, Heidelberg (2002). https://doi.org/10.1007/3-540-45986-3_10
24. Tao, R., Qiao, J.: Fast component tree computation for images of limited levels. IEEE Trans. Pattern Anal. Mach. Intell. 45(3), 3059–3071 (2023)

End-to-End Ultrametric Learning
for Hierarchical Segmentation

Raphael Lapertot[✉] , Giovanni Chierchia , and Benjamin Perret

LIGM, Univ Gustave Eiffel, CNRS, ESIEE Paris, 77454 Marne-la-Vallée, France
raphael.lapertot@esiee.fr

Abstract. Hierarchical image segmentation aims to capture the structure of objects of different sizes at different scales and helps to understand the scene. With the success of neural networks for image segmentation and the recent emergence of object and part segmentation datasets, the task of supervised learning of segmentation hierarchies naturally arises. In a previous work, we proposed a differentiable ultrametric layer that transforms any dissimilarity measure into an ultrametric distance equivalent to a hierarchical segmentation. In this paper, we study several loss functions for end-to-end learning of a neural network model predicting hierarchical segmentations. In particular, we propose a generalization of the Rand index for hierarchical segmentation and propose exact and approximate algorithms to compute it. We introduce new metrics to compare hierarchical segmentations, and we demonstrate the suitability of the proposed pipeline with several possible loss function combinations on a simulated hierarchical dataset.

Keywords: Image segmentation · Hierarchy · Ultrametric

1 Introduction

Image segmentation is the process of dividing an image into distinct regions that highlight relevant structures. One critical factor in this process is the selection of an appropriate scale, since it affects the level of detail visible in an image. Such difficulty can be avoided by using a hierarchical framework that generates consistent segmentations across multiple scales. This approach allows one to defer the decision regarding the scale until after the segmentation is complete.

With the success of neural networks in flat image segmentation [7,21,23,25] and the recent emergence of hierarchical segmentation datasets [4,8,9,14,19], the task of end-to-end supervised learning of hierarchical segmentation naturally arises. Notable efforts have been made to produce high quality hierarchical image segmentations, however, to the best of our knowledge, none of them

This work is supported by the French ANR grant ANR-20-CE23-0019, and was granted access to the HPC resources of IDRIS under the allocation 2023-AD011013101R1 made by GENCI.

propose loss functions for end-to-end supervised learning of hierarchical image segmentation. Some prior research has performed end-to-end supervised learning of flat image segmentations [1,16,24], while other studies have produced hierarchical segmentations, but not in an end-to-end fashion, using flat segmentation ground-truths [13,18,20]. HieraSeg [12] propose a pixel approach for hierarchical semantic segmentation, for which the hierarchy is on the semantic labels. Loss functions play a major role in the process as they guide the model towards meaningful segmentation hierarchies through the optimization process. Using an end-to-end pipeline has also shown to be advantageous.

This paper extends our previous works which proposed a differentiable ultrametric layer [5], and a pipeline for the supervised learning of ultrametrics [11]. We explore various loss functions including an adaptation of the Rand Index, a well-established metric for evaluating segmentation quality, to the hierarchical context for end-to-end optimization. We also introduce quantitative metrics to assess well-ordered hierarchical segmentations. Our main contributions are new loss functions for the supervised learning of hierarchical segmentation, corresponding algorithms for approximate or exact computation, and quantitative metrics for hierarchical segmentation assessments.

2   Model

Our primary objective is to establish an end-to-end supervised learning framework for hierarchical segmentation. Given the success of deep learning methodologies in the domain of image segmentation, it is natural to explore their applicability to this particular context. We frame the problem as a regression task operating on the edges of the 4-adjacency graph (or grid) $\mathcal{G}$ of the image, with pixels as vertices and edges linking each neighbor pixels.

2.1   Hierarchical Segmentation Using Graphs

Flat segmentation is often approached from a region-based perspective. From this point of view, a segmentation S associates each pixel v of the image with an associated region label s_v. It can also be seen from a boundary-based perspective. Let us consider the 4-adjacency graph of an image. Considering a segmentation S, the *cut* ϕ_S indicates whether two pixels belong to the same region in S or not: $\phi_S(v, v') = 1$ if $s_v \neq s_{v'}$ else 0 with v and v' two pixels.

Hierarchical segmentation can also be seen from a region point of view and a boundary point of view. In the first case, it is represented as a sequence of ordered segmentations, the next one being a refinement of the previous one. From the boundary perspective, a hierarchical image segmentation is represented as an ultrametric (dissimilarity) grid $(\mathcal{G}, w)$ (also called saliency map or ultrametric contour map), that is, a 4-adjacency graph $\mathcal{G}$ whose edges $e \in E$ are weighted by a dissimilarity value $w(e)$ that satisfies the ultrametric constraint ($\forall C \in \mathcal{C}, \forall e \in C, w(e) \leq \max_{e' \in C \setminus \{e\}} w(e')$, with $\mathcal{C}$ the set of cycles of $\mathcal{G}$).

In general, a dissimilarity grid $(\mathcal{G}, w)$ is not an ultrametric grid, but a simple way to transform it into an ultrametric grid is to compute its *subdominant* *ultrametric* $(\mathcal{G}, u_w)$, *i.e.*, the largest ultrametric grid which is smaller or equal than $(\mathcal{G}, w)$. In image analysis, computing the subdominant ultrametric of a dissimilarity grid can be seen as *removing borders with holes*. The subdominant ultrametric of a dissimilarity grid can also be defined thanks to the notion of *min-max paths*, *i.e.* paths that minimizes the maximal value along the path. More formally, we denote $\mathcal{P}_{v,v'}$ the set of paths from v to v' in a dissimilarity grid $(\mathcal{G}, w)$. The min-max path $P^*_{v,v'}$ from the pixel v to the pixel v' is

$$P^*_{v,v'} = \underset{P \in \mathcal{P}_{v,v'}}{\arg\min} \ \underset{e \in P}{\max} \ w(e). \tag{1}$$

The maximal edge of the min-max path is called the *min-max edge*, and will be denoted $mm_w(v, v')$. Finally, the subdominant ultrametric $d_w(v, v')$ of $(\mathcal{G}, w)$ on the edge $\{v, v'\}$ is equal to the weight of the min-max edge between v and v':

$$d_w(v, v') = w(mm_w(v, v')). \tag{2}$$

In this paper, we will consider hierarchies represented as ultrametric grid.

2.2 Ultrametric Network

We propose the design of a neural network that predicts an ultrametric grid from an input image. The pipeline is shown in Fig. 1 and unfolds as follows.

- The input image is given to a U-Net neural network. The network outputs a transformed image of identical dimensions, preserving spatial relationships.
- The output is transformed into edge-weights that correspond to the 4-adjacency graph of the original image.
- A sigmoid activation function binds these edge-weights within the range of 0 (no contour) to 1 (strong contour).
- An ultrametric layer converts the graph into an ultrametric graph, while preserving the differentiability of the computation [5].

In particular, the U-Net architecture shows versatility by accommodating different convolutional or transformer-based neural networks, ensuring adaptability to upcoming innovations. Our approach converts the U-Net's output in the pixel domain to edge-weights by taking the average weight of neighboring pixels.

3   Loss Functions

In this section, we introduce differentiable loss functions for comparing two ultrametric grids, one of which is predicted by the proposed ultrametric network. We give algorithms to compute them both approximately and exactly. These loss functions will be used for training the proposed ultrametric network.

In the following, we assume that any ground-truth ultrametric grid w is composed of a few different levels $\Lambda(w)$. For example, a ground-truth in an *objects-and-parts* dataset typically has three levels: the level 0 denotes the absence of frontier, the level 1 delineates objects, and the level 2 defines the parts of objects.

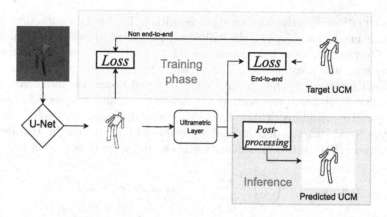

Fig. 1. Our proposed pipeline with optimization either end-to-end or not. In this figure, the U-Net outputs edge-weights (seen as a dissimilarity measure) of the 4-adjacency graph built on the input image. The loss function is computed either before or after (or both) the ultrametric layer that transforms the dissimilarity grid into an ultrametric one, which is equivalent to a hierarchical segmentation.

3.1 Quadratic Error

A simple approach is to consider the L2 distance between the ground-truth ultrametric grid (w) and the predicted edge-weights ($\hat{w}$), while mitigating the imbalance between the number of edges in the different levels of the ground-truth. This can be formulated as

$$L2(\hat{w}, w) = \frac{1}{|E|} \sum_{\lambda \in \Lambda(w)} \beta_\lambda \sum_{e \in E_\lambda} (\hat{w}(e) - w(e))^2, \tag{3}$$

where E denotes the set of edges within the 4-adjacency graph corresponding to the input image, E_λ represents the set of edges with a ground-truth value of λ, and $\beta_\lambda = 1 - \frac{|E_\lambda|}{|E|}$ serves as a class-balancing weight. The latter term is essential as there are typically a lot more non-edges than edges; when this term is not included, the non-edges are naturally privileged, and actual edges are harder to learn and detect. While this approach seems natural, it is sensitive to small geometric transformations such as translations, which is undesirable since the quality of a segmentation should not decrease significantly with small spatial variations. This loss can be applied after the ultrametric layer, offering an end-to-end learning pipeline. It can also be applied before the ultrametric layer, which can serve as a baseline even if the training is no longer end-to-end.

3.2 Hierarchical Rand Index

From a region-based perspective, the Rand Index for two segmentations S and $\hat{S}$ can be defined as follows

$$1 - RI(\hat{S}, S) = \binom{N}{2}^{-1} \sum_{v < v'} |\delta(s_v, s_{v'}) - \delta(\hat{s}_v, \hat{s}_{v'})|. \tag{4}$$

In this equation, N is the number of pixels, with $\binom{N}{2}^{-1}$ used for normalization, v and v' are pixels, and $\delta(x, y)$ if the indicator function which is 0 if $x = y$ and 1 of $x \neq y$. Overall, the Rand index measures the proportion of pixel pairs for which the two segmentations disagree. This measure increases when regions are split or merged incorrectly, while being robust to small geometric transformations. From a boundary-based perspective, the Rand Index can be expressed using the cuts $\phi_{\hat{S}}$ and ϕ_S similarly. This is from this boundary-based perspective that we relax the Rand Index to hierarchical segmentation. We define the Hierarchical Rand Index (HRI) of two dissimilarity grids w and $\hat{w}$ on $\mathcal{G}$ as

$$HRI(\hat{w}, w) = \binom{N}{2}^{-1} \sum_{v < v'} \beta_{d_w(v, v')} \left(d_w(v, v') - d_{\hat{w}}(v, v') \right)^2. \tag{5}$$

It computes the mean ultrametric distance error for every pixel pair of the graph. This measure keeps the same benefits as the Rand index, that is, it focuses on the incorrect merging and splitting of hierarchical regions. We use the same class-balancing weights β_λ as described previously.

3.3   Naive Algorithm for HRI

Let us focus on the computation of the HRI. Firstly, in an optimization framework, the loss function needs to be differentiable on the edges of the graph. To calculate the ultrametric distance between pairs of pixels, we exploit an interesting property of minimal spanning trees (or *MST* for short): each path in an MST of a graph is a min-max path. A binary partition tree (*BPT*) associated with the MST can be built during the Kruskal algorithm [15] which enables to efficiently find min-max edge between any pixel pair by querying their lowest common ancestor in this BPT, which can be done efficiently with a linear time preprocessing of the tree [3]. A naive algorithm for computing the HRI is detailed in Algorithm 1. This naive algorithm has a quadratic runtime w.r.t. the number of vertices as it browses each pixel pair.

A first approach to alleviate this problem is to approximate the HRI by considering a subset of M pixel pairs. This reduces the time complexity from $\mathcal{O}(N^2)$ to $\mathcal{O}(M + N \log N)$ but we only get an approximation of the correct result. In the context of a stochastic gradient descent algorithm, such kind of approximation in the loss function might be acceptable.

3.4   Optimized Algorithm for HRI

In this section, we propose a more efficient algorithm to compute the HRI. This algorithm leverages on the properties of min-max paths and the BPT. The idea is as follows: when computing the MST of a graph and the corresponding BPT using Kruskal's algorithm, when an edge is added to the BPT, it merges two subtrees. The amount of pixels in a tree will be called its *area*. The number of pixel pairs for which the added edge is the min-max edge can be deduced by

---

Algorithm 1. Naive hierarchical Rand index

1: **procedure** NAIVE-HRI($\mathcal{G}, \hat{w}, w$)   ▷ $\mathcal{O}(N^2)$
2:     Compute $BPT(\mathcal{G}, \hat{w})$ ▷ $\mathcal{O}(N \log N)$
3:     Compute $BPT(\mathcal{G}, w)$ ▷ $\mathcal{O}(N \log N)$
4:     Compute the class-balancing weight: $\forall \lambda \in \Lambda(w), \beta_\lambda = 1 - \frac{|E_\lambda|}{|E|}$ ▷ $\mathcal{O}(N)$
5:     **for** $(v, v') \in V^2$ with $v < v'$ **do**   ▷ $\mathcal{O}(N^2)$
6:         Find the predicted min-max edge $mm_{\hat{w}}(v, v')$ using $BPT(\mathcal{G}, \hat{w})$ ▷ $\mathcal{O}(1)$
7:         Get its dissimilarity value: $d_{\hat{w}}(v, v') \leftarrow \hat{w}(mm_{\hat{w}}(v, v'))$ ▷ $\mathcal{O}(1)$
8:         Find the ground-truth min-max edge $mm_w(v, v')$ using $BPT(\mathcal{G}, w)$ ▷ $\mathcal{O}(1)$
9:         Get its dissimilarity value: $d_w(v, v') \leftarrow w(mm_w(v, v'))$ ▷ $\mathcal{O}(1)$
10:        Compute error: $HRI(\hat{w}, w)_{v,v'} = \beta_{d_w(v,v')}(d_w(v, v') - d_{\hat{w}}(v, v'))^2$ ▷ $\mathcal{O}(1)$
11:    **end for**
12:    Return the mean HRI error
13: **end procedure**

---

looking at the areas of the trees it merges. Formally, for each edge $e \in MST(\mathcal{G}, \hat{w})$ and each ground-truth level $\lambda \in \Lambda(w)$, we define the area $A(e, \lambda)$ at e for the level λ as the number of pixel pairs whose min-max edge in $(\mathcal{G}, \hat{w})$ is e and whose ultrametric distance in the ground truth is λ:

$$A(e, \lambda) = \left| \{(v, v') \in V^2 \mid d_w(v, v') = \lambda, e = mm_{\hat{w}}(v, v')\} \right|. \tag{6}$$

The HRI can then be rewritten as follows

$$HRI(\hat{w}, w) = \binom{N}{2}^{-1} \sum_{\lambda \in \Lambda(w)} \sum_{e \in MST(\mathcal{G}, \hat{w})} A(e, \lambda)(\lambda - \hat{w}(e))^2. \tag{7}$$

In this new formulation which is equivalent to Eq. 5, we have replaced the sum over all pixel pairs by a double sum over the ground truth levels, which is small, and the MST edges, which is in the order of N the number of pixels.

Then, let's see how to calculate $A(e, \lambda)$. Assume that the finest ground-truth partition if composed of k regions labeled $\{1, ..., k\}$. We define the ultrametric distance between any two of these regions as

$$U(i, j) = d_w(v, v'). \tag{8}$$

where v (resp v') is a pixel of the region i (resp. j); note that the choice of v and v' does not matter as they are all at the same ultrametric distance. Let e be an edge of $MST(\mathcal{G}, \hat{w})$. This edge corresponds to a node n in the associated BPT. The number of pixels of label i contained in the subtree rooted at the node n is denoted $R(n, i)$. Let c_1^e and c_2^e be respectively the left and right children of the node of the BPT associated to the edge e. Note that, as e merges its left and right subtrees in the BPT, e is the min-max edge between any pixel in the subtrees rooted in c_1^e and in c_2^e. We can observe that, for any MST edge e and for any two different regions i and j, there are $R(c_1^e, i)R(c_2^e, j) + R(c_1^e, j)R(c_2^e, i)$ pairs of pixels from the regions i and j in the ground-truth between the subtrees rooted in c_1^e and c_2^e. Similarly, there are $R(c_1^e, i)R(c_2^e, i)$ pairs of pixels of the same

---

Algorithm 2. Optimized hierarchical Rand index

1: **procedure** OPTIMIZED-HRI$(\mathcal{G}, \hat{w}, w)$ ▷ $\mathcal{O}(N(\log N + k^2 + |\Lambda(w)|))$
2:      Compute $BPT'(\mathcal{G}, \hat{w})$ ▷ $\mathcal{O}(N \log N)$
3:      Compute $BPT'(\mathcal{G}, w)$ ▷ $\mathcal{O}(N \log N)$
4:      Compute the ultrametric distance of regions of the finest partition: U ▷ $\mathcal{O}(k^2)$
5:      Compute areas of regions per edge of $MST(\mathcal{G}, \hat{w})$: R   ▷ $\mathcal{O}(Nk)$
6:      Compute areas of regions pair per edge of $MST(\mathcal{G}, \hat{w})$: B   ▷ $\mathcal{O}(Nk^2)$
7:      Compute areas per ultrametric value and edge of $MST(\mathcal{G}, \hat{w})$: A ▷ $\mathcal{O}(N|\Lambda(w)|)$
8:      Compute $HRI = \sum_{\lambda \in \Lambda(w)} \sum_{e \in MST(\mathcal{G}, \hat{w})} A(e, \lambda)(\lambda - \hat{w}(e))^2$ ▷ $\mathcal{O}(N|\Lambda(w)|)$
9:      Return HRI error   ▷ $\mathcal{O}(1)$
10: **end procedure**

---

region i in the ground-truth between the subtrees rooted in c_1^e and c_2^e. Thus, the number of pixels pairs of region i and j, whose min-max edge is e, reads

$$B(e, i, j) = \begin{cases} R(c_1^e, i)R(c_2^e, j) + R(c_1^e, j)R(c_2^e, i) & \text{if } i \neq j \\ R(c_1^e, i)R(c_2^e, i) & \text{if } i = j. \end{cases} \tag{9}$$

Then, the area $A(e, \lambda)$ indicating for each edge e of the MST and for each ultrametric value $\lambda \in \Lambda(w)$, how many pixel pairs of ultrametric distance λ in the ground-truth have the min-max edge e in $\hat{w}$, can be rewritten as:

$$A(e, \lambda) = \sum_{i=1}^{k} \sum_{j=i}^{k} \left(1 - \delta(\lambda, U(i, j))\right) B(e, i, j). \tag{10}$$

In practice, R is a $(N-1) \times k$ matrix and B is a $(N-1) \times k^2$ matrix: they can be seen as attributes mapping a k (resp. k^2) vector to any internal node of the BPT. Both can be computed by browsing the BPT from the leaves to the root.

The complete HRI can be computed by Algorithm 2 with a runtime complexity of $\mathcal{O}(N(\log N + k^2 + |\Lambda(w)|))$. The fewer regions in the ground-truth, the faster the algorithm. Both Algorithms 1 and 2 yield the same result.

4   Experiments

We now evaluate the proposed loss functions. Specifically, we analyze the effectiveness of these loss functions for training a neural network in the context of hierarchical image segmentation.

4.1   Dataset

We use a custom-made *Humanoid* Dataset, which is a toy dataset consisting of a virtual humanoid moving in front of 6 cameras with different angles for 120 frames, resulting in 720 images of size 128×128. The ground-truths consists of Ultrametric Contour Maps (UCM) [2,6,11], where interpixels are added to the original image to represent the edges, resulting in an image twice as large as the input image. These UCMs have object-edges drawing the external contours, and part-edges drawing the contours of the parts (fore-arms, face, ...), with respectively value 1 and 0.5. When there are no edges, the value is 0. This dataset was generated using Blender. In our experiments, 4 cameras are used for training, 1 camera is used for validation, and the remaining camera is used for testing. Three samples of the Humanoid Dataset are displayed in Fig. 2.

4.2   Neural Network and Hyper-parameters

We use a ResNet18 pretrained on ImageNet for the backbone of our U-Net. We first freeze the encoder to only train the decoder for 300 epochs with a One Cycle learning rate scheduler [22] with a maximum learning rate of 0.001. We use an Adam optimizer with a weight decay of 0.01. Once this pre-training is done, we unfreeze the encoder and train it for an additional 100 epochs with a One Cycle learning rate scheduler with a maximum learning rate of 0.0002. We use a batch-size of 64 and simple data augmentations such as horizontal flips and slight random brightness, saturation and contrast modifications.

In terms of computation time, training the neural network as described above using four NVIDIA Tesla V100 GPUs, the partial algorithm took around 47 min to compute with one pixel pair, approximately 51 min with 16384 pixel pairs, whereas the optimized algorithm required approximately 1 h and 16 min. For reference, the control group with the L2 loss prior to the ultrametric layer took around 32 min to train. Therefore, in the experiments, we used 16384 random pixel pairs when computing the partial algorithm.

The network predictions are post-processed with an area filter removing 1-pixel regions [17]. The ultrametric grids are then converted to Ultrametric Contour Maps (UCM) [2,6,11], where interpixels are added to the original image to represent the edges, yielding an image twice as large as the input.

4.3   Metrics

For assessing hierarchical segmentation, we build on our previous work [11] using the Level Recovery Fraction (LRF) and the False Discovery Fraction (FDF). Let U_{pred} be a predicted UCM and U_{tar} be the corresponding ground-truth UCM. The threshold of U_{pred} at levels t is denoted by U_{pred_t}.

For any ground-truth level λ and any threshold level t, $LRF(\lambda, t)$ denotes the fraction of ground-edges of level λ which are matched to an edge in U_{pred_t}:

$$LRF(\lambda, t) = \frac{\left|\{e \mid \text{match}_{U_{pred_t}}(e) \text{ and } U_{tar}(e) = \lambda\}\right|}{\left|\{e \mid U_{tar}(e) = \lambda\}\right|}, \tag{11}$$

where $\text{match}_{U_{pred_t}}(e)$ is true if the edge e is matched with a corresponding edge in U_{pred_t} using a bipartite graph matching method.

For the False Discovery Fraction, it is however not immediate to associate a false positive edge in the prediction to a specific level of the ground-truth. For

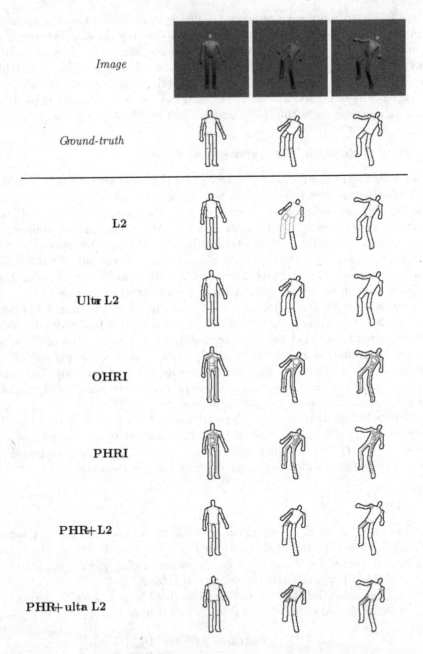

Fig. 2. Predictions of the neural networks on the Humanoid Dataset trained with different loss functions. When training with L2 alone, some contours are opened, resulting in edges that are less salient than expected after post-processing. When training with HRI alone, the object-edges are well detected and closed, but the part-edges are not well located. The other configurations produce closed edges that are well located.

any threshold level t, $FDF(t)$ is then defined as the fraction of prediction edges at the threshold t that cannot be matched with any edge of the ground-truth:

$$FDF(t) = \frac{|\{e \mid !\text{match}_{U_{pred_t}}(e) \text{ and } U_{pred_t}(e) = 1\}|}{|\{e \mid U_{pred_t}(e) = 1\}|}. \tag{12}$$

In a good prediction, the object boundaries will be recovered at a high threshold (1 in our setting), and the part boundaries will be recovered at a medium threshold (0.5 in our setting), resulting in two step-functions. The FDF, on the other hand, would be 0 for every threshold.

To summarize LRF curves, we propose to calculate the area enclosed between the ideal step curve and the observed curve within the LRF at each level of the hierarchy (LRF_{obj} for object boundaries, and LRF_{part} for part boundaries). Additionally, we compute the mean value of the FDF among the thresholds, giving $mFDF$. Smaller values for these metrics indicate improved performance. All in all, LRF_{obj} and LRF_{part} measure if detected edges are in the right order in the hierarchy, and $mFDF$ gives insights about false detections.

4.4   Results

We address the following problematic: how effective the loss functions are for training a neural network for hierarchical image segmentation? To this end, we train neural networks with various combinations of the proposed loss functions. We automatically combine multiple losses using learnable parameters [10,26]. In the following, $L2$ stands for the quadratic error loss applied before the ultrametric layer (not end-to-end) and serves as a control group, $UL2$ is the quadratic error loss applied after the ultrametric layer (end-to-end), and $OHRI$ and $PHRI$ are the hierarchical rand index computed with respectively the optimized algorithm and the partial algorithm (both end-to-end).

The metrics on the test set are shown in Fig. 3, and the predictions of those networks are displayed in Fig. 2. The Level Recovery Fraction (objects) plot shows the limits of the control group when computing the L2 prior on the ultrametric layer; the blue line representing object boundaries recovery shows that object borders are recovered at a medium threshold instead of a high threshold, resulting in a high LRF_{obj} error. This effect, visible in Fig. 2, is mostly due to open borders that are removed when computing the subdominant ultrametric. The aforementioned effect does not happen with the other loss functions.

Part boundaries are recovered at a medium threshold for most loss functions, except when training with HRI alone (green and red line). For the latter, part boundaries are recovered more linearly at a mid-low threshold (instead of the ideal step curve at a mid threshold), and wrongly recovered even at a high threshold, resulting in a high LRF_{part} error. The control group also recovers part boundaries at a lower threshold than other methods.

Finally, the False Discovery Fraction plot shows that training with HRI alone results in a lot of wrong boundaries at a medium to low threshold, mostly because it does not learn where to place the part borders as visible in Fig. 2.

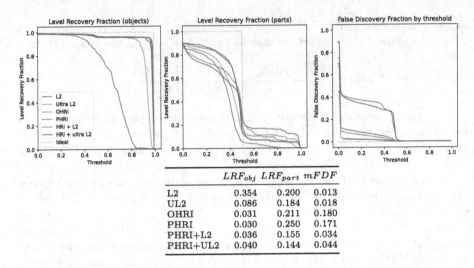

	LRF_{obj}	LRF_{part}	$mFDF$
L2	0.354	0.200	0.013
UL2	0.086	0.184	0.018
OHRI	0.031	0.211	0.180
PHRI	0.030	0.250	0.171
PHRI+L2	0.036	0.155	0.034
PHRI+UL2	0.040	0.144	0.044

Fig. 3. Experiment results. The errors are the area between the ideal case (dotted line) and the actual line for the three metrics. Lower errors means better results.

5   Conclusion

In this paper, we presented new loss functions and developed algorithms for both approximate and optimized computation. We demonstrated their effectiveness through a proof-of-concept experiment on a simplified dataset. Additionally, we introduced quantitative metrics for evaluating hierarchical segmentation. Future research will involve applying this method to larger and more complex hierarchical datasets to further evaluate its potential. Additionally, there is a need for further investigation into the selection of pixel pairs for the approximate computation of the HRI. A particularly intriguing and challenging direction for future research is the incorporation of a semantic dimension into the hierarchical segmentation pipeline.

References

1. Al-Huda, Z., Peng, B., Yang, Y., Algburi, R.N.A.: Object scale selection of hierarchical image segmentation with deep seeds. IET Image Process. **15**, 191–205 (2020)
2. Arbelaez, P., Maire, M., Fowlkes, C., Malik, J.: Contour detection and hierarchical image segmentation. IEEE TPAMI **33**(5), 898–916 (2010)
3. Bender, M.A., Farach-Colton, M.: The LCA problem revisited. In: Gonnet, G.H., Viola, A. (eds.) LATIN 2000. LNCS, vol. 1776, pp. 88–94. Springer, Heidelberg (2000). https://doi.org/10.1007/10719839_9
4. Chen, X., Mottaghi, R., Liu, X., Fidler, S., Urtasun, R., Yuille, A.: Detect what you can: detecting and representing objects using holistic models and body parts. In: IEEE CVPR (2014)
5. Chierchia, G., Perret, B.: Ultrametric fitting by gradient descent. In: NeurIPS, vol. 32. Curran Associates, Inc. (2019)

6. Cousty, J., Najman, L., Kenmochi, Y., Guimarães, S.: Hierarchical segmentations with graphs: quasi-flat zones, minimum spanning trees, and saliency maps. JMIV **60**(4), 479–502 (2018)

7. Elizar, E., Zulkifley, M.A., Muharar, R., Zaman, M.H.M., Mustaza, S.M.: A review on multiscale-deep-learning applications. Sensors **22**(19), 7384 (2022)

8. de Geus, D., Meletis, P., Lu, C., Wen, X., Dubbelman, G.: Part-aware panoptic segmentation. In: IEEE CVPR (2021)

9. He, J., et al.: PartImageNet: a large, high-quality dataset of parts. CoRR (2021)

10. Kendall, A., Gal, Y., Cipolla, R.: Multi-task learning using uncertainty to weigh losses for scene geometry and semantics (2018)

11. Lapertot, R., Chierchia, G., Perret, B.: Supervised learning of hierarchical image segmentation. In: Vasconcelos, V., Domingues, I., Paredes, S. (eds.) CIARP 2023. LNCS, vol. 14469, pp. 201–213. Springer, Cham (2023). https://doi.org/10.1007/978-3-031-49018-7_15

12. Li, L., Zhou, T., Wang, W., Li, J., Yang, Y.: Deep hierarchical semantic segmentation. In: IEEE CVPR, pp. 1236–1247 (2022)

13. Maninis, K., Pont-Tuset, J., Arbeláez, P., Gool, L.V.: Convolutional oriented boundaries: from image segmentation to high-level tasks. IEEE TPAMI **40**, 819–833 (2017)

14. Meletis, P., Wen, X., Lu, C., de Geus, D., Dubbelman, G.: Cityscapes-panoptic-parts and pascal-panoptic-parts datasets for scene understanding. CoRR (2020)

15. Najman, L., Cousty, J., Perret, B.: Playing with Kruskal: algorithms for morphological trees in edge-weighted graphs. In: Hendriks, C.L.L., Borgefors, G., Strand, R. (eds.) ISMM 2013. LNCS, vol. 7883, pp. 135–146. Springer, Heidelberg (2013). https://doi.org/10.1007/978-3-642-38294-9_12

16. Ôn Vû Ngoc, M., et al.: Introducing the boundary-aware loss for deep image segmentation. In: BMVC 2021 (2021)

17. Perret, B., Cousty, J., Guimarães, S.J.F., Kenmochi, Y., Najman, L.: Removing non-significant regions in hierarchical clustering and segmentation. PRL **128**, 433–439 (2019)

18. Pont-Tuset, J., Arbeláez, P., Barron, J., Marques, F., Malik, J.: Multiscale combinatorial grouping for image segmentation and object proposal generation. arXiv:1503.00848 (2015)

19. Ramanathan, V., et al.: PACO: parts and attributes of common objects (2023)

20. Ren, Z., Shakhnarovich, G.: Image segmentation by cascaded region agglomeration. In: IEEE CVPR, pp. 2011–2018 (2013)

21. Ronneberger, O., Fischer, P., Brox, T.: U-net: convolutional networks for biomedical image segmentation. In: Navab, N., Hornegger, J., Wells, W.M., Frangi, A.F. (eds.) MICCAI 2015. LNCS, vol. 9351, pp. 234–241. Springer, Cham (2015). https://doi.org/10.1007/978-3-319-24574-4_28

22. Smith, L.N., Topin, N.: Super-convergence: very fast training of residual networks using large learning rates. CoRR (2017)

23. Thisanke, H., Deshan, C., Chamith, K., Seneviratne, S., Vidanaarachchi, R., Herath, D.: Semantic segmentation using vision transformers: a survey (2023)

24. Wolf, S., Schott, L., Köthe, U., Hamprecht, F.: Learned watershed: end-to-end learning of seeded segmentation (2017)

25. Xie, E., Wang, W., Yu, Z., Anandkumar, A., Alvarez, J.M., Luo, P.: SegFormer: simple and efficient design for semantic segmentation with transformers (2021)

26. Zhang, Y., Wang, C., Wang, X., Zeng, W., Liu, W.: FairMOT: on the fairness of detection and re-identification in multiple object tracking. IJCV **129**(11), 3069–3087 (2021)

Out-of-Core Attribute Algorithms for Binary Partition Hierarchies

Josselin Lefèvre[1,2]([✉]), Jean Cousty[1], Benjamin Perret[1],
and Harold Phelippeau[2]

[1] LIGM, Univ Gustave Eiffel, CNRS, ESIEE Paris, 77454 Marne-la-Vallée, France
`josselin.lefevre@esiee.fr`
[2] Thermo Fisher Scientific, Bordeaux, France

Abstract. Binary Partition Hierarchies (BPHs) and Minimum Spanning Trees are key structures in hierarchical image analysis. However, the explosion in the size of image data poses a new challenge, as the memory available in conventional workstations becomes insufficient to execute classical algorithms. To address this problem, specific algorithms have been proposed for out-of-core computation of BPHs, where a BPH is actually represented by a collection of smaller trees, called a distribution, thus reducing the memory footprint of the algorithms. In this article, we address the problem of designing efficient out-of-core algorithms for computing classical attributes in distributions of BPHs, which is a necessary step towards a complete out-of-core hierarchical analysis workflow that includes tasks such as connected filtering and the generation of other representations such as hierarchical watersheds. The proposed algorithms are based on generic operations designed to propagate information through the distribution of trees, enabling the computation of attributes such as area, volume, height, minima and number of minima.

1 Introduction

Hierarchies are versatile representations that are useful in many image analysis and processing problems and are attracting increasing interest [15]. Among them, binary partition hierarchies [2,18] (BPHs) paired to minimum spanning trees are key structures for several (hierarchical) segmentation methods: in particular, it has been shown [2,14] that the BPH can be used to efficiently compute quasi-flat zone hierarchies [2,13] (also called α-trees), watershed hierarchies [2,12] or seed-based watersheds [10]. Efficient algorithms for building BPHs on standard size images are well established [14], but with the continuous improvement of image acquisition systems, the image resolutions are increasing dramatically, resulting in images that can reach dozens of gigabytes. In [3,11] the authors proposed algorithms to compute the BPH under the out-of-core constraint, *i.e.*, when the goal is to minimize the amount of memory required by the algorithms. In this framework, the BPH is spread into a collection of smaller local hierarchies called a distribution. Each local hierarchy is represented by a tree data structure that is small enough to fit in the main memory of a classical workstation.

S. Brunetti et al. (Eds.): DGMM 2024, LNCS 14605, pp. 298–311, 2024.
https://doi.org/10.1007/978-3-031-57793-2_23

In this article, we address the problem of the out-of-core computation of tree attributes, which is a necessary step to obtain a complete out-of-core hierarchical image analysis pipeline including for example: connected filters [17], attribute openings [1] or extinction values computation [19]. Several authors have already explored the issue of attribute computation in related contexts, such as incremental attribute computation [20] and parallel or distributed algorithms for hierarchical image analysis. In [4,7,9], the authors investigate distributed memory algorithms for computation of min and max trees to perform user-defined attribute filtering and multiscale analysis of terabytes images. In [5], the method was extended to allow a posteriori attribute computation on distributed component trees. In [6], the computation of minimum spanning trees of streaming images is considered whereas a parallel algorithm for the computation of quasi-flat zones hierarchies has been proposed in [8], allowing to efficiently implement interactive filtering segmentation of video.

In this study, we propose a generic scheme and detailed algorithms for computing regional attributes on the distribution of a BPH under the out-of-core constraint. These algorithms do not require any additional global data structure and only require having two trees of the distribution simultaneously in memory. We consider classical geometric attributes such as area, volume, height, or rightmost intersecting slice of a region, and topological attributes such as the Boolean attribute indicating whether a region is a regional minimum and the number of regional minima included in each region. The implementation of the method in C++ and Python based on the hierarchical graph processing library Higra [16] is available online https://github.com/PerretB/Higra-distributed.

This article is organized as follows. Section 2 gives the formal problem statement recalling the notion of BPH and distribution of BPH before introducing the distribution of an attribute. Section 3 explains the proposed data structures and presents a general scheme for out-of-core attribute computation. Section 4 presents uses of this scheme for calculating attributes such as area, volume, height, local minima and number of minima.

2   Out-of-Core Attribute Computation: Problem Statement

In this section, we give formal definitions of a BPH and of the distribution of a BPH, and we state the problem of computing an attribute over the distribution of a BPH.

2.1   Binary Partition Hierarchy by Altitude Ordering

Let us first recall the definition of a hierarchy of partitions. Then we define the binary partition hierarchy by altitude ordering.

Let V be a set. A *partition of* V is a set of pairwise disjoint subsets of V. Any element of a partition is called a *region* of this partition. The *ground* of a partition $\mathbf{P}$, denoted by $gr(\mathbf{P})$, is the union of the regions of $\mathbf{P}$. A partition whose ground is V is called a *complete partition of* V. Let $\mathbf{P}$ and $\mathbf{Q}$ be two partitions of V. We say that $\mathbf{Q}$ is a *refinement of* $\mathbf{P}$ if any region of $\mathbf{Q}$ is included in a region of $\mathbf{P}$. A *hierarchy on* V is a sequence $(\mathbf{P}_0, \ldots, \mathbf{P}_\ell)$ of partitions of V such that, for any λ in $\{0, \ldots, \ell - 1\}$, the partition $\mathbf{P}_\lambda$ is a refinement of $\mathbf{P}_{\lambda+1}$. Let $\mathcal{H} = (\mathbf{P}_0, \ldots, \mathbf{P}_\ell)$ be a hierarchy. The integer ℓ is called the *depth of* $\mathcal{H}$ and, for any λ in $\{0, \ldots, \ell\}$, the partition $\mathbf{P}_\lambda$ is called the λ-*scale of* $\mathcal{H}$. In the following, if λ an integer in $\{0, \ldots, \ell\}$, we denote by $\mathcal{H}[\lambda]$ the λ-scale of $\mathcal{H}$. For any λ in $\{0, \ldots, \ell\}$, any region of the λ-scale of $\mathcal{H}$ is also called a *region of* $\mathcal{H}$. The hierarchy $\mathcal{H}$ is *complete* if $\mathcal{H}[0] = \{\{x\} \mid x \in V\}$ and if $\mathcal{H}[\ell] = \{V\}$. We denote by $\mathcal{H}_\ell(V)$ the set of all hierarchies on V of depth ℓ, by $\mathcal{P}(V)$ the set of all partitions on V, and by $2^{|V|}$ the set of all subsets of V.

In the following, the symbol ℓ stands for any strictly positive integer.

We define a *graph* as a pair $G = (V, E)$ where V is a finite set and E is composed of ℓ unordered pairs of distinct elements in V. Each element of V is called a *vertex of* G, and each element of E is called an *edge of* G. The Binary Partition Hierarchy (BPH) by altitude ordering relies on a total order on E, denoted by $\prec$. Let k in $\{1, \ldots, \ell\}$, we denote by $u_k^\prec$ the k-th element of E for the order $\prec$. Let u be an edge in E, the *rank of* u *for* $\prec$, denoted by $r^\prec(u)$, is the unique integer k such that $u = u_k^\prec$. We set $\mathcal{B}[0] = \{\{x\} \mid x \in V\}$. The *partial binary partition hierarchy* $\mathcal{B}[k]$ *at rank* k is the hierarchy on V defined by $\mathcal{B}[k] = \left(\mathcal{B}[k-1] \setminus \{R_x, R_y\}\right) \cup \{R_x \cup R_y\}$ where $u_k^\prec = \{x, y\}$ and where R_x (resp. R_y) is the unique region of $\mathcal{B}[k-1]$ that contains x (resp. y). The partial binary partition hierarchy at rank ℓ is the *binary partition hierarchy by* $\prec$ and it is denoted by $\mathcal{B}^\prec$. The *rank* $r(R)$ of a region R of $\mathcal{B}^\prec$ is the lowest scale λ at which the region appears in the hierarchy and, if $r(R) > 0$, the *building edge* of R, denoted by $\mu^\prec(R)$, is the edge $\mu^\prec(R) = u_{r(R)}^\prec$ (*i.e.*, the edge that lead to "building" region R).

2.2 Distribution of Binary Partition Hierarchy on a Causal Partition

Intuitively, distributing a hierarchy consists in splitting it into a set of smaller hierarchies such that: 1) each smaller hierarchy corresponds to a *selection* of a subpart of the whole that intersects a slice of the graph and 2) the initial hierarchy can be reconstructed by "gluing" those smaller hierarchies.

Let V be a set. The operation *sel* is the map from $2^{|V|} \times \mathcal{P}(V)$ to $\mathcal{P}(V)$ which associates to any subset X of V and to any partition $\mathbf{P}$ of V the subset $\mathrm{sel}(X, \mathbf{P})$ of $\mathbf{P}$ which contains every region of $\mathbf{P}$ that contains an element of X. The operation *select* is the map from $2^{|V|} \times \mathcal{H}_\ell(V)$ in $\mathcal{H}_\ell(V)$ which associates to any subset X of V and to any hierarchy $\mathcal{H}$ on V the hierarchy $\mathrm{select}(X, \mathcal{H}) = (\mathrm{sel}(X, \mathcal{H}[0]), \ldots, \mathrm{sel}(X, \mathcal{H}[\ell]))$.

We are then able to define the distribution of a hierarchy thanks to the *select* operation. Let V be a set, let $\mathbf{P}$ be a complete partition on V and let $\mathcal{H}$ be a hierarchy on V. The distribution of $\mathcal{H}$ over $\mathbf{P}$ is the set $\delta_\mathcal{H} = \{\mathrm{select}(R, \mathcal{H}) \mid R \in \mathbf{P}\}$ and for any region R of $\mathbf{P}$, $\mathrm{select}(R, \mathcal{H})$ is called a *local hierarchy* of $\delta_\mathcal{H}$ In Fig. 1, the distribution $(\mathcal{H}_0, \mathcal{H}_1, \mathcal{H}_2)$ of the BPH $\mathcal{H}$ is computed over the causal partition $(\{a, d\}, \{b, e\}, \{c, f\})$.

In the following sections, we consider the special case of a distribution built on a 4-adjacency graph representing a 2d image divided into slices (this is not a limiting factor, and the method can easily be adapted to any regular grid graph such as 6 adjacency for 3d images). Let h and w be two integers representing the height and the width of an image. Then, the vertex set V is the Cartesian product $\{0, \cdots, h-1\} \times \{0, \cdots, w-1\}$ and the edge set E is given by the well-known 4-adjacency relation on V. Let k be a positive integer, the *causal partition of V into $k+1$ slices* is the sequence $(S_0, \ldots, S_k)$ such that for any t in $\{0, \cdots, k\}$, $S_t = \{(i,j) \in V \mid t \times \frac{w}{k} \leq i < (t+1) \times \frac{w}{k}\}$. Each element of the sequence $(S_0, \ldots, S_k)$ is called a *slice* and it can be seen that each slice can have up to two neighboring slices.

Important notation. *In this article, the symbols $G = (V, E)$, $\prec$, and $\mathcal{H}$ denote a 4-adjacency graph of ℓ edges, a total order on its edge set E, and the associated BPH by $\prec$, respectively. Furthermore, the symbol $\delta_{\mathcal{H}}$ denotes the distribution of $\mathcal{H}$ over the causal partition $(S_0, \cdots, S_k)$ of V into $k+1$ slices.*

2.3   Distribution of Attributes

An *attribute on $\mathcal{H}$* is a mapping A associating an attribute value $A(R)$ to every region R of $\mathcal{H}$. In this article, we consider attributes values that can be either Boolean or scalar and we are interested in computing attributes from the distribution of a BPH over a causal partition. To this end, given an attribute A on $\mathcal{H}$, we define the *distribution of A over $\delta_{\mathcal{H}}$* as the series $\delta_A = (A_{\mathcal{B}_0}, \cdots, A_{\mathcal{B}_k})$ such that for any i in $\{0, \ldots, k\}$ and any R in $\mathcal{B}_i$, we have $A(R) = A_{\mathcal{B}_i}(R)$. Hence, in this article, our main goal is to solve the following problem.

Problem. *Given the distribution $\delta_{\mathcal{H}}$ of the hierarchy $\mathcal{H}$, compute the distribution δ_A of an attribute A without explicitly computing $\mathcal{H}$ nor A while maintaining the out-of-core constraint, that is having a limited amount of memory at each computation steps.*

3   Propagate Algorithm: A Fundamental Brick for Out-of-Core Attribute Computation

In this section, we introduce *Propagate* Algorithm which is called as a fundamental step of all subsequent attribute computations. Since a region may contain pixels belonging to several slices, it is generally not possible to compute the attribute of a region using only the information available in a single local hierarchy of the distribution. Thus, the general out-of-core attribute computation scheme considers a first step of local computation followed by a second step consisting of merging local information, done with Propagate Algorithm. More precisely, the proposed computation proceeds as follows:

1. Compute a *partial attribute* value locally on each tree of the distribution;

2. *Propagate* partial attribute values from neighboring trees in the causal direction, *i.e.*, from slices of lower indices to those of higher indices. At the end of this step, the attribute values of the last tree of the distribution are correct;
3. Backpropagate the "correct" attribute values in the anti-causal direction, *i.e.*, from slices of higher indices to those of lower indices. At the end of this step, all the attribute values of all the trees in the distribution are correct.

In this scheme, each local hierarchy of the distribution is visited twice (once in the causal pass and once in the anti-causal pass) and at any step, we never need to have more than 2 neighboring hierarchies simultaneously in memory.

Before describing precisely *Propagate* Algorithm in Sect. 3.2, we first introduce the necessary data structures in Sect. 3.1. Then, the computation of specific attributes using *Propagate* Algorithm is addressed in Sect. 4.

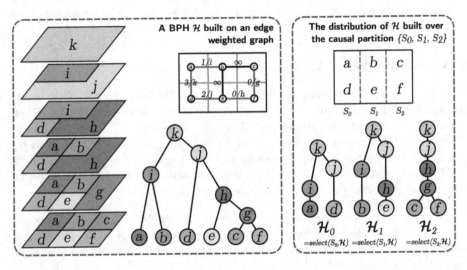

Fig. 1. Left, a BPH $\mathcal{H}$ built on an edge weighted graph (*index/weight*) and the visualisation of the nested series of partition. Red edges of the graph belong to the minimum spanning tree. Right, the distribution $\{\mathcal{H}_0, \mathcal{H}_1, \mathcal{H}_2\}$ of $\mathcal{H}$ over the causal partition $\{S_0, S_1, S_2\}$.

3.1   Data Structures

The data structures used in our algorithms are designed to contain only the necessary and sufficient information so that we never need to have all the data in the main memory at once. The tree-based data structure representing a local hierarchy $\mathcal{H}$ assumes that the regions, also called nodes of the hierarchy in this context, are indexed in a particular order and relies on three arrays: 1) a mapping of the indices from the local context (a given slice) to the global one (the whole graph) noted $\mathcal{H}$.map, 2) a parent array denoted by $\mathcal{H}$.par encoding the parent relation between the tree nodes, and 3) an array $\mathcal{H}.\mathtt{alt}$ giving, for each non-leaf-node of the tree, the weight of its corresponding building edge also called *altitude*.

More precisely, given a binary partition hierarchy $\mathcal{H}$ with n regions, every integer between 0 and $n-1$ is associated to a unique region of $\mathcal{H}$. Moreover, this indexing of the regions of $\mathcal{H}$ follows a topological order such that: 1) any leaf region is indexed before any non-leaf region; 2) two leaf regions $\{x\}$ and $\{y\}$ are sorted with respect to an arbitrary order on the element V (*e.g.*, the raster scan order in 2d) and 3) two non-leaf regions are sorted according to their altitude, *i.e.*, the order of their building edges for $\prec$. This order can be seen as an extension of the order $\prec$ on E to the set $V \cup E$ that enables 1) to efficiently browse the nodes of a hierarchy according to their scale of appearance in the hierarchy and 2) to efficiently match regions of V with the leaves of the hierarchy. By abuse of notation, this extended order is also denoted by $\prec$ in the following.

To keep track of the global context, a link between the indices in the local tree and the global indices in the whole graph is stored in the form of an array $\mathtt{map}$ which associates: 1) to the index i of any leaf region R, the vertex x of the graph G such that $R = \{x\}$, *i.e.* $\mathtt{map[i]=x}$; and 2) to the index i of any non-leaf region R, its building edge, *i.e.* $\mathtt{map[i]}=\mu^{\prec}(R)$.

The parent relation of the tree is stored in an array $\mathtt{par}$ such that $\mathtt{par[i]=j}$ if the region of index j is the parent of the region of index i. For the root node r, which has no parent, we set $\mathtt{par[r]=r}$.

The binary partition hierarchy is built for a particular ordering $\prec$ of the edges of G. In practice, this ordering is induced by weights computed over the edges of G. To this end, we store an array $\mathtt{alt}$ of size $|R^{\star}|$, *i.e.* the number of non-leaf regions, elements such that, for every region R in $R^{\star}$ of index i, $\mathtt{alt[i]}$ is the weight of the building edge $\mu^{\prec}(R)$ of region R, with $R^{\star}$ be the set of non-leaf regions R in $\mathcal{H}$. The edges can then be compared according to the following total order induced by the weights: we set $u \prec v$ if the weight of u is less than the one of v or if u and v have equal weights but u comes before v with respect to the raster scan order.

3.2   Attribute Propagation

Let us now give a precise description of PROPAGATE Algorithm (see Algorithm 1) that is a fundamental brick to compute regional attributes in an out-of-core manner. This algorithm allows for transforming a first version of an attribute, which can be seen as local to each slice, to a global version. To this end, the algorithm considers the regions which are replicated in two neighbouring hierarchies associated with successive slices and propagate their attribute values: 1) firstly, in a causal order (lines 1 to 3), from the slices of lower indices to the ones of higher indices; and 2) secondly, in an anti-causal order (lines 5 to 6), from the higher indices to the lower ones. During the causal propagation, a binary operator is used to merge the attribute values of regions replicated in consecutive slices S_i and S_{i+1}. This binary operator (denoted by $\oplus$ in Algorithm 1) is given as a parameter of the algorithm. In the following sections, depending on the kind of attributes, we consider the sum, the maximum, or the minimum of two values. During the anti-causal propagation, the attribute value of the region in the slice of higher index is copied as the attribute value of the same replicated region in the slice of lower index.

Algorithm 1: PROPAGATE

Params: The distribution $(\mathcal{B}_0, \ldots, \mathcal{B}_k)\}$ of a BPH; a series $(\texttt{attr}_{\mathcal{B}_0}, \ldots, \texttt{attr}_{\mathcal{B}_k})$ of attributes; and a binary operator $\oplus$

Result: A new series $(\texttt{attr}_{\mathcal{B}_0}^{\downarrow}, \ldots, \texttt{attr}_{\mathcal{B}_k}^{\downarrow})$ of attributes where values of regions replicated on consecutive slices have been propagated.

1 $\texttt{attr}_{\mathcal{B}_0}^{\uparrow} := \texttt{attr}_{\mathcal{B}_0}$

2 **foreach** i from 1 to k **do**

3 $\quad \texttt{attr}_{\mathcal{B}_i}^{\uparrow} := \text{MERGE}(\mathcal{B}_{i-1}, \mathcal{B}_i, \texttt{attr}_{\mathcal{B}_{i-1}}^{\uparrow}, \texttt{attr}_{\mathcal{B}_i}, \oplus)$

4 $\texttt{attr}_{\mathcal{B}_k}^{\downarrow} := \texttt{attr}_{\mathcal{B}_k}^{\uparrow}$

5 **foreach** i from $k-1$ to 0 **do**

6 $\quad \texttt{attr}_{\mathcal{B}_i}^{\downarrow} := \text{MERGE}(\mathcal{B}_{i+1}, \mathcal{B}_i, \texttt{attr}_{\mathcal{B}_{i+1}}^{\downarrow}, \texttt{attr}_{\mathcal{B}_i}^{\uparrow}, \lhd)$

7 **return** $\left(\texttt{attr}_{\mathcal{B}_0}^{\downarrow}, \ldots, \texttt{attr}_{\mathcal{B}_k}^{\downarrow} \right)$

In order to identify the regions that are replicated in two consecutive hierarchies $\mathcal{X}$ and $\mathcal{Y}$, PROPAGATE Algorithm calls the auxiliary function MERGE (see Algorithm 2). This function simultaneously browses the regions of the two hierarchies in increasing order. When two nodes x and y of, respectively, $\mathcal{X}$ and $\mathcal{Y}$ are found as occurrences of the same region (*i.e.*, when $\mathcal{X}.\text{map}[x] = \mathcal{Y}.\text{map}[y]$, see line 3 in Algorithm 2) the attribute values $\texttt{attr}_{\mathcal{X}}[x]$ and $\texttt{attr}_{\mathcal{Y}}[y]$ are merged and the result is stored as the new attribute value of y: $\texttt{attr}'_{\mathcal{Y}}(y) := \texttt{attr}_{\mathcal{X}}(x) \oplus \texttt{attr}_{\mathcal{Y}}(y)$. If *assign-first* operator, denoted by $\lhd$, is given to Merge as the merging operator (*i.e.*, if $\oplus = \lhd$), then the result $\texttt{attr}_{\mathcal{X}}(x) \oplus \texttt{attr}_{\mathcal{Y}}(y)$ is simply the value $\texttt{attr}_{\mathcal{X}}(x)$.

Algorithm 2: MERGE

Params: Two hierarchies $\mathcal{X}$ and $\mathcal{Y}$; two attributes $\texttt{attr}_{\mathcal{X}}$ and $\texttt{attr}_{\mathcal{Y}}$ associated with these hierarchies ; and a binary operator $\oplus$

Result: A new attribute $\texttt{attr}'_{\mathcal{Y}}$, update of $\texttt{attr}_{\mathcal{Y}}$

1 $x := 0; y := 0$ $\qquad$ // x iterates over $\mathcal{X}$ and y over $\mathcal{Y}$

2 **while** $x < |\mathcal{X}|$ **or** $y < |\mathcal{Y}|$ **do**

3 $\quad$ **if** $x < |\mathcal{X}|$ **and** $y < |\mathcal{Y}|$ **and** $\mathcal{X}.\text{map}[x] = \mathcal{Y}.\text{map}[y]$ **then**

4 $\quad\quad$ $\texttt{attr}'_{\mathcal{Y}}[y] := \texttt{attr}_{\mathcal{X}}[x] \oplus \texttt{attr}_{\mathcal{Y}}[y]$

5 $\quad\quad$ $x := x + 1; \ y := y + 1$

6 $\quad$ **else if** $\mathcal{X}.\text{map}[x] \prec \mathcal{Y}.\text{map}[y]$ **then** $x := x + 1$

7 $\quad$ **else** $y := y + 1; \texttt{attr}'_{\mathcal{Y}}[y] := \texttt{attr}_{\mathcal{Y}}[y]$

8 **return** $\texttt{attr}'_{\mathcal{Y}}$

Figure 2 illustrates the use of PROPAGATE Algorithm. More precisely, in Fig. 2b, an initial attribute value is mapped to each node (subscript) of the local hierarchies (*i.e.*, $(attr_{\mathcal{B}_0}, attr_{\mathcal{B}_1}, attr_{\mathcal{B}_2})$). The values of $(attr_{\mathcal{B}_0}^{\uparrow}, attr_{\mathcal{B}_1}^{\uparrow}, attr_{\mathcal{B}_2}^{\uparrow})$ obtained after the causal pass of PROPAGATE are then shown in Fig. 2c. In particular, red arrows indicate the mapping between replicated regions detected by MERGE as well as the computation which is made to obtain the value of the updated attribute (here $\oplus = +$ is given to MERGE). Finally, Fig. 2d shows the result of $(attr_{\mathcal{B}_0}^{\downarrow}, attr_{\mathcal{B}_1}^{\downarrow}, attr_{\mathcal{B}_2}^{\downarrow})$ obtained after the anticausal pass.

It can be observed that every node of each tree is browsed once during the execution of MERGE Algorithm. Thus, the overall time-complexity of MERGE Algorithm is linear with respect to the number of regions of $\mathcal{X}$ and $\mathcal{Y}$. Furthermore, given a causal partition with $k + 1$ slices, it can be seen that PROPAGATE performs exactly $2 \times k$ calls to MERGE and that the overall time complexity of PROPAGATE Algorithm is $O(k + N)$ where N is the sum of the numbers of regions of the local hierarchies $\mathcal{B}_0, \ldots, \mathcal{B}_k$.

4   Out-of-Core Attributes Algorithms

In this section, following the general scheme proposed in Sect. 3, we present out-of-core algorithms to compute common attributes in hierarchical analysis.

Rightmost Slice. We first consider *Rightmost slice* attribute that maps to every region R of $\mathcal{H}$, the highest index of a slice containing a vertex of R. Aside from being a simplest attribute whose values can be computed with the help of PROPAGATE, *Rightmost slice* attribute, denoted by `right`, will be used subsequently as a preprocessing necessary before computing more complex attributes such as minima number. More precisely, it is used when one needs to select a single representative of a region that is replicated over several slices. In such case, we arbitrarily pick the representative of a region R as the occurrence of R in local hierarchy $\mathcal{B}_i$ such that $\text{right}_{\mathcal{B}_i}(R) = i$.

To compute the *Rightmost slice* of each region, firstly, for each local hierarchy $\mathcal{B}_i$ in the distribution $\delta_{\mathcal{H}}$, we locally initialize an attribute $I_{\mathcal{B}_i}(R) = i$ and then we call PROPAGATE with the supremum operator.

---

1 **Function** *RightmostSlice($\delta_{\mathcal{H}} = \{\mathcal{B}_0, \ldots, \mathcal{B}_k\}$)*

2   **foreach** i from 0 to k **do**

3         $\text{right}_{\mathcal{B}_i}[R] := i$ for each region R of $\mathcal{B}_i$

4   **return** PROPAGATE($\left(\text{right}_{\mathcal{B}_0}, \ldots, \text{right}_{\mathcal{B}_k}\right), \delta_{\mathcal{H}}, \vee$)

Area. Let us now consider the area attribute that maps to every region of a hierarchy the number of pixels in that region. In the tree-based representation of a hierarchy, the area of a node n can be obtained recursively by setting $\mathbf{area}(n) = 1$ for every leaf n and $\mathbf{area}(n) = \sum_{c \in \mathbf{children}(n)} \mathbf{area}(c)$ for every non-leaf node n with $\mathbf{children}(n) = \{c \mid n = \mathbf{parent}(c)\}$. In the case of a distribution, for each local tree, we initialize the area of the nodes with the previous recursive formula (lines 2–5 below) followed by a call to PROPAGATE with the $+$ operator.

---

1   **Function** $Area(\delta_{\mathcal{H}} = \{\mathcal{B}_0, \ldots, \mathcal{B}_k\})$
2   **foreach** i from 0 to k **do**
3   Initialize **area** values with 1 for every leaf and 0 for every non-leaf
4   **foreach** non-root region R of $\mathcal{B}_i$ in topological order **do**
5           $\mathbf{area}_{\mathcal{B}_i}[\mathbf{parent}[R]]+ = \mathbf{area}_{\mathcal{B}_i}[R]$
6   **return** PROPAGATE$((\mathbf{area}_{\mathcal{B}_0}, \ldots, \mathbf{area}_{\mathcal{B}_k}), \delta_{\mathcal{H}}, +)$

Note that if we consider superpixels instead of pixels or the integral of a function over the domain, the algorithm remains valid provided an adaptation of its initialization at line 3.

Volume. In the tree-based representation of a hierarchy, the volume $V(n)$ of a node n can be obtained recursively by $V(n) = 0$ for every leaf n and $V(n) = A(n) \times |altitude(parent(n)) - altitude(n)| + \sum_{c \in children(n)} V(c)$ for every non-leaf node n. To compute the volume for a distribution, we firstly initialize the volume locally with the recursive formula. Similarly to the area, at this point, each region will lack the volumes of included regions that do not belong to the same local hierarchy. To correct this value, we propagate the partial volume by calling PROPAGATE with the $+$ operator.

---

1   **Function** $Volume(\delta_{\mathcal{H}} = \{\mathcal{B}_0, \ldots, \mathcal{B}_k\})$
2   **foreach** i from 0 to k **do**
3   Locally initialize $\mathbf{area}_{\mathcal{B}_i}$ *i.e.*, lines 3–5 of *Area*
4   **foreach** non-root region R of $\mathcal{B}_i$ in topological order **do**
5           $\mathbf{vol}_{\mathcal{B}_i}[R]+ = \mathbf{area}_{\mathcal{B}_i}[R] \times (\mathbf{alt}[\mathcal{B}_i.\mathbf{par}[R]] - \mathbf{alt}[R])$
6           $\mathbf{vol}_{\mathcal{B}_i}[\mathcal{B}_i.\mathbf{par}[R]]+ = \mathbf{vol}_{\mathcal{B}_i}[R]$
7   **return** PROPAGATE$((\mathbf{vol}_{\mathcal{B}_0}, \ldots, \mathbf{vol}_{\mathcal{B}_k}), \delta_{\mathcal{H}}, +)$

Height. The height of a region R of a BPH $\mathcal{H}$ is the difference between the altitude of the region R and the altitude of the lowest region included in R. From the tree based representation of a hierarchy, it can be computed with the following recursion: $H(n) = 0$ for every leaf n and $H(n) = altitude(parent(n)) - altitude(n) + \max\{H(c), c \in \mathbf{children}(n)\}$ for every non-leaf node n. In a distribution, after initializing the height locally, following the above recursion, it is possible that the children leading to maximize the height in the global hierarchy do not belong to the same local hierarchy. It is therefore necessary to "search" for these children within the distribution in order to maximize the height. To do so we call *Propagate* with the supremum operator.

---

1 **Function** $Height(\delta_{\mathcal{H}} = \{\mathcal{B}_0, \ldots, \mathcal{B}_k\})$
2   **foreach** i from 0 to k **do**
3   Initialize $\mathtt{heig}_{\mathcal{B}_i}$ to 0 for every node
4   **foreach** non-root region R of $\mathcal{B}_i$ in topological order **do**
5              $\mathtt{heig}_{\mathcal{B}_i}[\mathcal{B}_i.\mathtt{par}[R]] :=$
             $\max(\mathtt{heig}_{\mathcal{B}_i}[\mathcal{B}_i.\mathtt{par}[R]], \mathtt{alt}[\mathcal{B}_i.\mathtt{par}[\mathcal{B}_i.\mathtt{par}[R]]] -$
             $\mathtt{alt}_{\mathcal{B}_i}[\mathcal{B}_i.\mathtt{par}[R]] + \mathtt{heig}_{\mathcal{B}_i}[R])$
6   **return** PROPAGATE$((\mathtt{heig}_{\mathcal{B}_0}, \ldots, \mathtt{heig}_{\mathcal{B}_k}), \delta_{\mathcal{H}}, \vee)$

Note that this method can also be used to compute the topological height, *i.e.* the maximum length of a paths from a node to a descendant leaf node.

Minima. A regional minimum of a weighted graph is a set of vertices connected by edges of weight k and whose adjacent edges are all of strictly higher weight. In the tree-based representation of a BPH, a node n can be identified as a minimum if 1) the altitude of the parent of n is different from the one of n, and 2) the altitude of any non-leaf node included in the sub-tree rooted in n is equal to the altitude of n. We can formalize and test these two criteria in our framework. Firstly, let $L_{\mathcal{B}_i}(n) =$ true if $altitude(n) < altitude(parent(n))$ and false otherwise: this can be calculated locally, as only the altitude of the parents is required. The second criterion is defined recursively as $E^L_{\mathcal{B}_i}(R) = false$ for any leaf R and as $E^L_{\mathcal{B}_i}(R) = \bigwedge_{c \in children(R)} \neg L_{\mathcal{B}_i}(c)$ for an any non-leaf region R. This last criterion depends on the children of R, so we need to make sure it is satisfied for all children in the distribution. We can verify this by calling PROPAGATE with the infimum operator to compute a *global* criterion $E^G_{\mathcal{B}_i}$. Then, an occurrence of a region R in the distribution is a minimum if $M_{\mathcal{B}_i}(R) = L_{\mathcal{B}_i}(R) \wedge E^G_{\mathcal{B}_i}(R)$.

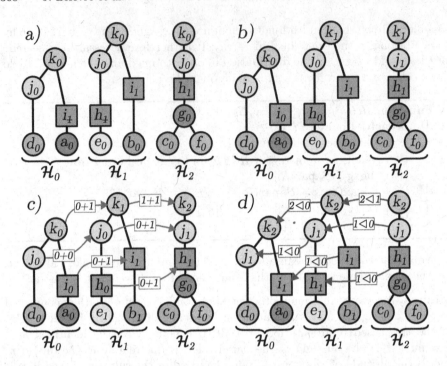

Fig. 2. Minima number attribute computation. The subscript of each node indicate its attribute value. **a.** The two local minima i and h (square node) have both two occurrences in the distribution. To ensure that only one occurrence of each minimum contributes, only the rightmost replica is set to 1, *e.g.*, region h in $\mathcal{H}_1$ count as 0 because its rightmost occurrence is in $\mathcal{H}_2$. **b.** Local initialisation of `minnb` on the basis of modified minima values. **c.** Causal pass with the operator $+$. The arrows link two occurrences of a region in neighbouring hierarchies. The arrow points to the region whose attribute is updated. **d.** Anti-causal pass, the values associated with rightmost occurrences of each region area back-propagated with $\lhd$.

Minima Number. In this part, we are interested in the number of minima included in each region of a hierarchy. In a tree, we can obtain it recursively by considering, $M(n) = 0$ for every leaf n, $M(n) = 1$ for every minimum n and $M(n) = \sum_{x \in \mathtt{childre}(n)} M(n)$ for any other node.

```
1  Function Minima(δ_H = {B_0,...,B_k})
2  |   foreach i from 0 to k do
3  |   |   foreach region R of B_i in topological order do
4  |   |   |   includesMin[R] := false; mini_{B_i}[R] := false
5  |   |   foreach non-leaf region R of B_i in topological order do
6  |   |   |   if alt[R] < alt[B_i.par[R]] and includesMin[R] = false then
7  |   |   |   |   mini_{B_i}[R] := true; includesMin[R] := true
8  |   |   |   includesMin[B_i.par[R]] := includesMin[R]
9  |   return PROPAGATE((mini_{B_0},...,mini_{B_k}), δ_H, ∧)
```

At first glance, the calculation of this attribute seems similar to that of area. For the latter, we propagate the sum value of the leaves, and no leaf can contribute more than once to the final result because a leaf belongs to one and only one local hierarchy. However, as the minima are defined on non-leaf regions, they can be replicated in several local hierarchies *e.g.*, on Fig. 2 the node h is replicated in $\mathcal{H}_1$ and $\mathcal{H}_2$. If this situation, if we use the same strategy as for area, a single minimum would contribute several times to the minima count. We must therefore ensure that, one and only one occurrence of a minimum contributes to the calculation of the attribute for the ancestors of thus minimum. As a solution, we use *RightmostSlice* attribute to arbitrarily discriminate the last occurrence of each region. We can therefore guarantee that only one of these occurrences will contribute to the local minimum account. In Fig. 2a), since the occurrence of the minimum h in $\mathcal{H}_1$ is not the rightmost, it does not contribute to the bottom-up sum contrary to the occurrence in $\mathcal{H}_2$. It should be noted that contrary to the previously introduced attributes, this one necessitate two calls to *Propagate* (minima and *RightmostSlice* can be computed simultaneously) rather than one.

```
1  Function NumberMinima(δ_H = {B_0,...,B_k})
2  |   right := RightmostSlice(δ_H); mini := Minima(δ_H)
3  |   foreach i from 0 to k do
4  |   |   Initialize nbmin_{B_i} to 0 for every node
5  |   |   foreach non-root region R of the each B_i in topological order do
6  |   |   |   if mini_{B_i}[R] = true and right_{B_i}[R] = i then
7  |   |   |   |   nbmin_{B_i}[R] := 1
8  |   |   |   nbmin_{B_i}[B_i.par[R]] += nbmin_{B_i}[R]
9  |   return PROPAGATE((nbmin_{B_0},...,nbmin_{B_k}), δ_H, +)
```

5   Discussion and Conclusion

In this paper, we proposed a global scheme for computing attributes for a distribution of BPH. This method is based on a linear complexity algorithm and requires having only the information about two adjacent regions in main memory at any step of the algorithm. We have given applications for computing common attributes such as area, volume, and local minima. It should be noted that the use of this methodology is not limited to the presented attributes, but can easily be used to compute other attributes such as bounding boxes or to determine watershed edges. In future work, we plan to study the time and memory consumption of this methodology and to extend it to the computation of more complex attributes such as extinction values or smallest common ancestors.

References

1. Breen, E.J., Jones, R.: Attribute openings, thinnings, and granulometries. Comput. Vis. Image Underst. **64**(3), 377–389 (1996)
2. Cousty, J., Najman, L., Perret, B.: Constructive links between some morphological hierarchies on edge-weighted graphs. In: Hendriks, C.L.L., Borgefors, G., Strand, R. (eds.) ISMM 2013. LNCS, vol. 7883, pp. 86–97. Springer, Heidelberg (2013). https://doi.org/10.1007/978-3-642-38294-9_8
3. Cousty, J., Perret, B., Phelippeau, H., Carneiro, S., Kamlay, P., Buzer, L.: An algebraic framework for out-of-core hierarchical segmentation algorithms. In: Lindblad, J., Malmberg, F., Sladoje, N. (eds.) DGMM 2021. LNCS, vol. 12708, pp. 378–390. Springer, Cham (2021). https://doi.org/10.1007/978-3-030-76657-3_27
4. Gazagnes, S., Wilkinson, M.H.F.: Distributed connected component filtering and analysis in 2D and 3D tera-scale data sets. IEEE TIP **30**, 3664–3675 (2021)
5. Gazagnes, S., Wilkinson, M.H.: Parallel attribute computation for distributed component forests. In: 2022 IEEE ICIP, pp. 601–605 (2022)
6. Gigli, L., Velasco-Forero, S., Marcotegui, B.: On minimum spanning tree streaming for hierarchical segmentation. PRL **138**, 155–162 (2020)
7. Götz, M., Cavallaro, G., Geraud, T., Book, M., Riedel, M.: Parallel computation of component trees on distributed memory machines. TPDS **29**, 2582–2598 (2018)
8. Havel, J., Merciol, F., Lefèvre, S.: Efficient tree construction for multiscale image representation and processing. JRTIP **16**, 1129–1146 (2019)
9. Kazemier, J.J., Ouzounis, G.K., Wilkinson, M.H.F.: Connected morphological attribute filters on distributed memory parallel machines. In: Angulo, J., Velasco-Forero, S., Meyer, F. (eds.) ISMM 2017. LNCS, vol. 10225, pp. 357–368. Springer, Cham (2017). https://doi.org/10.1007/978-3-319-57240-6_29
10. Lebon, Q., Lefèvre, J., Cousty, J., Perret, B.: Interactive segmentation with incremental watershed cuts. In: Vasconcelos, V., Domingues, I., Paredes, S. (eds.) CIARP 2023. LNCS, vol. 14469, pp. 189–200. Springer, Cham (2024). https://doi.org/10.1007/978-3-031-49018-7_14
11. Lefèvre, J., Cousty, J., Perret, B., Phelippeau, H.: Join, select, and insert: efficient out-of-core algorithms for hierarchical segmentation trees. In: Baudrier, É., Naegel, B., Krähenbühl, A., Tajine, M. (eds.) DGMM 2022. LNCS, vol. 13493, pp. 274–286. Springer, Cham (2022). https://doi.org/10.1007/978-3-031-19897-7_22
12. Meyer, F.: The dynamics of minima and contours. In: Maragos, P., Schafer, R.W., Butt, M.A. (eds.) Mathematical Morphology and its Applications to Image and Signal Processing. Computational Imaging and Vision, vol. 5, pp. 329–336. Springer, Boston (1996). https://doi.org/10.1007/978-1-4613-0469-2_38
13. Meyer, F., Maragos, P.: Morphological scale-space representation with levelings. In: Nielsen, M., Johansen, P., Olsen, O.F., Weickert, J. (eds.) Scale-Space 1999. LNCS, vol. 1682, pp. 187–198. Springer, Heidelberg (1999). https://doi.org/10.1007/3-540-48236-9_17

14. Najman, L., Cousty, J., Perret, B.: Playing with Kruskal: algorithms for morpho-logical trees in edge-weighted graphs. In: Hendriks, C.L.L., Borgefors, G., Strand, R. (eds.) ISMM 2013. LNCS, vol. 7883, pp. 135–146. Springer, Heidelberg (2013). https://doi.org/10.1007/978-3-642-38294-9_12
15. Passat, N., Kurtz, C., Vacavant, A.: Virtual special issue: "hierarchical represen-tations: new results and challenges for image analysis". PRL **138**, 201–203 (2020)
16. Perret, B., Chierchia, G., Cousty, J., Guimarães, S.J.F., Kenmochi, Y., Najman, L.: Higra: hierarchical graph analysis. SoftwareX **10**, 100335 (2019)
17. Salembier, P., Serra, J.: Flat zones filtering, connected operators, and filters by reconstruction. IEEE TIP **8**, 1153–1160 (1995)
18. Salembier, P., Garrido, L.: Binary partition tree as an efficient representation for image processing, segmentation, and information retrieval. TIP **9**(4), 561–576 (2000)
19. Silva, A.G., de Alencar Lotufo, R.: Efficient computation of new extinction values from extended component tree. PRL **32**(1), 79–90 (2011)
20. Silva, D.J., Alves, W.A., Hashimoto, R.F.: Incremental bit-quads count in compo-nent trees: theory, algorithms, and optimization. PRL **129**, 33–40 (2020)

Multi-scale Component-Tree: A Hierarchical Representation for Sparse Objects

Romain Perrin[1]([⊠])[iD], Aurélie Leborgne[1][iD], Nicolas Passat[2][iD], Benoît Naegel[1][iD], and Cédric Wemmert[1][iD]

[1] University of Strasbourg, ICube, Strasbourg, France
romain.perrin@unistra.fr
[2] University of Reims Champagne-Ardenne, CReSTIC, Reims, France

Abstract. Component-trees are hierarchical structures developed in the framework of mathematical morphology. They model images via the inclusion relationships between the connected components of their successive threshold sets. There exist many variants of component-trees, but to the best of our knowledge, none of them deals with the representation of the image at different scales. In this article, we propose such a variant of component-tree that tackles this issue, namely the Multi-Scale Component-Tree (MSCT). We describe an algorithmic scheme for building the MSCT from the standard computation of component-trees of the image, seen from its lowest to its highest scale. At each step, a local upscaling is performed on relevant parts of the image, corresponding to nodes of the MSCT which are selected according to a stability analysis. The last step builds elements which are part of the standard component-tree (at the highest, native scale of the image). The MSCT provides a compact, efficient representation of images compared to the standard (single-scale) component-tree. In particular, the MSCT is especially suited to analyse images containing sparse objects, which require to be represented at a high scale, vs. large background regions that can be losslessly represented at a lower scale. We illustrate the relevance of the MSCT in the context of cellular image segmentation.

Keywords: Component-tree · Multi-scale · Hierarchical representation · Segmentation · Mathematical morphology

1 Introduction

The component-tree [22] is a morphological graph-based model defined by considering the connected components of binary sets obtained from successive thresholdings of an image. Initially proposed in the field of statistics [10], the component-tree has been redefined in the theoretical framework of mathematical morphology and involved, in particular, in the development of morphological tools [4,22]. In particular, the component-tree allows to design connected operators [23] that can process images while preserving contour information.

This work was supported by the French Agence Nationale de la Recherche (ArtIC, Grant ANR-20-THIA-0006, and Grant ANR-23-CE45-0015).

The component-tree led to the development of various filtering and segmentation approaches, mainly by building upon two paradigms: the selection of nodes by attribute filtering [11] or the computation of optimal cuts within the tree [9]. The component-tree was then involved in various application fields, e.g. medical imaging [24], astronomy [1], agriculture imaging [3], microscopic imaging [14].

The popularity of the component-tree relies, on the one hand, on its ability to encode an image in a compact and lossless fashion, and on the other hand to the low complexity of its construction and handling. In particular, the component-tree can be built in quasi-linear time [1,5,16]. Many algorithms have been proposed for its construction, based e.g. on watershed-based, flooding-based or merging-based strategies (see [5] for a survey). More recently, efforts were geared towards its efficient construction in the case of very large images. Götz et al. [8] presented a distributed-memory parallel method for computation of component-trees, while Moschini et al. [15] worked on shared-memory parallel computation at extreme dynamic ranges. Gazagnes et al. [6,7] introduced novel parallel algorithms for max-tree construction on tera-scale images. Recently, Blin et al. [2] offered the first GPU implementation for building a component-tree.

Many variants of the component-tree were proposed over the last years. Some of them are trees, e.g. the hyperconnection tree [21] (that extends the component-tree to the case of hyperconnections), the multivalued component-tree [12] (that handles values endowed with hierarchical orders), the shaping paradigm [25] (that builds component-trees on component-trees), the complete tree of shapes [17] (that unifies the min- and max-trees and links them to the tree of shapes). Others are directed acyclic graphs, e.g. the component graph [19] (that handles values endowed with partial orders), the asymmetric hierarchies [20] (that handle non-symmetric adjacency links). Nonetheless, to the best of our knowledge, there has been no attempt to design variants of the component-tree that deal with the multiscale modeling of an image. The closest works on that topic deal with the notion of component-hypertrees [18], that provide a forest of component-trees induced by a family of increasing connectivities. In this context, the notion of scale was considered at the topological level.

In this article, we aim to design a variant of component-tree that handles the notion of multiple scales at the spatial level. This new tree is called the *Multi-Scale Component-Tree* (MSCT, for brief). By contrast with the standard component-tree, that models the image without taking into account the local informativeness, the MSCT aims to adapt the scale of modeling according to the carried information, with lower (resp. higher) scales where few (resp. many) details/structures of interest are available.

2   Component-Tree

Let $f : \mathbb{Z}^2 \to \mathbb{N}$ be a 2D image. In practice, this image is finite. Without loss of generality, we can them assume that it is defined on a square support $\mathbb{S}_n = [\![0, 2^n - 1]\!]^2 \subset \mathbb{Z}^2$ ($n \in \mathbb{N}$), and then composed of $N = 2^{2n}$ pixels.

We can also assume that it takes its values in a finite subset $\mathbb{V} \subset \mathbb{N}$, that can be chosen as $\mathbb{V} = [\![0, m-1]\!]$. The image f being finite, we assume that $f(x) = 0$ for any $x \notin \mathbb{S}_n$. We endow $\mathbb{N}$ (and thus $\mathbb{V}$) with the standard order $\leqslant$. We endow $\mathbb{Z}^2$ (and thus $\mathbb{S}_n$) with a connectivity framework inherited from the standard adjacencies in digital topology.

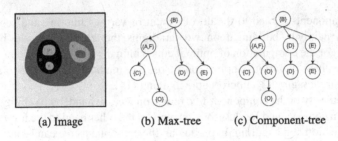

(a) Image   (b) Max-tree   (c) Component-tree

Fig. 1. The max-tree (b) and the component-tree (c) of a grey-level image (a).

For any nonempty subset $X \subseteq \mathbb{Z}^2$, we note $CC(X) \subset 2^{\mathbb{Z}^2}$ the set of the connected components (i.e. the maximal connected subsets) of X.

Let $\lambda \in \mathbb{V}$. We note the upper threshold set of f at value λ as $[f \geqslant \lambda] = \{x \in \mathbb{Z}^2 \mid f(x) \geqslant \lambda\}$. The family $\{CC([f \geqslant \lambda])\}_{\lambda \in \mathbb{V}}$ is increasing. Let $\lambda \in \mathbb{V}$ ($\lambda \neq 0$) and $X_{\lambda+1} \in CC([f \geqslant \lambda + 1])$ be a connected component of the threshold set of f at value $\lambda + 1$. There exists a unique $X_\lambda \in CC([f \geqslant \lambda])$ such that $X_{\lambda+1} \subseteq X_\lambda$.

This hierarchical organisation of the connected components can be represented in an inclusion tree, which is called the max-tree. (By considering the dual order $\geqslant$ on $\mathbb{N}$, one may define the dual min-tree).

More formally, the max-tree is defined as the Hasse diagram of the partially ordered set $(\bigcup_{\lambda \in \mathbb{V}} CC([f \geqslant \lambda]), \subseteq)$. Its root (i.e. its maximum) is the set $\mathbb{Z}^2$, which is the unique connected component of $[f \geqslant 0]$. Its leaves (i.e. its minimal elements) are the flat zones of locally maximal value in the image. An example of max-tree is illustrated in Fig. 1(b) for the image depicted in Fig. 1(a). One may note that a set may be a connected component for many successive threshold sets. While the max-tree only models such an element once, we may also consider this connected component at each value where it appears, leading to a multiset of connected components instead of a set. This paradigm provides a less compact version of the max-tree, that is sometimes called the component-tree, illustrated in Fig. 1(c).

3   Multi-scale Component-Tree

In this section, we explain how to build the Multi-Scale Component-Tree. The purpose of this construction process is to represent background and/or non-relevant parts of the image within flat zones at the lowest scales, while representing relevant/fine detailed parts of the image within flat zones at the highest scales. The proposed method, summarized in Algorithm 1 and Fig. 2, revolves around three steps:

1. downsampling of the gray-scale image (Sect. 3.1 and Algorithm 1, line 2);
2. definition of the *Base Component-Tree* at the lowest scale (Sect. 3.2 and Algorithm 1, line 3);
3. iterative upsampling that promotes significant regions from one scale to the next (Sect. 3.3 and Algorithm 1, lines 4–10).

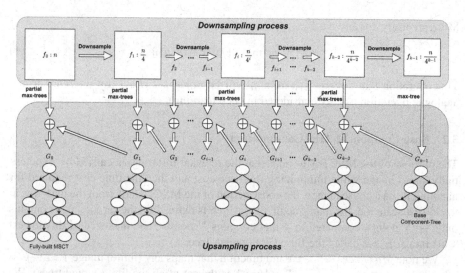

Fig. 2. Illustration of the MSCT construction (see Sect. 3 and Algorithm 1).

Algorithm 1: MSCT construction.

Data: $f : \mathbb{S}_n \to \mathbb{V}$ (gray-scale image), $k \in \mathbb{N}^\star$ number of scales
Result: $G = (V, E)$ Multi-Scale Component-Tree of f

1 **begin**
2     $F \longleftarrow \{f_0, \ldots, f_{k-1}\}$ with $f_0 = f$ and $\forall i \in [\![1, k-1]\!], f_i \longleftarrow MaxPool(f_{i-1})$
3     $G \longleftarrow G_{k-1} \longleftarrow MaxTree(f_{k-1})$
4   **for** i *from* $k - 2$ *down to* 0 **do**
5        $C_i \longleftarrow NodeSelection(G, i)$
6   **foreach** $N \in C_i$ **do**
7           $G_i(N) \longleftarrow PartialMaxTree(f_i, N)$
8           $Merge(G, G_i(N), N)$
9   **end**
10   **end**
11 **end**

3.1 Downsampling Process

The input of the construction procedure is an image $f : \mathbb{S}_n \to \mathbb{V}$ such as defined in Sect. 2. From f, we define a set of downsampled images $F = \{f_i : \mathbb{S}_{n-i} \to \mathbb{V}\}_{i=0}^{k-1}$ where k is the number of scales ($1 \leqslant k \leqslant n$). For each $i \in [\![0, k-1]\!]$, the image f_i is the i-th downsampled version of f, i.e. the image at scale $\frac{1}{2^i}$.

We have $f_0 = f$. For each $i \in [\![0, k-1]\!]$, the image f_{i+1} is obtained from the image f_i by collapsing the 2×2 sets of points of $\mathbb{S}_{n-i}$ into one point of $\mathbb{S}_{n-(i-1)}$. The max-tree is an image model which is especially well-fitted for modeling images where the relevant information is related to the local maxima in the image. Based on this fact, we assume that in the MSCT, it is relevant to preserve the areas of greatest values vs. a background of lower values. As a consequence, we use a maximum policy to define f_{i+1} from f_i.

More precisely, for any $x \in \mathbb{S}_{n-(i+1)}$, we set

$$f_{i+1}(x) = \max_{0 \leqslant a,b,\leqslant 1} \{f_i(2x + (a, b))\} \tag{1}$$

This operation that defines f_{i+1} from f_i will be noted *MaxPool* (by reference to the analogue operation usually considered in deep learning).

3.2 Base Component-Tree Computation

The purpose of the MSCT is to represent the relevant information carried by a (potentially large) image while minimizing its space cost and then the time cost required for handling it. As a consequence, the construction of the MSCT starts from the max-tree at the lowest scale, i.e. the max-tree of f_{k-1}, which is defined on the set $\mathbb{S}_{k-1}$ that contains only $\frac{1}{4^{k-1}}N$ points, compared to f that contains N points. The max-tree of this lowest scale image f_{k-1} is called the Base Component-Tree.

We note $MaxTree : \mathbb{V}^{\mathbb{Z}^2} \to \mathbb{T}$ the function that maps any (finite) image $f : \mathbb{Z}^2 \to \mathbb{V}$ onto its max tree $G = (V, E) \in \mathbb{T}$ (where $\mathbb{T}$ is the set of all the finite rooted trees). We recall that the computational cost of optimal algorithms that implement $MaxTree$ is $O(n \log n)$ where n is the number of points of the image support [5].

Here, the Base Component-Tree $G_{k-1} = (V_{k-1}, E_{k-1}) \in \mathbb{T}$ is computed from the image $f_{k-1} : \mathbb{S}_{k-1} \to \mathbb{V}$, and the induced time cost is then $\frac{N}{4^{k-1}} \log \frac{N}{4^{k-1}}$.

3.3 Upsampling Process

The MSCT $G = (V, E)$ is initialized as the Base Component-Tree G_{k-1} of f_{k-1}. At this stage, the whole image f, including both informative and non-informative parts, is represented at the lowest scale. The purpose is now to modify this tree G in order to represent the informative regions of the image at higher scales, according to their degree of relevance.

This process is carried out iteratively, scale by scale, from the lower to the higher (Algorithm 1, line 4). At each iteration/scale $i \in [\![0, k-2]\!]$, a set of nodes $C_i \subseteq V$ is selected from the tree G. These nodes are those assumed as containing a relevant information that motivates the computation of a local max-tree on their associated region. The procedure of selecting these nodes in G at scale i (noted $NodeSelection(G, i)$ in Algorithm 1, line 5) is described in the next subsection.

For each selected node $N \in C_i$ (Algorithm 1, line 6), the max-tree of the image f_i restricted to the region $N \subseteq \mathbb{S}_{n-i}$ is computed. This "local" max-tree computation, noted $PartialMaxTree(f_i, N)$ is nothing but the computation of a standard max-tree on a given (connected) region N. Note in particular that algorithmically, the behaviour of $MaxTree(f_{k-1})$ (Algorithm 1, line 3) is the same as $PartialMaxTree(f_k, \mathbb{S}_{n-k-1})$ (In our algorithmic scheme, we used a version of Najman and Couprie's algorithm [16][1]). Once this new partial max-tree $G_i(N) = (V_i(N), E_i(N))$ is computed, it must be embedded in the MSCT G. This embedding and its side effects on the structure of G (Algorithm 1, line 8) are detailed at the end of this section.

---

[1] Although we only consider integer values, Najman and Couprie's algorithm uses Tarjan's Union-Find method and is able to efficiently process floating-point values as well.

Nodes Selection for Local Upsampling. At each iteration/scale i, a set C_i of nodes is selected for an upsampling procedure. The choice of the most relevant nodes is carried out based on a priority score assigned to each node $N \in V$ of the current MSCT $G = (V, E)$. This score is computed based on the notion of Maximally Stable Extremal Regions (MSER) [13]. The MSER stability value of a node $N \in V$, that belongs to a threshold set at value $v \in \mathbb{V}$ is defined as $MSER(N) = \frac{|N_{-\varDelta}| - \sum_p |N^p_{+\varDelta}|}{|N|}$ where $N_{-\varDelta} \in V$ is the ancestor node at value $v - \varDelta$ such that $N \subseteq N_{-\varDelta}$, and the $N^p_{+\varDelta} \in V$ are all the descendant nodes of N at value $v + \varDelta$, i.e. such that $N \supseteq N^p_{+\varDelta}$. The nodes of V are then sorted by decreasing stability.

The set of selected nodes is defined by choosing the nodes by decreasing stability in the list. The selected nodes have to be non-overlapping, which means that for any two distinct nodes $N_1, N_2 \in C_i$ we have neither $N_1 \subseteq N_2$ nor $N_2 \subseteq N_1$. Indeed, each of these nodes will become the root of a partial tree that will be embedded in the MSCT G. It is then required that none of these new trees be part of another. As a consequence the ancestors and descendants of the selected nodes are progressively discarded from the list. The selection ends once the list is empty.

Local Upsampling. For each node $N \in V$ selected from the above process, a max-tree $G_i(N) = (V_i(N), E_i(N))$ of the image f_i restricted to the support N is computed. This new max-tree $G_i(N)$ is then dedicated to replace the current subtree of G starting at node N. However, the node N which is the root of the subtree to be replaced is defined at a given value $v \in \mathbb{V}$, and its parent node N' is defined at a value $v' \in \mathbb{V}$ with $v' < v$. By contrast, the new partial tree built in the region N has its root at a value lower than v. In this context, the tree $G_i(N) = (V_i(N), E_i(N))$ has to be split into two parts, that must be processed distinctly. On the one hand, all the nodes $N_u \in V_i(N)$ that are defined for a value $u \in \mathbb{V}$ with $u \leqslant v'$ have to be merged with the ancestor node $N'_u \in V$ at value u, which is the ancestor of N in G. The "new" version of the node N'_u then corresponds to $N'_u \cup N_u$. On the other hand, all the nodes $N_u \in V_i(N)$ such that their parent node $N_w \in V_i(N)$ is defined for a value $w \leqslant v'$ now become the root of a partial tree of f_i restricted to N_u, and this root N_u has to be connected to the node $N_{v'}$. In other words, the node N is replaced by a forest extracted from its max-tree $G_i(N)$, that is connected to the parent node of N, whereas the remainder of the tree is absorbed by the branch of G located between N and the root. This policy leads in particular to a structure of MSCT where each node may encode pixels at different scales. An example of such replacement is illustrated in Fig. 3.

4 Object Segmentation Using the MSCT

The built MSCT $G = (V, E)$ now contains a hierarchy of flat-zones with pixels at different scales. The MSCT can be used, in particular, for segmentation tasks. A possible use-case is cell segmentation, where high contrast denotes the presence of a bio-marker (foreground) while low contrast denotes its absence (background). We propose a segmentation method illustrated by Algorithm 2 which involves three steps. First, a set of nodes of interest is extracted from the MSCT (line 3). Second, for each node of interest,

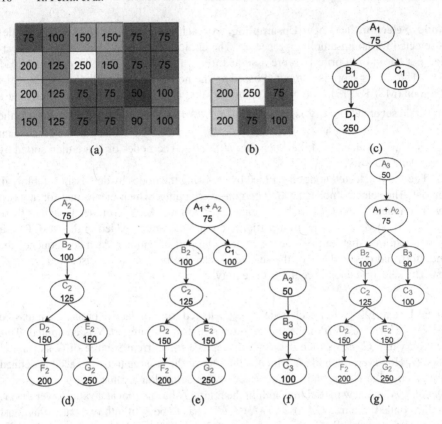

Fig. 3. Example of MSCT sub-tree replacement during upsampling. (a) is the input image and (b) is the downsampled image at half scale. (c) is the max-tree of (b) also referred to as the base component-tree of the MSCT. (d) is the partial max-tree of the red L-shaped area of (a). (e) is the merging of the partial max-tree (d) on the MSCT (c). (f) is the partial max-tree of the blue square-shaped area of (b). (g) is the merging of the MSCT (e) with the partial max-tree (f) and is also the fully-built MSCT. It is important to note that (f) introduces gray-levels that do not exist in the MSCT, even changing the root of the MSCT (e) (levels 50 and 90) prior to being merged with (f). (Color figure online)

its local sub-tree is filtered (line 4) according to a maximum MSER value. The filtered nodes are intermediate clusters containing either objects or clusters of touching objects. Third, each cluster is cleaned to remove the background around the object(s) (line 6) and then undergoes a watershed to separate the cluster into individual objects (line 7).

4.1 Nodes Selection for Segmentation

The retrieval of objects of interest is analogous to the upsampling process described in Sect. 3.3, given that the extracted flat-zones inherently encompass these objects. Subsequently, the same function is employed to define a set of disjoint sub-trees, facilitating the separation into distinct objects (line 3).

Algorithm 2: MSCT segmentation

Data: $G = (V, E)$ MSCT, $MSER_{max} \in \mathbb{N}$ maximum MSER value, $f : \mathbb{Z}^2 \rightarrow \mathbb{N}$ input
image, $n \in \mathbb{N}$ subdivision factor

Result: $P = \{P_0, \ldots, P_k\}$

1 **begin**
2      $P \leftarrow \emptyset$
3      $C_0 \leftarrow UpsampleNodeSelection(G)$
4      $C'_0 \leftarrow FilterTree(G, C_0, MSER_{max})$
5   **for** $c \in C'_0$ **do**
6          $c' \leftarrow FillHoles(Otsu(GaussianFilter(C_0)))$
7          $S \leftarrow Watershed(f, c, UltimateErosion(c'))$
8          $P \leftarrow P \cup S$
9   **end**
10 **end**

4.2 MSCT Filtering

A first cluster separation can be performed by filtering nodes of the MSCT. A simplified version of the MSCT sub-tree is computed. It consists of reducing each branch to a single node representing a local flat-zone and being assigned the minimum MSER value among all nodes of the branch. This simplified tree is filtered using a maximum MSER value criterion (line 4). The function *KeepNode* (Eq. (2)) is recursively called on each node of the sub-tree starting at its root. Leaves of the filtered sub-tree are the intermediate clusters.

$$KeepNode(n, v) = \begin{cases} True, & \text{if } n.nChildren = 0 \wedge n.mser \leq v \\ True, & \text{if } n.children > 0 \wedge n.mser \leq v \wedge \forall n' \in \\ & n.children, n'.mser \leq v \\ False, & \text{otherwise} \end{cases} \quad (2)$$

4.3 Object Segmentation

Each intermediate cluster $c \in C'_0$ is processed separately (lines 5–9). In a first step, the intermediate cluster is cleaned to retrieve the exact contour of its underlying object(s) and get rid of background pixels (line 6). The presence of these background pixels is a side effect of the upsampling steps due to the maximum nature of the downsampling operations. They are removed using a Gaussian filter followed by a Otsu thresholding and a hole filling operation. In a second step, the number of individual objects inside an intermediate cluster is estimated by performing an ultimate erosion on its previously cleaned flat-zone (line 7). The centroid of each ultimate erosion connected component is then used to initialize a watershed algorithm resulting in a complete partition of said intermediate cluster. The set of all watershed partitions of all intermediate clusters P is the final segmentation result, with each partition being a connected component representing one individual object.

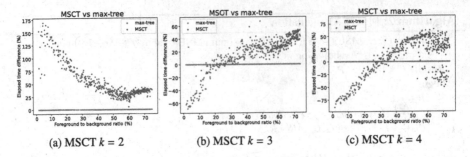

(a) MSCT $k = 2$ (b) MSCT $k = 3$ (c) MSCT $k = 4$

Fig. 4. Computing-time differences for several downsampling factors. The MSCT (blue curves) are shown relative to the max-trees (red curves). (Color figure online)

5   Experimental Results

5.1   Implementation

To assess the validity of the Multi-Scale Component-Tree and its ability to produce satisfactory segmentations of cellular images, we have implemented it in Python. In this section, we discuss important choices regarding algorithms and global parameters.

The main parameter to build the MSCT is the number of scales k that also defines the number of downsampling and upsampling steps. It defines the size of the image f_{k-1} used to build the Base Component-Tree G_{k-1}. The MSER parameter δ is computed relative to the tree height at scale i (cardinal of the support set of f_i) using a parameter $MSER_{height}$. Node selection is constrained by two parameters: the maximum area nodes is set to A_{max} while the maximum stability of nodes is set to $MSER_{select}$. Regarding the segmentation, $MSER_{filter}$ sets the maximum stability of nodes when filtering the tree.

5.2   Computational Cost Evaluation

The computation time required to build the MSCT depends on multiple factors: the size of the input image, the number of objects or clusters of objects and their respective size. An experiment (Fig. 4) is made by generating synthetic images of a fixed size composed of a variable number of bright objects on a dark background. We can observe that the MSCT computation outperforms the regular max-tree when 30% of the image surface consists of objects, using a downsampling factor $k > 2$ (Figs. 4(b) and 4(c)). When using a downsampling of $k = 2$, the size of the downsampled image f_1 upon which the Base Component-Tree is computed is only $\frac{N}{4}$ and the cost of choosing and computing partial max-trees outweights the benefit of working with downsampled images (Fig. 4a). This illustrates the effectiveness of building the Base Component-Tree (Sect. 3.2) on a low scale version of the image especially when said image is sparse.

5.3   Space Complexity Analysis

To complete the second experiment in Sect. 5.2, the number of created nodes and stored pixels have been measured for growing numbers of objects in the image. We can observe

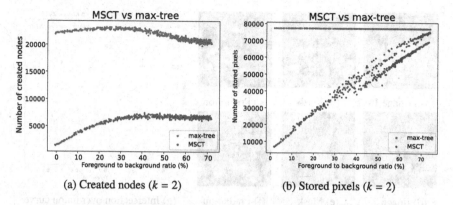

(a) Created nodes ($k = 2$)   (b) Stored pixels ($k = 2$)

Fig. 5. Space complexity measures.

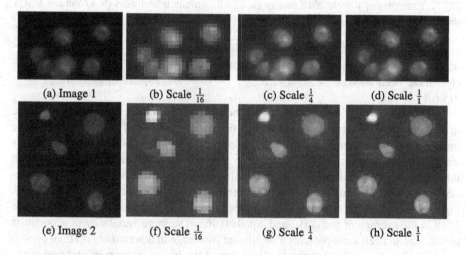

(a) Image 1   (b) Scale $\frac{1}{16}$ (c) Scale $\frac{1}{4}$ (d) Scale $\frac{1}{1}$

(e) Image 2   (f) Scale $\frac{1}{16}$ (g) Scale $\frac{1}{4}$ (h) Scale $\frac{1}{1}$

Fig. 6. Examples of images with their reconstruction after each upsampling step.

that the number of created nodes is significantly lower for the MSCT (blue curve in Fig. 5(a)) compared to a regular max-tree (red curve in Fig. 5(a)). The number of pixels is roughly linear with respect to the sparsity of the image for the MSCT (blue curve in Fig. 5(b)) as opposed to constant and equal to the image size for the max-tree (red curve in Fig. 5(b)). The reason behind the low number of pixels resides in the multi-scale nature of those pixels as only highly-contrasted flat-zones are promoted from low scale to high scale during the upsampling process (Sect. 3.3).

5.4 Kaggle 2018

Our method has been tested on the Kaggle 2018 Data Science Bowl dataset[2]. It contains gray-scale images of various sizes, blurriness, cell types and sizes. Only gray-scale

---

[2] www.kaggle.com/competitions/data-science-bowl-2018.

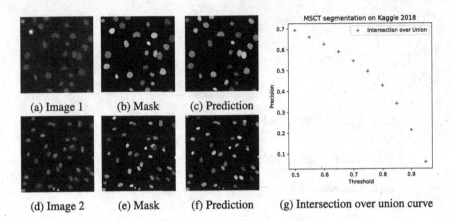

(a) Image 1   (b) Mask   (c) Prediction

(d) Image 2   (e) Mask   (f) Prediction   (g) Intersection over union curve

Fig. 7. Examples of Kaggle 2018 images (a, d) with ground truth masks (b, e) and the MSCT segmentation results (c, f).

images with bright objects on dark background are considered. The following parameters have been used: $k = 3$, $MSER_{height} = 10\%$, $A_{max} = 10\%$, $MSER_{select} = 1$, $MSER_{filter} = 1$.

Some examples of input images and their reconstruction at different steps of the upsampling process are illustrated in Fig. 6. A total number of $n = 547$ gray-scale images have been segmented using the method described in Sect. 4. Figure 7 shows some segmentation results. Kaggle uses an intersection over union (IoU) metric with thresholds between 0.50 and 0.95. A prediction is deemed a true positive if the intersection IoU score of its mask with the ground truth mask is at least equal to the threshold. Figure 7(g) shows the curve of precision per threshold. The global score for Kaggle 2018 is the average of the curve and equals 0.465. The large variability in acquisition techniques and cell sizes presents a significant challenge in determining optimal parameters. Depending on the specific configuration of the image under consideration, these methods frequently result in either under-segmentation or over-segmentation.

6   Discussion

We have introduced the concept of the Multi-Scale Component-Tree (MSCT) and presented an algorithm for its construction on gray-scale images. Our results demonstrate the MSCT ability in storing flat-zones across various scales and its enhanced capability to distinguish between flat-zones encompassing objects of interest and those constituting the background, especially when compared to its single-scale counterpart. The efficiency of the MSCT has been further exemplified through its application in a cellular segmentation task. More complex filtering schemes could be applied prior to the segmentation step. Other attributes could be used as well such as the border gradient, complexity or compactness. When dealing with non-increasing attributes, shaping could be employed, that is building a component-tree of the MSCT, filtering it and reconstructing the MSCT. In terms of efficient building, the upsampling step could benefit from parallel computing as all selected sub-trees are mutually disjoint by virtue of the selection function. Indeed, partial max-trees could be computed in parallel on different sets of pixels and only the replacement step in the MSCT has to be performed sequentially. Other segmentation methods could be defined to avoid resorting to use watershed like an ellipse fitting model that might offer improvements for the given examples.

References

1. Berger, C., Géraud, T., Levillain, R., Widynski, N., Baillard, A., Bertin, E.: Effective component tree computation with application to pattern recognition in astronomical imaging. In: ICIP, pp. 41–44 (2007)
2. Blin, N., Carlinet, E., Lemaitre, F., Lacassagne, L., Geraud, T.: Max-tree computation on GPUs. IEEE Trans. Parallel Distrib. Syst. **33**, 3520–3531 (2022)
3. Bosilj, P., Duckett, T., Cielniak, G.: Connected attribute morphology for unified vegetation segmentation and classification in precision agriculture. Comput. Ind. **98**, 226–240 (2018)
4. Breen, E.J., Jones, R.: Attribute openings, thinnings, and granulometries. Comput. Vis. Image Underst. **64**(3), 377–389 (1996)
5. Carlinet, E., Géraud, T.: A comparative review of component tree computation algorithms. IEEE Trans. Image Process. **23**, 3885–3895 (2014)
6. Gazagnes, S., Wilkinson, M.H.F.: Distributed component forests in 2-D: hierarchical image representations suitable for tera-scale images. Int. J. Pattern Recogn. Artif. Intell. **33**, 1940012 (2019)
7. Gazagnes, S., Wilkinson, M.H.F.: Distributed connected component filtering and analysis in 2D and 3D tera-scale data sets. IEEE Trans. Image Process. **30**, 3664–3675 (2021)
8. Götz, M., Cavallaro, G., Géraud, T., Book, M., Riedel, M.: Parallel computation of component trees on distributed memory machines. IEEE Trans. Parallel Distrib. Syst. **29**, 2582–2598 (2018)
9. Guigues, L., Cocquerez, J.P., Le Men, H.: Scale-sets image analysis. Int. J. Comput. Vis. **68**, 289–317 (2006)
10. Hartigan, J.A.: Statistical theory in clustering. J. Classif. **2**, 63–76 (1985)
11. Jones, R.: Connected filtering and segmentation using component trees. Comput. Vis. Image Underst. **75**, 215–228 (1999)
12. Kurtz, C., Naegel, B., Passat, N.: Connected filtering based on multivalued component-trees. IEEE Trans. Image Process. **23**, 5152–5164 (2014)
13. Matas, J., Chum, O., Urban, M., Pajdla, T.: Robust wide-baseline stereo from maximally stable extremal regions. Image Vis. Comput. **22**, 761–767 (2004)
14. Meyer, C., Baudrier, É., Schultz, P., Naegel, B.: Combining max-tree and CNN for segmentation of cellular FIB-SEM images. In: Kerautret, B., Colom, M., Krähenbühl, A., Lopresti, D., Monasse, P., Perret, B. (eds.) RRPR 2022. LNCS, vol. 14068, pp. 77–90. Springer, Cham (2023). https://doi.org/10.1007/978-3-031-40773-4_7
15. Moschini, U., Meijster, A., Wilkinson, M.H.F.: A hybrid shared-memory parallel max-tree algorithm for extreme dynamic-range images. IEEE Trans. Pattern Anal. Mach. Intell. **40**, 513–526 (2018)
16. Najman, L., Couprie, M.: Building the component tree in quasi-linear time. IEEE Trans. Image Process. **15**, 3531–3539 (2006)
17. Passat, N., Mendes Forte, J., Kenmochi, Y.: Morphological hierarchies: a unifying framework with new trees. J. Math. Imaging Vis. **65**(5), 718–753 (2023)
18. Passat, N., Naegel, B.: Component-hypertrees for image segmentation. In: Soille, P., Pesaresi, M., Ouzounis, G.K. (eds.) ISMM 2011. LNCS, vol. 6671, pp. 284–295. Springer, Heidelberg (2011). https://doi.org/10.1007/978-3-642-21569-8_25

19. Passat, N., Naegel, N.: Component-trees and multivalued images: structural properties. J. Math. Imaging Vis. **49**, 37–50 (2014)
20. Perret, B., Cousty, J., Tankyevych, O., Talbot, H., Passat, N.: Directed connected operators: asymmetric hierarchies for image filtering and segmentation. IEEE Trans. Pattern Anal. Mach. Intell. **37**, 1162–1176 (2015)
21. Perret, B., Lefèvre, S., Collet, C., Slezak, É.: Hyperconnections and hierarchical representations for grayscale and multiband image processing. IEEE Trans. Image Process. **21**, 14–27 (2012)
22. Salembier, P., Oliveras, A., Garrido, L.: Anti-extensive connected operators for image and sequence processing. IEEE Trans. Image Process. **7**, 555–570 (1998)
23. Salembier, P., Serra, J.: Flat zones filtering, connected operators, and filters by reconstruction. IEEE Trans. Image Process. **4**, 1153–1160 (1995)
24. Wilkinson, M.H.F., Westenberg, M.A.: Shape preserving filament enhancement filtering. In: Niessen, W.J., Viergever, M.A. (eds.) MICCAI 2001. LNCS, vol. 2208, pp. 770–777. Springer, Heidelberg (2001). https://doi.org/10.1007/3-540-45468-3_92
25. Xu, Y., Géraud, T., Najman, L.: Connected filtering on tree-based shape-spaces. IEEE Trans. Pattern Anal. Mach. Intell. **38**, 1126–1140 (2016)

An Approach to Colour Morphological Supremum Formation Using the LogSumExp Approximation

Marvin Kahra[1]([✉]), Michael Breuß[1], Andreas Kleefeld[2,3], and Martin Welk[4]

[1] Institute for Mathematics, Brandenburg University of Technology
Cottbus-Senftenberg, 03046 Cottbus, Germany
{marvin.kahra,breuss}@b-tu.de

[2] Forschungszentrum Jülich GmbH, Jülich Supercomputing Centre,
Wilhelm-Johnen-Straße, 52425 Jülich, Germany
a.kleefeld@fz-juelich.de

[3] Faculty of Medical Engineering and Technomathematics, University of Applied
Sciences Aachen, Heinrich-Mußmann-Straße 1, 52428 Jülich, Germany

[4] UMIT TIROL – Private University for Health Sciences and Health Technology,
Eduard-Wallnöfer-Zentrum 1, 6060 Hall/Tyrol, Austria
martin.welk@umit-tirol.at

Abstract. Mathematical morphology is a part of image processing that has proven to be fruitful for numerous applications. Two main operations in mathematical morphology are dilation and erosion. These are based on the construction of a supremum or infimum with respect to an order over the tonal range in a certain section of the image. The tonal ordering can easily be realised in grey-scale morphology, and some morphological methods have been proposed for colour morphology. However, all of these have certain limitations.

In this paper we present a novel approach to colour morphology extending upon previous work in the field based on the Loewner order. We propose to consider an approximation of the supremum by means of a log-sum exponentiation introduced by Maslov. We apply this to the embedding of an RGB image in a field of symmetric 2×2 matrices. In this way we obtain nearly isotropic matrices representing colours and the structural advantage of transitivity. In numerical experiments we highlight some remarkable properties of the proposed approach.

Keywords: mathematical morphology · colour image · matrix-valued image · symmetric matrix · transitivity

1 Introduction

Mathematical morphology is a theory used to analyse spatial structures in images. Over the decades it has developed into a very successful field of image processing, see e.g. [16–18] for an overview. Morphological operators basically

© The Author(s), under exclusive license to Springer Nature Switzerland AG 2024
S. Brunetti et al. (Eds.): DGMM 2024, LNCS 14605, pp. 325–337, 2024.
https://doi.org/10.1007/978-3-031-57793-2_25

consist of two main components. The first of these is the structuring element (SE), which is characterised by its shape, size and position. These in turn can be divided into two types of SEs, flat and non-flat, cf. [9]. A flat SE basically defines a neighbourhood of the central pixel where appropriate morphological operations are performed, while a non-flat SE also contains a mask with finite values that are used as additive offsets. The SE is usually implemented as a window sliding over the image. The second main component is used to perform a comparison of values within a SE. Two basic operations in mathematical morphology are dilation and erosion, where a pixel value is set to the maximum and minimum, respectively, of the discrete image function within the SE. Many morphological filtering procedures of practical interest, such as opening, closing or top hats, can be formulated by combining dilation and erosion operations. Since dilation and erosion are dual operations, it is often sufficient to restrict oneself to one of the two when constructing algorithms.

Let us briefly discuss how the operations of dilation and erosion can be applied to colour morphology, as it is the underlying concept for our further considerations. As already mentioned, one important operation in morphology is to perform a comparison of the tonal values within the SE. For the simpler application areas such as grey value morphology, one can act directly on complete lattices in order to obtain a total order of the colour values, cf. [4,18]. In the case of colour morphology, this is no longer the case, as there is no total order of the colour values. For this reason, corresponding half-orders and different basic structures are used, cf. [3]. The first approach that could be used for this is to regard each colour channel of an image as an independent image and to perform grey value morphology on each of them. The other approach uses a vector space structure where each colour is considered as a vector in an underlying colour space. This can take the form of [1,2] or [13]. We will take the latter approach, but use symmetric matrices instead of vectors, and order the elements by means of a half-order, namely the Loewner order (see [6]).

To calculate the basic operations of colour morphology, the supremum or infimum is determined instead of the maximum or minimum. There are also different approaches on how to choose the supremum of symmetric matrices. To give some examples, we mention here the nuclear norm, the Frobenius norm and the spectral norm. For a comparison of these norms we refer to the work [21] by Welk, Kleefeld and Breuß.

In this paper, however, we want to consider a different approach, namely the approximation of the supremum by means of a log-sum exponentiation. The reason why we follow this approach is that it is an approximation which already gave promising results in the work by Kahra, Sridhar and Breuß (see [12]) for grey scale images in connection with a fast Fourier transform. This approach was also used for colour images in [20], but only in the above described sense of a channel-wise structuring of colour images. Another connection worth mentioning is with the paper of Burgeth, Welk, Feddern and Weickert (see [8]), where root and power were used for symmetric positive semidefinite matrices instead of logarithm and exponential function. However, our approach here has the advantage of transitivity, which is an important property that is necessary in grey-scale morphology to define the basic concepts of dilation/erosion (see [10]).

This paper will be the first step for transferring the LogSumExp approach to colour morphology with tonal vectors/matrices. The goal is to present a first introduction to the characterisation of this approach for tonal value matrices and a comparison with some of the other already existing approaches.

2 General Definitions

2.1 Colour Morphology

In order to make this paper self-contained, we briefly recall some basics of mathematical morphology in general and colour morphology in particular.

For this purpose, we first consider a two-dimensional, discrete image domain $\Omega \subseteq \mathbb{Z}^2$ and a single-channel grey-scale image, which is explained by a function $f : \Omega \to [0, 255]$. In the case of non-flat morphology, the SE can be represented as a function $b : \mathbb{Z}^2 \to \mathbb{R} \cup \{-\infty\}$ with

$$b(\boldsymbol{x}) := \begin{cases} \beta(\boldsymbol{x}) & , \ \boldsymbol{x} \in B_0, \\ -\infty & , \ \text{otherwise}, \end{cases} \quad B_0 \subset \mathbb{Z}^2,$$

where B_0 is a set centred at the origin. In the case of a flat filter (flat morphology), it is simply the special case $\beta(\boldsymbol{x}) = 0$. The most elementary operations of mathematical morphology are dilation and erosion. They are defined as

$$(f \oplus b)(\boldsymbol{x}) := \max_{\boldsymbol{u} \in \mathbb{Z}^2} \{f(\boldsymbol{x} - \boldsymbol{u}) + b(\boldsymbol{u})\}$$

$$(f \ominus b)(\boldsymbol{x}) := \min_{\boldsymbol{u} \in \mathbb{Z}^2} \{f(\boldsymbol{x} + \boldsymbol{u}) - b(\boldsymbol{u})\}.$$

With these two operations, many other operations can be defined that are of great interest in practice, for example opening $f \circ b = (f \ominus b) \oplus b$ and closing $f \bullet b = (f \oplus b) \ominus b$.

We now come to our real area of interest, namely colour morphology. This is similar to the grey-scale morphology already shown, with the difference that we no longer have just one channel, but three. There are many useful formats for expressing this, see [19]. A classic approach in this sense is the channel-by-channel processing of an image with the RGB colour model, see e.g. [20] for a recent example of channel-wise scheme implementation. However, instead of RGB vectors, we will use symmetric, positive semi-definite 2×2 matrices. For this we assume that the colour values are already normalised to the interval $[0, 1]$. First, we transfer this vector into the HCL colour space by means of $M = \max\{r, g, b\}$, $m = \min\{r, g, b\}$, $c = M - m$, $\ell = \frac{1}{2}(M + m)$ and

$$h = \begin{cases} \frac{g-b}{6c} \quad \text{mod } 1 & , \ \text{if } M = r, \\ \frac{b-r}{6c} + \frac{1}{3} \quad \text{mod } 1 & , \ \text{if } M = g, \\ \frac{r-g}{6c} + \frac{2}{3} \quad \text{mod } 1 & , \ \text{if } M = b. \end{cases}$$

Then we replace the luminance ℓ with $\tilde{\ell} = 2\ell - 1$ and consider the quantities c, $2\pi h$ and $\tilde{\ell}$ as radial, angular and axial coordinates of a cylindrical coordinate system, respectively. Since the transformation shown here maps each colour from the RGB colour space one-to-one to a colour in the HCL bi-cone, this represents a bijection onto the bi-cone, which in turn is interpreted with Cartesian coordinates by $x = c \cdot \cos(2\pi h)$, $y = c \cdot \sin(2\pi h)$ and $z = \tilde{\ell}$. Finally, we map these Cartesian coordinates onto a symmetric matrix as follows:

$$A := \frac{\sqrt{2}}{2} \begin{pmatrix} z - y & x \\ x & z + y \end{pmatrix},$$

where the complete transformation process is a bijective mapping, see [6].

2.2 The Loewner Order and the Decision of a Minimiser

Since there is no classical ordering of the elements in $\mathbb{R}^3$ or $\mathbb{R}^{2\times 2}$, we need to ask when an element is larger or smaller than another element. To answer this question, we need to use a weaker definition of an ordering relation, i.e. a half-order. For this reason, we resort to the promising Loewner order and the colour morphological processing based on this, which was already presented by Burgeth and Kleefeld (see [7]).

For two symmetric matrices A and B, the Loewner order $\leq_L$ or $<_L$ is defined as

$$A <_L B \ (A \leq_L B) \iff B - A \text{ is positive (semi-)definite.}$$

The problem with this semi-order is that it is not a lattice order [5] and therefore it is not possible to find a unique matrix maximum or supremum. To get around this problem, one needs another property to select a uniquely determined matrix maximum from the convex set of symmetric matrices $\mathcal{U}(\mathcal{X})$ that are upper bounds in the Loewner sense for the multi-set $\mathcal{X} = \{X_1, X_2, \ldots, X_n\}$ of given data matrices of the set $\mathrm{Sym}(2)$ of symmetric real 2×2 matrices with

$$\mathcal{U}(\mathcal{X}) := \{Y \in \mathrm{Sym}(2) : X \leq_L Y \ \ \forall X \in \mathcal{X}\},$$

see [21]. For representation of this property, we use the function $\varphi : \mathcal{U}(\mathcal{X}) \to \mathbb{R}$, which is convex and Loewner-monotone, i.e.

$$\varphi(A) \leq \varphi(B) \iff A \leq_L B.$$

Furthermore, φ should have a unique minimiser in $\mathcal{U}(\mathcal{X})$, then we can define the φ-supremum of $\mathcal{X}$ as said minimiser:

$$\mathrm{Sup}_{\varphi}(\mathcal{X}) := \arg\min_{Y \in \mathcal{U}(\mathcal{X})} \varphi(Y).$$

The matrix supremum introduced in the paper [8] is based on the calculation of the trace and is therefore also called trace-supremum. That is, one has $\varphi(Y) = \mathrm{tr}\, Y$ and we get as supremum:

$$\mathrm{Sup}_{\mathrm{tr}}(\mathcal{X}) := \arg\min_{Y \in \mathcal{U}(\mathcal{X})} \mathrm{tr}\, Y.$$

Based on the corresponding norms, the Frobenius supremum

$$\mathrm{Sup}_2(\mathcal{X}) := \underset{Y \in \mathcal{U}(\mathcal{X})}{\arg\min} \sum_{X \in \mathcal{X}} \|Y - X\|_2$$

and the spectral supremum

$$\mathrm{Sup}_\infty(\mathcal{X}) := \underset{Y \in \mathcal{U}(\mathcal{X})}{\arg\min} \sum_{X \in \mathcal{X}} |\lambda_1(Y - X)|,$$

where $\lambda_1(A)$ denotes the largest eigenvalue of A, were derived in [21]. Note that, in the case of positive semidefinite matrices, all three norms are Schatten norms $\|\cdot\|_p$ for $p \in \{1, 2, \infty\}$.

3 Mathematical Background of the Log-Exp-Supremum

In this section, we will give a brief introduction to the construction of a characterisation for the so-called log-exp-supremum and summarise some of its properties, but will refer to a future paper for the proofs.

Let the multi-set $\mathcal{X} = \{X_1, \ldots, X_n\}$, $n \in \mathbb{N}$, of symmetric real 2×2 matrices represent our colour values in the considered neighbourhood. Then, we define the **log-exp-supremum (LES)** as

$$S := \mathrm{Sup}_{\mathrm{LE}}(\mathcal{X}) := \lim_{m \to \infty} \left(\frac{1}{m} \log \sum_{i=1}^{n} \exp(m X_i) \right) \tag{1}$$

for the dilation and erosion follows by duality. For the calculation of the LES, we will use a characterisation of the form

$$S = \lambda_1 u_1 u_1^{\mathrm{T}} + \mu v_1 v_1^{\mathrm{T}}, \tag{2}$$

where λ_1 is the largest eigenvalue of the matrices of $\mathcal{X}$ with the corresponding normalised eigenvector u_1, v_1 is the normalised eigenvector perpendicular to u_1 and μ is the next largest eigenvalue of $\mathcal{X}$ with an eigenvector not aligned with u_1 if λ_1 is unique, otherwise $\mu = \lambda_1$ holds. Note that μ does not have to be the corresponding eigenvalue to the eigenvector v_1. The characterisation is based on the spectral decomposition of the input X_i:

$$X_i = \lambda_i u_i u_i^{\mathrm{T}} + \mu_i v_i v_i^{\mathrm{T}}, \quad \lambda_i \geq \mu_i, \quad \langle u_i, v_i \rangle = 0, \quad |u_i| = 1 = |v_i|,$$
$$u_i = (c_i, s_i)^{\mathrm{T}}, \quad v_i = (-s_i, c_i)^{\mathrm{T}}, \quad c_i = \cos(\varphi_i), \quad s_i = \sin(\varphi_i), \tag{3}$$
$$\varphi_i \in \left[-\frac{\pi}{2}, \frac{\pi}{2} \right], \quad i = 1, \ldots, n,$$

where $\lambda_i, \mu_i \in \mathbb{R}$ are the eigenvalues of X_i and $u_i \perp v_i$ are the corresponding unit eigenvectors. Additionally, we used for the Eq. (2) that without loss of generality the matrix X_1 has (one of) the largest eigenvalues of $\mathcal{X}$. The proof exploits the spectral decomposition (3) and certain properties of the Rayleigh product (see [11]) and symmetric 2×2 matrices. However, we will not go into the proof of the LES characterisation (2) here, as it would strain the scope of this paper too much. Instead, we intend to do this in a separate paper and refer to the results of our experiments for now.

We now turn our attention to some of the properties of (1). First of all, we need to clarify whether it is actually a supremum in the Loewner sense. In fact, it is easy to show that it is an upper bound:

Lemma 1. *The LES S according to (1) is an upper bound in the Loewner sense for the given matrices $\mathcal{X}$.*

Proof. We certainly have

$$\sum_{i=1}^{n} \exp(m\boldsymbol{X}_i) \geq_L \exp(m\boldsymbol{X}_j) \quad \forall j \in \{1, \dots, n\},$$

which in combination with the fact that the logarithm is an operator-monotone function (see [14]), i.e. $\boldsymbol{A} \leq_L \boldsymbol{B} \implies \log \boldsymbol{A} \leq_L \log \boldsymbol{B}$ for symmetric positive definite matrices $\boldsymbol{A}, \boldsymbol{B}$, results in $\boldsymbol{S} \geq_L \boldsymbol{X}_j$ for all $j \in \{1, \dots, n\}$. $\square$

We already mentioned in the previous section that it is necessary to have another property to select a uniquely determined maximum from the convex set $\mathcal{U}(\mathcal{X})$. This should be represented by a function φ, which should have a unique minimiser in $\mathcal{U}(\mathcal{X})$. Unfortunately, there is no such total ordering function φ on $\mathcal{U}(\mathcal{X})$ for which $\boldsymbol{S}$ is the unique minimiser. However, by defining the **p-power upper bound cone** as

$$\mathcal{U}_p(\mathcal{X}) := (\mathcal{U}(\mathcal{X}^p))^{\frac{1}{p}} = \left\{ \boldsymbol{Y} \in \mathrm{Sym}(2) : \boldsymbol{Y}^p \in \mathcal{U}(\mathcal{X}^p) \right\} \tag{4}$$

for the element-wise application of the p-th power to the multi-set $\mathcal{X}$ in form of $\mathcal{X}^p := \{\boldsymbol{X}^p : \boldsymbol{X} \in \mathcal{X}\}$, we can slightly modify the problem to obtain a unique minimiser. To be exact, we denote the intersection of all p-power upper bound cones as the **super-upper bound cone**

$$\mathcal{U}_*(\mathcal{X}) := \bigcap_{p>0} \mathcal{U}_p(\mathcal{X}) \tag{5}$$

and use this cone to find the minimiser instead of $\mathcal{U}(\mathcal{X})$. In fact, following the reasoning in [8], one sees that $\boldsymbol{S} \in \mathcal{U}_p(\mathcal{X})$ holds for any $p > 0$, and therefore $\boldsymbol{S} \in \mathcal{U}_*(\mathcal{X})$. By declaring the function $\varphi : \mathrm{Sym}(2) \rightarrow \mathbb{R}^2$ with $\varphi(\boldsymbol{A}) = (\lambda_1, \lambda_2)$, where $\lambda_1 \geq \lambda_2$ are the eigenvalues of $\boldsymbol{A} \in \mathrm{Sym}(2)$, and endowing $\mathbb{R}^2$ with the lexicographic order

$$(a, b) \prec (a', b') :\Longleftrightarrow (a < a' \vee (a = a' \wedge b \leq b')) \tag{6}$$

one can show that the function φ is Loewner-monotone, convex and $\boldsymbol{S}$ is its unique minimiser in $\mathcal{U}_*(\mathcal{X})$. In addition, one can show the following lemma for transitivity, which we will not do here, but refer to our experiments.

Lemma 2. *The LES (1) is transitive, i.e. for multi-sets $\mathcal{X}$ and $\mathcal{Y}$ of symmetric 2×2 matrices one has*

$$\mathrm{Sup}_{LE}(\mathcal{X} \cup \mathcal{Y}) = \mathrm{Sup}_{LE}\left(\{\mathrm{Sup}_{LE}(\mathcal{X}), \mathrm{Sup}_{LE}(\mathcal{Y})\}\right). \tag{7}$$

4 Experiments

We divide this section into three parts with four experiments. First, we will demonstrate that our new method yields correct results in the Loewner sense, as did the previous Loewner methods. Then we will illustrate how the approximations of different morphological operations affect it. Our main focus will be on the almost isotropic character and especially the transitivity compared to previous approaches. For this, we will apply some elementary flat morphological operations to synthetic and natural colour images of different sizes. For the experiments we are using various structuring elements: a 3×3 square, a 5×5 square and a 9×9 square, all centred on the middle pixel.

4.1 Correctness of the Loewner Order

For this we consider a simple 30×30 image divided into blue (RGB $= (0,0,1)$) and green (RGB $= (0,1,0)$) and perform a dilation and an erosion on it with the 9×9 SE, see Fig. 1. To better distinguish the images, we have placed a black frame around them. The white border in between does not belong to the image.

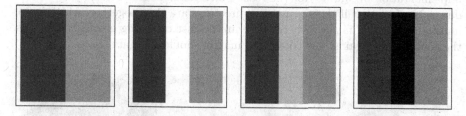

Fig. 1. Comparison of different dilation and erosion methods of an 30×30 image with a 9×9 SE. **From left to right:** Original image, dilation with the new method, channel-wise dilation and erosion with our method or channel-wise. (Colour figure online)

For the erosion, both our method and a channel-by-channel erosion with the corresponding Matlab function result in black. However, if we look at the dilation for these two methods, the channel-by-channel approach gives RGB $= (0,1,1)$ or in the bi-cone $(-1,0,0)$ while ours is white. The issue is that the symmetric matrix belonging to the blue-green colour is no longer larger than blue and green in the Loewner sense, because the blue-green cone does not include the green and blue cone. Our white, however, retains this property. The reason for the white colour is related to the construction of our method and will therefore be discussed in more detail in our future work.

4.2 Application of Different Morphological Operations

We will use the 64×64 pepper image and apply dilation, erosion, closing and opening with the 3×3 SE according to our approach, see Fig. 2. We immediately notice that most of the colours fade. As a result of log-sum exponentiation, we basically only compare the leading eigenvalues with different eigenvectors, and as m increases, the leading terms become more dominant, which results in higher eigenvalues and thus brighter colours. This is especially true if the two largest eigenvalues in the neighbourhood under consideration are already close to each other, which often happens with smaller SEs and natural images. If the two largest eigenvalues λ_1 and $\lambda_2 = \lambda_1 + \varepsilon$ for a sufficiently small $\varepsilon > 0$ have different eigenvectors, we obtain with our usual notation

$$S = \lambda_2 u_2 u_2^{\mathrm{T}} + \lambda_1 v_2 v_2^{\mathrm{T}} = (\lambda_1 + \varepsilon) u_2 u_2^{\mathrm{T}} + \lambda_1 v_2 v_2^{\mathrm{T}} = \lambda_1 I + \varepsilon u_2 u_2^{\mathrm{T}},$$

where we set $u_2 = (1,0)^{\mathrm{T}}$ and $v_2 = (0,1)^{\mathrm{T}}$. For example, the exact dilation in Fig. 2 has an average difference between the largest and second largest eigenvalue of 0.0268 per pixel, while the largest and smallest eigenvalues of the image are 0.5574 and -0.7071. To illustrate this further, we have included in Fig. 2 an approximate dilation with a scaling factor of $m = 69$, chosen because this is the largest factor that Matlab can compute without error. The clear difference to the converged dilation is possibly due to two reasons. Firstly, the factor is far from the limit value, but this cannot be changed, at least with the current implementation in Matlab, for the reason mentioned above. Secondly, it may be due to the fact that if the eigenvector directions are close together, which is not unusual in natural images where there are no fast or strong colour transitions, the convergence against the converged matrix should be quite slow.

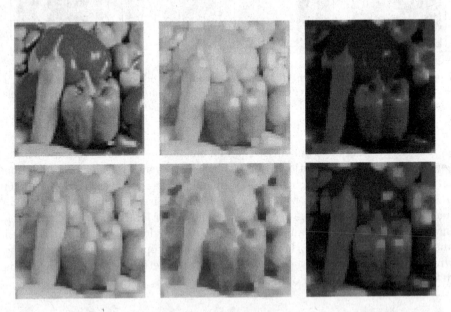

Fig. 2. Illustration of exact filtering results obtained with the new LES method for different operations of a 64×64 pepper image with a 3×3 SE. **Top: From left to right:** Original image, exact dilation and exact erosion. **Bottom: From left to right:** Approximate dilation ($m = 69$), exact closing and exact opening. (Colour figure online)

4.3   Transitivity

For this experiment we again use the 64×64 pepper image and apply a dilation
with the 5×5 mask to it and compare the result with applying a dilation with the
3×3 mask twice for our method and Burgeth's and Kleefeld's trace optimisation
method, see Fig. 3. By comparison, we mean a channel-by-channel comparison
of the two images in the form of a difference. From this, we take the absolute
values. Our method should give the exact same image due to transitivity and
Burgeth's and Kleefeld's method [6] should give a slightly different image.

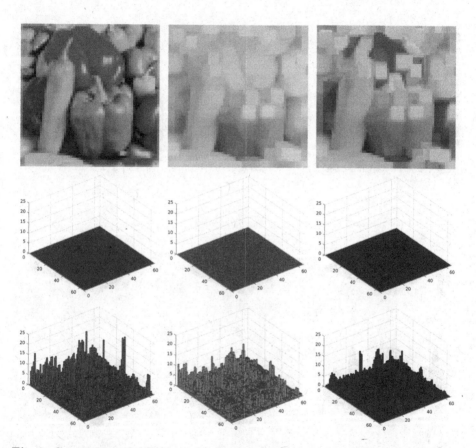

Fig. 3. Comparison of differences in the channel-wise deviations between the exact
filtering results of the dilation with a 5×5 SE and applying a dilation with the 3×3
SE twice using our new method and the method of Burgeth and Kleefeld. **Top: From
left to right:** Original image, dilation with the 5×5 SE by using our method and
Burgeth's and Kleefeld's method. **Mid: From left to right:** Differences in our method
with absolute values in the channels red, green and blue. **Bottom: From left to right:**
Differences in Burgeth's and Kleefeld's method with absolute values in the channels
red, green and blue. (Colour figure online)

In fact, with our method we do not see any differences in the colour channels, i.e. the difference is zero. In contrast, Burgeth's and Kleefeld's method shows clear deviations, which occur across the entire image in all channels. In particular, the red colour channel shows strong deviations of up to around 20 tonal values.

To examine this difference more closely, we look at how the error behaves when we double the number of dilations. This means we compare a dilation with a 9×9 SE with a fourfold dilation with a 3×3 SE, see Fig. 4. Because of this doubling, the doubling of operations, we would expect the error for the corresponding methods to also double.

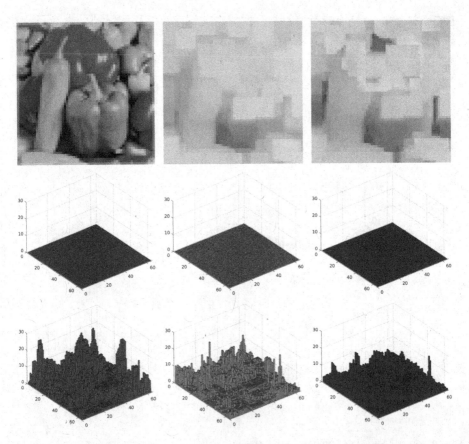

Fig. 4. Comparison of differences in the channel-wise deviations between the exact filtering results of the dilation with a 9×9 SE and fourfold dilation with the 3×3 SE using our new method and the method of Burgeth and Kleefeld. **Top: From left to right:** Original image, dilation with the 9×9 SE by using our method and Burgeth's and Kleefeld's method. **Mid: From left to right:** Differences in our method with absolute values in the channels red, green and blue. **Bottom: From left to right:** Differences in Burgeth's and Kleefeld's method with absolute values in the channels red, green and blue. (Colour figure online)

We see a continuation of the previous behaviour in the individual channels, which means that the transitivity of our method is preserved and that errors still occur everywhere with Burgeth's and Kleefeld's method, but particularly in the red colour channel. We see a further increase in the size of the errors. To measure the error more accurately, we use the Frobenius norm, which we apply to the channel-by-channel difference of the images, and then add this up for the individual channels. In this way, we obtain the value 336.9 for the first case of a twofold dilation and the value 686.5 for the second case of a fourfold dilation, which is almost twice the error.

For a complexity comparison between the methods mentioned, we consider the average time for the dilation of the 64×64 pepper image, which is made up of ten attempts. Our approach takes about 2.525 s, the approach of Burgeth and Kleefeld about 7.480 s and the channel-wise approach with the dilation function of Matlab 0.140 s. So we see that our non-optimised algorithm is already significantly faster than Burgeth's and Kleefeld's, but of course slower than the channel-by-channel approach, which could produce false colours.

5 Conclusion and Future Work

In this work, we have used the colour morphology used by Burgeth and co-authors, see [6–8], using Loewner order to introduce a new method for determining a "maximum colour" in the sense of dilation. We applied the log-sum exponentiation to approximate the supremum, which was introduced by Maslov [15], to the symmetric 2×2 matrices, which were created using a bijective mapping from the HCL̃ bi-cone. By using the spectral decomposition of these matrices, we were able to specify an approximation that depended only on them, which means that it depends only on the input data. We conducted several experiments on synthetic and natural images to demonstrate Loewner order compliance, how different morphological operations affect our approach, and identified some key factors that influence the approximation. In particular, we have shown that our approach fulfils the property of transitivity, which is important for lattices and orders, in contrast to Burgeth's and Kleefeld's method [7] or other methods derived from it, cf. [21].

In future work, we will take a closer look at the construction briefly introduced here and prove the properties presented. In this regard, we will particularly examine the minimality of the approach in details and provide a relaxation for the LES (1) so that it even depends continuously on the input data.

References

1. Angulo, J.: Morphological color processing based on distances. Application to color denoising and enhancement by centre and contrast operators. In: Proceedings of VIIP, pp. 314–319. Citeseer (2005)
2. Aptoula, E., Lefèvre, S.: On lexicographical ordering in multivariate mathematical morphology. Pattern Recogn. Lett. **29**(2), 109–118 (2008). https://doi.org/10.1016/j.patrec.2007.09.011

3. Barnett, V.: The ordering of multivariate data. J. R. Stat. Soc. Ser. A (Gen.) **139**(3), 318–344 (1976). https://doi.org/10.2307/2344839

4. Blusseau, S., Velasco-Forero, S., Angulo, J., Bloch, I.: Adaptive anisotropic morphological filtering based on co-circularity of local orientations. Image Process. On Line **12**, 111–141 (2022). https://doi.org/10.5201/ipol.2022.397

5. Borwein, J., Lewis, A.: Convex Analysis and Nonlinear Optimization. Springer, New York (2000). https://doi.org/10.1007/978-0-387-31256-9

6. Burgeth, B., Kleefeld, A.: An approach to color-morphology based on Einstein addition and Loewner order. Pattern Recogn. Lett. **47**, 29–39 (2014). https://doi.org/10.1016/j.patrec.2014.01.018

7. Burgeth, B., Kleefeld, A.: Morphology for color images via Loewner order for matrix fields. In: Hendriks, C.L.L., Borgefors, G., Strand, R. (eds.) ISMM 2013. LNCS, vol. 7883, pp. 243–254. Springer, Heidelberg (2013). https://doi.org/10.1007/978-3-642-38294-9_21

8. Burgeth, B., Welk, M., Feddern, C., Weickert, J.: Morphological operations on matrix-valued images. In: Pajdla, T., Matas, J. (eds.) ECCV 2004. LNCS, vol. 3024, pp. 155–167. Springer, Heidelberg (2004). https://doi.org/10.1007/978-3-540-24673-2_13

9. Haralick, R.M., Sternberg, S.R., Zhuang, X.: Image analysis using mathematical morphology. IEEE Trans. Pattern Anal. Mach. Intell. **9**(4), 532–550 (1987). https://doi.org/10.1109/TPAMI.1987.4767941

10. Heijmans, H.J., Ronse, C.: The algebraic basis of mathematical morphology I. Dilations and erosions. Comput. Vis. Graph. Image Process. **50**(3), 245–295 (1990). https://doi.org/10.1016/0734-189X(90)90148-O

11. Horn, R.A., Johnson, C.R.: Matrix Analysis. Cambridge University Press, Cambridge (2012)

12. Kahra, M., Sridhar, V., Breuß, M.: Fast morphological dilation and erosion for grey scale images using the Fourier transform. In: Elmoataz, A., Fadili, J., Quéau, Y., Rabin, J., Simon, L. (eds.) SSVM 2021. LNCS, vol. 12679, pp. 65–77. Springer, Cham (2021). https://doi.org/10.1007/978-3-030-75549-2_6

13. Lezoray, O., Elmoataz, A., Meurie, C.: Mathematical morphology in any color space. In: 14th International Conference of Image Analysis and Processing – Workshops, pp. 183–187. IEEE (2007). https://doi.org/10.1109/ICIAPW.2007.33

14. Löwner, K.: Über monotone Matrixfunktionen. Math. Z. **38**(1), 177–216 (1934). https://doi.org/10.1007/BF01170633

15. Maslov, V.P.: On a new superposition principle for optimization problems. Russian Math. Surv. **42**(3), 39–48 (1987)

16. Najman, J., Talbot, H.: Mathematical Morphology: From Theory to Applications. ISTE-Wiley, London-Hoboken (2010). https://doi.org/10.1002/9781118600788

17. Roerdink, J.B.T.M.: Mathematical morphology in computer graphics, scientific visualization and visual exploration. In: Soille, P., Pesaresi, M., Ouzounis, G.K. (eds.) ISMM 2011. LNCS, vol. 6671, pp. 367–380. Springer, Heidelberg (2011). https://doi.org/10.1007/978-3-642-21569-8_32

18. Serra, J., Soille, P.: Mathematical Morphology and Its Applications to Image Processing, vol. 2. Springer, Dordrecht (2012)

19. Sharma, G., Bala, R.: Digital Color Imaging Handbook. CRC Press, Boca Raton (2017)

20. Sridhar, V., Breuss, M., Kahra, M.: Fast approximation of color morphology. In: Bebis, G., et al. (eds.) ISVC 2021. LNCS, vol. 13018, pp. 488–499. Springer, Cham (2021). https://doi.org/10.1007/978-3-030-90436-4_39
21. Welk, M., Kleefeld, A., Breuß, M.: Quantile filtering of colour images via symmetric matrices. Math. Morphol. Theory Appl. 1(1), 136–174 (2016). https://doi.org/10.1515/mathm-2016-0008

Image Segmentation by Hierarchical Layered Oriented Image Foresting Transform Subject to Closeness Constraints

Luiz Felipe Dolabela Santos[1], Felipe Augusto de Souza Kleine[1,2], and Paulo André Vechiatto Miranda[1]([⊠]) [iD]

[1] Institute of Mathematics and Statistics, University of São Paulo, São Paulo, SP, Brazil
pmiranda@ime.usp.br
[2] Researcher at the IPT - Institute for Technological Research of the State of São Paulo, São Paulo, Brazil

Abstract. In this work, we address the problem of image segmentation, subject to high-level constraints expected for the objects of interest. More specifically, we define closeness constraints to be used in conjunction with geometric constraints of inclusion in the Hierarchical Layered Oriented Image Foresting Transform (HLOIFT) algorithm. The proposed method can handle the segmentation of a hierarchy of objects with nested boundaries, each with its own expected boundary polarity constraint, making it possible to control the maximum distances (in a geodesic sense) between the successive nested boundaries. The method is demonstrated in the segmentation of nested objects in colored images with superior accuracy compared to its precursor methods and also when compared to some recent click-based methods.

Keywords: Multi-object segmentation · Oriented Image Foresting Transform · Layered graphs · Closeness constraints

1 Introduction

Image segmentation, such as to extract an object from a background, is one of the most fundamental and challenging problems in image processing, computer vision, and image analysis. It is a fundamental procedure widely used in the area of autonomous vehicles and robotics for object detection and content recognition [33,35], disease identification and etiology study in the field of medicine and biological image analysis [15,24,26], crop area measurement for agriculture [4,11], and many other applications [1,29,34].

Thanks to Conselho Nacional de Desenvolvimento Científico e Tecnológico – CNPq – (Grant 407242/2021-0, 313087/2021-0, 166631/2018-3), CAPES (88887.136422/2017-00), FAPESP (2014/12236-1, 2014/50937-1) and IPT (Institute for Technological Research).

S. Brunetti et al. (Eds.): DGMM 2024, LNCS 14605, pp. 338–349, 2024.
https://doi.org/10.1007/978-3-031-57793-2_26

Given that segmentation can be seen as the task of partitioning an input image into objects/regions of interest by assigning distinct labels to their composing pixels, it can be interpreted as a classification problem at the pixel level from the perspective of machine learning-based methods. Hence, an important taxonomy of image segmentation concerns how the methods consider the usage of training data. Unsupervised segmentation does not require prior training aside from some parameters to adjust [28], self-supervised segmentation uses unlabeled training data [27], supervised segmentation uses labeled data for training [31], and semi-supervised segmentation uses both labeled and unlabeled data for training [36]. For example, supervised segmentation by deep learning methods have been shown to be suitable for general semantic or instance segmentation of natural images [12].

Segmentation methods can also be divided, according to the level of user intervention, into interactive and automatic methods. Interactive methods are guided by user actions that provide clues about the desired outcome, thus reducing possible existing ambiguities that would otherwise lead to an ill-posed problem. Interactive methods can be subdivided according to the type of user constraint provided, such as boundary-based methods by anchor points [21], minimum bounding box by extreme points [16,25] or by grabcut-style [32], and region-based methods by drawing scribbles (markers/seeds) [13], which include the class of click-based methods as a particular case.

Optimization frameworks in graphs by combinatorial analysis can handle image segmentation as a graph partition problem subject to soft and hard constraints. In this context, *Oriented Image Foresting Transform* (OIFT) [22] is a region-based method by optimal cuts in directed weighted graphs, which lies in the intersection of Generalized Graph Cut and General Fuzzy Connectedness frameworks, supporting several high-level priors for object segmentation, including global properties such as connectedness [17,18], shape constraints [3,20,23], boundary polarity [19], maximum allowable size [5], and hierarchical constraints [13], which allow the customization of the segmentation to a given target object [14]. These generic priors in general scenarios represent more abstract concepts, potentially covering an unlimited amount of cases, that are difficult for supervised methods to learn and extrapolate from only limited annotated databases. Hence, it is typically impractical to train networks for this purpose, even though there are works on how to design loss functions that favor the segmentation to have the same topology as the ground truth [10]. Indeed, in the case of more specific areas, such as medical imaging, there is an effort in segmenting objects based on known priors to achieve better performance [7]. Therefore, due to their complementary forces, the integration of machine learning techniques with methods built on the strong formalism of graph partitions has become a very relevant research topic [2,9].

In this work, we propose a region-based method which accepts as priors a hierarchy of nested objects with minimum and maximum distance constraints, and each object can have, encoded in the weights of the arcs of its respective layer (see Sect. 3.2), its own boundary polarity configuration, among other particular restrictions (e.g., shape constraints). In this way, it extends the *Hierarchical Layered Oriented Image Foresting Transform* (HLOIFT) [13], subject to the inclusion constraint of nested objects, to support a novel closeness constraint.

The reminder of the paper is structured as follows: Sect. 2 describes the notations and definitions for our algorithm. Section 3 reviews the two previous works (OIFT and HLOIFT) which will be extended in this paper. In Sect. 4, we present the closeness constraint and our new method, which produces an optimal energy segmentation subject to the proposed constraint. Finally, in Sects. 5 and 6, we show the experimental results and state our conclusions.

2   Notations and Definitions

A weighted digraph G is a triple $(\mathcal{N}, \mathcal{A}, \omega)$, where $\mathcal{N}$ is a nonempty set of vertices or nodes, $\mathcal{A}$ is a set of ordered pairs of distinct vertices called arcs or directed edges, and $\omega : \mathcal{A} \to \mathbb{R}$ represents the weights associated with the arcs. The digraph G is symmetric if for any of its arcs $(s,t) \in \mathcal{A}$, the pair (t,s) is also an arc of G, but we can have $\omega(s,t) \neq \omega(t,s)$. For a given graph $G = (\mathcal{N}, \mathcal{A}, \omega)$, a path $\pi = \langle t_1, t_2, \ldots, t_n \rangle$ is a sequence of adjacent nodes (*i.e.*, $(t_i, t_{i+1}) \in \mathcal{A}$, $i = 1, 2, \ldots, n-1$) with no repeated vertices ($t_i \neq t_j$ for $i \neq j$). The set of nodes along a path π can be indicated as $\mathcal{V}(\pi) = \{t_1, \ldots, t_n\}$. A path $\pi_t = \langle t_1, t_2, \ldots, t_n = t \rangle$ is a path with terminus at a node t. A path is *trivial* when $\pi_t = \langle t \rangle$. A path $\pi_t = \pi_s \cdot (s,t)$ indicates the extension of a path π_s by an arc (s,t).

A *predecessor map* is a function P that assigns to each node t in $\mathcal{N}$ either some other adjacent node in $\mathcal{N}$, or a distinctive marker *nil* not in $\mathcal{N}$—in which case t is said to be a *root* of the map. A *spanning forest* is a predecessor map which contains no cycles—i.e., one which takes every node to *nil* in a finite number of iterations. For any node $t \in \mathcal{N}$, a spanning forest P defines a path π_t^P recursively as $\langle t \rangle$ if $P(t) = nil$, and $\pi_s^P \cdot (s,t)$ if $P(t) = s \neq nil$ (Fig. 1a).

3   Background

3.1   Oriented Image Foresting Transform (OIFT)

In the case of binary segmentation, dividing $\mathcal{N}$ into background O_0 and object $O_1 = \mathcal{N} \setminus O_0$, we consider two non-empty disjoint seed sets $\mathcal{S}_0, \mathcal{S}_1 \subset \mathcal{N}$ indicating hard constraints for background and object, respectively, such that $\mathcal{S}_0 \subset O_0$ and $\mathcal{S}_1 \subset O_1$. The object O_1 is identified with its *labeling* $X \colon \mathcal{N} \to \{0,1\}$, so that $O_1 = \{v \in \mathcal{N} \colon X(v) = 1\}$.

Let $\mathcal{U}(\mathcal{S}_0, \mathcal{S}_1) = \{X \in \mathcal{X} \colon X(t) = i \text{ for all } t \in \mathcal{S}_i, i \in \{0,1\}\}$ be the universe of all segmentations satisfying seed constraints, where $\mathcal{X}$ is the set of all labeling functions from $\mathcal{N}$ to $\{0,1\}$. Consider the graph-cut measure ε_{min} defined as

$$\varepsilon_{min}(X) = \min\{\omega(s,t) \colon (s,t) \in \mathcal{A} \text{ and } X(s) > X(t)\}. \tag{1}$$

The OIFT segmentation X gives, subject to seed constraints, a global optimum solution by maximizing ε_{min}. That is, $\varepsilon_{min}(X) = \max\limits_{X' \in \mathcal{U}(\mathcal{S}_0, \mathcal{S}_1)} \varepsilon_{min}(X')$.

OIFT can be build upon the *Image Foresting Transform* framework (IFT) [8] by considering appropriate IFT parameters in a connected and symmetric digraph G, as described in [22], where we usually take $\mathcal{N}$ to be the image domain $\mathcal{I}$ (i.e., the set of pixels in $\mathbb{Z}^2$) and $\mathcal{A}$ is defined by 8-neighborhood for 2D images.

Next, we present the extension of OIFT to multiple objects in layered graphs from [13], where each layer is a graph at the level of pixels.

3.2   Hierarchical Layered OIFT (HLOIFT)

Let $\mathcal{L} = \{1, \ldots, m\}$ denote a set of labels, where each element $i \in \mathcal{L}$ is associated with a corresponding object $O_i \subset \mathcal{I}$ to be segmented of a total of m objects. The HLOIFT graph associated with $\mathcal{L}$ and an image will be defined on the set of nodes $\mathcal{N} = \mathcal{L} \times \mathcal{I}$, so that each node t is now a pair (i, v), $i \in \mathcal{L}$ and $v \in \mathcal{I}$. For each node $t = (i, v)$, we use $\lambda \colon \mathcal{N} \to \mathcal{L}$ and $p \colon \mathcal{N} \to \mathcal{I}$ to denote the projections onto the first and second coordinates, that is, $\lambda(t) = i$ and $p(t) = v$. The HLOIFT segmentation will be identified with a binary variable $X \colon \mathcal{N} \to \{0, 1\}$, where, for $i \in \mathcal{L}$, the ith object O_i and the background O_0 are defined, respectively, as

$$O_i = \{v \in \mathcal{I} \colon X(i, v) = 1\} \quad \text{and} \quad O_0 = \mathcal{I} \setminus \bigcup_{i \in \mathcal{L}} O_i. \tag{2}$$

Each object/background object O_i, $i \in \mathcal{L} \cup \{0\}$, will be identified with a corresponding set $\mathcal{S}_i \subset \mathcal{I}$ of seeds, aiming for $\mathcal{S}_i \subseteq O_i$.

The first step of HLOIFT is to create a set of m layers, where each layer $\mathcal{H}_i$, $i \in \mathcal{L}$, represents a single corresponding object O_i. A layer $\mathcal{H}_i = (\mathcal{N}_i, \mathcal{A}_i, \omega_i)$ is a weighted digraph, where $\mathcal{N}_i = \{i\} \times \mathcal{I}$ and each node $t = (i, v) \in \mathcal{N}_i$ corresponds to the image pixel $p(t) = v \in \mathcal{I}$, such that $\mathcal{N} = \mathcal{L} \times \mathcal{I} = \bigcup_{i \in \mathcal{L}} \mathcal{N}_i$. We define the set of intra-layer arcs $\mathcal{A}_i$ on $\mathcal{H}_i$, $i = 1, \ldots, m$, as $(s, t) \in \mathcal{A}_i$ if, and only if, $p(s)$ and $p(t)$ are neighboring pixels. Regarding the weight function ω_i, it is highly customizable and, ideally, it should highlight the desired boundaries for O_i as clearly as possible, being possible to incorporate higher level priors in its definition whenever it is appropriate [13], as will be explained in Sect. 5.

In the second step, HLOIFT generates a *hierarchical layered* weighted digraph $\mathcal{H}$ as the union of all layered graphs $\mathcal{H}_i$, $i = 1, \ldots, m$, with additional *inter-layer* arcs connecting only some of the distinct layers. These inter-layer arcs represent a prior knowledge on any pair (O_i, O_j) of objects given by a hierarchy, such that either $O_i \cap O_j = \emptyset$ (exclusion relation), or one of them is properly contained in the other (inclusion relation). Here, we consider only the inclusion relation, represented as a function $h \colon \mathcal{L} \to \mathcal{L}$, so that $h(i) = j$ if, and only if, O_j is the smallest of the objects properly containing O_i. In this work, we consider a set of m objects with nested boundaries, such that $h(i) = i + 1$, $i = 1, \ldots, m-1$. If $h(i) = j$, then we will refer to O_j as the *parent* of O_i and in such a case, we define $\omega(t, s) = \infty$ and $\omega(s, t) = -\infty$, for all $s = (i, v) \in \mathcal{N}_i$ and $t = (j, u) \in \mathcal{N}_j$ such that $\|u - v\| \leq \rho$, where the distance parameter $\rho \geq 0$ indicates the minimum distance between the boundaries of the parent-offspring pair of objects. More specifically, for the parent-offspring pair (O_j, O_i) (i.e., with $h(i) = j$) we assume that $u \in O_j$ whenever there exists an $v \in O_i$ with $\|u - v\| \leq \rho$.

Finally, in the last step, a modified OIFT algorithm is applied over $\mathcal{H}$ to compute the segmentation map $X: \mathcal{N} \to \{0, 1\}$. However, in the case where only the inclusion relation is considered, as done in this work, this algorithm becomes the regular OIFT (Sect. 3.1) over the graph $\mathcal{H}$, using as object and background seeds, the sets $\mathcal{S}_1^* = \{(i, v) \in \mathcal{N}: v \in \mathcal{S}_i\}$ and $\mathcal{S}_0^* = \{t \in \mathcal{N}: p(t) \in \mathcal{S}_0\}$.

In HLOIFT, for a given valid solution of m objects satisfying the seed and inclusion constraints, the energy of O_i in layer $\mathcal{H}_i$, $i \in \mathcal{L}$, is given by:

$$e(O_i) = \min_{(s,t) \in \mathcal{A}_i} \{\omega_i(s, t) : p(s) \in O_i \text{ and } p(t) \notin O_i\}, \tag{3}$$

and the final energy of the set of m objects is given by:

$$e(O_1, \ldots, O_m) = \min_{i \in \mathcal{L}} e(O_i), \tag{4}$$

such that $e(O_1, \ldots, O_m) = \varepsilon_{min}(X)$, as defined by Eq. 1.

4   HLOIFT with Closeness Constraints (HLOIFT-CC)

Let P be a map of predecessors defined in an image graph with vertices in $\mathcal{I}$, representing a tree rooted in a pixel internal to the nested objects $O_1, \ldots, O_m$ of a hierarchy, with only inclusion relations. Here, we consider P computed as indicated for the *Geodesic Star Convexity* (GSC) constraint in [20].

If O_i satisfies the GSC constraint with respect to P (indicated as O_i is GSC), then for all $u, v \in \mathcal{I}$, such that $u = P(v)$, we have that $X(i, v) = 1 \implies X(i, u) = 1$ and $X(i, u) = 0 \implies X(i, v) = 0$ (Fig. 1). The aim of this section is to define an optimal algorithm with maximum energy by Eq. 4, such that O_i is GSC for $i \in \mathcal{L}$ (GSC satisfaction can be easily achieved by changing some weights of intra-layer arcs, as explained in [20] and Sect. 5), and some pairs of objects also satisfy the proposed closeness constraint. Given two objects O_i and O_j, the closeness constraint consists of a binary relation $O_i \xrightarrow{\text{clo}} O_j$ defined as follows:

Definition 1 (Closeness constraint). *For $L \geq 0$ and a forest given by a predecessor map $P: \mathcal{I} \to \mathcal{I} \cup \{nil\}$, an object $O_i \subset \mathcal{I}$ satisfies the closeness constraint with respect to $O_j \subset \mathcal{I}$ with parameter L (indicated as $O_i \xrightarrow{\text{clo}} O_j$) provided for all $s \in O_j$ there exists $t \in O_i$ such that $D(s) - D(t) \leq L$ and $t \in \mathcal{V}(\pi_s^P)$, where for all $q \in \mathcal{I}$, $D(q) = 0$ if $P(q) = nil$, and $D(q) = D(p) + \|q - p\|$ if $P(q) = p \neq nil$, with the symbol $\|\cdot\|$ denoting the L2-norm. Violations of the constraint are indicated by the notation $O_i \xcancel{\xrightarrow{\text{clo}}} O_j$.*

Note that the resulting binary relation is not symmetric. For the symmetric case, we use the notation $O_i \xleftrightarrow{\text{clo}} O_j$ to indicate that $O_i \xrightarrow{\text{clo}} O_j$ and $O_j \xrightarrow{\text{clo}} O_i$. Note that when we have the inclusion relation, such that O_j is the parent of O_i (i.e., $h(i) = j$), then $O_i \xrightarrow{\text{clo}} O_j \implies O_i \xleftrightarrow{\text{clo}} O_j$.

For the sake of simplicity, we present the algorithm considering the closeness constraint only between layers 1 and 2. Figure 2 presents an example of the

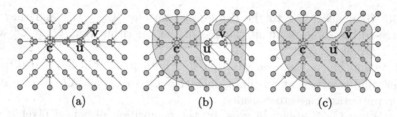

Fig. 1. (a) Example of a spanning forest P with 8-connected adjacency, where the arrows (in magenta) point to the predecessor of each node. A path $\pi_v^P = \pi_u^P \cdot (u, v)$ (in blue) is stored backwards in P, where $P(v) = u$ is the predecessor node of v. (b) GSC constraint violation for O_i (in green). (c) Updated example where O_i is GSC. (Color figure online)

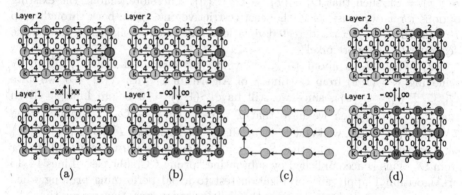

Fig. 2. HLOIFT-CC example for $\rho = 0$ and $L = 1$. (a) The graph $\mathcal{H}$ with the input seeds, using $\mathcal{S}_0 = \{p(J)\} = \{p(j)\}$ and $\mathcal{S}_1 = \{p(F)\}$ (i.e., $\mathcal{S}_0^* = \{j, J\}$ and $\mathcal{S}_1^* = \{F\}$). Note that background seeds affect all layers, while object seeds only directly affect their own layers (Lines 3–6 of Algorithm 1). (b) HLOIFT result has energy $e(O_1, O_2) = \min\{5, 8\} = 5$, but $O_1 \overset{\text{clo}}{\not\to} O_2$. (c) The predecessor map P considered for HLOIFT-CC, defining a tree rooted at pixel $p(F)$. (d) HLOIFT-CC has $e(O_1, O_2) = \min\{3, 4\} = 3$.

HLOIFT-CC algorithm. Note that among all solutions satisfying $O_1 \overset{\text{clo}}{\to} O_2$, $h(1) = 2$ and seeds, HLOIFT-CC obtains the one with the highest energy.

During the execution of the algorithm, the m objects ($i \in \mathcal{L}$) are gradually computed through the growth of their internal regions as:

$$\mathcal{O}_i = \{v \in \mathcal{I} : X(i, v) = 1 \text{ and } S(i, v) = 1\} \tag{5}$$

where $S : \mathcal{N} \to \{0, 1\}$ is an array of status, which indicates with value 1 the nodes that have already been processed. In a complementary way, objects can also be obtained gradually by complementing the growth of background seeds:

$$\begin{aligned} \mathcal{B}_i &= \{v \in \mathcal{I} : X(i, v) = 0 \text{ and } S(i, v) = 1\} \\ \mathcal{O}_i' &= \mathcal{I} \setminus \mathcal{B}_i \end{aligned} \tag{6}$$

Note that, according to the definitions, we have $\mathcal{O}_i \subseteq \mathcal{O}'_i$ and they will converge to the same result at the end of the algorithm (i.e., $\mathcal{O}_i = \mathcal{O}'_i$ when $S(t) = 1$ for all $t \in \mathcal{N}$).

The algorithm is based on ensuring that both pairs $(\mathcal{O}_1, \mathcal{O}_2)$ and $(\mathcal{O}'_1, \mathcal{O}'_2)$ contain valid solutions at all times, such that $\mathcal{O}_1 \overset{clo}{\to} \mathcal{O}_2$ and $\mathcal{O}'_1 \overset{clo}{\to} \mathcal{O}'_2$. There are two important cases to consider:

(1) When $\mathcal{O}_2$ is about to grow by the acquisition of a new pixel $p(s) = v$, by removing $s = (2, v)$ from Q (Line 8 of Algorithm 1) with $X(s) = 1$, we have $\mathcal{O}_2 \cup \{v\} \subseteq \mathcal{O}'_2$, since we still have $S(2, v) = 0$ before Line 9. Given that $\mathcal{O}'_1 \overset{clo}{\to} \mathcal{O}'_2$, then we have that $\mathcal{O}'_1 \overset{clo}{\to} \mathcal{O}_2 \cup \{v\}$. Therefore, when $\mathcal{O}_1 \overset{clo}{\not\to} \mathcal{O}_2 \cup \{v\}$ (i.e., $\mathcal{O}_1 \overset{clo}{\not\to} \{v\}$), we know that there is $x \in \mathcal{O}'_1 \setminus \mathcal{O}_1$ (note that $S(1, x) = 0$), such that $\mathcal{O}_1 \cup \{x\} \overset{clo}{\to} \mathcal{O}_2 \cup \{v\}$. Therefore, among the existing options for x, we must select the least restrictive one, going back through map P, in order to correct the detected violation. This is accomplished by subroutine "conquer_CC_support_pixel".

(2) When $\mathcal{B}_1$ is about to grow by the acquisition of a new pixel v, by removing $s = (1, v)$ from Q (Line 8 of Algorithm 1) with $X(s) = 0$, we have $\mathcal{O}_1 \subseteq \mathcal{I} \setminus (\mathcal{B}_1 \cup \{v\})$, since we still have $S(1, v) = 0$ before Line 9. Given that $\mathcal{O}_1 \overset{clo}{\to} \mathcal{O}_2$, then we have that $\mathcal{I} \setminus (\mathcal{B}_1 \cup \{v\}) \overset{clo}{\to} \mathcal{O}_2$. Therefore, when $\mathcal{I} \setminus (\mathcal{B}_1 \cup \{v\}) \overset{clo}{\not\to} \mathcal{O}'_2$, we know that for all $x \in \mathcal{O}'_2$ such that $\mathcal{I} \setminus (\mathcal{B}_1 \cup \{v\}) \overset{clo}{\not\to} \{x\}$ we have that $x \notin \mathcal{O}_2$ (note that $S(2, x) = 0$). Hence, we must remove all such x from $\mathcal{O}'_2$. This is accomplished by subroutine "prune_CC_subtrees". Lines 14–15 of Algorithm 1 apply an optimization test to avoid performing pruning when pruning has already been done for the node's predecessor, thus saving time.

Algorithm 1 – HLOIFT Algorithm Subject to Closeness Constraints

INPUT: Hierarchical layered digraph $\mathcal{H} = (\mathcal{N}, \mathcal{A}, \omega)$, builded from an image according to the tree h of hierarchical constraints of inclusion, such that $h(i) = i + 1, i = 1, \ldots, m - 1$, and with the minimal distance parameter ρ embedded in its inter-layer arcs; the seed sets $(\mathcal{S}_0, \ldots, \mathcal{S}_m)$; the predecessor map $P \colon \mathcal{I} \to \mathcal{I} \cup \{nil\}$; and closeness parameter L.

OUTPUT: The binary map $X \colon \mathcal{N} \to \{0, 1\}$ identifying segmentation given by (2).

AUXILIARY: Priority queue Q, variable tmp, the cost map $V \colon \mathcal{N} \to [-\infty, \infty]$, and an array of status $S \colon \mathcal{N} \to \{0, 1\}$, where $S(t) = 1$ for processed nodes and $S(t) = 0$ for unprocessed nodes.

1. **For each** $t \in \mathcal{N}$ and $i \in \mathcal{L}$ **do**
2. $\quad$ Set $S(t) \leftarrow 0$ and $V(t) \leftarrow \infty$;
3. $\quad$ **If** $p(t) \in \mathcal{S}_0$ **then**
4. $\quad\quad$ $V(t) \leftarrow -\infty$, $X(t) \leftarrow 0$, and insert t in Q;
5. $\quad$ **If** $p(t) \in \mathcal{S}_i$ and $\lambda(t) = i$ **then**
6. $\quad\quad$ $V(t) \leftarrow -\infty$, $X(t) \leftarrow 1$, and insert t in Q.
7. **While** $Q \neq \emptyset$ **do**
8. $\quad$ Remove s from Q such that $V(s)$ is minimum;
9. $\quad$ Set $S(s) \leftarrow 1$;
10. $\quad$ **If** $X(s) = 1$, **then**
11. $\quad\quad$ **If** $\lambda(s) = 2$, **then**

12.   L   L   *conquer_CC_support_pixel(p(s), L, P, X, S, V, Q);*
13.   **Else If** $\lambda(s) = 1$, **then**
14.   Set $q \leftarrow (1, P(p(s)))$;
15.   **If** $\neg [X(q) = 0 \ and \ S(q) = 1]$, **then**
16.   L   L   *prune_CC_subtrees(p(s), L, P, X, S, V, Q).*
17.   **For each** $(s, t) \in \mathcal{A}$ such that $S(t) = 0$ **do**
18.   **If** $X(s) = 1$ **then** $tmp \leftarrow \omega(s, t)$
19.   **Else** $tmp \leftarrow \omega(t, s)$;
20.   **If** $tmp < V(t)$, **then**
21.   Set $V(t) \leftarrow tmp$;
22.   Set $X(t) \leftarrow X(s)$;
23.   L   L   L   **If** $t \notin Q$ **then** *insert t in Q;*

Algorithm 2 – CONQUER_CC_SUPPORT_PIXEL

INPUT: Pixel $v \in \mathcal{I}$, parameter L, maps P,X,S,V, and priority queue Q.

1. Set $t \leftarrow v$ and $q \leftarrow P(v)$;
2. **While** $\neg [S(1, q) = 1 \ and \ X(1, q) = 1]$ and $D(v) - D(q) \leq L$ **do**
3.   L   Set $t \leftarrow q$ and $q \leftarrow P(q)$;
4. **If** $D(v) - D(q) > L$, **then**
5.   Set $u \leftarrow (1, t)$;
6.   **If** $S(u) = 0$, **then**
7.   Set $X(u) \leftarrow 1$ and $V(u) \leftarrow -\infty$;
8.   L   L   **If** $u \notin Q$ **then** *insert u in Q;*

Algorithm 3 – PRUNE_CC_SUBTREES

INPUT: Pixel $r \in \mathcal{I}$, parameter L, maps P,X,S,V, and priority queue Q.
AUXILIARY: A stack T, and variable cst.

1. Set $cst \leftarrow D(P(r))$ and insert r in T;
2. **While** $T \neq \emptyset$ **do**
3.   Remove u from the top of T;
4.   **For each** neighboring pixel v of u **do**
5.   **If** $u = P(v)$, **then**
6.   Set $z \leftarrow (2, v)$;
7.   **If** $D(v) - cst > L$ and $S(z) \neq 1$, **then**
8.   Set $X(z) \leftarrow 0$ and $V(z) \leftarrow -\infty$;
9.   L   **If** $z \notin Q$ **then** *insert z in Q;*
10.   **Else If** $\neg [S(z) = 1 \ and \ X(z) = 0]$, **then**
11.   L   L   L   *Insert v in T.*

5   Experimental Results

We conducted experiments, comparing HLOIFT-CC with its precursors as well as with a state-of-the-art click-based method. In the following, RITM denotes the click-based interactive segmentation from [30] with its models trained on a combination of COCO and LVIS, as available in[1]. SAM denotes *Segment Anything Model*, a new AI model from Meta AI[2]. HLOIFT denotes the standard method described in Sect. 3.2 and HLOIFT-GSC also includes the GSC constraint embedded in its intra-layer arcs [20], as explained below.

HLOIFT-CC, HLOIFT-GSC and HLOIFT consider the **boundary polarity priors** embedded in its intra-layer arcs as a soft constraint, in order to explore the object-contour orientations, while HLOIFT-CC and HLOIFT-GSC also consider the GSC constraint as a hard constraint. For this purpose, we use in our experiments ω_i as defined by the following formula:

$$\omega_i(s,t) = \begin{cases} -\infty & \text{if } P(p(s)) = p(t) \text{ and } O_i \text{ is GSC} \\ \|I(t) - I(s)\| \times (1 + \alpha_i) & \text{else if } l_s > l_t \\ \|I(t) - I(s)\| \times (1 - \alpha_i) & \text{else if } l_s < l_t \\ \|I(t) - I(s)\| & \text{otherwise} \end{cases} \quad (7)$$

where $I(s) = (l_s, a_s, b_s)$ and $I(t) = (l_t, a_t, b_t)$ are the colors of pixels $p(s)$ and $p(t)$ in CIELAB color space, the symbol $\|\cdot\|$ denotes the vector norm, and α_i is a polarity parameter. In this setting, each object O_i has its own constant $\alpha_i \in [-1, 1]$, so that we can favor the segmentation of O_i with transitions from *bright to dark* pixels with $\alpha_i > 0$, or the opposite orientation with $\alpha_i < 0$. In the experiments, we consider $|\alpha_i| = 50\%$ and with the sign of α_i configured according to the expected transitions for the objects of interest.

We performed quantitative experiments for colored images of 640×480 pixels divided into three databases (DB1, DB2 and DB3) with 30, 20 and 16 images, respectively, representing flat objects with nested boundaries under different viewpoints. Sample images of the three bases are shown in Fig. 3. For each image, four segmentations were performed for different locations of the internal seed of the child object, leading to a total of 264 single-click segmentations per method. Our databases and source code are available to the community[3].

Table 1 presents the experimental results. The proposed method for $\rho = 1.5$ and $L = 40$ achieved an average accuracy greater than 99% for both O_1 and O_2 for all bases. In the case of RITM (or SAM), its only generated binary mask is compared with the two nested objects present in the ground truth, but for neither of them was its accuracy higher than that obtained by HLOIFT-CC.

Regarding the computational time of the experiments, the mean execution time per image of 640×480 pixels with two layers ($m = 2$) was 380.8 ms to compute HLOIFT, 444.5 ms for HLOIFT-GSC and 504.0 ms for HLOIFT-CC, in an Intel Core i5-10210U CPU @ 1.60 GHz×8. Therefore, we have an increase of only 13% compared to the HLOFT-GSC time. A worst-case upper bound for HLOIFT-CC from Algorithm 1 is $O(M + NT)$, when Q uses bucket sorting, where T is the mean size of the pruned subtrees by Algorithm 3, which have their depth limited by L, N and M are the number of nodes and arcs in $\mathcal{H}$ (Fig. 4).

---

[1] https://github.com/SamsungLabs/ritm_interactive_segmentation/blob/master.
[2] https://segment-anything.com/.
[3] https://github.com/pauloavmiranda/HLOIFT_CC.

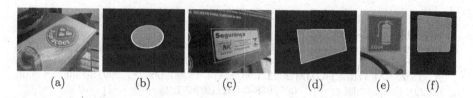

(a)   (b)   (c)   (d)   (e)   (f)

Fig. 3. Sample images and their respective ground truths (O_2 is the union of the gray and white regions while O_1 is the gray region only). (a-b) DB1. (c-d) DB2. (e-f) DB3.

Table 1. The mean accuracy values by the Dice coefficient and their respective standard deviation values for the different evaluated methods and databases.

Data	Obj	HLOIFT-CC	HLOIFT-GSC	HLOIFT	RITM	SAM
DB1	O_1	$(99.71 \pm 0.06)\%$	$(74.69 \pm 34.04)\%$	$(51.97 \pm 41.51)\%$	$(66.73 \pm 21.56)\%$	$(51.71 \pm 44.32)\%$
	O_2	$(99.72 \pm 0.08)\%$	$(93.13 \pm 14.54)\%$	$(91.21 \pm 14.14)\%$	$(71.75 \pm 22.25)\%$	$(53.84 \pm 46.84)\%$
DB2	O_1	$(99.85 \pm 0.06)\%$	$(71.09 \pm 44.34)\%$	$(29.75 \pm 44.81)\%$	$(79.42 \pm 24.01)\%$	$(32.44 \pm 41.90)\%$
	O_2	$(99.81 \pm 0.09)\%$	$(97.22 \pm 4.72)\%$	$(90.86 \pm 16.01)\%$	$(82.17 \pm 24.54)\%$	$(32.53 \pm 42.03)\%$
DB3	O_1	$(99.84 \pm 0.04)\%$	$(83.71 \pm 35.80)\%$	$(46.79 \pm 44.43)\%$	$(93.68 \pm 1.16)\%$	$(56.51 \pm 44.10)\%$
	O_2	$(99.86 \pm 0.05)\%$	$(99.17 \pm 1.89)\%$	$(96.47 \pm 3.40)\%$	$(95.36 \pm 1.16)\%$	$(57.34 \pm 44.98)\%$

(a) Input seed (b) HLOIFT-CC (c) HLOIFT-GSC   (d) HLOIFT   (e) RITM

Fig. 4. Example of segmentation results for the different methods for Fig. 3c.

6   Conclusion

In this work, we proposed the HLOIFT-CC method that successfully incorporated the closeness constraint in HLOIFT with inclusion relations. As attested in the experiments, by the simple specification of the expected high-level constraints of the desired nested objects, HLOIFT-CC can be quickly employed in new applications without the need to have a previously built training base.

As future work, we intend to include size constraints in HLOIFT-CC through the use of differential OIFT [5,6]. We also intend to extend HLOIFT-CC to include the exclusion relation between sibling objects in conjunction with closeness constraints and to adapt it to other graphs apart from a grid of pixels only.

References

1. Barreto, T.L., et al.: Classification of detected changes from multitemporal high-res Xband SAR images: intensity and texture descriptors from superpixels. IEEE J. Sel. Top. Appl. Earth Obs. Remote Sens. **9**(12), 5436–5448 (2016)
2. Belém, F., et al.: Fast and effective superpixel segmentation using accurate saliency estimation. In: Baudrier, É., Naegel, B., Krähenbühl, A., Tajine, M. (eds.) DGMM

2022. LNCS, vol. 13493, pp. 261–273. Springer, Cham (2022). https://doi.org/10.1007/978-3-031-19897-7_21

3. Braz, C.M., Santos, L.F.D., Miranda, P.A.V.: Graph-based image segmentation with shape priors and band constraints. In: Baudrier, É., Naegel, B., Krähenbühl, A., Tajine, M. (eds.) DGMM 2022. LNCS, vol. 13493, pp. 287–299. Springer, Cham (2022). https://doi.org/10.1007/978-3-031-19897-7_23

4. Cai, W., Wei, Z., Song, Y., Li, M., Yang, X.: Residual-capsule networks with threshold convolution for segmentation of wheat plantation rows in UAV images. Multimedia Tools Appl. **80**, 32131–32147 (2021)

5. Condori, M.A.T., Miranda, P.A.V.: Differential oriented image foresting transform segmentation by seed competition. In: Baudrier, É., Naegel, B., Krähenbühl, A., Tajine, M. (eds.) DGMM 2022. LNCS, vol. 13493, pp. 300–311. Springer, Cham (2022). https://doi.org/10.1007/978-3-031-19897-7_24

6. Condori, M.A., Miranda, P.A.: Differential oriented image foresting transform and its applications to support high-level priors for object segmentation. J. Math. Imaging Vis. **65**, 802–817 (2023)

7. Conze, P.H., Andrade-Miranda, G., Singh, V.K., Jaouen, V., Visvikis, D.: Current and emerging trends in medical image segmentation with deep learning. IEEE Trans. Radiat. Plasma Med. Sci. **7**, 545–569 (2023)

8. Falcão, A., Stolfi, J., Lotufo, R.: The image foresting transform: theory, algorithms, and applications. IEEE Trans. Pattern Anal. Mach. Intell. **26**(1), 19–29 (2004)

9. Farabet, C., Couprie, C., Najman, L., LeCun, Y.: Learning hierarchical features for scene labeling. IEEE Trans. Pattern Anal. Mach. Intell. **35**, 1915–1929 (2013)

10. Hu, X., Fuxin, L., Samaras, D., Chen, C.: Topology-Preserving Deep Image Segmentation. Curran Associates Inc., Red Hook (2019)

11. Jin, X., Che, J., Chen, Y.: Weed identification using deep learning and image processing in vegetable plantation. IEEE Access **9**, 10940–10950 (2021)

12. Kirillov, A., et al.: Segment anything. arXiv preprint arXiv:2304.02643 (2023)

13. Leon, L.M., Ciesielski, K.C., Miranda, P.A.: Efficient hierarchical multi-object segmentation in layered graphs. Math. Morphol. Theory Appl. **5**(1), 21–42 (2021). https://doi.org/10.1515/mathm-2020-0108

14. Lézoray, O., Grady, L.: Image Processing and Analysis with Graphs: Theory and Practice. CRC Press, Boca Raton (2012)

15. Magadza, T., Viriri, S.: Deep learning for brain tumor segmentation: a survey of state-of-the-art. J. Imaging **7**(2), 19 (2021)

16. Maninis, K.K., Caelles, S., Pont-Tuset, J., Gool, L.V.: Deep extreme cut: from extreme points to object segmentation. In: IEEE Conference on Computer Vision and Pattern Recognition, pp. 616–625 (2018)

17. Mansilla, L.A.C., Miranda, P.A.V.: Oriented image foresting transform segmentation: connectivity constraints with adjustable width. In: 29th SIBGRAPI Conference on Graphics, Patterns and Images, pp. 289–296, October 2016

18. Mansilla, L.A.C., Miranda, P.A.V., Cappabianco, F.A.M.: Oriented image foresting transform segmentation with connectivity constraints. In: 2016 IEEE International Conference on Image Processing (ICIP), pp. 2554–2558, September 2016

19. Mansilla, L., Miranda, P.: Image segmentation by oriented image foresting transform: handling ties and colored images. In: 18th International Conference on Digital Signal Processing, Greece, pp. 1–6, July 2013

20. Mansilla, L., Miranda, P.: Image segmentation by oriented image foresting transform with geodesic star convexity. In: 15th International Conference on Computer Analysis of Images and Patterns (CAIP), York, UK, vol. 8047, pp. 572–579, August 2013

21. Miranda, P., Falcao, A., Spina, T.: Riverbed: a novel user-steered image segmentation method based on optimum boundary tracking. IEEE Trans. Image Process. **21**(6), 3042–3052 (2012)

22. Miranda, P., Mansilla, L.: Oriented image foresting transform segmentation by seed competition. IEEE Trans. Image Process. **23**(1), 389–398 (2014)

23. de Moraes Braz, C., Miranda, P.A., Ciesielski, K.C., Cappabianco, F.A.: Optimum cuts in graphs by general fuzzy connectedness with local band constraints. J. Math. Imaging Vis. **62**, 659–672 (2020)

24. Nigri, E., Ziviani, N., Cappabianco, F., Antunes, A., Veloso, A.: Explainable deep CNNs for MRI-based diagnosis of Alzheimer's disease. In: 2020 International Joint Conference on Neural Networks (IJCNN), pp. 1–8. IEEE (2020)

25. Oliveira, D.E.C., Demario, C.L., Miranda, P.A.V.: Image segmentation by relaxed deep extreme cut with connected extreme points. In: Lindblad, J., Malmberg, F., Sladoje, N. (eds.) DGMM 2021. LNCS, vol. 12708, pp. 441–453. Springer, Cham (2021). https://doi.org/10.1007/978-3-030-76657-3_32

26. Phuah, C.L., Chen, Y., Strain, J.F., Yechoor, N., Laurido-Soto, O.J., Ances, B.M., et al.: Association of data-driven white matter hyperintensity spatial signatures with distinct cerebral small vessel disease etiologies. Neurology **99**(23), e2535–e2547 (2022)

27. Rani, V., Nabi, S.T., Kumar, M., Mittal, A., Kumar, K.: Self-supervised learning: a succinct review. Arch. Comput. Methods Eng. **30**(4), 2761–2775 (2023)

28. Raza, K., Singh, N.K.: A tour of unsupervised deep learning for medical image analysis. Curr. Med. Imaging **17**(9), 1059–1077 (2021)

29. Sampath, A., Bijapur, P., Karanam, A., Umadevi, V., Parathodiyil, M.: Estimation of rooftop solar energy generation using satellite image segmentation. In: 2019 IEEE 9th International Conference on Advanced Computing (IACC), pp. 38–44 (2019)

30. Sofiiuk, K., Petrov, I.A., Konushin, A.: Reviving iterative training with mask guidance for interactive segmentation. In: 2022 IEEE International Conference on Image Processing (ICIP), pp. 3141–3145. IEEE (2022)

31. Vercio, L.L., et al.: Supervised machine learning tools: a tutorial for clinicians. J. Neural Eng. **17**(6), 062001 (2020)

32. Xu, N., Price, B., Cohen, S., Yang, J., Huang, T.: Deep GrabCut for object selection. In: Proceedings of the British Machine Vision Conference (BMVC), pp. 182.1–182.12. BMVA Press, September 2017

33. Yasuda, Y.D., Cappabianco, F.A., Martins, L.E.G., Gripp, J.A.: Automated visual inspection of aircraft exterior using deep learning. In: Anais Estendidos do XXXIV Conference on Graphics, Patterns and Images, pp. 173–176. SBC (2021)

34. Yu, X., Ye, X., Gao, Q.: Pipeline image segmentation algorithm and heat loss calculation based on gene-regulated apoptosis mechanism. Int. J. Press. Vessels Pip. **172**, 329–336 (2019)

35. Zeng, R., Wen, Y., Zhao, W., Liu, Y.J.: View planning in robot active vision: a survey of systems, algorithms, and applications. Comput. Visual Media **6**, 225–245 (2020)

36. Zhang, M., Zhou, Y., Zhao, J., Man, Y., Liu, B., Yao, R.: A survey of semi-and weakly supervised semantic segmentation of images. Artif. Intell. Rev. **53**, 4259–4288 (2020)

Discrete and Combinatorial Topology

Elastic Analysis of Augmented Curves and Constrained Surfaces

Esfandiar Nava-Yazdani[✉][iD]

Zuse Institute Berlin, Berlin, Germany
navayazdani@zib.de
https://www.zib.de/members/navayazdani

Abstract. The square root velocity transformation is crucial for efficiently employing the elastic approach in functional and shape data analysis of curves. We study fundamental geometric properties of curves under this transformation. Moreover, utilizing natural geometric constructions, we employ the approach for intrinsic comparison within several classes of surfaces and augmented curves, which arise in the real world applications such as tubes, ruled surfaces, spherical strips, protein molecules and hurricane tracks.

Keywords: Elastic shape analysis · Tube · Manifold-valued · Ruled surface · Hurricane track

1 Introduction

Metric comparison of curves is a core task in a wide range of application areas such as morphology, image and shape analysis, computer vision, action recognition and signal processing. Thereby, a Riemannian structure is highly desirable, since it naturally provides powerful tools, beneficial for such applications.

In the recent years, the use of Riemannian metrics for the study of sequential data, such as shapes of curves, trajectories given as longitudinal data or time series, has rapidly grown. In elastic analysis of curves, one considers deformations caused from both bending and stretching. A Riemannian metric, which quantifies the amount of those deformations is called elastic (cf. [18,19]). Therein, in contrast to landmark-based approaches (cf. [12,21,22]), one considers whole continuous curves instead of finite number of curve-points. Consequently, the underlying spaces are infinite dimensional and computational cost becomes a significant issue. The square root velocity (SRV) framework provides a convenient and numerically efficient approach for analysing curves via elastic metrics and has been widely used in the recent years (cf. [3,4,11,14] and the comprehensive work [24]).

In many applications the curves are naturally manifold-valued. For instance, Lie groups such as the Euclidean motion group, or more generally, symmetric spaces including the Grassmannian and the Hadamard-Cartan manifold of positive definite matrices are widely used in modelling of real world applications.

S. Brunetti et al. (Eds.): DGMM 2024, LNCS 14605, pp. 353–363, 2024.
https://doi.org/10.1007/978-3-031-57793-2_27

Extensions of SRV framework from euclidean to general manifold-valued data can be found in [9, 13, 25–27].

Our contributions are the following. We expose for plane curves the behaviour of speed and curvature under the SRV transformation and geometric invariants. Moreover, we apply the elastic approach to augmented curves, determining certain classes of surfaces, tubes, ruled surfaces and spherical strips, as well as hurricane tracks considered with their intensities. We recall that with distance and geodesic at hand, significant ingredients of statistical analysis such as mean and principal geodesic components as well as approximation and modelling concepts such as splines can be computed.

This paper is organized as follows. Section 2, presents the Riemannian setting and notations. Section 3 is devoted to applications. Therein, we consider time series, for which in addition to spatial data, auxiliary information give rise to augmented curves and some classes of surfaces generated by them. Thereby, we apply the elastic approach to both euclidean and spherical trajectories. Future prospects and concluding remarks are presented in Sect. 4.

For the convenience of those readers primary interested in the applications, we mention that, advanced parts and details from differential geometry, presented in Sect. 2, can be skipped. Thereby, the essential point is the use of a framework (SRV) for computation of shortest paths on the spaces of curves and their shapes.

2   Riemannian Framework

2.1   Preliminaries

For the background material on Riemannian geometry, we refer to [8] and [10]. Let (M, g) be a finite dimensional Riemannian manifold and $\mathcal{M}$ the Fréchet manifold of smooth immersed curves from $\mathcal{D}$ in M, where $\mathcal{D}$ denotes either the unit circle S^1 or the unit interval $I := [0, 1]$ for closed or open curves respectively. Moreover, we denote the group of orientation preserving diffeomorphisms on $\mathcal{D}$ by $Diff^+$. The following reparametrization invariance is crucial for a Riemannian metric G on $\mathcal{M}$:

$$G_{c \circ \varphi}(h \circ \varphi, k \circ \varphi) = G_c(h, k),$$

for any $c \in \mathcal{M}$, $h, k \in T_c \mathcal{M}$ and $\varphi \in Diff^+$. The above equivariance ensures that the induced distance function satisfies the following, which is often desirable in applications:

$$d(c_0 \circ \varphi, c_1 \circ \varphi) = d(c_0, c_1),$$

for any two curves c_0 and c_1 in $\mathcal{M}$. Similarly, denoting the isometry group of M by $Isom(M)$ and the tangent map of $F \in Isom(M)$ by TF, the invariance

$$G_{F \circ c}(TF \circ h, TF \circ k) = G_c(h, k),$$

ensures that

$$d(F \circ c_0, F \circ c_1) = d(c_0, c_1).$$

With the above invariances, we can divide out the spaces $Isom(M)$ and $Diff^+$, and consider the natural induced distance d^S on the quotient space

$$\mathcal{S} = \mathcal{M}/(Diff^+ \times Isom(M))$$

given by

$$d^S([c_0], [c_1]) = \inf \{d(c_0, f \circ c_1 \circ \varphi) : \varphi \in Diff^+, f \in Isom(M)\}$$
$$= \inf \{d(f \circ c_0 \circ \varphi, c_1) : \varphi \in Diff^+, f \in Isom(M)\}.$$

In the context of shape analysis of curves, $\mathcal{M}$ and $\mathcal{S}$ are called the pre-shape and shape space, respectively. Note that the order of quotient operations does not matter, since the left action of $Isom(M)$ and the right action of $Diff^+$ commute. $\mathcal{M}/Diff^+$ is the space of unparametrized curves and its inherited distance reads

$$\inf \{d(c_0, c_1 \circ \varphi) : \varphi \in Diff^+\}.$$

We remark that particular essential challenges are due to the fact that some basic concepts and results from finite dimensional differential geometry such as Hopf-Rinow theorem, do not carry over to the infinite dimensional case. Now, let ∇ be the Levi-Civita connection of M and denote the arc length parameter, speed and unit tangent of c by θ, ω and T respectively. Thus, we have $\omega = |\dot{c}|$, $d\theta = \omega dt$ and $T = \frac{\dot{c}}{\omega}$, where dot stands for derivation with respect to the parameter t.

Due to a remarkable result in [16] the geodesic distance induced by the simplest natural choice, the L^2-metric

$$G_c^{L^2}(h, k) = \int_{\mathcal{D}} g_c(h, k) \, d\theta,$$

always vanishes. Consequently, some stronger Sobolev metrics have been considered in several works including [5, 7, 17]. They are given by

$$G_c(h, k) = \sum_{i=0}^{n} \int_{\mathcal{D}} a_i g_c(\nabla_T^i h, \nabla_T^i k) \, d\theta,$$

with a_1 non-vanishing and all a_i non-negative, distinguish the curves. We consider first order metrics with constant coefficients. We remark that the coefficients a_i can be chosen such that the metric is scale invariant, which is a desired property for some applications in shape analysis. A family of certain weighted Sobolev-type metrics, the so-called elastic metrics, based on a decomposition of derivatives of the vector fields into normal and tangent components, has been introduced in [18,19]:

$$G_c^{a,b}(h, k) = \int_{\mathcal{D}} a g_c((\nabla_T h)^\top, (\nabla_T k)^\top) + b g_c((\nabla_T h)^\perp, (\nabla_T k)^\perp) \, d\theta,$$

with $4b \geq a > 0$. In this work, we use the square root velocity (SRV) framework, which allows for a convenient and computationally efficient elastic approach. The main tool in this framework is the square root velocity transformation, which for euclidean M reads

$$q : c \mapsto \frac{\dot{c}}{\sqrt{|\dot{c}|}}.$$

It isometrically maps curves modulo translations, with the metric $G^{1,1/4}$ to $\mathcal{M}$ with the flat L^2-metric given by

$$G^0(v, w) = \int_D g(v(t), w(t)) dt.$$

This metric is frequently called (cf. [2, 6, 15]) flat, to emphasize its footpoint independence. Note that the elastic metric $G^{1,1}$ corresponds to the first order Sobolev metric with $a_0 = 0$ and $a_1 = 1$. We remark, that for plane curves, the work [23] has extended the SRV transformation to general parameters $a, b > 0$. For further reading on the SRV framework and applications in shape analysis, we refer to [11, 14] (numerical aspects), the survey [6] and particularly, the comprehensive work [24].

2.2   Plane Curves

A natural question that arises is, how essential geometric characteristics of a curve behave under the SRV transformation. In the following, we provide an answer for speed and curvature in the case of plane curves. Let $M = \mathbb{R}^2$, $\tilde{c} := q(c)$ and denote the curvature of c by κ. Note that $\tilde{c}$ does not need to be an immersion.

Proposition 1. *Denoting the speed of $\tilde{c}$ by $\tilde{\omega}$, we have*

$$\tilde{\omega} = \sqrt{\frac{\dot{\omega}^2}{4\omega} + \omega^3 \kappa^2}. \tag{1}$$

Moreover, $\tilde{c}$ is an immersion if and only if κ and $\dot{\omega}$ have no common zeros. In this case,

$$\tilde{\kappa}\tilde{\omega} = \kappa\omega + \dot{\varphi}, \tag{2}$$

where $\tilde{\kappa}$ denotes the curvature of $\tilde{c}$ and

$$\varphi := \arctan\left(\frac{2\omega^2 \kappa}{\dot{\omega}}\right).$$

Proof. Let N denote the unit normal of c. With the shorthand notations $\alpha := \sqrt{\omega}$ and $\beta := \alpha^3 \kappa$, a straightforward application of the Frenet equations $\dot{T} = \omega\kappa N$ and $\dot{N} = -\omega\kappa T$, yield

$$\dot{\tilde{c}} = \dot{\alpha}T + \beta N,$$

$$\ddot{\tilde{c}} = (\ddot{\alpha} - \frac{\beta^2}{\alpha})T + (\dot{\beta} + \frac{\dot{\alpha}\beta}{\alpha})N.$$

Thus, we have

$$\tilde{\omega} = \sqrt{\dot{\alpha}^2 + \beta^2},$$

immediately implying (1). Obviously, zeros of $\tilde{\omega}$ are common zeros of κ and $\dot{\omega}$. Thus, $\tilde{c}$ is an immersion if and only if κ and $\dot{\omega}$ have no common zeros. In this case, $\tilde{\kappa}$ and $\varphi = \arctan(\beta/\dot{\alpha}) = \arctan\left(\frac{2\omega^2\kappa}{\tilde{\omega}}\right)$ are well-defined and

$$\tilde{\kappa}\tilde{\omega}^3 = \tilde{\omega}^2\beta/\alpha + \dot{\alpha}\dot{\beta} - \ddot{\alpha}\beta,$$

which immediately implies the curvature formula (2).

Next, we apply the proposition to study some geometric quantities, which are invariant under the SRV transformation. For closed curves, integrating the curvature above formula over $\mathcal{D} = S^1$ (note that in this case, $\tilde{\omega} > 0$ almost everywhere), we see that the SRV transformation preserves the total curvature and particularly the turning number. Moreover, $\kappa\omega$ is preserved if and only if $\kappa = a\frac{d}{dt}\left(\frac{1}{\omega}\right)$ with a constant a.

Clearly, with κ and ω at hand, utilizing Frenet equations, we can compute c up to rigid motions. The following explicit solution is an immediate application of the above proposition. In light of the above proposition, immersed curves, which are mapped to straight lines, can easily be determined as follows.

Example 1. Let a, b, A be constants with ab, $A > 0$, $\omega(t) = A/\sin^2(at + b)$ and $\kappa = a/\omega$. A straightforward computation, utilizing the curvature formula (2), implies $\tilde{\kappa} = 0$.

2.3   Curves in Homogeneous Spaces

For the background material on Lie groups and homogeneous spaces, we refer to [10]. The works [13,27] provide extensions of the SRV framework for euclidean curves to the case of general manifolds. The former has high computational cost, while the latter, transported SRV, depends on a reference point and also suffers from distortion or bias caused by holonomy effects. We use the natural extension to homogeneous spaces exposed in [9,26]. For reader's convenience, we sketch the core ingredients of the approach and refer to the mentioned works for details and some applications.

Let M be a homogeneous space, i.e., $M = H/K$, where K is a closed Lie subgroup of a Lie Group H. Let $\|\cdot\|$ denote the induced norm by a left invariant metric on H, L the tangent map of the left translation, and $Imm(\mathcal{D}, H)$ the space of immersed curves from $\mathcal{D}$ to H. The SRV transformation is given by $Q(\alpha) = (\alpha(0), q(\alpha))$, where

$$q(\alpha) = \frac{L_{\alpha^{-1}}\dot{\alpha}}{\sqrt{\|\dot{\alpha}\|}}$$

Here, $\alpha^{-1}(t)$ denotes the inverse element of $\alpha(t)$ in H and $\mathcal{H}$ the Lie algebra of H. The map Q is a bijection from $\mathcal{M}(\mathcal{D}, H)$ onto $H \times L^2(\mathcal{D}, \mathcal{H})$. Now, M can be equipped with the Riemannian metric given by the pullback of the product metric of $H \times L^2(\mathcal{D}, \mathcal{H})$ using the map Q and horizontal lifting. Let c_1 and c_2 be immersed curves in M with horizontal lifts α_1 and α_2 respectively. The induced distance on M reads

$$d(c_1, c_2) = inf \left\{ \sqrt{d_H^2(\alpha_1(0), \alpha_2(0)x) + \|q(\alpha_1) - Ad_{x^{-1}}(q(\alpha_2)\|_{L^2}^2} : x \in K \right\}.$$

3   Applications

Frequently, besides spatiotemporal data, represented by a curve γ in a manifold M, there are additional or auxiliary information associated with the curve, thus with the same time-correspondence. These can jointly with γ be comprised and represented as a so-called augmented curve $\tilde{\gamma}$ in a higher dimensional manifold $\tilde{M}$. In some applications, the curve $\tilde{\gamma}$ uniquely determines a submanifold N of M via a natural construction. An important example is provided, when $\tilde{M}$ is a submanifold of the tangent bundle of M, where the auxiliary information is represented as a vector field along γ and the construction is given by the Riemannian exponential map. Significant special cases occur, when M is $\mathbb{R}^3$ or the unit two-sphere S^2 and N a surface. In the next two subsections, we consider certain classes of surfaces in $\mathbb{R}^3$, which often arise in applications and are determined by augmented curves in $\mathbb{R}^4$. In the last two subsections, we consider certain spherical regions as well as hurricane tracks together with their intensities. In both cases, we utilize the Riemannian distance from Subsect. 2.3 to $S^2 \times \mathbb{R}$, which is a homogeneous space (recall that S^2 can be identified with $SO(3)/SO(2)$).

For our example applications, we present geodesic paths representing deformations, minimizing the elastic energy within the SRV framework. We remark, that in a Riemannian setting, distance and geodesics are essential Building blocks for many major issues in the morphology and shape analysis, such as computation of mean and test statistics as well as principal component or geodesic analysis. Moreover, besides statistical analysis, also some methods for clustering and classification use Riemannian metrics and geodesics.

For the code implementing our approach, which particularly includes Riemannian optimization for the computation of geodesic paths, we utilized our publicly available python package https://github.com/morphomatics, introduced in [1].

3.1   Tubes

A tube or canal surface c is a one-parameter family of circles, whose centers constitute a regular curve γ such that the circles are perpendicular to γ. More precisely, denoting the radii of the circles by r,

$$c(s, .) = \gamma + r(N \cos s + B \sin s), \ 0 \leq s \leq 2\pi,$$

where N and B are the normal and binormal of the curve $\gamma = \gamma(t)$, $t \in \mathcal{D}$, resp. Due to the unique correspondence of c to (γ, r), comparison of tubes reduces to comparison of curves in $\mathbb{R}^4$. Figure 1 shows some examples of shortest paths of tubes. Real world applications include a variety of fields such as examination of vein, pipes, capsules and plant roots. Clearly, tubes include surfaces of revolution.

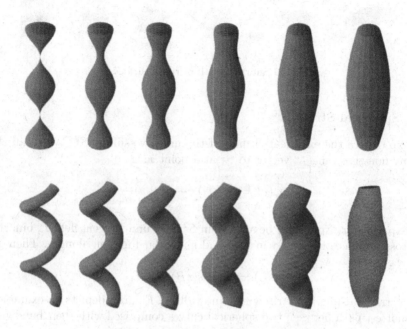

Fig. 1. Two shortest paths of tubes

3.2 Ruled Surfaces

A ruled surface is formed by moving a straight line segment (possibly with varying length) along a base curve. More precisely, let γ be a curve in $\mathbb{R}^3$ and v a unit vector field along γ. Then

$$c(s, .) = \gamma + sv, \ s \in I,$$

parametrizes a ruled surface generated by (γ, v). Figure 2 depicts an example, where each surface consists of straight line segments connecting the blue (for better visibility) curves γ and $\gamma + v$. The class of ruled surfaces includes many prominent surfaces such as cone, cylinder, helicoid (a minimal surface) and Möbius strip. They arise in manufacturing (construction by bending a flat sheet), cartography, architecture and biochemistry (secondary and tertiary structure of protein molecules).

Fig. 2. Shortest path of ruled surfaces

3.3 Spherical Strips

Let exp denote the exponential map of the unit two-sphere S^2. We recall that for any non-zero tangent vector to S^2 at a point x:

$$\exp_x(v) = \cos(|v|)x + \sin(|v|)\frac{v}{|v|}$$

and $\exp_x(0) = x$. Now, let γ be a curve in S^2 with unit tangent field T, binormal B (cross product of T and γ in $\mathbb{R}^3$), and r a scalar function along γ. Then, the map c given by

$$c(s,.) := \exp_\gamma s(rB), \ s \in I,$$

parametrizes a spherical strip with bandwidth r. Figure 3 depicts an example of the shortest path between two spherical curves comprised with their bandwidth functions visualised as strips.

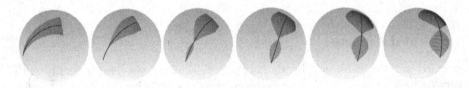

Fig. 3. Shortest path of spherical strips

3.4 Hurricane Tracks

Hurricanes belong to the most extreme natural phenomena and can cause major impacts regarding environment, economy, etc. Intensity of a hurricane is determined by the maximum sustained wind (maxwind), monotonically classifying the storms into categories (due to Saffir-Simpson wind scale; for instance, maxwind $\geq$137 knots corresponds to category 5). Due to their major impacts on economy, human life and environment, as well as extreme variability and complexity, hurricanes have been studies in a large number of works. For our example, we used

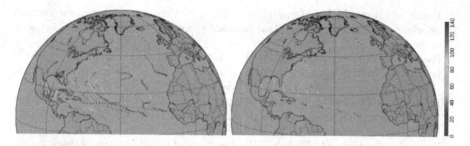

Fig. 4. 2010 Atlantic hurricane tracks (left) and the shortest path between two of them with color-coded maximum sustained wind (in knots). (Color figure online)

the HURDAT 2 database provided by the U.S. National Oceanic and Atmospheric Administration publicly available on https://www.nhc.noaa.gov/data/, supplying latitude, longitude, and maxwind on a 6 h base of Atlantic hurricanes.

We represent the tracks as discrete trajectories in S^2. For further details and comparison with other approaches, we refer to [24,25] and the recent work [20]. The latter, also provides statistical analysis and a classification of hurricane tracks in terms of their intensities. Figure 4 illustrates this data set with a visualization of the 2010 hurricane tracks and a shortest path, where the intensities, considered as auxiliary information, are color-marked.

4 Conclusion

In this paper, we analysed the behaviour of speed and curvature under the square root velocity framework for elastic approach to plane curves. Moreover, we applied an extension of this framework to homogeneous Spaces, to metrically compare augmented curves and special surfaces, generated by those curves, using a natural construction via the Riemannian exponential map. Our approach, allows for computationally efficient determination of geodesic paths in the shape spaces of the respective classes of surfaces. Our example applications include tubes, ruled surfaces, spherical strips and hurricane tracks. Future work includes further real world applications, particularly concerning statistical analysis of longitudinal data such as comparison of group wise trends within a hierarchical model as well as classification and prediction.

Acknowledgements. This work was supported through the German Research Foundation (DFG) via individual funding (project ID 499571814).

References

1. Ambellan, F., Hanik, M., von Tycowicz, C.: Morphomatics: geometric morphometrics in non-Euclidean shape spaces (2021). https://doi.org/10.12752/8544. https://morphomatics.github.io/

362   E. Nava-Yazdani

2. Bauer, M., Bruveris, M., Marsland, S., Michor, P.: Constructing reparametrization invariant metrics on spaces of plane curves. Differential Geometry (2012). https://arxiv.org/pdf/1207.5965.pdf
3. Bauer, M., Bruveris, M., Charon, N., Møller-Andersen, J.: A relaxed approach for curve matching with elastic metrics. ESAIM: Control Optim. Calc. Var. **25**, (March 2018). https://doi.org/10.1051/cocv/2018053
4. Bauer, M., Bruveris, M., Harms, Philipp Michor, P.W.: Soliton solutions for the elastic metric on spaces of curves. Discret. Contin. Dyn. Syst. A **38**, 1161–1185 (2018). https://doi.org/10.3934/dcds.2018049
5. Bauer, M., Bruveris, M., Michor, P.W.: Overview of the geometries of shape spaces and diffeomorphism groups. J. Math. Imaging Vis. **50**(1–2), 60–97 (2014)
6. Bauer, M., Charon, N., Klassen, E., Brigant, A.L.: Intrinsic Riemannian metrics on spaces of curves: theory and computation. arXiv preprint (2020). https://arxiv.org/abs/2003.05590
7. Bauer, M., Harms, P., Michor, P.W., et al.: Sobolev metrics on the manifold of all Riemannian metrics. J. Differ. Geom. **94**(2), 187–208 (2013)
8. do Carmo, M.P.: Riemannian Geometry. Mathematics: Theory and Applications, 2nd edn. Birkhäuser, Boston (1992)
9. Celledoni, E., Eidnes, S., Schmeding, A.: Shape analysis on homogeneous spaces: a generalised SRVT framework. In: Celledoni, E., Di Nunno, G., Ebrahimi-Fard, K., Munthe-Kaas, H.Z. (eds.) Abelsymposium 2016. AS, vol. 13, pp. 187–220. Springer, Cham (2018). https://doi.org/10.1007/978-3-030-01593-0_7
10. Gallot, S., Hullin, D., Lafontaine, J.: Riemannian Geometry. Universitext, 3rd edn. Springer, Berlin (2004)
11. Huang, W., Gallivan, K.A., Srivastava, A., Absil, P.A.: Riemannian optimization for registration of curves in elastic shape analysis. J. Math. Imaging Vis. **54**(3), 320–343 (2016)
12. Kendall, D., Barden, D., Carne, T., Le, H.: Shape and Shape Theory. Wiley, New York (1999)
13. Le Brigant, A.: Computing distances and geodesics between manifold-valued curves in the SRV framework. J. Geom. Mech. **9**(2), (2017)
14. Liu, W., Srivastava, A., Zhang, J.: Protein structure alignment using elastic shape analysis. In: Proceedings of the First ACM International Conference on Bioinformatics and Computational Biology, pp. 62–70 (2010)
15. Michor, P., Mumford, D., Shah, J., Younes, L.: A metric on shape space with explicit geodesics. Atti Accad. Naz. Lincei Cl. Sci. Fis. Mat. Natur. Rend. Lincei (9) Mat. Appl. **19**, (July 2007). https://doi.org/10.4171/RLM/506
16. Michor, P.W., Mumford, D.: Vanishing geodesic distance on spaces of submanifolds and diffeomorphisms. Doc. Math. **10**, 217–245 (2005)
17. Michor, P.W., Mumford, D.: An overview of the Riemannian metrics on spaces of curves using the Hamiltonian approach. Appl. Comput. Harmon. Anal. **23**(1), 74–113 (2007)
18. Mio, W., Srivastava, A., Joshi, S.H.: On shape of plane elastic curves. Int. J. Comput. Vis. **73**, 307–324 (2006)
19. Mio, W., Srivastava, A., Joshi, S.: On shape of plane elastic curves. Int. J. Comput. Vis. **73**, 307–324 (2007). https://doi.org/10.1007/s11263-006-9968-0
20. Nava-Yazdani, E., Ambellan, F., Hanik, M., von Tycowicz, C.: Sasaki metric for spline models of manifold-valued trajectories. Comput. Aided Geom. Des. **104**, 102220 (2023). https://doi.org/10.1016/j.cagd.2023.102220

21. Nava-Yazdani, E., Hege, H.C., Sullivan, T.J., von Tycowicz, C.: Geodesic analysis in Kendall's shape space with epidemiological applications. J. Math. Imaging Vis. 1–11 (2020)
22. Nava-Yazdani, E., Hege, H.C., von Tycowicz, C.: A hierarchical geodesic model for longitudinal analysis on manifolds. J. Math. Imaging Vis. **64**(4), 395–407 (2022). https://doi.org/10.1007/s10851-022-01079-x
23. Needham, T., Kurtek, S.: Simplifying transforms for general elastic metrics on the space of plane curves. SIAM J. Imaging Sci. **13**(1), 445–473 (2020). https://doi.org/10.1137/19M1265132
24. Srivastava, A., Klassen, E.P.: Functional and Shape Data Analysis, vol. 1. Springer, New York (2016)
25. Su, Z., Klassen, E., Bauer, M.: The square root velocity framework for curves in a homogeneous space. In: 2017 IEEE Conference on Computer Vision and Pattern Recognition Workshops (CVPRW), pp. 680–689 (2017)
26. Su, Z., Klassen, E., Bauer, M.: Comparing curves in homogeneous spaces. Differ. Geom. Appl. **60**, 9–32 (2018)
27. Zhang, Z., Su, J., Klassen, E., Le, H., Srivastava, A.: Rate-invariant analysis of covariance trajectories. J. Math. Imaging Vis. **60**, 1306–1323 (2018)

Morse Frames

Gilles Bertrand[ID] and Laurent Najman[✉][ID]

Univ Gustave Eiffel, CNRS, LIGM, 77454 Marne-la-Vallée, France
{gilles.bertrand,laurent.najman}@esiee.fr

Abstract. In the context of discrete Morse theory, we introduce Morse frames, which are maps that associate a set of critical simplexes to each simplex of a given complex. The main example of Morse frames are the Morse references. In particular, Morse references allow computing Morse complexes, an important tool for homology. We highlight the link between Morse references and gradient flows. We also propose a novel presentation of the annotation algorithm for persistent cohomology, as a variant of a Morse frame. Finally, we propose another construction, that takes advantage of the Morse reference for computing the Betti numbers of a complex in mod 2 arithmetic.

Keywords: Homology · Cohomology · Discrete Morse Theory

1 Introduction

In this paper, we aim at developing new concepts and algorithm schemes for computing topological invariants for simplicial complexes, such as cycles, cocycles and Betti numbers (Sect. 2). In [2], one of the authors of the present paper, introduces a novel, sequential, presentation of discrete Morse theory [8], termed *Morse sequences*. In Sect. 3, we introduce *Morse frames*, that are maps that associate a set of critical simplexes to each simplex. These maps allow adding information to Morse sequences, so that we can compute cycles and cocycles that detect "holes". The main example of Morse frames is called the Morse reference (Sect. 4), and is a by-product of Morse sequences. We discuss the link between reference maps and gradient flows. This leads us to the Morse complex. We then see (Sect. 5) that Morse frames allows for a novel presentation of annotations [6] for computing persistent cohomology. Then, inspired by the annotation technique, we propose (Sect. 6) an efficient construction for computing Betti numbers in mod 2 arithmetic. We then discuss (Sect. 7) how to implement the notions presented in the paper. Finally, we conclude the paper.

2 Simplicial Complexes, Homology, and Cohomology

2.1 Simplicial Complexes

Let K be a finite family composed of non-empty finite sets. The family K is a *(simplicial) complex* if $\sigma \in K$ whenever $\sigma \neq \emptyset$ and $\sigma \subseteq \tau$ for some $\tau \in K$.

S. Brunetti et al. (Eds.): DGMM 2024, LNCS 14605, pp. 364–376, 2024.
https://doi.org/10.1007/978-3-031-57793-2_28

An element of a simplicial complex K is *a simplex of K*, or *a face of K*. A *facet of K* is a simplex of K that is maximal for inclusion. The *dimension* of $\sigma \in K$, written $dim(\sigma)$, is the number of its elements minus one. If $dim(\sigma) = p$, we say that σ is a *p-simplex*. We denote by $K^{(p)}$ the set of all p-simplexes of K.

We recall the definitions of the collapses/expansions operators [15].

Let K, L be simplicial complexes. Let $\sigma \in K^{(p)}$, $\tau \in K^{(p+1)}$. The couple (σ, τ) is a *free pair for K*, or a *free p-pair for K*, if τ is the only face of K that contains σ. Thus, τ is necessarily a facet of K. If (σ, τ) is a free (p-)pair for K, then $L = K \setminus (\sigma, \tau)$ is *an elementary (p-)collapse of K*, and K is *an elementary (p-)expansion of L*. We say that K *collapses onto L*, or that L *expands onto K*, if there exists a sequence $\langle K = K_0, \ldots, K_k = L \rangle$, such that K_i is an elementary collapse of K_{i-1}, $i \in [1, k]$.

2.2   Homology and Cohomology

Let K be a simplicial complex. We write $K[p]$ for the set composed of all subsets of $K^{(p)}$. Also, we set $K^{(-1)} = \emptyset$ and $K[-1] = \{\emptyset\}$. Each element of $K[p]$, $p \geq -1$, is a *p-chain of K*. The symmetric difference of two elements of $K[p]$ endows $K[p]$ with the structure of a vector space over the field $\mathbb{Z}_2 = \{0, 1\}$. The set $K^{(p)}$ is a basis for this vector space. Within this structure, a chain $c \in K[p]$ is written as a sum $\sum_{\sigma \in c} \sigma$, the chain $c = \emptyset$ being written 0. The sum of two chains is obtained using the modulo 2 arithmetic.

Let K be a simplicial complex. As we are dealing with a finite simplicial complex, boundary and coboundary operators can be defined as operators on $K[p]$. If $\sigma \in K^{(p)}$, with $p \geq 0$, we set:
$$\partial(\sigma) = \{\tau \in K^{(p-1)} \mid \tau \subset \sigma\} \text{ and } \delta(\sigma) = \{\tau \in K^{(p+1)} \mid \sigma \subset \tau\}.$$
The *boundary operator* $\partial_p : K[p] \to K[p-1]$, $p \geq 0$, is such that, for each $c \in K[p]$, $\partial_p(c) = \sum_{\sigma \in c} \partial(\sigma)$, with $\partial_p(\emptyset) = 0$.
The *coboundary operator* $\delta^p : K[p] \to K[p+1]$, $p \geq -1$, is such that, for each $c \in K[p]$, $\delta^p(c) = \sum_{\sigma \in c} \delta(\sigma)$, with $\delta^p(\emptyset) = 0$.
For each $p \geq 0$, we have $\partial_p \circ \partial_{p+1} = 0$ and $\delta^p \circ \delta^{p-1} = 0$.
We define four subsets of $K[p]$, $p \geq 0$, which are vector spaces over $\mathbb{Z}_2$:

- the set $Z_p(K)$ of *p-cycles of K*, $Z_p(K)$ is the kernel of ∂_p;
- the set $B_p(K)$ of *p-boundaries of K*, $B_p(K)$ is the image of ∂_{p+1};
- the set $Z^p(K)$ of *p-cocycles of K*, $Z^p(K)$ is the kernel of δ^p;
- the set $B^p(K)$ of *p-coboundaries of K*, $B^p(K)$ is the image of δ^{p-1}.

Figure 1 depicts an annulus, with various cycles and cocyles, coloured in blue. In Fig. 1a, we see a 1-cycle that is the 1-boundary of the two pink triangles. In Fig. 1b, we have a 1-cycle that is not a 1-boundary. Such a cycle detects a "hole" by "contouring" it. In Fig. 1c, we see a 1-cocycle which is the 1-coboundary of the four pink points. In Fig. 1d, we have a 1-cocycle that is not a 1-coboundary. Such a cocycle detects a "hole" by "cutting" the annulus.

We also define the following quotient vector spaces:

- $H_p(K) = Z_p(K) \setminus B_p(K)$, which is the pth *homology vector space of K*;
- $H^p(K) = Z^p(K) \setminus B^p(K)$, which is the pth *cohomology vector space of K*.

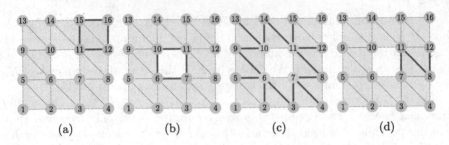

Fig. 1. An annulus, with various cycles and cocyles. See text for details. (Color figure online)

An element h in $H_p(K)$ is such that $h = z + B_p(K)$ for some $z \in Z_p(K)$. We write $h = [z]_p$, which is the *homology class of the cycle z*.

Similarly, an element h in $H^p(K)$ is such that $h = z + B^p(K)$ for some $z \in Z^p(K)$. We write $h = [z]^p$, which is the *cohomology class of the cocycle z*.

Let $\beta_p(K) = dim(H_p(K))$ and $\beta^p(K) = dim(H^p(K))$. We have $\beta_p(K) = \beta^p(K)$ (See [7, Sec. V.1]). The number $\beta_p(K) = \beta^p(K)$ is the pth *Betti number (mod 2) of* K.

3   Morse Sequences and Morse Frames

Let us first introduce the two following basic operators [15].

Let K, L be simplicial complexes. If $\sigma \in K$ is a facet of K, and if $L = K \setminus \{\sigma\}$, we say that L is *an elementary perforation of K*, and that K is *an elementary filling of L*.

The notion of a "Morse sequence" [2] is defined by simply considering expansions and fillings of a simplicial complex.

Definition 1. *Let K be a simplicial complex. A* Morse sequence (on K) *is a sequence* $\overrightarrow{W} = \langle \emptyset = K_0, \ldots, K_k = K \rangle$ *of simplicial complexes such that, for each* $i \in [1, k]$, K_i *is either an elementary expansion or an elementary filling of* K_{i-1}.

Let $\overrightarrow{W} = \langle K_0, \ldots, K_k \rangle$ be a Morse sequence. For each $i \in [1, k]$:

- If K_i is an elementary filling of K_{i-1}, we write $\hat{\sigma}_i$ for the simplex σ such that $K_i = K_{i-1} \cup \{\sigma\}$. We say that the face σ is *critical for* $\overrightarrow{W}$.
- If K_i is an elementary expansion of K_{i-1}, we write $\hat{\sigma}_i$ for the free pair (σ, τ) such that $K_i = K_{i-1} \cup \{\sigma, \tau\}$. We say that $\hat{\sigma}_i$, σ, τ, are *regular for* $\overrightarrow{W}$.

We write $\widehat{W} = \langle \hat{\sigma}_1, \ldots, \hat{\sigma}_k \rangle$, and we say that $\widehat{W}$ is a *(simplex-wise)* Morse sequence. Clearly, $\overrightarrow{W}$ and $\widehat{W}$ are two equivalent forms. We shall pass from one of these forms to the other without notice.

There are several ways to obtain a Morse sequence $\overrightarrow{W}$ from a given complex K. The two following schemes are basic ones to achieve this goal:

1. *The increasing scheme.* We build $\overrightarrow{W}$ from the left to right. Starting from $\emptyset$, we obtain K by iterative expansions and fillings. We say that this scheme is *maximal* if we make a filling only if no expansion can be made.

2. *The decreasing scheme.* We build $\overrightarrow{W}$ from the right to the left. Starting from K, we obtain $\emptyset$ by iterative collapses and perforations. We say that this scheme is *maximal* if we make a perforation only if no collapse can be made.

See [2, Section 7] for a discussion of the differences between these schemes.

Definition 2. *The* gradient vector field *of a Morse sequence* $\overrightarrow{W}$ *is the set of all regular pairs for* $\overrightarrow{W}$. *We say that two Morse sequences* $\overrightarrow{W}$ *and* $\overrightarrow{V}$ *on a given complex* K *are* equivalent *if they have the same gradient vector field.*

It is worth mentioning that there is no loss of generality when using Morse sequences as a presentation of gradient vector fields. In fact, we can prove that the gradient vector field of an arbitrary Morse function may be seen as the gradient vector field of a Morse sequence (see [2]).

Let $\overrightarrow{W}$ be a Morse sequence on K. We write $\ddot{W} = \{\sigma \in K \mid \sigma$ is critical for $\overrightarrow{W}\}$, $\ddot{W}^{(p)} = \{\sigma \in K^{(p)} \mid \sigma$ is critical for $\overrightarrow{W}\}$, and $\ddot{W}^{(-1)} = \emptyset$. For each $p \geq -1$, we write $\ddot{W}[p]$ for the set composed of all subsets of $\ddot{W}^{(p)}$. An element $c \in \ddot{W}[p]$ is a *p-chain of* $\ddot{W}$. We have $\ddot{W}[p] \subseteq K[p]$.

A Morse frame is simply a map which assigns, to each p-simplex of K, a certain set of critical p-simplexes.

Definition 3. *Let* $\overrightarrow{W}$ *be a Morse sequence on a simplicial complex* K. *We say that* Υ *is a* (Morse) frame *on* $\overrightarrow{W}$ *if* Υ *is a map such that:*
$$\Upsilon : \sigma \in K^{(p)} \mapsto \Upsilon(\sigma) \in \ddot{W}[p].$$
If Υ *is a Morse frame on* $\overrightarrow{W}$, *we also denote by* Υ *the map:*
$$\Upsilon : c \in K[p] \mapsto \Upsilon(c) \in \ddot{W}[p], \text{ where } \Upsilon(c) = \sum_{\sigma \in c} \Upsilon(\sigma) \text{ and } \Upsilon(\emptyset) = 0.$$

4   The Morse Reference

A Morse complex is a basic tool for efficiently computing simplicial homology using discrete Morse theory. Since a Morse complex is built solely on critical complexes, its dimension is generally much smaller than the one of the original complex. In this section, we introduce two frames which allow simplifying the construction of a Morse complex.

4.1   Reference and Co-reference

Definition 4. *Let* $\overrightarrow{W}$ *be a Morse sequence and let* Υ', Υ'' *be two Morse frames on* $\overrightarrow{W}$ *such that, for each critical simplex* σ *of* $\overrightarrow{W}$, *we have* $\Upsilon'(\sigma) = \Upsilon''(\sigma) = \{\sigma\}$. *We say that* Υ' *is the* (Morse) reference *of* $\overrightarrow{W}$ *if, for each regular pair* (σ, τ) *of* $\overrightarrow{W}$, *we have* $\Upsilon'(\tau) = 0$ *and* $\Upsilon'(\sigma) = \Upsilon'(\partial(\tau) \setminus \{\sigma\})$.
We say that Υ'' *is the* (Morse) co-reference *of* $\overrightarrow{W}$ *if, for each regular pair* (σ, τ) *of* $\overrightarrow{W}$, *we have* $\Upsilon''(\sigma) = 0$ *and* $\Upsilon''(\tau) = \Upsilon''(\delta(\sigma) \setminus \{\tau\})$. *If* Υ *is the reference of* $\overrightarrow{W}$, *we write* Υ^* *for the co-reference of* $\overrightarrow{W}$.

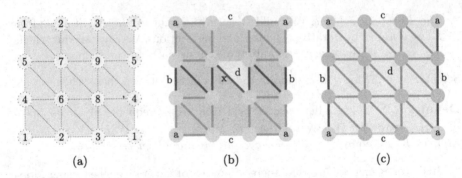

Fig. 2. (a) A torus. Points with the same label are identified. (b) A Morse reference map. (c) A Morse co-reference map. See text for details. (Color figure online)

Thus, if $\varUpsilon$ is the Morse reference of $\overrightarrow{W}$ then, for each regular pair (σ, τ) of $\overrightarrow{W}$, we have $\varUpsilon(\partial(\tau)) = 0$ and $\varUpsilon^*(\delta(\sigma)) = 0$.

Let $\overrightarrow{W}$ be a Morse sequence on K and let $\widehat{W} = \langle \hat{\sigma}_1, \ldots, \hat{\sigma}_k \rangle$. We see that a Morse reference $\varUpsilon$ of $\overrightarrow{W}$ may be computed by scanning the sequence $\widehat{W}$ from the left to the right. Also, a Morse co-reference $\varUpsilon^*$ of $\overrightarrow{W}$ may be computed by scanning $\widehat{W}$ from the right to the left. The uniqueness of $\varUpsilon$ and $\varUpsilon^*$ is a consequence of these constructions. As a limit case, observe that:

- If τ is a facet of K, then we have $\varUpsilon(\tau) = 0$ whenever τ is not critical.
- If σ is a 0-simplex of K, then we have $\varUpsilon^*(\sigma) = 0$ whenever σ is not critical.

Also, it can be checked that the references and co-references of two Morse sequences are equal whenever these sequences are equivalent in the sense given in Definition 2. The converse is, in general, not true.

Figure 2a depicts a two-dimensional torus. We first illustrate, in Fig. 2b, the Morse reference of a Morse sequence $\overrightarrow{T}$ on this torus obtained by a maximal increasing scheme (depicted in detail in [2, Fig. 1]). In this figure, any simplex σ in grey is such that $\varUpsilon(\sigma) = 0$. At the first step, the first critical simplex $\sigma_1 = a$ is coloured in pink, with $\varUpsilon(\sigma_1) = a$. After all the possible expansions from a, we have $\varUpsilon(\sigma) = a$ (in pink) for all σ of dimension 0. At the next stage, we introduce a first 1-critical simplex b (in blue), and we have $\varUpsilon(b) = b$; this leads to $\varUpsilon(\sigma) = b$ for all simplexes σ of dimension 1 highlighted in blue. We then introduce a second critical 1-simplex, c (in green), and we have $\varUpsilon(c) = c$; this leads to $\varUpsilon(\sigma) = c$ for all simplexes σ of dimension 1 in green. At the penultimate step of the Morse sequence, we have a free pair (σ, τ), and $\varUpsilon(\sigma) = x$, in purple, where $x = \varUpsilon(\partial(\tau) \setminus \{\sigma\}) = b + c$. The ultimate step of the Morse sequence is the critical 2-simplex d, and we have $\varUpsilon(d) = d$, highlighted in yellow.

Conversely, in Fig. 2c, by scanning $\overrightarrow{T}$ from right to left, we obtain its Morse co-reference map. In this figure, any simplex σ in grey is such that $\varUpsilon^*(\sigma) = 0$. Starting with the critical 2-simplex d, after several steps in the sequence, we have $\varUpsilon^*(\sigma) = d$ for all simplexes σ of dimension 2, highlighted in yellow. We then have $\Upsilon^*(\sigma) = c$ for all simplexes σ of dimension 1 highlighted in green, and $\Upsilon^*(\sigma) = b$ for all simplexes σ of dimension 1 highlighted in blue. Finally, we have $\Upsilon^*(\sigma_1) = a$ for the last simplex of dimension 0, which is critical, and is coloured in pink.

4.2   Gradient Paths, Co-Gradient Paths and Gradient Flows

References are closely related to the notion of a gradient path. In the following, we recall the classical definition of such a path. We also introduce the notion of a co-gradient path, which arises naturally from the definition of a co-reference.

Let $\overrightarrow{W}$ be a Morse sequence on K.

1. Let $\pi = \langle \sigma_0, \tau_0, \ldots, \sigma_{k-1}, \tau_{k-1}, \sigma_k \rangle$, $k \geq 0$, be a sequence with $\sigma_i \in K^{(p)}$, $\tau_i \in K^{(p+1)}$. We say that π is a *gradient path in* $\overrightarrow{W}$ *(from* σ_0 *to* σ_k *)* if, for any $i \in [0, k-1]$, the pair (σ_i, τ_i) is regular for $\overrightarrow{W}$ and $\sigma_{i+1} \in \partial(\tau_i)$, with $\sigma_{i+1} \neq \sigma_i$. The path π is *trivial* if $k = 0$, that is, if $\pi = \langle \sigma_0 \rangle$ with $\sigma_0 \in K^{(p)}$.
2. Let $\pi = \langle \tau_0, \sigma_1, \tau_1, \ldots, \sigma_k, \tau_k \rangle$, $k \geq 0$, be a sequence with $\tau_i \in K^{(p)}$, $\sigma_i \in K^{(p-1)}$. We say that π is a *co-gradient path in* $\overrightarrow{W}$ *(from* τ_0 *to* τ_k *)* if, for any $i \in [1, k]$, the pair (σ_i, τ_i) is regular for $\overrightarrow{W}$ and $\tau_{i-1} \in \delta(\sigma_i)$, with $\tau_i \neq \tau_{i-1}$. The path π is *trivial* if $k = 0$, that is, if $\pi = \langle \tau_0 \rangle$ with $\tau_0 \in K^{(p)}$.

Proposition 1 below can be proved by induction, by considering the two scanning processes of $\widehat{W}$ that are mentioned above. Theorem 1 reflects an important duality relation between the reference and the co-reference of a Morse sequence. It can be proved by changing the extremities of gradient and co-gradient paths.

Proposition 1. *Let $\overrightarrow{W}$ be a Morse sequence on K and Υ be the reference of $\overrightarrow{W}$. Let $\sigma, \nu \in K^{(p)}$ such that ν is critical for $\overrightarrow{W}$.*

1. *We have $\nu \in \Upsilon(\sigma)$ if and only if the number of gradient paths from the simplex σ to the critical simplex ν is odd.*
2. *We have $\nu \in \Upsilon^*(\sigma)$ if and only if the number of co-gradient paths from the critical simplex ν to the simplex σ is odd.*

Theorem 1. *Let $\overrightarrow{W}$ be a Morse sequence on K and Υ be the reference of $\overrightarrow{W}$. Let $\sigma \in K^{(p)}$ and $\tau \in K^{(p+1)}$ be two simplexes that are both critical for $\overrightarrow{W}$. We have $\sigma \in \Upsilon(\partial(\tau))$ if and only if $\tau \in \Upsilon^*(\delta(\sigma))$.*

An important concept in discrete Morse theory is the one of gradient flows [8], by which, using Forman's own words [9], *loosely speaking, a simplex flows along the gradient paths* for infinite time (see also [9] for the dual concept). See [14, Def. 8.6] for a precise definition. Gradient flows are a basic ingredient for setting the fundamental property of a Morse complex, that is, the equality of homology between a complex and its Morse complex. In fact, there is a deep link between co-references of a Morse sequence and gradient flows. For the sake of space, we give only an informal presentation of this relation, which may be checked by the interested readers. If τ is a p-simplex of a complex K, the gradient flow which starts from τ is obtained:

1. By considering regular pairs (σ', τ'), with $\sigma' \in \partial(\tau)$. Such a pair may be seen as the beginning of a co-gradient path that starts at τ.
2. By considering some p-simplexes that are in the boundary of a $(p+1)$-simplex ν, such that (τ, ν) is regular.

If τ is a critical simplex, then the case 2. cannot happen. Also, this case cannot happen for τ', since τ' belongs to the regular pair (σ', τ'). By induction, the gradient flow starting at a critical simplex corresponds exactly to co-gradient paths. In a dual manner, the gradient flow ending at a critical simplex corresponds to gradient paths. Thus, if $\sigma, \tau \in K^{(p)}$ and if τ is critical for $\overrightarrow{W}$, then:

The simplex σ is in the gradient flow starting at τ if and only if $\tau \in \Upsilon^*(\sigma)$.
The simplex σ is in the gradient flow ending at τ if and only if $\tau \in \Upsilon(\sigma)$.

It is interesting to compare Υ and Υ^* with the analogous constructions in smooth Morse theory. A gradient flow associates a critical simplex to a chain which is invariant under the flow. According to [9], this chain is the discrete analogue of the *unstable (or descending) cell associated to a critical point of a smooth Morse function*, and it is obtained with Υ^*. Forman [9] also studies the dual of the flow, the coflow. The coflow maps a critical simplex to a chain that is invariant under the coflow. This chain plays *the role of the stable (or ascending) cell associated to a critical point of a smooth Morse function*, and it is obtained thanks to Υ.

4.3   The Morse Complex

Now, let us consider a boundary map that is restricted to the critical simplexes. This map may be easily built with a Morse reference.

Let Υ be the reference of $\overrightarrow{W}$. If $\sigma \in \ddot{W}^{(p)}$, we set $d(\sigma) = \Upsilon(\partial(\sigma))$.
We denote by d_p the map:
$$d_p : c \in \ddot{W}[p] \mapsto d_p(c) \in \ddot{W}[p-1], \text{ where } d_p(c) = \Upsilon(\partial_p(c)).$$
Thus, we have $d_p(c) = \sum_{\sigma \in c} d(\sigma)$ with $d_p(\emptyset) = 0$.

Theorem 2. *Let Υ be the reference of a Morse sequence $\overrightarrow{W}$ on K.*
For each $c \in K[p]$, we have $d_p(\Upsilon(c)) = \Upsilon(\partial_p(c))$.

Proof. Let $\overrightarrow{W} = \langle \emptyset = K_0, ..., K_k = K \rangle$ be a Morse sequence on K, and let $\widehat{W} = \langle \hat{\sigma}_1, ..., \hat{\sigma}_k \rangle$. We consider the statement (S_i): For each $c \in K_i[p]$, we have $d_p(\Upsilon(c)) = \Upsilon(\partial_p(c))$. We have $K_0[p] = \{\emptyset\}$. Thus (S_0) holds.
Suppose (S_{i-1}) holds with $0 \le i - 1 \le k - 1$. Let $c \in K_i[p]$.

1) Suppose $\hat{\sigma}_i = \sigma$, with $\sigma \in \ddot{W}$. If $\sigma \notin c$, then we are done. Otherwise, we have $c = c' \cup \{\sigma\}$, with $c' \in K_{i-1}[p]$.
   We have $\partial_p(c) = \partial_p(c') + \partial(\sigma)$. Thus $\Upsilon(\partial_p(c)) = \Upsilon(\partial_p(c')) + \Upsilon(\partial(\sigma))$.
   By the induction hypothesis and by the definition of $d(\sigma)$, we obtain $\Upsilon(\partial_p(c)) = d_p(\Upsilon(c')) + d(\sigma)$. Therefore $\Upsilon(\partial_p(c)) = d_p(\Upsilon(c')) + d_p(\Upsilon(\{\sigma\})) = d_p(\Upsilon(c))$.
2) Suppose $\hat{\sigma}_i = (\sigma, \tau)$ is a free pair. If $\sigma \notin c$ and $\tau \notin c$, then we are done.

2.1) Suppose $\sigma \in c$. Let $c' = c + \partial_{p+1}(\tau)$. We have $\Upsilon(c') = \Upsilon(c) + \Upsilon(\partial_{p+1}(\tau)) = \Upsilon(c)$. We also have $\partial_p(c') = \partial_p(c) + \partial_p(\partial_{p+1}(\tau)) = \partial_p(c)$.
But $c' = (c \setminus \{\sigma\}) + c''$, with $c'' = \{\eta \in \partial_{p+1}(\tau) \mid \eta \neq \sigma\}$. Thus $c' \in K_{i-1}[p]$.
By the induction hypothesis, it follows that $d_p(\Upsilon(c')) = \Upsilon(\partial_p(c'))$. By the previous equalities, we obtain $d_p(\Upsilon(c)) = \Upsilon(\partial_p(c))$.

2.2) Suppose $\tau \in c$. Let $c = c' \cup \{\tau\}$, with $c' \in K_{i-1}[p]$. Since $\Upsilon(\tau) = 0$, we obtain $\Upsilon(c) = \Upsilon(c')$. Furthermore $\Upsilon(\partial_p(c)) = \Upsilon(\partial_p(c')) + \Upsilon(\partial(\tau)) = \Upsilon(\partial_p(c'))$. By the induction hypothesis, we have $d_p(\Upsilon(c')) = \Upsilon(\partial_p(c'))$. Therefore $d_p(\Upsilon(c)) = \Upsilon(\partial_p(c))$. □

The two following results are direct consequences of Theorem 2.

Proposition 2. *Let Υ be the reference of a Morse sequence $\overrightarrow{W}$ on K. For any $c, c' \in K[p]$ we have $\Upsilon(\partial_p(c)) = \Upsilon(\partial_p(c'))$ whenever $\Upsilon(c) = \Upsilon(c')$.*

Proof. Let $c, c' \in K[p]$ with $\Upsilon(c) = \Upsilon(c')$. Thus, $d_p(\Upsilon(c)) = d_p(\Upsilon(c'))$. By Theorem 2, we have $\Upsilon(\partial_p(c)) = \Upsilon(\partial_p(c'))$. □

Proposition 3. *If $\overrightarrow{W}$ is a Morse sequence, then the maps d_p are boundary operators. That is, we have $d_p \circ d_{p+1} = 0$.*

Proof. Let $\sigma \in \ddot{W}^{(p+1)}$. We have $d_{p+1}(\{\sigma\}) = d(\sigma) = \Upsilon(\partial(\sigma)) = \Upsilon(\partial_{p+1}(\{\sigma\}))$. By Theorem 2, we have $d_p(\Upsilon(\partial_{p+1}(\{\sigma\}))) = \Upsilon(\partial_p(\partial_{p+1}(\{\sigma\}))) = \Upsilon(0) = 0$. Thus $d_p \circ d_{p+1}(\{\sigma\}) = 0$, which gives the result by linearity. □

Since $d_p \circ d_{p+1} = 0$, the couple $(\ddot{W}[p], d_p)$ satisfies the definition of a *chain complex* [12]. We say that $(\ddot{W}[p], d_p)$ is the *Morse (chain) complex of* $\overrightarrow{W}$. This notion of a Morse complex is equivalent to the classical one given in the context of discrete Morse theory. This fact may be verified using [14, Theorem 8.31], Proposition 1, and the very definition of the differential d_p.

Dual results for Theorem 2, Proposition 2 and Proposition 3 can be written by considering Υ^* instead of Υ.

In the following, we denote by $H_p(\ddot{W})$ (resp. $H^p(\ddot{W})$) the pth homology (resp. cohomology) vector space corresponding to the Morse complex of $\overrightarrow{W}$. By Theorem 2, the map Υ is a *chain map* [12] from the chain complex $(K[p], \partial_p)$ to the chain complex $(\ddot{W}[p], d_p)$. Hence, Υ induces a linear map between $H_p(K)$ and $H_p(\ddot{W})$; see [12]. Furthermore, we have the following.

Theorem 3 (from [8]). *For each $p \geq 0$, the vector spaces $H_p(K)$ and $H_p(\ddot{W})$ are isomorphic.*

5   Annotations

If σ is a p-simplex in a complex K, an annotation for σ, as introduced in [4], is a length g binary vector, where g is the rank of the homology group $H_p(K)$.

These annotations, when summed up for simplexes in a given cycle, provide a way to determine the homology class of this cycle. The following definition is an adaptation for Morse sequences of this notion. The main difference is that we annotate each simplex with a subset of the critical simplexes of the sequence, instead of a vector.

Let $\overrightarrow{W}$ be a Morse sequence on K. We say that a Morse frame Υ on $\overrightarrow{W}$ is an *annotation on* $\overrightarrow{W}$ if Υ satisfies the three conditions:

C1: For each $\sigma \in K^{(p)}$, we have $\Upsilon(\sigma) \subseteq \ddot{V}^{(p)}$ where $\ddot{V}^{(p)}$ is a subset of $\ddot{W}^{(p)}$;
C2: For each p, we have $Card(\ddot{V}^{(p)}) = \beta_p(K)$;
C3: For any cycles $z, z' \in Z_p(K)$, we have $\Upsilon(z) = \Upsilon(z')$ if and only if their homology classes are such that $[z]_p = [z']_p$.

Let Υ be a frame on $\overrightarrow{W}$. If $\tau \in \ddot{W}^{(p)}$, we set $\Upsilon^{\sharp}(\tau) = \{\sigma \in K^{(p)} \mid \tau \in \Upsilon(\sigma)\}$. The following proposition, derived from [6], indicates that an annotation may be seen as a way to determine a cohomology basis of the complex.

Proposition 4 (adapted from [6]). *Let $\overrightarrow{W}$ be a Morse sequence on K. A frame Υ on $\overrightarrow{W}$ is an annotation on $\overrightarrow{W}$ if and only if Υ satisfies the conditions C1, C2, and the following condition C4.*

C4: *The set of chains $\{\Upsilon^{\sharp}(\tau) \mid \tau \in \ddot{V}^{(p)}\}$ is a set of cocycles whose cohomology classes $\{[\Upsilon^{\sharp}(\tau)]^p \mid \tau \in \ddot{V}^{(p)}\}$ constitute a basis of $H^p(K)$.*

We give a construction for obtaining an annotation. Again, it is an adaptation for a Morse sequence of the one given in [5] and [6]. Three cases are considered:

1. If a critical simplex is added, and if the annotation of the boundary of this simplex is trivial, then a new cycle is created. The label associated to this simplex is composed solely of the simplex itself.
2. If a critical simplex is added, and if the annotation of the boundary of this simplex is not trivial, then a cycle is removed. This is done by selecting one label in the annotation of the boundary of this simplex, and by removing this label from all the previous annotations.
3. If a free pair is added, we propagate the labels of the annotations to this pair, according to the simple rule of Definition 4.

See [5] and [6] for the validity of this construction for the cases 1 and 2 The validity for the case 3 is an easy consequence of the definition of a free pair.

Let $\overrightarrow{W} = \langle K_0, \dots, K_k \rangle$ be a Morse sequence and $\widehat{W} = \langle \hat{\sigma}_1, \dots, \hat{\sigma}_k \rangle$. We write $\overrightarrow{W}_i = \langle K_0, \dots, K_i \rangle$, $i \in [0, k]$. We consider the sequence $\langle \Upsilon_0, \dots, \Upsilon_i \rangle$, $i \in [0, k]$, such that Υ_i is a frame for $\overrightarrow{W}_i$, with $\Upsilon_0(\emptyset) = 0$ and:

1. If $\hat{\sigma}_i = \sigma_i$ and $\Upsilon_{i-1}(\partial(\sigma_i)) = 0$, then Υ_i is such that $\Upsilon_i(\sigma_i) = \sigma_i$ and $\Upsilon_i(\tau) = \Upsilon_{i-1}(\tau)$ otherwise.
2. If $\hat{\sigma}_i = \sigma_i$ and $\Upsilon_{i-1}(\partial(\sigma_i)) \neq 0$, then we select an arbitrary critical face $\nu \in \Upsilon_{i-1}(\partial(\sigma_i))$. The map Υ_i is such that $\Upsilon_i(\sigma_i) = 0$, $\Upsilon_i(\tau) = \Upsilon_{i-1}(\tau) + \Upsilon_{i-1}(\partial(\sigma_i))$ if $\nu \in \Upsilon_{i-1}(\tau)$, and $\Upsilon_i(\tau) = \Upsilon_{i-1}(\tau)$ otherwise.

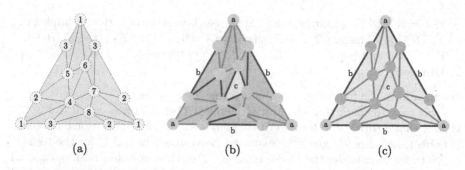

Fig. 3. (a) A dunce hat. Points with the same label are identified. (b) A Morse reference map. (c) A Morse co-reference map. See text for details. (Color figure online)

3. If $\hat{\sigma}_i = (\sigma_i, \tau_i)$, then Υ_i such that $\Upsilon_i(\tau_i) = 0$, $\Upsilon_i(\sigma_i) = \Upsilon_{i-1}(\partial(\tau_i) + \sigma_i)$, and $\Upsilon_i(\tau) = \Upsilon_{i-1}(\tau)$ otherwise.

Under the above construction, each frame Υ_i is an annotation on $\overrightarrow{W_i}$. Let $\ddot{V}_i^{(p)} = \{\sigma \in K_i \mid \Upsilon_i(\sigma) = \sigma\}$, $\ddot{V}_i^{(p)}$ is composed of critical faces for $\overrightarrow{W_i}$. For each $\sigma \in K_i^{(p)}$, we have $\Upsilon_i(\sigma) \subseteq \ddot{V}_i^{(p)}$. Furthermore, for each p, we have $Card(\ddot{V}_i^{(p)}) = \beta_p(K_i)$.

The interested reader can check that the reference map of the torus, given in Fig. 2b, is indeed an annotation. Here, the above case 2 does not happen. This case corresponds to the cell c from Fig. 3b.

6 Computing Betti Numbers with the Morse Reference

In the construction described in Sect. 5, we have to remove a label from all previous annotations. We now present another construction that reduces the amount of operations required for this task. The basic idea is to use the information given by the reference of a Morse sequence, and to remove labels only for some faces which are in the boundary of critical simplexes. Thus, annotations are not computed for all simplexes, but this construction allows us to obtain the Betti numbers of the complex.

Let $\overrightarrow{W}$ be a Morse sequence on a simplicial complex K, and let Υ be a Morse frame on $\overrightarrow{W}$. We say that Υ is *perfect* if each Betti number $\beta_p(K)$ is exactly equal to the number of critical p-simplexes σ in $\overrightarrow{W}$ such that $\Upsilon(\sigma) = \{\sigma\}$.

A key observation is the following. The Morse reference Υ of $\overrightarrow{W}$ is perfect if, and only if, for any critical simplex σ in $\overrightarrow{W}$, we have $\Upsilon(\partial(\sigma)) = 0$. In the next construction, we take advantage of this observation to iteratively remove suitable pairs of critical simplexes from the image of Υ.

Let $\overrightarrow{W} = \langle \emptyset = K_0, ..., K_k = K \rangle$ be a Morse sequence on a complex K, and let Υ be the Morse reference of $\overrightarrow{W}$. We write $\widehat{W} = \langle \hat{\sigma}_1, ..., \hat{\sigma}_k \rangle$. We set:

$$\ddot{W}^+ = \ddot{W} \cup \{\tau \in K \mid \tau \in \partial(\sigma) \text{ for some } \sigma \in \ddot{W}\}$$

We consider the sequence of frames $\Upsilon_0, ..., \Upsilon_k$ such that $\Upsilon_0 = \Upsilon$ and:

1. If $\hat{\sigma}_i = \sigma_i$ and $\Upsilon_{i-1}(\partial(\sigma_i)) \neq 0$, then we select an arbitrary critical simplex $\nu \in \Upsilon_{i-1}(\partial(\sigma_i))$. The map Υ_i is such that $\Upsilon_i(\sigma_i) = 0$, $\Upsilon_i(\tau) = \Upsilon_{i-1}(\tau) + \Upsilon_{i-1}(\partial(\sigma_i))$ if $\tau \in \ddot{W}^+$ and $\nu \in \Upsilon_{i-1}(\tau)$, $\Upsilon_i(\tau) = \Upsilon_{i-1}(\tau)$ otherwise.
2. Otherwise, we have $\Upsilon_i = \Upsilon_{i-1}$.

We then have the following: the Morse frame Υ_k is perfect.

It is easy to check that the Morse reference Υ of the torus, given in Fig. 2b, is such that $\Upsilon(\partial(\sigma)) = 0$ for all critical simplexes σ. Thus, this Morse reference is perfect, and directly gives the expected Betti numbers $(1,2,1)$ for the torus.

Now, let us consider the Morse reference Υ of the dunce hat that is depicted in Fig. 3b (see the corresponding Morse sequence in [2, Fig. 2]). The last critical simplex c in the sequence is such that $\Upsilon(\partial(c)) = b + b + b = b$. Thus, there exist at this last step some edges annotated with b in $\ddot{W}^+$. We remove both b and c from the set of annotations of $\ddot{W}^+$, having for effect to "kill" the blue cocycle (or dually, to "kill" the blue cycle in Fig. 3c). We then retrieve the expected Betti numbers $(1,0,0)$ for the dunce hat. This result confirms that the dunce hat is acyclic, although the Morse sequence contains 3 critical simplexes, a, b and c.

7   Implementing Morse Frames

In the literature, specific, independent algorithms are designed for computing the gradient vector field [1,11,13], the Morse complex [10,13], or the Betti numbers [6]. The framework of Morse frames shows that we can compute the gradient vector field and the Morse complex simultaneously, in only one pass, and the Betti numbers in two passes.

As long as we can check whether a pair is free in constant time (e.g., for example with cubical complexes and the use of a mask to check the neighborhood of a simplex), the complexity of a Morse sequence is $\mathcal{O}(dn)$, where d is the dimension of the complex, and n the number of its simplexes. When we compute a Morse reference map, we need to maintain a list of labels for each simplex, each label corresponding to a critical simplex. This leads to a complexity in $\mathcal{O}(dcn)$, where c is the number of critical simplexes. The Morse reference has a memory complexity of $\mathcal{O}(cn)$. In contrast, algorithms that compute a Morse complex, such as [10,13], claim a cubic worst-case complexity for $d = 3$ (because they have to run several times on each gradient path); furthermore, such algorithms can only be applied *after* obtaining a gradient vector field.

The framework of Morse frames allows retrieving the concept of annotations [6]. The current implementation of annotations [3] can be described, in our language, as a Morse sequence where all simplexes are critical, i.e., with only fillings. A key point for efficiency of this implementation [3], is the ordering of the simplexes: a heuristic is used to try preventing the creation of unnecessary cycles. Morse frames show that, with a simple change of heuristic (using, for example, a maximal increasing scheme), the annotation algorithm can take advantage of gradient fields. The ordering of the simplexes provided by such a scheme, avoids the creation of unnecessary cycles, by using expansions and fillings, instead of only fillings.

Section 6 provides an algorithm for computing the Betti numbers in mod 2 arithmetic, that is inspired by annotations. This algorithm uses the reference map and only considers the set of critical simplexes and their boundary.

8   Conclusion

This paper introduces Morse frames, that are based on a novel presentation of discrete Morse theory, called Morse sequences. Morse frames allow for adding information to a Morse sequence, associating a set of specific simplexes to each simplex. The main example of Morse frames, the Morse reference, offers substantial utility in the context of homology. In particular, together with its dual, the Morse co-reference, they provide the discrete analogue of ascending/stable and descending/unstable cell associated to a critical point of a smooth Morse function. Significantly, the Morse reference allows retrieving the Morse chain complex. Using Morse frames, we give a novel presentation of the annotation algorithm. Inspired by these annotations, we describe an efficient scheme for computing Betti numbers in mod 2 arithmetic.

On the theoretical side, for future work, we aim at providing a proof of Theorem 3, that will rely only on the Morse reference. We also intend to compute persistence with the Morse reference, and to extend our framework to other fields than the mod 2 arithmetic. On a more practical level, we also want to test the proposed algorithms, and to compare their efficiency with respect to the state-of-the-art.

References

1. Benedetti, B., Lutz, F.H.: Random discrete Morse theory and a new library of triangulations. Exp. Math. **23**(1), 66–94 (2014)
2. Bertrand, G.: Morse sequences. In: Rinaldi, S. (ed.) DGMM 2024. LNCS, vol. 14605, pp. 377–389. Springer, Cham (2024). https://hal.science/hal-04227281
3. Boissonnat, J.D., Dey, T.K., Maria, C.: The compressed annotation matrix: an efficient data structure for computing persistent cohomology. Algorithmica **73**(3), 607–619 (2015)
4. Busaryev, O., Cabello, S., Chen, C., Dey, T.K., Wang, Y.: Annotating simplices with a homology basis and its applications. In: Fomin, F.V., Kaski, P. (eds.) SWAT 2012. LNCS, vol. 7357, pp. 189–200. Springer, Heidelberg (2012). https://doi.org/10.1007/978-3-642-31155-0_17
5. De Silva, V., Vejdemo-Johansson, M.: Persistent cohomology and circular coordinates. In: Proceedings of the Twenty-Fifth Annual Symposium on Computational Geometry, pp. 227–236 (2009)
6. Dey, T.K., Fan, F., Wang, Y.: Computing topological persistence for simplicial maps. In: Proceedings of the Thirtieth Annual Symposium on Computational Geometry, pp. 345–354 (2014)
7. Edelsbrunner, H., Harer, J.: Computational Topology - An Introduction. American Mathematical Society (2010)
8. Forman, R.: Witten-Morse theory for cell complexes. Topo **37**(5), 945–979 (1998)

9. Forman, R.: Discrete Morse theory and the cohomology ring. Trans. Am. Math. Soc. **354**(12), 5063–5085 (2002)
10. Fugacci, U., Iuricich, F., De Floriani, L.: Computing discrete Morse complexes from simplicial complexes. Graph. Models **103**, 101023 (2019)
11. Harker, S., Mischaikow, K., Mrozek, M., Nanda, V.: Discrete Morse theoretic algorithms for computing homology of complexes and maps. Found. Comput. Math. **14**, 151–184 (2014)
12. Hatcher, A.: Algebraic Topology. Cambridge University Press, Cambridge (2002)
13. Robins, V., Wood, P.J., Sheppard, A.P.: Theory and algorithms for constructing discrete Morse complexes from grayscale digital images. IEEE Trans. Pattern Anal. Mach. Intell. **33**(8), 1646–1658 (2011)
14. Scoville, N.A.: Discrete Morse Theory, vol. 90. American Mathematical Society (2019)
15. Whitehead, J.H.C.: Simplicial spaces, nuclei and m-groups. Proc. Lond. Math. Soc. **2**(1), 243–327 (1939)

Morse Sequences

Gilles Bertrand[✉][iD]

Univ Gustave Eiffel, CNRS, LIGM, 77454 Marne-la-Vallée, France
gilles.bertrand@esiee.fr

Abstract. We introduce the notion of a Morse sequence, which provides a simple and effective approach to discrete Morse theory. A Morse sequence is a sequence composed solely of two elementary operations, that is, expansions (the inverse of a collapse), and fillings (the inverse of a perforation). We show that a Morse sequence may be seen as an alternative way to represent the gradient vector field of an arbitrary discrete Morse function. We also show that it is possible, in a straightforward manner, to make a link between Morse sequences and different kinds of Morse functions. At last, we introduce maximal Morse sequences, which formalize two basic schemes for building a Morse sequence from an arbitrary simplicial complex.

Keywords: Discrete Morse theory · Expansions and collapses · Fillings and perforations · Simplicial complex

1 Introduction

Discrete Morse theory, developed by Robin Forman [10,11], studies the topology of objects using functions that assign values to their cells of different dimensions. A discrete Morse function detects some special cells, called critical cells, which capture the essential topological features of the object.

In this paper, we present an approach where, instead of a Morse function, a sequence of elementary operators is used for a simple representation of an object. This sequence, that we called *a Morse sequence*, is composed solely of two elementary operations, that is, expansions (the inverse of a collapse), and fillings (the inverse of a perforation). These operations correspond exactly to the ones introduced by Henry Whitehead [20]. After some basic definitions and two meaningful examples (Sects. 2, 3, and 4), we show that a Morse sequence is an alternative way to represent the gradient vector field of an arbitrary discrete Morse function (Sect. 5). We also show that it is possible to recover immediately, from a Morse sequence, different kinds of Morse functions (Sect. 6). At last, we introduce maximal Morse sequences, which formalize two basic schemes for building a Morse sequence from an arbitrary simplicial complex (Sect. 7).

2 Basic Definitions

Let K be a finite family composed of non-empty finite sets. The family K is a *(simplicial) complex* if $\sigma \in K$ whenever $\sigma \neq \emptyset$ and $\sigma \subseteq \tau$ for some $\tau \in K$.

S. Brunetti et al. (Eds.): DGMM 2024, LNCS 14605, pp. 377–389, 2024.
https://doi.org/10.1007/978-3-031-57793-2_29

An element of a simplicial complex K is *a simplex of K* or *a face of K*. A *facet of K* is a simplex of K that is maximal for inclusion. The *dimension* of $\sigma \in K$, written $dim(\sigma)$, is the number of its elements minus one. If $dim(\sigma) = p$, we say that σ is a *p-simplex*. The *dimension of K*, written $dim(K)$, is the largest dimension of its simplices, the *dimension of $\emptyset$*, the void complex, being defined to be -1. We denote by $K^{(p)}$ the set composed of all p-simplexes of K.

If $\sigma \in K^{(p)}$ we set $\partial(\sigma) = \{\tau \in K^{(p-1)} \mid \tau \subset \sigma\}$, which is the *boundary of σ*.

We recall some basic definitions related to the collapse operator [20].

Let K be a complex and let σ, τ be two distinct faces of K. The couple (σ, τ) is a *free pair for K* if τ is the only face of K that contains σ. Thus, the face τ is necessarily a facet of K. If (σ, τ) is a free pair for K, then $L = K \setminus \{\sigma, \tau\}$ is *an elementary collapse of K*, and K is *an elementary expansion of L*. We say that *K collapses onto L*, or that *L expands onto K*, if there exists a sequence $\langle K = M_0, ..., M_k = L \rangle$ such that M_i is an elementary collapse of M_{i-1}, $i \in [1, k]$. The complex K is *collapsible* if K collapses onto a *vertex*, that is, onto a complex of the form $\{\{a\}\}$. We say that K is *(simply) homotopic to L*, or that K and L are *(simply) homotopic*, if there exists a sequence $\langle K = M_0, ..., M_k = L \rangle$ such that M_i is an elementary collapse or an elementary expansion of M_{i-1}, $i \in [1, k]$. The complex K is *(simply) contractible* if K is simply homotopic to a vertex.

3   Morse Sequences

Let us start first with the definition of perforations and fillings.

Let K, L be simplicial complexes. If $\sigma \in K$ is a facet of K and if $L = K \setminus \{\sigma\}$, we say that L is *an elementary perforation of K*, and that K is *an elementary filling of L*.

These transformations were introduced by Whitehead in a seminal paper [20]. Combined with collapses and expansions, it has been shown that we obtain four operators that correspond to the homotopy equivalence between two simplicial complexes (Th. 17 of [20]). See also [6] which provides another kind of equivalence based on a variant of these operators.

In this paper, we introduce the notion of a "Morse sequence" by simply considering expansions and fillings of a simplicial complex.

Definition 1. *Let K be a simplicial complex. A* Morse sequence (on K) *is a sequence $\overrightarrow{W} = \langle \emptyset = K_0, ..., K_k = K \rangle$ of simplicial complexes such that, for each $i \in [1, k]$, K_i is either an elementary expansion or an elementary filling of K_{i-1}.*

Let $\overrightarrow{W} = \langle K_0, ..., K_k \rangle$ be a Morse sequence. For each $i \in [1, k]$:

- If K_i is an elementary filling of K_{i-1}, we write $\hat{\sigma}_i$ for the simplex σ such that $K_i = K_{i-1} \cup \{\sigma\}$, we say that the face σ is *critical for $\overrightarrow{W}$*.
- If K_i is an elementary expansion of K_{i-1}, we write $\hat{\sigma}_i$ for the free pair (σ, τ) such that $K_i = K_{i-1} \cup \{\sigma, \tau\}$, we say that $\hat{\sigma}_i$, σ, τ, are *regular for $\overrightarrow{W}$*.

We write $\widehat{W} = \langle \hat{\sigma}_1, ..., \hat{\sigma}_k \rangle$, and we say that $\widehat{W}$ is a *(simplex-wise) Morse sequence*. Clearly, $\overrightarrow{W}$ and $\widehat{W}$ are two equivalent forms. We shall pass from one of these forms to the other without notice.

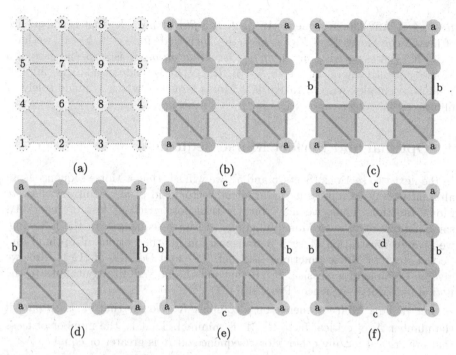

Fig. 1. A Morse sequence on the torus. (a) A triangulation, points with the same label are identified. (b) The sequence begins with the critical 0-simplex a. Elementary expansions are added to the sequence until we obtain a maximal expansion from a. (c) The critical 1-simplex b is added to the sequence. (d) A maximal expansion from b is done. (e) The second critical 1-simplex c is added, and a maximal expansion from c is done. (f) The critical 2-simplex d is added.

Observe that, if $\overrightarrow{W} = \langle K_0, ..., K_k \rangle$ is a Morse sequence, with $k \geq 1$, then K_1 is necessarily a filling of $\emptyset$. Thus, K_1 is necessarily a vertex. That is, K_1 is made of a single 0-simplex that is critical for $\overrightarrow{W}$.

Figure 1 presents an example of a Morse sequence $\overrightarrow{W}$ on a torus T. There are different ways to obtain a Morse sequence. In Fig. 1, we apply the following strategy. We build $\overrightarrow{W}$ from the left to the right. Starting from $\emptyset$, we obtain T by iterative elementary expansions and fillings. Also, we make maximal expansions, that is, we make a filling only if no elementary expansion can be made.

Remark 1. Let $\widehat{W} = \langle \hat{\sigma}_1, ..., \hat{\sigma}_k \rangle$ be a Morse sequence and let $\hat{\sigma}_i, \hat{\sigma}_j, j > i$, be two consecutive critical faces of $\widehat{W}$, that is, $\hat{\sigma}_{i+1}, ..., \hat{\sigma}_{j-1}$ are regular pairs. Then, as a direct consequence of the definition of a Morse sequence, the complex X_{j-1} collapses onto X_i. This property is the core of a fundamental theorem, called *the collapse theorem*, which makes the link between the basic definitions of discrete Morse theory and discrete homotopy (See Theorem 3.3 of [9] and Theorem 4.27 of [19]). In a certain sense, we can say that Morse sequences provide an introduction to discrete Morse theory by starting from this property.

Remark 2. Any Morse sequence $\overrightarrow{W}$ on K is a *filtration on* K, that is a sequence of nested complexes $\langle \emptyset = K_0, ..., K_k = K \rangle$ such that, for $i \in [0, k-1]$, we have $K_i \subseteq K_{i+1}$; see [8]. Also any *simplex-wise filtration on* K is a special case of a Morse sequence where, for $i \in [0, k-1]$, $K_{i+1} \setminus K_i$ is made of a single simplex. That is, a simplex-wise filtration is a Morse sequence which is made solely of fillings; all faces of K are critical for such a sequence.

4   Optimal and Perfect Morse Sequences

In the next two sections (Sects. 6 and 5), we will see that a Morse sequence is an alternative way to represent the gradient vector field of any arbitrary discrete Morse function. Thus, we may directly transpose, without loss of generality, some notions relative to Morse functions to Morse sequences. In the following, we give an illustration of such a transposition for the notions of optimal and perfect discrete Morse functions (see Def. 2.87 and Def. 4.6 of [19]). Also, we give an exemple of a classical result that may be proved directly thanks to the notion of a Morse sequence (Proposition 1).

Let $\overrightarrow{W}$ be a Morse sequence on a complex K. We say that $\overrightarrow{W}$ is *optimal* if the number N of critical faces for $\overrightarrow{W}$ is minimal. That is, the number of faces that are critical for any other Morse sequence on K is greater or equal to N. If $dim(K) = d$, the *Morse vector of* $\overrightarrow{W}$ is the vector $\vec{c}(\overrightarrow{W}) = (c_0, \ldots, c_p, \ldots, c_d)$ where c_p is the number of p-simplexes that are critical for $\overrightarrow{W}$. We denote by $\vec{b}(K)$ the vector $\vec{b}(K) = (b_0, \ldots, b_p, \ldots, b_d)$ where b_p is the *p*th Betti number (mod. 2) of K (see [13]). We also use the notations $c_p(\overrightarrow{W})$ and $b_p(K)$ when $\overrightarrow{W}$ and K are not clear from the context.

We say that a Morse sequence $\overrightarrow{W}$ on K is *perfect* if $\vec{c}(\overrightarrow{W}) = \vec{b}(K)$. In other words, a Morse sequence $\overrightarrow{W}$ on K is perfect if each number b_p of "p-dimensional holes of K" is equal to the number c_p of critical p-simplexes of $\overrightarrow{W}$.

Suppose a complex K is collapsible. Then we have $\vec{b}(K) = (1, 0, \ldots, 0)$. Also, we easily see that K admits a Morse sequence which has a single critical face, this face being a 0-simplex. For this sequence, we have $\vec{c}(\overrightarrow{W}) = (1, 0, \ldots, 0)$, thus K admits a perfect discrete Morse sequence.

Now, let us consider a complex that is contractible but not collapsible. The dunce hat [21] is a basic example of such a complex. In Fig. 2, a Morse sequence $\overrightarrow{W}$ for a triangulation D of the dunce hat is given; the same strategy as above has been used. We see that, in this example, we have $\vec{c}(\overrightarrow{W}) = (1, 1, 1)$. But, by contractibility of D, we have $\vec{b}(D) = (1, 0, 0)$. This leads to the question: Is it possible to have a perfect Morse sequence for D?

We have the answer to this question by simply reading the definition of a Morse sequence: If a sequence $\overrightarrow{W}$ on K has a single critical simplex (therefore, a 0-simplex), then clearly the complex K is collapsible.

Thus – thanks to the notion of a Morse sequence – we have a straightforward proof of the following classical result (see Prop. 4.10 of [19], see also [2] and [3]).

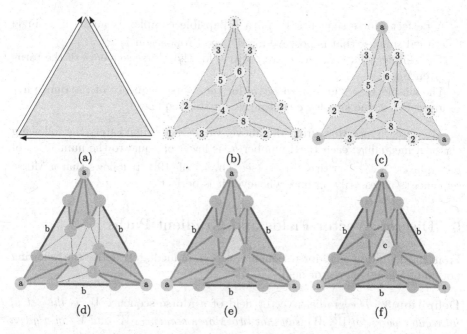

Fig. 2. A Morse sequence on the dunce hat. (a) the dunce hat, the three edges of the triangle have to be identified with the arrows. (b) A triangulation of the dunce hat. (c) The sequence begins with the critical 0-simplex a. c) A maximal expansion from a is done, then the 1-critical simplex b is added. (e) A maximal expansion from b. (f) The critical 2-simplex c is added.

Proposition 1. *Let K be a complex with $\vec{b}(K) = (1, 0, \cdots, 0)$. The complex K admits a perfect discrete Morse sequence if and only if K is collapsible.*

To conclude this section, we underline a fundamental link between Morse sequences and homology. Let $K = L \cup \{\sigma\}$ be an elementary filling of L, with $\sigma \in K^{(p)}$, $p \geq 1$. It is well-known that the addition of σ will either increase $b_p(L)$ by 1 or decrease $b_{p-1}(L)$ by 1 (but not both), all other Betti numbers being unaffected; see Lemma 3.36 of [19]. Also, it is well-known that, if K is an elementary expansion of L, then all Betti numbers are unaffected. This leads us to the following definition where each critical simplex is either positive or negative.

Definition 2. *Let $\vec{W} = \langle K_0, ..., K_k \rangle$ be a Morse sequence and $\widehat{W} = \langle \hat{\sigma}_1, ..., \hat{\sigma}_k \rangle$. Let $\hat{\sigma}_i = \sigma_i$ be a critical p-simplex for $\vec{W}$. We say that σ_i is positive for $\vec{W}$ if $i = 1$ or if $b_p(K_i) = b_p(K_{i-1}) + 1$. We say that σ_i is negative for $\vec{W}$ if $i \geq 2$ and if $b_{p-1}(K_i) = b_{p-1}(K_{i-1}) - 1$, with $p \geq 1$.*

We check at once that a Morse sequence $\vec{W}$ is perfect if and only if all critical simplexes for $\vec{W}$ are positive for $\vec{W}$. For example:

- A perfect Morse sequence $\overrightarrow{W}$ on a collapsible complex is made of a single critical simplex that is positive for $\overrightarrow{W}$ (see Proposition 1).
- The simplexes a, b, c, and d are positive for the Morse sequence of the torus given Fig. 1.
- The simplexes a and b are positive for the Morse sequence of the dunce hat given Fig. 2; the simplex c is negative for this sequence.

We observe also that, from the above, we deduce immediately the following classical inequality: each Betti number b_p is lower or equal to the number c_p of p-simplexes that are critical for $\overrightarrow{W}$ (see Th. 4.1 of [19]). It follows that a Morse sequence is necessarily optimal whenever it is perfect.

5 Discrete Vector Fields and Gradient Paths

From the definition of a Morse sequence, we can immediately derive the following notion of a gradient vector field.

Definition 3. *The* gradient vector field *of a Morse sequence* $\overrightarrow{W}$ *is the set of all regular pairs for* $\overrightarrow{W}$. *We say that two Morse sequences* $\overrightarrow{W}$ *and* $\overrightarrow{V}$ *on a given complex* K *are* equivalent *if they have the same gradient vector field.*

Let us recall the definitions of a discrete vector field and a p-gradient path, see Definitions 2.43 and 2.46 of [19].

Let K be a complex and V be a set of pairs (σ, τ), with $\sigma, \tau \in K$ and $\sigma \in \partial\tau$. We say that V is a *(discrete) vector field on* K if each simplex of K is in at most one pair of V. We say that $\sigma \in K$ is *critical for* V if σ is not in a pair of V.

Let V be a vector field on a complex K. A *p-gradient path in V (from* σ_0 *to* σ_k*)* is a sequence $\pi = \langle \sigma_0, \tau_0, \sigma_1, \tau_1, ..., \sigma_{k-1}, \tau_{k-1}, \sigma_k \rangle$, with $k \geq 0$, composed of faces $\sigma_i \in K^{(p)}$, $\tau_i \in K^{(p+1)}$ such that, for all $i \in [0, k-1]$, (σ_i, τ_i) is in V, $\sigma_{i+1} \subset \tau_i$, and $\sigma_{i+1} \neq \sigma_i$. This sequence π is said to be *trivial* if $k = 0$, that is, if $\pi = \langle \sigma_0 \rangle$; otherwise, if $k \geq 1$, we say that π is *non-trivial*. Also, the sequence π is *closed* if $\sigma_0 = \sigma_k$. We say that a vector field V is *acyclic* if V contains no non-trivial closed p-gradient path.

Now, let $\overrightarrow{W}$ be a Morse sequence on K. Then, the gradient vector field of $\overrightarrow{W}$ is clearly a vector field. We say that a p-gradient path in this vector field is a *p-gradient path in* $\overrightarrow{W}$.

In the sequel of this section, we show that a Morse sequence may be seen as an alternative way to represent the gradient vector field of an arbitrary discrete Morse function. A classical result of discrete Morse theory states that a discrete vector field V is the gradient vector field of a discrete Morse function if and only if V is acyclic (Theorem 2.51 of [19]). Thus, in order to achieve this goal, we establish the equivalence between gradient vector fields of Morse sequences and acyclic vector fields (Theorem 1). Before, we introduce the notion of a maximal p-gradient path. Such a path allows us to extract, in the top dimension of a complex K, either a critical simplex or a free pair for K (Lemma 1). This formalizes an incremental deconstruction of the complex, which is usually given with certain Morse functions, see Remark 13 of [1].

Let V be a vector field on K and let $\pi = \langle \sigma_0, \tau_0, ..., \sigma_{k-1}, \tau_{k-1}, \sigma_k \rangle$ be a p-gradient path in V. We say that a pair of simplexes (η, ν) is an *extension of* π *(in V)* if $\langle \eta, \nu, \sigma_0, \tau_0, ..., \sigma_{k-1}, \tau_{k-1}, \sigma_k \rangle$ or if $\langle \sigma_0, \tau_0, ..., \sigma_{k-1}, \tau_{k-1}, \sigma_k, \nu, \eta \rangle$ is a p-gradient path in V. We say that π is *maximal (in V)* if π has no extension in V. If V is acyclic, it can be checked that, for any $p \geq 0$, there exists a maximal p-gradient path in V. To see this point, we can pick an arbitrary (possibly trivial) p-gradient path and extend it iteratively with extensions. If V is acyclic, we obtain a maximal path after a finite number of extensions.

Lemma 1 (deconstruction). *Let V be an acyclic vector field on K, with $dim(K) = d$. Then, at least one of the following holds:*

1) There exists a facet τ of K, with $dim(\tau) = d$, that is critical for V.
2) There exists a pair (σ, τ) in V, with $dim(\tau) = d$, that is a free pair for K.

Proof. If K has a d-face that is critical for V, then we are done. Suppose there is no such faces in K. If $d = 0$, then all the 0-faces of K are critical, thus we must have $d \geq 1$. Let τ be an arbitrary d-face of K. Since τ is not critical, there exists a pair (σ, τ) that is in V. Since $d \geq 1$, there is a face $\sigma' \in K$ such that $\pi' = \langle \sigma, \tau, \sigma' \rangle$ is a $(d-1)$ gradient path in V. By iteratively extending π' with extensions we obtain a maximal $(d-1)$-gradient path in V that is non-trivial. Let $\pi = \langle \sigma_0, \tau_0, ..., \sigma_{k-1}, \tau_{k-1}, \sigma_k \rangle$ be such a path, we have $k \geq 1$. If (σ_0, τ_0) is a free pair for K, then we are done. Otherwise, σ_0 must be a subset of a d-simplex ν, with $\nu \neq \tau_0$. By our hypothesis ν is not critical for V. Since ν is a facet for K, there must exist a $(d-1)$-simplex η, $\eta \neq \sigma_0$, such that (η, ν) is in V. In this case, the path $\pi' = \langle \eta, \nu, \sigma_0, \tau_0, ..., \sigma_{k-1}, \tau_{k-1}, \sigma_k \rangle$ would be a $(d-1)$-gradient path in V. Thus, the path π would not be maximal, a contradiction: the pair (σ_0, τ_0) must be a free pair for K. $\square$

Theorem 1. *Let K be a simplicial complex. A vector field V on K is acyclic if and only if V is the gradient vector field of a Morse sequence on K.*

Proof. i) Let $\overrightarrow{W} = \langle \emptyset = K_0, ..., K_i, ..., K_l = K \rangle$ be a Morse sequence on K and let V be the gradient vector field of $\overrightarrow{W}$. For each $\sigma \in K$, let $\rho(\sigma)$ be the index i such that $\sigma \in K_i$ and $\sigma \notin K_{i-1}$. Now, let $\pi = \langle \sigma_0, \tau_0, \sigma_1, \tau_1, ..., \sigma_{k-1}, \tau_{k-1}, \sigma_k \rangle$, $k \geq 1$, be a non-trivial p-gradient path in V. For all $i \in [0, k-1]$, (σ_i, τ_i) is in V, thus $\rho(\sigma_i) = \rho(\tau_i)$. Since $\sigma_{i+1} \subset \tau_i$ and since $\overrightarrow{W}$ is a filtration, we have $\rho(\sigma_{i+1}) \leq \rho(\tau_i)$. Since $\sigma_{i+1} \neq \sigma_i$ the pair (σ_{i+1}, τ_i) is not a regular pair for $\overrightarrow{W}$, thus we have $\rho(\sigma_{i+1}) < \rho(\tau_i)$. It follows that, for all $i \in [0, k-1]$, we have $\rho(\sigma_{i+1}) < \rho(\sigma_i)$. This gives $\rho(\sigma_k) < \rho(\sigma_0)$. It means that $\sigma_k \neq \sigma_0$, and so the path π cannot be closed. Consequently the vector field V is acyclic.
ii) Let V be an acyclic vector field on K, with $dim(K) = d$.

1) Suppose there exists a facet τ of K, with $dim(\tau) = d$, that is critical for V. Let $K' = K \setminus \{\tau\}$ and $V' = V$. Then, the set V' is also an acyclic vector field on the complex K'.

2) Suppose there exists a pair (σ, τ) in V, with $dim(\tau) = d$, that is a free pair for K. Clearly, the set $V' = V \setminus \{(\sigma, \tau)\}$ is also an acyclic vector field on the complex $K' = K \setminus \{\sigma, \tau\}$.

By 1), 2), and by Lemma 1, we can build inductively two sequences $\overleftarrow{W} = \langle K = K_0, ..., K_k = \emptyset \rangle$ and $\langle V = V_0, ..., V_k = \emptyset \rangle$ such that, for each $i \in [1, k]$:

– either K_i is an elementary perforation of K_{i-1} and $V_i = V_{i-1}$,
– or $K_i = K_{i-1} \setminus \{\sigma, \tau\}$ is an elementary collapse of K_{i-1} and $V_i = V_{i-1} \setminus \{(\sigma, \tau)\}$.

By considering the inverse of $\overleftarrow{W}$ we obtain the sequence $\overrightarrow{W} = \langle K_k = K'_0, ..., K'_k = K_0 \rangle$, which is such that, for each $i \in [1, k]$, either K'_i is an elementary expansion of K'_{i-1}, or K'_i is an elementary filling of K'_{i-1}. In other words, $\overrightarrow{W}$ is a Morse sequence on $K_0 = K$; the gradient field of $\overrightarrow{W}$ is precisely V, as required.   □

6 Morse Functions and Morse Sequences

Discrete Morse theory is classically introduced through the concept of a discrete Morse function. In this section we show that it is possible, in a straightforward manner, to make a link between Morse sequences and these functions.

We first introduce the notion of a Morse function on a Morse sequence $\overrightarrow{W}$.

Definition 4. *Let $\overrightarrow{W}$ be a Morse sequence on K and $\widehat{W} = \langle \hat{\sigma}_1, ..., \hat{\sigma}_k \rangle$. A map $f : K \rightarrow \mathbb{Z}$ is a Morse function on $\overrightarrow{W}$ whenever f satisfies the two conditions:*

1) If $\hat{\sigma}_i = \sigma_i$ is critical for $\overrightarrow{W}$ and $\sigma \in \partial(\sigma_i)$, then $f(\sigma_i) > f(\sigma)$.
2) If $\hat{\sigma}_i = (\sigma_i, \tau_i)$ is regular for $\overrightarrow{W}$, then $f(\sigma_i) \geq f(\tau_i)$.

Now, we can check that the following definition of a Morse function on a simplicial complex K is equivalent to the classical one [10,11].

Let K be a simplicial complex and let $f : K \rightarrow \mathbb{Z}$ be a map on K. Let V be the set of all pairs (σ, τ), with $\sigma, \tau \in K$, such that $\sigma \in \partial(\tau)$ and $f(\sigma) \geq f(\tau)$. If each $\nu \in K$ is in at most one pair in V, we say that f is a *Morse function on K*, and V is the *gradient vector field of f*. We say that two Morse functions on K are *equivalent* if they have the same gradient vector field.

Let f be a Morse function on K, and V be the gradient vector field of f. From the above definition, the set V is a discrete vector field on K. If $\pi = \langle \sigma_0, \tau_0, \sigma_1, \tau_1, ..., \sigma_{k-1}, \tau_{k-1}, \sigma_k \rangle$ is a p-gradient path in V, we see that we must have $f(\sigma_i) \geq f(\tau_i)$, and also $f(\tau_i) > f(\sigma_{i+1})$. Thus, $f(\sigma_0) > f(\sigma_k)$ whenever $k \geq 1$. It means that V contains no non-trivial closed p-gradient path. In other words, we have the classical result:

Proposition 2. *If f is a Morse function on K, then the gradient vector field of f is an acyclic vector field.*

Let $\overrightarrow{W}$ be a Morse sequence on K. We see that a Morse function on $\overrightarrow{W}$ is indeed a Morse function on K, the gradient vector field of this Morse function is precisely the gradient vector field of $\overrightarrow{W}$. Conversely, by Proposition 2 and by Theorem 1, if f is a Morse function on K, then there exists a Morse sequence $\overrightarrow{W}$ on K which has the same gradient vector field as f. It is easy to check that f is also a Morse function on $\overrightarrow{W}$. This leads us to the following result.

Theorem 2. *If f is a Morse function on K, then there exists a Morse sequence $\overrightarrow{W}$ on K such that f is a Morse function on $\overrightarrow{W}$. Furthermore, a Morse function g on K is equivalent to f if and only if g is a Morse function on $\overrightarrow{W}$.*

We introduce hereafter a particular kind of Morse function. Since a Morse sequence is a filtration, the following function f is indeed a Morse function on $\overrightarrow{W}$.

Definition 5. *Let $\overrightarrow{W}$ be a Morse sequence on K and $\widehat{W} = \langle \hat{\sigma}_1, \ldots, \hat{\sigma}_k \rangle$. The canonical Morse function of $\overrightarrow{W}$ is the map $f \colon K \to \mathbb{Z}$ such that:*

1) If $\hat{\sigma}_i = \sigma_i$ is critical for $\overrightarrow{W}$, then $f(\sigma_i) = i$.
2) If $\hat{\sigma}_i = (\sigma_i, \tau_i)$ is regular for $\overrightarrow{W}$, then $f(\sigma_i) = f(\tau_i) = i$.

As a consequence of Theorem 2, any Morse function on K is equivalent to a canonical Morse function.

We note that a canonical Morse function f is *flat*, that is, we have $f(\sigma) = f(\tau)$ whenever (σ, τ) is in the gradient vector field of f (Definition 4.14 of [19]). Also f is *excellent*, that is, all values of the critical simplexes are distinct (Definition 2.31 of [19]). In fact, a canonical Morse function has the three properties which define a basic Morse function (see [4] and also Definition 2.3 of [19]).

Let $f \colon K \to \mathbb{Z}$ be a map on K. We say that f is a *basic Morse function* if f satisfies the properties:

1) *monotonicity*: we have $f(\sigma) \leq f(\tau)$ whenever $\sigma \subseteq \tau$;
2) *semi-injectivity*: for each $i \in \mathbb{Z}$, the cardinality of $f^{-1}(i)$ is at most 2;
3) *genericity*: if $f(\sigma) = f(\tau)$, then either $\sigma \subseteq \tau$ or $\tau \subseteq \sigma$.

We observe that, if f is a basic Morse function on K, then we can build a Morse sequence $\overrightarrow{W}$ if we pick the simplexes of K by increasing values of f. For each i, $f^{-1}(i)$ gives a critical simplex if the cardinality of $f^{-1}(i)$ is one, and $f^{-1}(i)$ gives a regular pair if the cardinality of $f^{-1}(i)$ is two.

Let f and g be two basic Morse functions on K. We say that f and g are *strongly equivalent* if f and g induce the same order on K. That is, we have $f(\sigma) \leq f(\tau)$ if and only if $g(\sigma) \leq g(\tau)$.

With the above scheme for building a Morse sequence from a basic Morse function, we obtain the following result.

Proposition 3. *Let f be a basic Morse function on K. There exists one and only one Morse sequence $\overrightarrow{W}$ such that the canonical Morse function of $\overrightarrow{W}$ is strongly equivalent to f.*

7   Maximal Morse Sequences

Building a gradient vector field from a complex is a fundamental issue in discrete Morse theory. This problem is equivalent to building a Morse sequence $\overrightarrow{W}$ from a complex K. Clearly, the two following schemes are two basic ways to achieve this goal:

1) *The increasing scheme.* We build $\overrightarrow{W}$ from the left to the right. Starting from $\emptyset$, we obtain K by iterative expansions and fillings. We say that this scheme is *maximal* if we make a filling only if no expansion can be made.

2) *The decreasing scheme.* We build $\overrightarrow{W}$ from the right to the left. Starting from K, we obtain $\emptyset$ by iterative collapses and perforations. We say that this scheme is *maximal* if we make a perforation only if no collapse can be made.

Clearly, any Morse sequence may be obtained by an increasing scheme and any Morse sequence may be obtained by a decreasing scheme. By Theorem 1, it means that an arbitrary acyclic vector field may be obtained by each of these two schemes.

Now, let us focus our attention on maximal increasing and maximal decreasing schemes. The purpose of these two schemes is to try to minimize the number of critical simplexes. Thus, a filling or a perforation is made only if there is no other choice. The examples given in Fig. 1 and Fig. 2 are instances of a maximal increasing scheme.

First, it can be seen that the algorithm *Random Discrete Morse*, proposed by Benedetti and Lutz in [5], corresponds exactly to a maximal decreasing scheme. See this paper for many details of the algorithm (computational complexity, implementation in GAP, comparison with other algorithms, lower bounds for discrete Morse vectors...). See also Sect. 2.3 and Algorithm 1 in [19].

Also, there is a link between a maximal increasing scheme and *coreduction based algorithms* [12,16,18]. As mentioned in [12], a coreduction is not feasible on a simplicial complex. In fact, the coreduction algorithm presented in [12] may be formalized with a Morse sequence through the following definition.

Definition 6. *Let K be a simplicial complex. A* coreduction sequence (on K) *is a sequence* $\overrightarrow{C} = \langle K = C_0, ..., C_k = \emptyset \rangle$ *such that the sequence* $\overrightarrow{W} = \langle \emptyset = K_0 = K \setminus C_0, ..., K_k = K \setminus C_k = K \rangle$ *is a Morse sequence.*

In other words, a sequence $\overrightarrow{C} = \langle K = C_0, ..., C_k = \emptyset \rangle$ is a coreduction sequence if, for each $i \in [1, k]$, $K \setminus C_i$ is either an elementary expansion or an elementary filling of $K \setminus C_{i-1}$. It may be checked that the notion of a coreduction presented in [12] fully agrees with the above definition. It follows that the corresponding maximal coreduction algorithm may be seen as a maximal increasing scheme if we simply build a filtration with the simplexes that are removed by such an algorithm; see Sect. 5 of [12].

Thus, Morse sequences allow us to retrieve two methods for building a gradient vector field, which try to minimize the number of critical simplexes. Equivalently, they try to find optimal Morse sequences. It is worth mentioning that this problem is, in general, NP-hard [17]. Therefore, these methods do not, in general, give optimal results.

Now, let $\overrightarrow{W} = \langle \emptyset = K_0, ..., K_k = K \rangle$ be a Morse sequence on K. We write $\overleftarrow{W}$ for the inverse of the sequence $\overrightarrow{W}$, that is, we have $\overleftarrow{W} = \langle K = K_k, ..., K_0 = \emptyset \rangle$.

Thus, if $\overrightarrow{W}$ is a Morse sequence, $\overleftarrow{W}$ is a sequence $\langle L_0, ..., L_k \rangle$ such that, for each $i \in [1, k]$, L_i is either an elementary collapse or an elementary perforation of L_{i-1}. The following definition is a formal presentation of maximal increasing and decreasing schemes. See also [1, Definition 11] for an alternative formalization of a maximal decreasing scheme based on basic Morse functions.

Definition 7. Let $\overrightarrow{W} = \langle \emptyset = K_0, ..., K_k = K \rangle$ be a Morse sequence on K. For any $i \in [0, k]$, we say that K_i is maximal for $\overrightarrow{W}$ (resp. maximal for $\overleftarrow{W}$) if no elementary expansion (resp. collapse) of K_i is a subset of K.

1) We say that $\overrightarrow{W}$ is maximal if, for any $i \in [1, k]$, the complex X_{i-1} is maximal for $\overrightarrow{W}$ whenever X_i is critical for $\overrightarrow{W}$.
2) We say that $\overleftarrow{W}$ is maximal if, for any $i \in [0, k - 1]$, the complex X_{i+1} is maximal for $\overleftarrow{W}$ whenever X_i is critical for $\overleftarrow{W}$.

Perhaps surprisingly, there exist some significant differences between these two schemes, in particular in regard to the number of critical simplexes that are obtained.

The complex of Fig. 3, already considered in [5] and [12], illustrates this difference. The complex K in this example is a graph, that is, we have $dim(K) \leq 1$. In (a) and (b), the results that may be produced by a maximal decreasing scheme and by a maximal increasing scheme are given; the corresponding Morse vectors are $(2, 3)$ and $(1, 2)$. This last vector corresponds to the Betti numbers of K. It can be seen that the result in (a) cannot be obtained by a maximal increasing scheme. Actually, the following result is easy to check.

Proposition 4. Let K be a complex, and let $\overrightarrow{W}$ be a Morse sequence on K that is maximal. If K is a graph, then $\overrightarrow{W}$ is perfect.

Now, let us consider the complex K depicted in Fig. 4. This complex, given by Hachimori in [14, 15], is a slight modification of the dunce hat. We observe that K is collapsible, therefore $\vec{b}(K) = (1, 0, 0)$. We also observe that K contains a single free pair, which is $(\{1, 3\}, \{1, 3, 4\})$. Thus, any collapse sequence must begin with this pair. Now, we see that we can build a spanning tree on K that contains the edge $\{1, 3\}$. It is possible that a complex which is built in the first steps of a maximal increasing scheme of K contains this edge. This edge will prevent further expansions of the sequence from recovering the full complex K. Such a sequence $\overrightarrow{W}$ is depicted in Fig. 4: $\overrightarrow{W}$ is not perfect. It can be seen that this cannot happen with a maximal decreasing scheme. In fact, we have the following result.

Proposition 5. Let K be a complex, with $dim(K) = 2$, and let $\overrightarrow{W}$ be a Morse sequence on K such that $\overleftarrow{W}$ is maximal. If K is collapsible, then $\overrightarrow{W}$ is perfect.

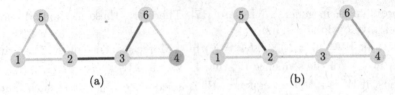

Fig. 3. A 1-dimensional complex. The results of a maximal decreasing scheme (a), and a maximal increasing scheme (b). See text for details.

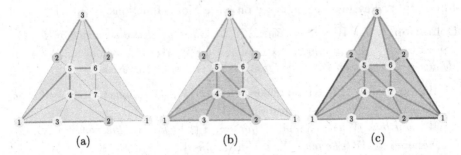

Fig. 4. A Morse sequence obtained by a maximal increasing scheme on Hachimori's example. (a): Starting from the critical 0-simplex 2, we create a maximal spanning tree that contains the edge $\{1,3\}$. (b): We then make all possible expansions. At this point, we have to select a critical 1-simplex, the edge $\{1,2\}$. (c): We continue with expansions, until this is no longer possible. We then have to add the critical 2-simplex $\{3,5,6\}$. See text for a discussion.

8   Conclusion

In this paper, we introduce the notion of a Morse sequence for a simple presentation of some basic ingredients of discrete Morse theory:

- The collapse theorem becomes a property that is contained in the very definition of a Morse sequence;
- The link between Morse sequences and different kinds of Morse functions is straightforward;
- A Morse sequence may represent the gradient vector field of an arbitrary discrete Morse function;
- Maximal Morse sequences formalize two basic schemes for building the gradient vector fields of an arbitrary simplicial complex.

Morse sequences are not only interesting by themselves, they also offer new perspectives for exploring the topology of simplicial complexes. For example, adding information to Morse sequences leads to novel schemes for computing topological invariant such as cycles, cocycles, and Betti numbers. This can be achieved by defining *Morse frames*, which are maps that associate a set of critical simplexes to each simplex of the complex. See the companion paper [7] where this approach is explored.

References

1. Adiprasito, K.A., Benedetti, B., Lutz, F.H.: Extremal examples of collapsible complexes and random discrete Morse theory. Discrete Comput. Geom. **57**, 824–853 (2017)
2. Ayala, R., Fernández-Ternero, D., Vilches, J.A.: Perfect discrete Morse functions on 2-complexes. Pattern Recogn. Lett. **33**(11), 1495–1500 (2012)
3. Ayala, R., Fernández-Ternero, D., Vilches, J.A.: Perfect discrete Morse functions on triangulated 3-manifolds. In: Ferri, M., Frosini, P., Landi, C., Cerri, A., Di Fabio, B. (eds.) CTIC 2012. LNCS, vol. 7309, pp. 11–19. Springer, Heidelberg (2012). https://doi.org/10.1007/978-3-642-30238-1_2
4. Benedetti, B.: Smoothing discrete Morse theory. Ann. Sc. Norm. Super. Pisa Cl. Sci. **XVI**, 335–368 (2016)
5. Benedetti, B., Lutz, F.H.: Random discrete Morse theory and a new library of triangulations. Exp. Math. **23**(1), 66–94 (2014)
6. Bertrand, G.: Completions, perforations and fillings. In: Lindblad, J., Malmberg, F., Sladoje, N. (eds.) DGMM 2021. LNCS, vol. 12708, pp. 137–151. Springer, Cham (2021). https://doi.org/10.1007/978-3-030-76657-3_9
7. Bertrand, G., Najman, L.: Morse frames. In: Rinaldi, S. (ed.) DGMM 2024. LNCS, vol. 14605, pp. 364–376. Springer, Cham (2024). https://hal.science/hal-04217818
8. Dey, T.K., Wang, Y.: Computational Topology for Data Analysis. C. U. P. (2022)
9. Forman, R.: Morse theory for cell complexes. Adv. Math. **134**, 90–145 (1998)
10. Forman, R.: Witten-Morse theory for cell complexes. Topology **37**(5), 945–979 (1998)
11. Forman, R.: A user's guide to discrete Morse theory. Séminaire Lotharingien Combin. [Electronic Only] **48**, B48c–35 (2002)
12. Fugacci, U., Iuricich, F., De Floriani, L.: Computing discrete Morse complexes from simplicial complexes. Graph. Models **103**, 101023 (2019)
13. Giblin, P.: Graphs, Surfaces and Homology. Cambridge University Press, Cambridge (2010)
14. Hachimori, M.: Simplicial complex library. https://infoshako.sk.tsukuba.ac.jp/%7Ehachi/math/library/nonextend_eng.html. Accessed 23 Sept 2023
15. Hachimori, M.: Combinatorics of constructible complexes. Ph.D. thesis, Tokyo University (2000)
16. Harker, S., Mischaikow, K., Mrozek, M., Nanda, V.: Discrete Morse theoretic algorithms for computing homology of complexes and maps. Found. Comput. Math. **14**, 151–184 (2014)
17. Joswig, M., Pfetsch, M.E.: Computing optimal Morse matchings. SIAM J. Discrete Math. **20**, 11–25 (2006)
18. Mrozek, M., Batko, B.: Coreduction homology algorithm. Discrete Comput. Geom. **41**(1), 96–118 (2009)
19. Scoville, N.A.: Discrete Morse Theory, vol. 90. American Mathematical Society (2019)
20. Whitehead, J.H.C.: Simplicial spaces, nuclei and m-groups. Proc. Lond. Math. Soc. **2**(1), 243–327 (1939)
21. Zeeman, E.C.: On the dunce hat. Topology **2**, 341–358 (1964)

1-Attempt and Equivalent Thinning on the Hexagonal Grid

Kálmán Palágyi[(✉)]

Department of Image Processing and Computer Graphics, University of Szeged,
Szeged, Hungary
palagyi@inf.u-szeged.hu

Abstract. Thinning in 2D is an iterative object reduction to produce
centerlines of discrete binary objects. A thinning algorithm is 1-attempt
if whenever a border point is not deleted in the actual iteration step,
it belongs to the resulting centerline. Parallel thinning algorithms alter
all deletable points simultaneously, while sequential ones traverse object
points in the current picture, and delete the actually visited one if it
is designated as deletable. A pair of thinning algorithms are equivalent
if they produce the same centerline for any input picture. This paper
presents the very first 1-attempt, equivalent, and topology-preserving
pair of parallel and sequential thinning algorithms acting on the nonconventional hexagonal grid. It is also illustrated that 1-attempt property
involves a remarkable speed up.

Keywords: Digital topology · Skeletons · Hexagonal grid · 1-Attempt
thinning · Equivalent thinning · Topology preservation

1 Introduction

Digital binary pictures (or *picture*s in short) assign a value of black or white
to each point of the given *digital space* (or *grid*) [9]. A *reduction* is an operation that transforms a picture only by changing a set of black points to white
ones, which is referred to as *deletion* [4]. *Parallel reduction*s can delete a set of
black points simultaneously, while *sequential reduction*s traverse the potentially
deletable black points of a picture, and focus on the actually visited single point
for possible deletion at a time [4].

For 2D pictures, *skeletonization* means extraction of *centerlines* (i.e., a stick-
like 1D representation of discrete objects). There is a fairly general agreement
that skeletonization plays a key role in a broad range of problems in image
processing and computer vision [20, 21]. *Thinning* is a frequently applied skeletonization approach: border points that satisfy certain geometric and topological
constraints are deleted in iteration steps, and the entire process is repeated until
stability is reached (i.e., no more points are deleted). Parallel thinning algorithms comprise parallel reductions, and sequential ones consist of sequential
reductions.

S. Brunetti et al. (Eds.): DGMM 2024, LNCS 14605, pp. 390–401, 2024.
https://doi.org/10.1007/978-3-031-57793-2_30

In the conventional and time-consuming implementation of thinning algorithms, the deletability of all border points in the actual picture is to be checked. That is why Palágyi and Németh introduced the concept of 1-*attempt thinning* [18]. In the case of a 1-attempt algorithm, if an examined border point 'survives' the current iteration step, it is 'immortal' (i.e., it cannot be deleted in the remaining thinning phases, and it is a safe element in the resulting centerline).

In [14], the author introduced the concept of *equivalent reductions* (*equivalent thinning algorithms*) and *equivalent deletion rules*. Two reductions (thinning algorithms) are said to be equivalent if they produce the same result for each input picture. A deletion rule is called equivalent if it yields a pair of equivalent parallel and sequential reductions.

A crucial issue in thinning algorithms is to ensure *topology preservation* [9]. It is generally agreed that a thinning algorithm is absolutely useless if it does not preserve the topology.

It is the common practice that 2D digital pictures are sampled on the regular square grid, since these pictures can be naturally stored in usual 2D arrays. The disadvantage of the square grid is that two grid elements can share an edge or only a vertex. The importance of the hexagonal grid shows an upward tendency due to its advantages of geometric and topologic properties [10,11].

In this paper, we present a pair of parallel and sequential thinning algorithms on the hexagonal grid. Instead of comparing the new algorithms with existing ones [2,3,5,6,8,22,23], three mathematical properties of the proposed algorithms are proved. We show that our algorithms are 1-attempt, equivalent, and topology-preserving. As far as we know they are the very first thinning algorithms being both 1-attempt and equivalent.

2   Basic Notions and Results

In this paper, we use the fundamental concepts of digital topology as reviewed by Kong and Rosenfeld in their seminal work [9].

A $(6,6)$ picture on the *hexagonal grid* $\mathcal{H}$ is a quadruple $(\mathcal{H}, 6, 6, B)$, where elements of $\mathcal{H}$ are the *points* of the picture; $B \subseteq \mathcal{H}$ denotes the set of *black points*; each point in $\mathcal{H} \setminus B$ is a *white point*; 6-*adjacency* is assigned to both B and $\mathcal{H} \setminus B$. Two points (i.e., regular hexagons) are 6-*adjacent* if they share an edge, and let $N(p)$ denote the set of six points that are 6-adjacent to point p, see Fig. 1a.

Since the studied adjacency relation is symmetric, its reflexive-transitive closure induces equivalence relations on B and $\mathcal{H} \setminus B$, and the generated equivalence classes are called *black components* (or *objects*) and *white components*, respectively.

A black point p is an *interior point* for B, if $N(p) \subset B$ (i.e., all the six points being 6-adjacent to p are black), and a black point is a *border point* for B, if it is not an interior point.

A reduction is *topology-preserving* if each object in the input picture contains exactly one object in the output picture, and each white component in the output

Fig. 1. Six elements in $N(p) = \{q_0, q_1, \ldots, q_5\}$ are 6-adjacent to the central point p (a). Two base matching templates for characterizing simple points in $(6,6)$ pictures on the hexagonal grid (b)-(c), in which each black template position matches a black point; each white element matches a white point; each position depicted in gray matches either a black or a white point. Notice that the simpleness of the central point p depends only on $N(p)$.

picture contains exactly one white component in the input picture. Note that it is valid only for reductions acting on 2D pictures.

A single black point is *simple* if its deletion preserves the topology. There is a fairly general agreement: a sequential reduction is topology-preserving if it deletes only simple points [9].

Kardos and Palágyi gave an illustrative characterization of simple points in $(6,6)$ pictures by matching templates [7]. The two base matching templates shown in Fig. 1b-c and their rotated versions match all simple points.

3   The Proposed Pair of Thinning Algorithms

In this section, we present a pair of parallel and sequential thinning algorithms **P** (see Algorithm 1) and **S** (see Algorithm 2).

Algorithm 1: parallel thinning algorithm **P**

Input: picture $(\mathcal{H}, 6, 6, X)$
Output: picture $(\mathcal{H}, 6, 6, Y)$
$Y \leftarrow X$
repeat
    // specifying the constraint set
    $C = \{\, p \mid p$ is an interior point in $Y \,\}$
    // collecting deletable points
    $D \leftarrow \{\, p \in Y \setminus C \mid p$ is deletable by deletion rule R for $Y \,\}$
    // deleting simultaneously
    $Y \leftarrow Y \setminus D$
until $D = \emptyset$;

In both algorithms **P** and **S**, the kernels of the **repeat** cycles correspond to one iteration step, and at the beginning of the iterations, the set black points Y of the current image is classified into two (disjoint) subsets: the *constraint set*

Algorithm 2: sequential thinning algorithm **S**

Input: picture $(\mathcal{H}, 6, 6, X)$
Output: picture $(\mathcal{H}, 6, 6, Y)$
$Y \leftarrow X$
repeat
    // specifying the constraint set
    $C = \{ p \mid p$ is an interior point in $Y \}$
    // traversal according to an arbitrary total ordering of $Y \setminus C$
    deletion $\leftarrow$ **false**
    foreach $p \in Y \setminus C$ **do**
        if *p is deletable by deletion rule R for Y* **then**
            // instant deletion
            $Y \leftarrow Y \setminus \{p\}$
            deletion $\leftarrow$ **true**
until *deletion* = **false**;

$C \subset Y$ containing interior points to be preserved, and points in $Y \setminus C$ are tested for possible deletion.

Both algorithms act on the hexagonal grid, and they use the same deletion rule R:

Definition 1. *A point $p \in B$ is deletable by deletion rule R for B from picture $(\mathcal{H}, 6, 6, B)$ (or deletable in short), if at least one of the 18 matching templates depicted in Fig. 2 matches it. Otherwise, point p is called non-deletable.*

Let us notice some useful properties of deletion rule R:

Proposition 1. *Only border points are deletable.*

On the one hand, we do not examine the deletability of the interior points (i.e., elements in the constraint set C, see Algorithms 1 and 2), and on the other hand, central point p is 6-adjacent to at least one white template position in all 18 matching templates depicted in Fig. 2.

Proposition 2. *All gray template positions of matching templates shown in Fig. 2 coincide with black points if the parallel algorithm P is performed.*

It can be readily seen by careful examination of the matching templates shown in Fig. 2. Template positions marked '$\star$' match interior points (i.e., elements in the constraint set C), and all points that are 6-adjacent to an interior point are black. Since each ignored gray position is 6-adjacent to a template position marked '$\star$', Proposition 2 holds.

Sequential reductions (and sequential thinning algorithms) generally suffer from the drawback that various visiting orders of border points yield different results. In Sect. 5, we show that our sequential thinning algorithm **S** is order-independent, and both algorithms **P** and **S** produce the same centerline for any input picture (i.e., they are equivalent).

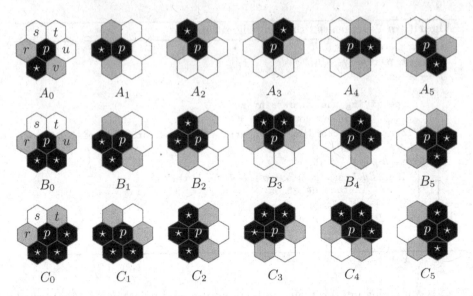

Fig. 2. The set of 18 matching templates associated with deletion rule R. Notations: each black position matches a black point; each white element matches a white point; p indicates the central element of a pattern; each position marked '$\star$' matches an element in the constraint set C. The template positions depicted in gray are not taken into consideration at all, they only play a role in the proof of Theorem 1.

Figure 3 is to illustrate an important property of the proposed algorithms: they cannot produce one point thin centerlines for any objects. It is usual in parallel thinning algorithms that use a single symmetric deletion rule like R.

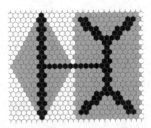

Fig. 3. The produced 2-point thick centerline superimposed on a small synthetic object.

Our algorithms were tested for numerous objects of different shapes (sampled on the hexagonal grid). For reasons on scope, we selected just five illustrative test images, see Fig. 4.

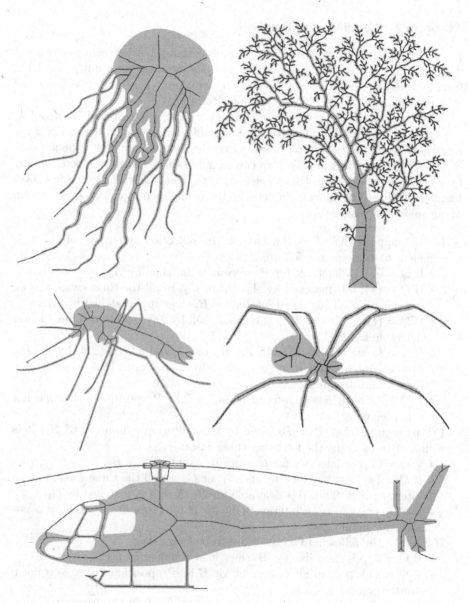

Fig. 4. Centerlines produced by our algorithms **P** and **S** for the selected five test images.

4   1-Attempt Property

Now we will show that our parallel thinning algorithm **P** is 1-attempt (i.e., if whenever a border point is not deleted in the actual iteration step, it cannot be deleted in the remaining iterations).

Theorem 1. *Algorithm* **P** *is* 1-*attempt.*

Proof. Let us prove it by contradiction. Assume that a border point $p \in B$ is not deleted from picture $(\mathcal{H}, 6, 6, B)$ in the previous iteration step, but p is deleted from picture $(\mathcal{H}, 6, 6, B \setminus D)$ in the current iteration, where $D \subset B$ is the set of all points that are deleted by the previous iteration.

Without loss of generality, we can suppose that p is matched by template $T \in \{A_0, B_0, C_0\}$ for $B \setminus D$, since the remaining templates shown in Fig. 2 are rotated versions of these three ones. Since the deletability of p depends only on the set of six points in $N(p)$, we can examine the (sub)set of deleted points $Q = D \cap N(p)$. It is obvious that all elements of Q coincide with white elements of template T. Then the following points can be readily seen by careful examination of our matching templates:

- Let us suppose that $T = A_0$. Due to the reflection symmetry of A_0, it is sufficient to examine the following cases:
  - If $Q = \emptyset$, p is deletable for B, since it is matched by A_0.
  - If $Q = \{s\}$, s is matched by A_0, B_1, or C_2. In all the three cases, r is an interior point. Thus p is deletable for B, since it is matched by B_1.
  - If $Q = \{t\}$ or $Q = \{s, t\}$, t is not matched by any template. Thus t is not deletable for B.
  - If $Q = \{s, u\}$, s is matched by A_0, B_1, or C_2, and u is matched by A_0, B_0, or C_0. Consequently, r and v are both interior points. Thus p is deletable for B, since it is matched by C_1.
  - If $Q = \{s, t, u\}$, p is an interior point for B by Proposition 2 (i.e., p is not a border point).
- Let us assume that $T = B_0$. Due to the reflection symmetry of B_0, it is sufficient to examine the following three cases:
  - If $Q = \emptyset$, p is deletable for B, since it is matched by B_0.
  - If $Q = \{s\}$, s is matched by A_0, B_1, or C_2. In all the three cases, r is an interior point. Thus p is deletable for B, since it is matched by C_1.
  - If $Q = \{s, t\}$, p is an interior point for B by Proposition 2 (i.e., p is not a border point).
- If $T = C_0$, the following two cases are to be examined:
  - If $Q = \emptyset$, p is deletable for B, since it is matched by C_0.
  - If $Q = \{s\}$, p is an interior point for B by Proposition 2 (i.e., p is not a border point).

In all cases, $p \in B$ is either not a border point, or it is deletable for B, or at least one element in Q is not deletable for B. Thus we arrived at a contradiction. It means that there was not any non-deletable border point in the previous iteration step, that can be deleted in the current iteration. Thus the proof by contradiction is completed.   □

The author proposed a computationally efficient implementation scheme for both sequential and parallel thinning algorithms [12]. A linked list stores all border points in the actual picture, its elements are examined for possible deletion, and updated after deletion. In the case of 1-attempt algorithms, non-deletable border points can be removed immediately from the list of border points, since the repeated evaluation of their deletability has become obsolete.

The advantage of the 1-attempt implementation scheme over the mentioned computationally efficient one is illustrated in Table 1 for the five test images shown in Fig. 4. Both implementations under comparison were run on a usual PC. Notice that the attainable speed-up is absolutely data-dependent.

Table 1. Computation times (in msec.) of algorithm **P** for the five test images shown in Fig. 4.

Test image	Size	Number of object points	Number of skeletal points	'conventional' comp. time C	'1-attempt' comp. time A	Speed-up C/A
	696 × 990	186 004	12 652	69.91	13.95	**4.29**
	576 × 430	87 994	25 801	32.13	7.84	**4.10**
	1656 × 537	278 825	10 953	57.67	20.09	**2.87**
	797 × 730	98 824	7 219	19.49	7.95	**2.45**
	488 × 344	25 826	3 305	5.01	2.24	**2.23**

Note that there exist some 1-attempt parallel thinning algorithms acting on the traditional square grid [16–18].

5   Equivalency

In this section, we show that the proposed thinning algorithms **P** and **S** are equivalent (i.e., they produce the same centerline for any picture).

Let us recall the author's previously reported results:

Theorem 2. [14] *Let us assume that a point $q \in B$ is deletable by deletion rule $\mathcal{R}$ for the set of black points B. Then $\mathcal{R}$ is equivalent (i.e., it yields a pair of*

*parallel and sequential reductions) if the deletability of any point $p \in B \setminus \{q\}$
does not depend on the 'color' of q:*

1. *if p is deletable by $\mathcal{R}$ for B, p is deletable by $\mathcal{R}$ for $B \setminus \{q\}$;*
2. *if p is not deletable by $\mathcal{R}$ for B, p is not deletable by $\mathcal{R}$ for $B \setminus \{q\}$.*

Theorem 3. [14] *A deletion rule yields an order-independent sequential reduction if and only if it is equivalent.*

We are now ready to state the following theorem:

Theorem 4. *Deletion rule R (assigned to both algorithms* **P** *and* **S** *) is equivalent.*

Proof. Let us suppose that $q \in B$ is deleted by R from picture $(\mathcal{H}, 6, 6, B)$. Then the following two points are to be shown for any $p \in B \setminus \{q\}$:

1. If p is deletable for B, p is deletable for $B \setminus \{q\}$.
2. If p is not deletable for B, p is not deletable for $B \setminus \{q\}$.

Let us examine the 18 matching templates shown in Fig. 2. Since the deletability of p depends only on the set of six points in $N(p)$, we can assume that $q \in N(p)$.

1. In this case, both points p and q are in B and are deletable. Let us assume that p is matched by a template $T \in \{A_0, \ldots, C_5\}$ (i.e., p coincides with the central position of T). Then deletable (black) point q only can coincide with a gray position of T. Hence q is not tested by R, implying that p is deletable for $B \setminus \{q\}$.
2. We will prove this point by contraposition. It is assumed that p is deletable for $B \setminus \{q\}$, and it is matched by a template T. We are to show that p is deletable for B. (In other words, q is white, it was deleted before, now, it is investigated what happens if q is 're-added'.)

   Without loss of generality, we can suppose that p is matched by template $T \in \{A_0, B_0, C_0\}$ for $B \setminus \{q\}$, since the remaining templates are rotated versions of these three ones. By Proposition 1, deletable point q for B may not coincide with a template position marked '$\star$'. Thus it is sufficient to assume that q coincides with a white template position of T. The following points can be readily seen by careful examination of our matching templates.
   - Let us suppose that $T = A_0$. Due to the reflection symmetry of A_0, it is sufficient to examine the following two cases:
     - If q coincides with s, q may be matched by templates A_0, B_1, or C_2. In all the three cases, p is matched by template B_1 for B.
     - If q coincides with t, q is not matched by any templates. Since q is deletable for B, we arrive at a contradiction.
   - Let us suppose that $T = B_0$. Due to the reflection symmetry of B_0, it is sufficient to examine just one case:
     If q coincides with s, q may be matched by templates A_0, B_1, or C_2. In all the three cases, p is matched by template C_1 for B.

- Let us suppose that $T = C_0$. In this case q coincides with s, and q may be matched by templates A_0, A_4, B_1, B_4, C_2, or C_4.
  - If q may be matched by templates A_0, B_1, or C_2, p is matched by template C_1.
  - If q may be matched by templates A_4, B_4, or C_4, p is matched by template C_5.

Since the deletability of p does not depend on the 'color' of the deletable point q, deletion rule R is equivalent by Theorem 2. □

We can state the following easy consequences of the theorems recalled and proved in this section:

- Sequential thinning algorithm **S** is order-independent (i.e., Algorithm 2 produces the same centerline for arbitrary total ordering of border points).
- The proposed sequential and parallel algorithms (i.e., **P** and **S**) are equivalent.

Lastly, we mention that there are some equivalent thinning algorithms on the traditional square grid as well [13,19].

6   Topology Preservation

Topology preservation is a crucial issue in thinning algorithms. In this section, we show that the proposed pair of parallel and sequential algorithms meet this requirement.

Theorem 5. *Both algorithms* **P** *and* **S** *are topology-preserving.*

Proof. Recall that a single black point is simple if its deletion preserves the topology, and a sequential reduction is topology-preserving if it deletes only simple points.

It is obvious by a careful examination of matching templates depicted in Fig. 2 and Fig. 1b-c that deletion rule R deletes only simple points. (Note that the 18 templates in Fig. 2 cause deletion of certain simple points, not of any simple point.)

Since the iteration steps of algorithm **S** (as sequential reductions) only delete simple points, our sequential algorithm **S** is topology-preserving.

By Theorem 4, algorithms **P** and **S** are equivalent. Consequently, parallel algorithm **P** is also topology-preserving. □

The topological correctness of our parallel algorithm **P** can also be verified in an alternative way.

Bertrand introduced the notion of a *P-simple set*, whose simultaneous deletion preserves the topology:

Definition 2. [1] *Let B be the set of black points in an arbitrary picture. A set of black points $X \subset B$ is a P-simple set for B if for any point $x \in X$ and any set of points $Y \subseteq X \setminus \{x\}$, x is a simple point for $B \setminus Y$.*

Theorem 6. [1] *A reduction that deletes solely P-simple sets is topology-preserving.*

The author proved the following theorem:

Theorem 7. [15] *If a reduction with equivalent deletion rule deletes only simple points, that reduction deletes solely P-simple sets.*

Since deletion rule R is equivalent (see Theorem 4) and deletes only simple points, it is topology-preserving by Theorems 7 and 6.

7 Conclusions

In this work, we have shown that it is possible to construct 1-attempt, equivalent, and topology-preserving sequential and parallel thinning algorithms on the nonconventional hexagonal grid. It is also illustrated that a thinning algorithm is considerably faster if we know that it fulfills the 1-attempt property.

Acknowledgements. This research was supported by project TKP2021-NVA-09. Project no. TKP2021-NVA-09 has been implemented with the support provided by the Ministry of Innovation and Technology of Hungary from the National Research, Development and Innovation Fund, financed under the TKP2021-NVA funding scheme.

References

1. Bertrand, G.: On P-simple points. Compte Rendu de l'Académie des Sciences de Paris, Série Math. **321**, 1077–1084 (1995)
2. Beucher, S.: From non connected to homotopic skeletons in multidimensional digital space. In: Proceedings of the International Symposium on Mathematical Morphology and Its Applications to Signal Processing, pp. 139–144 (1993)
3. Deutsch, E.S.: Thinning algorithms on rectangular hexagonal and triangular arrays. Commun. ACM **15**, 827–837 (1972). https://doi.org/10.1145/361573.361583
4. Hall, R.W.: Parallel connectivity-preserving thinning algorithms. In: Kong, T.Y., Rosenfeld, A. (eds.) Topological Algorithms for Digital Image Processing, pp. 145–179. Elsevier Science, Amsterdam (1996). https://doi.org/10.1016/S0923-0459(96)80014-0
5. Kardos, P., Palágyi, K.: On topology preservation for hexagonal parallel thinning algorithms. In: Aggarwal, J.K., Barneva, R.P., Brimkov, V.E., Koroutchev, K.N., Korutcheva, E.R. (eds.) IWCIA 2011. LNCS, vol. 6636, pp. 31–42. Springer, Heidelberg (2011). https://doi.org/10.1007/978-3-642-21073-0_6
6. Kardos, P., Palágyi, K.: Hexagonal parallel thinning algorithms based on sufficient conditions for topology preservation. In: Proceedings of the 3rd International Symposium of Computational Modeling on Objects Presented in Images: Fundamentals, Methods, and Applications, pp. 63–68 (2012). https://doi.org/10.1201/b12753-12

7. Kardos, P., Palágyi, K.: On topology preservation in triangular, square, and hexagonal grids. In: Proceedings of the 8th International Symposium on Image and Signal Processing and Analysis, pp. 782–787 (2013). https://doi.org/10.1109/ISPA.2013.6703844

8. Kardos, P., Palágyi, K.: Topology-preserving hexagonal thinning. Int. J. Comput. Math. **90**, 1607–1617 (2013). https://doi.org/10.1080/00207160.2012.724198

9. Kong, T.Y., Rosenfeld, A.: Digital topology: introduction and survey. Comput. Vis. Graph. Image Process. **48**, 357–393 (1989). https://doi.org/10.1016/0734-189X(89)90147-3

10. Lee, M., Jayanthi, S.: Hexagonal Image Processing: A Practical Approach. Springer, London (2005)

11. Marchand-Maillet, S., Sharaiha, Y.M.: Binary Digital Image Processing: A Discrete Approach. Academic Press, London (2000). https://doi.org/10.1117/1.1326456

12. Palágyi, K.: A 3D fully parallel surface-thinning algorithm. Theor. Comput. Sci. **406**, 119–135 (2008). https://doi.org/10.1016/j.tcs.2008.06.041

13. Palágyi, K.: Equivalent 2D sequential and parallel thinning algorithms. In: Barneva, R.P., Brimkov, V.E., Šlapal, J. (eds.) IWCIA 2014. LNCS, vol. 8466, pp. 91–100. Springer, Cham (2014). https://doi.org/10.1007/978-3-319-07148-0_9

14. Palágyi, K.: Equivalent sequential and parallel reductions in arbitrary binary pictures. Int. J. Pattern Recogn. Artif. Intell. **28**, 1460009 (2014). https://doi.org/10.1142/S021800141460009X

15. Palágyi, K.: How sufficient conditions are related for topology-preserving reductions. Acta Cybernetica **23**, 939–958 (2018). https://doi.org/10.14232/actacyb.23.3.2018.14

16. Palágyi, K., Németh, G.: 1-attempt subfield-based parallel thinning. In: Proceedings of the 12th International Symposium on Image and Signal Processing and Analysis, pp. 306–311 (2021). https://doi.org/10.1109/ISPA52656.2021.9552163

17. Palágyi, K., Németh, G.: 1-attempt 4-cycle parallel thinning algorithms. In: Proceedings of the 11th International Conference on Pattern Recognition Applications and Methods, pp. 229–236 (2022). https://doi.org/10.5220/0010819700003122

18. Palágyi, K., Németh, G.: 1-attempt parallel thinning. J. Comb. Optim. **44**, 2395–2409 (2022). https://doi.org/10.1007/s10878-021-00744-y

19. Palágyi, K., Németh, G., Kardos, P.: Topology-preserving equivalent parallel and sequential 4-subiteration 2D thinning algorithms. In: Proceedings of the 9th International Symposium on Image and Signal Processing and Analysis, pp. 306–311 (219–224). https://doi.org/10.1109/ISPA.2015.7306077

20. Saha, P.K., Borgefors, G., Sanniti di Baja, G.: Skeletonization: Theory, Methods and Applications. Academic Press, London (2017). https://doi.org/10.1016/B978-0-08-101291-8.00017-1

21. Siddiqi, K., Pizer, S.M.: Medial Representations: Mathematics, Algorithms and Applications. Springer, Dordrecht (2008). https://doi.org/10.1007/978-1-4020-8658-8

22. Staunton, R.C.: An analysis of hexagonal thinning algorithms and skeletal shape representation. Pattern Recogn. **29**, 1131–1146 (1999). https://doi.org/10.1016/0031-3203(94)00155-3

23. Staunton, R.C.: One-pass parallel hexagonal thinning algorithm. In: IEE Proceedings of Vision, Image and Signal Processing, vol. 148, pp. 45–53 (2001). https://doi.org/10.1049/ip-vis:20010076

A Survey on 2D Euclidean Curve Classes in Discrete Geometry with New Results

Étienne Le Quentrec$^{(\boxtimes)}$ ⓘ, Étienne Baudrier$^{(\boxtimes)}$ ⓘ, and Clément Jacquot

ICube-UMR 7357, 300 Bd Sébastien Brant - CS 10413, 67412 Illkirch Cedex, France
{elequentrec,baudrier}@unistra.fr

Abstract. Classes of curves like par-regularity, μ-reach, locally turn boundedness, quasi-regularity (and their generalizations) have been defined so as to guarantee geometrical or topological properties under discretization. An overview of their inter-relations is given. A focus is made on the Locally Turn Bounded (LTB) curves, a class having good discretization properties. It has already been shown that being LTB implies quasi-regularity. In this paper, it is shown that the LTB curves have a positive μ-reach. Moreover, we show that a LTB curve having a Lipschitz turn is par-regular.

1 Introduction

The discretization process plays an essential role in discrete geometry. Of course, part of discrete geometry takes advantage of the discrete nature of the data to propose efficient algorithms, but some discrete objects are the discretization of continuous objects or are detected in digital images from the Euclidean world. If the process of discretization in itself is modeled in several ways, the mathematical modeling of the continuous objects represented is also an important element. Indeed, the discretization process is accompanied by a loss of information between the continuous form and the discrete object obtained. To limit this loss of information, it is necessary to consider a restricted class of continuous objects not containing artifacts erased by the discretization process. This ensures that certain topological-geometric properties are preserved before and after discretization. Indeed, the conservation of topological-geometric properties may be required for the reconstruction of continuous objects from a discretization [4], the conservation of topological properties under rigid transformation [16], affine transformation [17], for image manipulation purposes or even data augmentation [20].

The set of continuous shapes is vast and contains curves far away from those of digital images. Conversely, the first classes considered (polygonal shapes, convex shapes) do not always cover the shapes encountered in the applications. In addition, they include curves with undesirable properties (e.g. polygons with very acute angles). It is then interesting to define a class of continuous forms different from these first classes whose discretization properties can be established. The history of this research dates back to at least the beginning of computing

S. Brunetti et al. (Eds.): DGMM 2024, LNCS 14605, pp. 402–414, 2024.
https://doi.org/10.1007/978-3-031-57793-2_31

[5] and continues today [10,15]. The fact that this research is still active is due to the difficulty of characterizing (usefully from a computer point of view) continuous curves so as to establish properties related to the discretization which can be global (like topological ones). We present a detailed state of the art on the main classes of curves defined in 2D as well as on the links between them in Sect. 2. In particular, we have already established the equivalence between the LTB curves with Lipschitz turn and the curves with positive reach, more precisely, it is shown in [10, Theorem 2] that par-regular curves are LTB with Lipschitz turn (Par-regularity is equivalent to a positive reach [7]); and the reciprocal is shown in [12]. A new relation is established in Sect. 3, where it is shown that the LTB curves have a strictly positive μ-reach.

2 Existing Notions on Continuous Shapes

The objective of this section is to introduce various classes of shapes whose complexity is limited enough to ensure certain properties in their discretization. Some necessary notions are recalled hereafter.

Let A be a subset of $\mathbb{R}^2$, A^c is its complement, $\overset{\circ}{A}$ denotes the biggest open set included in A (for the usual topology), $\bar{A}$, defined as the smallest closed set containing A, is the topological closure of A and $\partial A = \bar{A} \setminus \overset{\circ}{A}$ its topological boundary. Let $B(c,r) := \{x \in \mathbb{R}^2 \mid d(x,c) < r\}$ and $\bar{B}(c,r) := \{x \in \mathbb{R}^2 \mid d(x,c) \leq r\}$ where d is the Euclidean distance, denote the open and the closed ball of radius r and center c. The dilation of radius r of a set S denoted $S \oplus B(0,r)$ is the union of all the open balls of radius r with center in S and the respective erosion $S \ominus B(0,r)$ is the union of all center points of open balls included in S. The morphological opening with a structuring element being an open ball of radius r is defined as $S \circ B(0,r) := (S \ominus B(0,r)) \oplus B(0,r)$.

2.1 Par-Regularity

One of the earliest concepts of complexity control used in discrete geometry is par-regularity. First used to prove topology preservation results under digitization, this notion has later been applied to geometric estimation. It was independently introduced in 1982 by Pavlidis (referred to as *compatibility assumptions* in [18, Definition 7.4]) and by Serra (referred to as the *regular model* in [19, Section 5.C]). In Pavlidis' formulation, a binary image and a grid with spacing h are considered compatible if, on one hand, there exists a number $d > \sqrt{2}h$ such that for every point on the boundary of each connected component of the Euclidean shape S, there exists a tangent circle with a diameter of d contained entirely within S. On the other hand, the same condition is also satisfied for the complement of S (note that it is not specified in [18] whether S is an open or closed set). In Serra's formulation, the regular model consists of compact sets S stable under morphological opening and closing operations by a closed ball of radius r, as defined by:

$$S = (S \ominus \bar{B}(0,r)) \oplus \bar{B}(0,r) = (S \oplus \bar{B}(0,r)) \ominus \bar{B}(0,r). \qquad (1)$$

Gross and Latecki proposed a third definition of par-regularity based on segments of radius r orthogonal to the boundary of S in 1995 [6]. In 1998, they equivalently redefined par-regularity in a more readily usable statement, consistent with the compatibility assumptions.

Definition 1 (Par-regularity [8, Section 2]).
Let S be a closed subset of $\mathbb{R}^2$.

- *A closed disk $\bar{B}(c,r)$ is called the* inner osculating disk *(respectively, outer osculating disk) of radius r of S at point x if $\partial S \cap \partial B(c,r) = \{x\}$ and $\bar{B}(c,r) \subset \overset{\circ}{S} \cup \{x\}$ (respectively, if $\partial S \cap \partial B(c,r) = \{x\}$ and $\bar{B}(c,r) \subset S^c \cup \{x\}$).*
- *A shape S is said to be* par(r)-regular *if, for every point $x \in \partial S$, there exist an inner and an outer osculating disks of radius r at point x (see Fig. 1).*

Even if they are close, the different definitions are not equivalent. A closed ball of radius r is r-regular, but it is not par(r)-regular, it is only par(r')-regular for all radius $r' < r$. The par-regularity constrains the curvature of the curve and polygons are not par-regular. It also prohibits thinning of the shape and its complement. For a grid of step $h \leq \sqrt{2}r$, the Gauss digitization of a par(r)-regular shape S is homeomorphic to S [18, Theorem 7.1]. Moreover, under this hypothesis, the well-composedness has been established for three discretizations (supercover, surface and inner Jordan discretization [8, Theorem 8]. Lachaud and Thibert established in [7] the link between par-regularity and the reach defined by Federer.

Definition 2 (Reach [5, Section 4.1]). *Let S be a subset of $\mathbb{R}^2$. Noting, $\text{Unp}(S)$ the set of points p of $\mathbb{R}^2$ having a unique point of S at a minimum distance of p, the reach of S $\text{reach}(S)$ is then the upper bound of the positive numbers r such that*

$$S \oplus B(0,r) \subset \text{Unp}(S).$$

In other words $\text{Unp}(S)$ is the set on which the projection ξ of a point x on its nearest neighbor on S is uniquely defined. The reach of S is the greatest distance r for which any point less than S away has a uniquely defined projection $\xi(x)$ on S.

Property 1 ([7, Proposition 2], [5, Theorem 4.8, Rem. 4.20]). Let S be a compact domain of $\mathbb{R}^2$. The reach of ∂S is positive if and only if ∂S is a curve of class $C^{1,1}$ (which denotes the continuously differentiable functions on $\mathbb{R}^2$ with uniformly Lipschitz-continuous derivative).

Property 2 ([7, Lemma 1]). Let S be a compact subset of $\mathbb{R}^2$. The reach of ∂S is greater than or equal to r if and only if S is par(r')-regular for all $r' < r$.

Par-regularity is a widespread assumption in discrete geometry, several geometric estimation results have been proved (for instance the estimation result

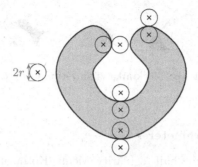

Fig. 1. Par(r)-regularity requires that there exist an inner osculating disk and an outer osculating disk of radius r at each point on the boundary of the form S (in blue). (Color figure online)

[7, Theorem 4] is achieved through the almost injective projection ξ from the digitized set onto the original continuous shape which is shown in [7, Theorem 3]). However it does not allow polygons to be included. We list, in the rest of this section, different generalizations of the par-regularity introduced in discrete geometry.

2.2 Half-Regularity

Noting the exclusion of forms S containing angular points from the family of par-regular forms, Pavlidis suggests a method for reconstructing a par-regular form from any form S ([18, Definition 7.4]). Stelldinger attempts to formalize this reconstruction (the morphological opening) to transform a set of forms, called half-regular, into par-regular forms before being discretized and thus be able to guarantee the preservation of the topology of the discretization of the morphological opening of S.

Definition 3 (Half-regularity, [21, Definition 5]). *Let S be a set and let S' be a component of $S \setminus (S \circ \bar{B}(0,r))$. Let n be the number of open disks included in S of radius r touching S'. These disks are called* bounding disks *of S' and*

- *if $n = 0$, S' is called an r-spot,*
- *if $n = 1$, S' is called an r-tip,*
- *if $n > 1$, S' is called an r-waist*

(see Fig. 2). A set S is said to be r-half-regular if for any point on the boundary, there exists an interior open osculating disk or an exterior open osculating disk of radius r included inside S, respectively outside S and if neither S nor S^c have a s-waist for $s \leq r$.

Polygons can be half-regular. If the half-regular shapes permit a reconstruction process, the contained r-spots still allow large peaks that are erased by the discretization process. Hence geometric properties of such shapes are not preserved, therefore there is no geometric estimation results from discretization of half-regular shapes.

Fig. 2. From left to right: the black area is an r-tip, an r-waist and r-spots (source [21, Figure 1]).

2.3 Stability of Parameter r

Generalizing the notion of half-regularity, Meine, Köthe and Stelldinger propose the notion of stability making it possible to reconstruct the boundaries ∂P of a partition of the plane P from points sampling ∂P with a controlled positioning error [14, Theorem 11]. Note that a *partition* designs here the union of a finite set of points and of disjoint arcs between pairs of those points such that the arcs have only there extremities in the finite point set [14, Definition 1].

Definition 4 (Stability of parameter r, [14, Definition 2]). *A partition P of the plane is said to be r-stable if its boundary ∂P can be dilated by a disk of radius s without changing its homotopy type for all $s \leq r$ (see Fig. 3).*

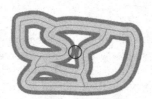

Fig. 3. [14, Figure 1]. The dark gray lines correspond to the ∂P boundaries of the partition of $\mathbb{R}^2$. The light gray tube (respectively dark gray tube) corresponds to the dilation of ∂P by r (respectively by $r' > r$). The set ∂P is r-stable because the homotopy type is not modified, but not r'-stable (a new connected component is created in the complement).

As for half-regularity, polygons can be r-stable and there is no result of geometric estimation from discretization of r-stable shapes.

If the definition of r-stability is very general, it seems difficult to verify concretely on a shape and does not seem to simplify the problem of reconstruction with homeomorphism close to the shape from a sampling of points: verifying the preservation of the homotopy type for dilations of ∂P would not be easier than guaranteeing the preservation of the topology of the reconstruction of the partition P from a sampling of points of ∂P. Also, the authors introduce the concept of (θ, d)-*peaks* which they prefer to use in the context of discrete geometry [14, Theorem 16]. Moreover, for certain shapes S, not having (θ, d)-peaks, [14, Theorem 16] allows us to extract from the edge of the discretization of a subset S of $\mathbb{R}^2$ of sequences of points sampling ∂S.

Definition 5 $((\theta, d)$**-peaks, [14, Definition 15]).** *Let ∂P be the boundary of a partition of the plane P. Two points $x_1, x_2 \in \partial P$ delimit an (θ, d)-peak if the distance between x_1 and x_2 is less than d and if there exists no arc of ∂P between x_1 and x_2 containing only points y such that $\widehat{x_1 y x_2} \geq \theta$.*

However this notion alone is not sufficient to guarantee that the form is r-stable, it is also necessary to assume that each element of the partition contains a disk of a certain size [14, Theorem 16]. It would then not be sufficient to exclude thinning in continuous forms.

2.4 Reach of Parameter μ

Introduced in computational geometry, the μ-reach is a generalization of the notion of reach (and, therefore, of the notion of par-regularity) to polygonal curves. This notion makes it possible to guarantee a reconstruction of the homotopy type from points sampling a curve with a certain μ-reach and provide stability results of curvature measures on the dilation of noisy manifold [2, Theorems 3 and 4]. The definition of the μ-reach is based on a generalization of the distance gradient (see [5, Theorem 4.8.3]).

Definition 6 (μ-reach, [3, Definition 4.3]).

- *Let K be a compact subset of $\mathbb{R}^2$. The generalized gradient ∇_K is the gradient of the distance of x to K, $\mathrm{d}(x, K)$, extended to the medial axis of K in the following way:*

$$\nabla_K(x) := \frac{x - c(x)}{\mathrm{d}(x, K)}$$

with $c(x)$ the center of the smallest disk $\sigma_K(x)$ containing all the nearest neighbors of x belonging to K (see Fig. 4).
- *A point x is said μ-critical if $\|\nabla_K(x)\| \leq \mu$.*
- *The μ-reach $r_\mu(K)$ is the infimum of distances between K and the topological closure of the set of μ-critical points.*

Remark 1. If there is only one nearest point, the generalized gradient is equal to the classical gradient and its norm is 1. In all cases, $\|\nabla_K(x)\| \leq 1$ and therefore one chooses the parameter $\mu \leq 1$ in general.

Remark 2 ([3, Section 4.2]). Note that $\|\nabla_K(x)\| = \cos(\theta(x))$ where $\theta(x)$ is the angle at which x sees its nearest neighbors on K (see Fig. 4). Thus the μ-reach is the distance to K, below which each point sees all of its nearest neighbors on K at an angle less than $2 \arccos(\mu)$. So for $\mu = 1$, we obtain $r_1(K) = \mathrm{reach}(\mathcal{C})$.

Here again, the class of shapes having a positive μ-reach is large enough to include polygons. Theorem 4.6 [3] allows us to reconstruct a set K up to homotopy equivalence with a set K' sampling it and of μ-strictly positive reach (see Fig. 5). Nevertheless, there are not yet results of geometric estimation in a discrete geometry frame.

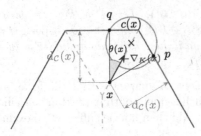
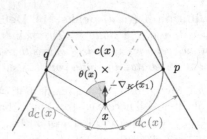

(a) If the nearest neighbors p and q on K of a point x are close to each other, the norm of the generalized gradient is close to 1.

(b) If the nearest neighbors p and q on K of point x are far from each other, the norm of the generalized gradient is close to 0.

Fig. 4. K a compact set. $-\nabla_K(x)$: generalization of the distance gradient to $\mathcal{C}$. $\Gamma_K(x)$: smallest disk containing the closest points to x on K.

Fig. 5. [3, Figure 5] The black points sample an equilateral triangle K: the set of black points K' is close, in the sense of the Hausdorff distance, to the triangle K. The theorem [3, Theorem 4.6] predicts values for which the dilation of the set of points K' is homotopically equivalent to the dilation of K (thick red boundary). (Color figure online)

2.5 Quasi-regularity

In order to generalize the topology preservation results on shapes with noisy boundaries, Ngo et al. extend the definition of par-regularity given by Jean Serra (Eq. (1)) by allowing the border of the shape to oscillate within a margin around its erosion by a ball.

Definition 7 (Quasi-regularity, [15, Definition 3]). *Let S be a bounded, simply connected subset of $\mathbb{R}^2$. The set S is quasi(r)-regular if S satisfies the following conditions:*

- *$S \ominus \bar{B}(0,r)$ is non-empty and connected,*
- *$S^c \ominus \bar{B}(0,r)$ is connected,*
- *$S \subset S \ominus \bar{B}(0,r) \oplus \bar{B}(0,\sqrt{2}r)$,*
- *$S^c \subset S^c \ominus \bar{B}(0,r) \oplus \bar{B}(0,\sqrt{2}r)$*

(see Fig. 6).

The quasi-regularity is based on morphological properties. It includes polygons with a sufficient thickness and whose acute angles are small enough. This notion allows some topological notions to be preserved through discretization. Thus, the Gauss digitization of a quasi(1)-regular shape is well-composed [15]. Due to its definition, it does not give control on the fine variations on the shape border and there is no discrete geometric estimation results for quasi-regularity.

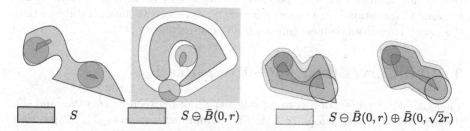

Fig. 6. [15, Figure 5]. From left to right: in the first figure, the shape S is not quasi-regular because $S \ominus \bar{B}(0, r)$ is not connected; in the second figure, the shape S is not quasi-regular because $S^c \ominus \bar{B}(0, r)$ is not connected; in the third figure, the form S is not quasi-regular because S is not included in $S \ominus \bar{B}(0, r) \oplus \bar{B}(0, \sqrt{2}r)$; in the fourth figure, the shape S satisfies all the conditions to be quasi-regular.

2.6   Locally Turn Boundedness

In order to consider both polygons and regulars curves, we introduce in [9] the notion of *Locally turn-bounded* curve to control the complexity of the curve and hence enables us to control the loss of geometric and topological information due to the digitization process [10]. The "complexity" of the curve is formally defined as its turn (a generalization of the integral of curvature).

Definition 8 (Turn [1, Section 5.1]).

- *The turn $\kappa(P)$ of a polygon $P = [x_i]_{i \in \mathbb{Z}/N\mathbb{Z}}$ is defined by:*

$$\kappa(P) := \sum_{i \in \mathbb{Z}/N\mathbb{Z}} \angle(x_i - x_{i-1}, x_{i+1} - x_i)$$

where $\angle(x_i - x_{i-1}, x_{i+1} - x_i) \in [0, \pi)$ is the absolute value of the angle between the vectors $x_i - x_{i-1}$ and $x_{i+1} - x_i$.
- *The turn $\kappa(\mathcal{C}) \in \mathbb{R}^+ \cup \{+\infty\}$ of a Jordan curve $\mathcal{C}$ is the supremum of its inscribed polygon.*

The turn can be understood as a generalization of the curvature to C^0-regular curves [1, Section 5.3] and as a lack of convexity [1, Theorem 5.1.5].

We defined in [9] the LTB curves by bounding locally the turn.

Definition 9 (LTB curves [9]**).** *A Jordan curve $\mathcal{C}$ is (θ, δ)-locally turn-bounded $((\theta, \delta)$-LTB) if, for any two points a and b in $\mathcal{C}$ such that the Euclidean distance $d(a, b)$ is less than δ, the turn of one of the arcs of the curve $\mathcal{C}$ delimited by a and b is less than or equal to θ.*

Polygons can be LTB and the Gauss discretization of a $(\frac{\pi}{2}, \delta)$-LTB shape is well-composed for a grid step $h < \frac{\delta}{\sqrt{2}}$ [10, Proposition 10, Corollary 4]. LTB control on the curvature makes it possible to have results on Multigrid convergence for length estimation [13, Theorems 2 and 3]. Moreover we have established that the Local Turn Boundedness implies the Quasi-regularity [11].

3   LTB Curves Have a Positive μ-Reach

The goal of current section is to establish the link between LTB curves and the μ-reach. We first need Property 3 that is a simplified version of [11, Proposition 2]. The application of Property 3 in the proof is illustrated Fig. 7.

Property 3. Let $\mathcal{C}_1$ a simple arc of ends p_1 and p_2. Let A be the topological closure of one of the bounded connected component of $\mathbb{R}^2 \setminus (\mathcal{C}_1 \cup [p_1, p_2])$. Let $\mathcal{C}_2$ a simple arc in A of ends p_1 and p_2 such that

– the arc $\mathcal{C}_2$ does not intersect the open segment (p_1, p_2),
– the interior of the connected component of the Jordan curve $\mathcal{C}_2 \cup [p_1, p_2]$ is convex

then

$$\kappa(\mathcal{C}_2) \leq \kappa(\mathcal{C}_1).$$

Theorem 1 (LTB curves have positive μ-reach). *Let $\theta \in [0, \pi]$ and $\delta > 0$. Let $\mathcal{C}$ be (θ, δ)-LTB curve.*
Then

$$\delta \leq 2r_{\cos \frac{\theta}{2}}(\mathcal{C}).$$

Proof. The proof is illustrated Fig. 7. Let $\theta \in [0, \pi]$ and $\mu := \cos \frac{\theta}{2}$. Let $\epsilon > 0$. By the very definition of the μ-reach (Definition 6), $r_\mu(\mathcal{C}) := \inf_{y \in \overline{M_\mu}(\mathcal{C})} d_{\mathcal{C}}(y)$ where $M_\mu(\mathcal{C})$ is the set of μ-critical point. Then there exist $\bar{\rho} \in [r_\mu(\mathcal{C}), r_\mu(\mathcal{C}) + \frac{\epsilon}{2})$ and $\bar{x} \in \overline{M_\mu}(\mathcal{C})$ such that $d_{\mathcal{C}}(\bar{x}) = \bar{\rho}$. Hence there exists $x \in M_\mu(\mathcal{C})$ such that $d_{\mathcal{C}}(x) \in [\bar{\rho}, \bar{\rho} + \frac{\epsilon}{2})$. Thus there exist $\rho \in [r_\mu(\mathcal{C}), r_\mu(\mathcal{C}) + \epsilon)$ and $x \in \mathbb{R}^2$ such that $\|\nabla_{\mathcal{C}}(x)\| \leq \mu$ and $d_{\mathcal{C}}(x) = \rho$. Let $\theta_x := 2\arccos(\|\nabla_{\mathcal{C}}(x)\|)$. Notice that $\theta \leq \theta_c$ since $\|\nabla_{\mathcal{C}}(x)\| \leq \mu = 2\cos \frac{\theta}{2}$. By definition of the generalized gradient $\|\nabla_{\mathcal{C}}(x)\|$ (Definition 6), there exist two distinct nearest points p_1, p_2 of $\mathcal{C}$, such that $\widehat{p_1 x p_2} = \theta_x$. And by definition of ρ, $d_{\mathcal{C}}(\hat{x}) = \rho$. Then $d(p_1, p_2) = 2\rho \sin(\frac{\theta_x}{2})$. Since $r_\mu(\mathcal{C}) \leq \rho < r_\mu(\mathcal{C}) + \epsilon$,

$$2\rho \sin\left(\frac{\theta}{2}\right) \leq d(p_1, p_2) < 2(r_\mu(\mathcal{C}) + \epsilon) \sin\left(\frac{\theta_x}{2}\right) \leq 2(r_\mu(\mathcal{C}) + \epsilon).$$

Since $d_{\mathcal{C}}(x) = \rho$, the open disk $B(x, \rho)$ does not intersect $\mathcal{C}$. Let $\mathcal{C}_2$ the arc of the circle $\partial B(x, \rho)$ with ends p_1 and p_2 and whose angle is θ_x. Let $\mathcal{C}_1$ be the arc of $\mathcal{C}$ with ends p_1 and p_2 such that $\mathcal{C}_2$ is included in the compact set bounded by $\mathcal{C}_1 \cup [p_1, p_2]$. Moreover by Property 3, $\kappa(\mathcal{C}_1) \geq \kappa(\mathcal{C}_2) \geq \theta_x$. Let $\mathcal{C}_2^c$ be the complementary closed arc of $\mathcal{C}_2$ (that is $\mathcal{C}_2^c := \partial B(c, r) \setminus \mathcal{C}_2$). Since $\theta_x \geq \theta$, $\kappa(\mathcal{C}_1) \geq \theta$. For the same reason, the complementary arc $\mathcal{C}_1^c$ of $\mathcal{C}_1$ ($\mathcal{C}_1^c := \mathcal{C} \setminus \mathcal{C}_1$) has its turn greater or equal to $2\pi - \theta_x$. And since $\theta \leq \pi$ and $\theta_x \leq \pi$, $\kappa(\mathcal{C}_1^c) \geq \theta$.

Then by definition of LTB curves (Definition 9), $\delta \leq d(p_1, p_2)$. Then for any $\epsilon > 0$,

$$\delta \leq d(p_1, p_2) \leq 2(r_\mu(\mathcal{C}) + \epsilon)$$

Finally,

$$\delta \leq 2r_\mu(\mathcal{C}).$$

Or equivalently, $\delta \leq 2r_{\cos \frac{\theta}{2}}(\mathcal{C})$.

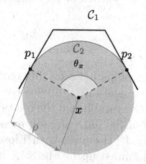

Fig. 7. Since the blue subarc $\mathcal{C}_1$ of $\mathcal{C}$ circumvents the arc $\mathcal{C}_2$, its turn is no lesser than those of $\mathcal{C}_2$, that is θ_x. (Color figure online)

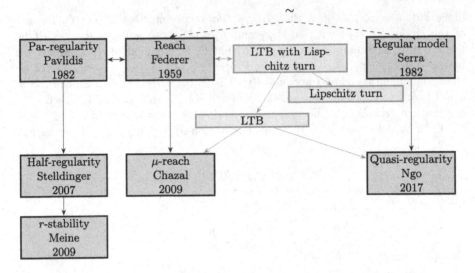

Fig. 8. Each arrow indicates a generalization of a notion and the notions in green boxes are the work of the authors (in the current or previous articles). (Color figure online)

4   Conclusion

We have explored the relationships between different classes of shapes used in discrete geometry in order to guarantee topological or geometric preservation properties. In particular we have established the relationship between LTB curves and the μ-reach. In addition of the links already presented, the technical report [12, Theorem 3.1] shows that LTB curves with Lipschitz turn are par-regular. All these relationships are summed up in Fig. 8. A last notion could also have been compared with the other: the *square-regularity* which naturally arose when we proved that LTB curves are quasi-regular [11, Proposition 3]. It consists on replacing the disks by square in definition of par-regularity. But no result of topological or geometric preservation under digitization has been proved for this class of shapes. All the definitions of the compared notions are straightforwardly generalized in higher dimension, but most results obtained in the two dimensional plane are no longer valid in the three-dimensional space. For instance, as shown by the counterexample [22, Figure 4], the digitization of very smooth shapes can be non well-composed. Considering LTB curves, the definition could also be straightforwardly extended to surfaces embedded in a three-dimensional space: between any two points a and b at distance less than δ of the surface there exists a path on the surface of turn less or equal to $\frac{\pi}{2}$. Two main challenges appears in three dimension: the choice of points a and b and the existence of infinitely many paths joining the points a and b (whereas there are only two in the two-dimensional case).

References

1. Alexandrov, A.D., Reshetnyak, Y.G.: General Theory of Irregular Curves. Mathematics and Its Applications, vol. 29. Springer, Dordrecht (1989). https://doi.org/10.1007/978-94-009-2591-5
2. Chazal, F., Cohen-Steiner, D., Lieutier, A., Thibert, B.: Stability of curvature measures. Comput. Graph. Forum **28**(5), 1485–1496 (2009). https://doi.org/10.1111/j.1467-8659.2009.01525.x
3. Chazal, F., Cohen-Steiner, D., Lieutier, A.: A sampling theory for compact sets in Euclidean space. Discr. Comput. Geom. **41**(3), 461–479 (2009)
4. Coeurjolly, D., Lachaud, J.O., Gueth, P.: Digital surface regularization with guarantees. IEEE Trans. Vis. Comput. Graph. **27**(6), 2896–2907 (2021)
5. Federer, H.: Curvature measures. Trans. Am. Math. Soc. **93**(3), 418–418 (1959)
6. Gross, A., Latecki, L.: Digitizations preserving topological and differential geometric properties. Comput. Vis. Image Underst. **62**(3), 370–381 (1995)
7. Lachaud, J.O., Thibert, B.: Properties of gauss digitized shapes and digital surface integration. J. Math. Imaging Vis. **54**(2), 162–180 (2016)
8. Latecki, L.J., Conrad, C., Gross, A.: Preserving topology by a digitization process. J. Math. Imaging Vis. **29** (1998)
9. Le Quentrec, E., Mazo, L., Baudrier, E., Tajine, M.: Local turn-boundedness: a curvature control for a good digitization. In: Couprie, M., Cousty, J., Kenmochi, Y., Mustafa, N. (eds.) 21st IAPR International Conference on Discrete Geometry for Computer Imagery, Paris, France (2019)
10. Le Quentrec, É., Mazo, L., Baudrier, É., Tajine, M.: Local turn-boundedness: a curvature control for continuous curves with application to digitization. J. Math. Imaging Vis. **62**, 673–692 (2020)
11. Le Quentrec, É., Mazo, L., Baudrier, É., Tajine, M.: Locally turn-bounded curves are quasi-regular. In: Lindblad, J., Malmberg, F., Sladoje, N. (eds.) DGMM 2021. LNCS, vol. 12708, pp. 202–214. Springer, Cham (2021). https://doi.org/10.1007/978-3-030-76657-3_14
12. Le Quentrec, E., Mazo, L., Baudrier, É., Tajine, M.: LTB curves with Lipschitz turn are par-regular. Research Report, Laboratoire ICube, université de Strasbourg (2021)
13. Le Quentrec, É., Mazo, L., Baudrier, É., Tajine, M.: Monotonic sampling of a continuous closed curve with respect to its gauss digitization: application to length estimation. J. Math. Imaging Vis. **64**, 869–891 (2022)
14. Meine, H., Köthe, U., Stelldinger, P.: A topological sampling theorem for Robust boundary reconstruction and image segmentation. Discr. Appl. Math. **157**(3), 524–541 (2009)
15. Ngo, P., Passat, N., Kenmochi, Y., Debled-Rennesson, I.: Convexity invariance of voxel objects under rigid motions. In: 2018 24th International Conference on Pattern Recognition (ICPR), pp. 1157–1162. IEEE, Beijing (2018)
16. Passat, N., Kenmochi, Y., Ngo, P., Pluta, K.: Rigid motions in the cubic grid: a discussion on topological issues. In: Couprie, M., Cousty, J., Kenmochi, Y., Mustafa, N. (eds.) DGCI 2019. LNCS, vol. 11414, pp. 127–140. Springer, Cham (2019). https://doi.org/10.1007/978-3-030-14085-4_11
17. Passat, N., Ngo, P., Kenmochi, Y., Talbot, H.: Homotopic affine transformations in the 2D cartesian grid. J. Math. Imaging Vis. **64**(7), 786–806 (2022)
18. Pavlidis, T.: Algorithms for Graphics and Image Processing. Springer, Heidelberg (1982). https://doi.org/10.1007/978-3-642-93208-3

19. Serra, J.P.: Image Analysis and Mathematical Morphology. Academic Press, London, New York (1982)
20. Shorten, C., Khoshgoftaar, T.M.: A survey on image data augmentation for deep learning. J. Big Data **6**(1), 1–48 (2019)
21. Stelldinger, P.: Digitization of non-regular shapes. In: Ronse, C., Najman, L., Decencière, E. (eds.) Mathematical Morphology: 40 Years On. Computational Imaging and Vision, vol. 30, pp. 269–278. Springer, Dordrecht (2005). https://doi.org/10.1007/1-4020-3443-1_24
22. Stelldinger, P., Latecki, L., Siqueira, M.: Topological equivalence between a 3D object and the reconstruction of its digital image. IEEE Trans. Pattern Anal. Mach. Intell. **29**(1), 126–140 (2007)

Mathematical Morphology and Digital
Geometry for Applications

Counting Melanocytes with Trainable h-Maxima and Connected Component Layers

Xiaohu Liu[1,2]($\boxtimes$) , Samy Blusseau[2] , and Santiago Velasco-Forero[2]

[1] Shanghai Jiaotong University, Shanghai, China
[2] Mines Paris, PSL University, Centre for Mathematical Morphology,
Fontainebleau, France
xiaohu-francois.liu@etu.minesparis.psl.eu

Abstract. Bright objects on a dark background, such as cells in microscopy images, can sometimes be modeled as maxima of sufficient dynamic, called h-maxima. Such a model could be sufficient to count these objects in images, provided we know the dynamic threshold that tells apart actual objects from irrelevant maxima. In this paper we introduce a neural architecture that includes a morphological pipeline counting the number of h-maxima in an image, preceded by a classical CNN which predicts the dynamic h yielding the right number of objects. This is made possible by geodesic reconstruction layers, already introduced in previous work, and a new module counting connected components. This architecture is trained end-to-end to count melanocytes in microscopy images. Its performance is close to the state of the art CNN on this dataset, with much fewer parameters (1/100) and an increased interpretability.

1 Introduction

Cell counting is a crucial step in biological experiments and medical diagnosis to provide both quantitative and qualitative information on cells. The process can be tedious, low-efficiency and prone to subjective errors, especially when cell clustering occurs and cells show high variance in shapes and contrasts. Automation offers a quick and accurate estimation of cell quantity in a sample. In this study, we focus on the cell counting task to quantify melanocyte population on fluorescent images with *Tyrosinase-related protein one* (TRP1) as melanocytic marker (see left hand image of Fig. 1). Previous studies [3,4] used deep learning with a U-Net architecture, which associates each input image with a density map, the integral of which yields the cell count. That method achieves satisfactory results but with almost two million parameters while cells, although diverse in shapes and often overlapping, seem to be well approximated by regional maxima over a dark background. Furthermore, the density map is ambiguous regarding what

---

S. Velasco-Forero—This work was granted access to the HPC resources of IDRIS under the allocation 2023-AD011012212R2 made by GENCI.

the model considers as a cell, especially in clusters, where the location of cells is often lost. This can be a limitation for users willing to check a posteriori the reliability of the count returned by the model, especially as the deep learning model is a black box and its decisions are hardly explainable.

Therefore, this is an interesting case study where it seems that a much simpler model, built on *a priori* knowledge, could achieve similar results with higher interpretability, regarding for example geometrical and contrast criteria used to recognize cells. In this paper, we propose to model cells as regional maxima with sufficient dynamic and size. The dynamic of an extremum is, simply speaking, its depth (see Fig. 2), and well known morphological methods exist to select extrema with dynamic larger than a threshold h, called h-maxima or h-minima [8]. Hence a simple morphological pipeline is presented, consisting in counting the h-maxima of a size-filtered version of the input image, for a given dynamic h. Since the image resolution is constant in the dataset, as well as the cell size, this approach only depends on the choice of the dynamic parameter, which needs however to be adapted to each image in order to achieve accurate counting. Thus, we take advantage of recent work [13] in which geodesic reconstruction, and therefore the computation of h-maxima, have been implemented as neural layers, allowing to train end-to-end a pipeline where a small Convolutional Neural Network (CNN) predicts the optimal dynamic h for the subsequent morphological layer.

The contribution of this paper is threefold. Firstly, a new end-to-end differentiable pipeline combining a trainable CNN and an h-maxima operator for cell counting is introduced. Secondly, we define a layer based on geodesical reconstruction, that counts connected components in binary images and is compatible with automatic differentiation [6]. Thirdly, we use a new joint loss function composed of differences on morphological geodesic reconstructions, and a term penalizing miscounts with respect to the true number of cells. This paper is structured as follows. The morphological method to count cells is presented in Sect. 2. A deep learning architecture based on the morphological method is developed in Sect. 3. Experiments are conducted in Sect. 4, the analysis of results in Sect. 5. Conclusions are drawn in Sect. 6.

2   Morphological Pipeline

In this section we present a simple morphological algorithm for cell counting in images where individual cells are well approximated by regional extrema of significant dynamic, like in Fig. 1. We assume images to be mappings of the discrete domain $\Omega := ([0, M-1] \times [0, N-1]) \cap \mathbb{N}^2, M, N \in \mathbb{N}^*$ to the discrete set of values $\mathcal{V} := [0,1] \cap \epsilon\mathbb{N}$, where $0 < \epsilon < 1$. We will denote by $\mathcal{F}(\Omega, \mathcal{V})$ the set of such functions. The algorithm takes as input an image f, as parameter a number $h \in \mathcal{V}$ and returns an estimated number n_c of cells after running the following steps:

1. Apply an alternate filter $\tilde{f} \leftarrow \varphi_B \circ \gamma_B(f)$ where φ_B and γ_B are respectively the closing and opening by B, the unit square structuring element

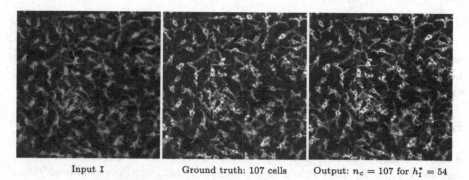

Input I   Ground truth: 107 cells   Output: $n_c = 107$ for $h_I^* = 54$

Fig. 1. Example of cell counting by identifying cells to h-maxima, in the TRP1 melanocyte dataset. From left to right: input RGB image (resized); ground truth mask for the 107 cells (in red); detection results using the best h for the image: $h_I^* = 54$. Note that despite the perfect count, there are some false positives and false negatives (e.g. just below the top right hand corner). (Color figure online)

2. Compute the h-extended maxima $M_h \leftarrow \text{EMAX}_h(\tilde{f})$, which is a binary image
3. Count the number of connected components of M_h: $n_c \leftarrow \text{CCC}(M_h)$

The definitions required for the key operator EMAX_h are recalled in the following paragraphs.

Morphological Geodesic Reconstruction. Given two functions $f, g \in \mathcal{F}(\Omega, \mathcal{V})$ such that $f \leq g$ the *geodesic dilation* of f under g, noted $\delta_g^{(1)}(f)$, is defined by [5,8]

$$\delta_g^{(1)}(f) := \delta_{\text{B}}(f) \wedge g \tag{1}$$

where $\wedge$ denotes the point-wise minimum operation, B denotes the unitary structuring element defined according to the pixel connectivity (either four or eight). Recall that $\delta_{\text{B}}(f)(x) = \max_{b \in B} f(x + b)$ for all $x \in \Omega$, therefore δ_{B} is an extensive operator, $\delta_{\text{B}}(f) \geq f$, since $0 \in \text{B}$. The geodesic dilation can be iterated and for $p \geq 1$ we note $\delta_g^{(p+1)}(f) := \delta_g^{(1)}\big(\delta_g^{(p)}(f)\big)$.

The *reconstruction by dilation* of f with respect to g, $\text{REC}^\delta(f, g)$ is then:

$$\text{REC}^\delta(f, g) := \text{REC}_g^\delta(f) := \delta_g^{(k)}(f) \tag{2}$$

where k is the first integer such that $\delta_g^{(k)}(f) = \delta_g^{(k+1)}(f)$, hence the dilation iterates until stable.

Regional Maxima by Reconstruction. The reconstruction by dilation in (2) can be used to extract regional maxima [14]. Let us consider an image $f \in \mathcal{F}(\Omega, \mathcal{V})$. $M \subset \Omega$ is a *regional maximum* at level t if M is connected, $f(x) = t$ $\forall x \in M$, and for any $y \in \Omega \setminus M$ with a neighbor in M, $f(y) < t$. It is well-known (refer to Section 6.3.3 in [8]) that the set of all regional maxima of f, denoted by $\text{RMAX}(f)$ is recovered by:

$$\forall x \in \Omega, \quad \text{RMAX}(f)(x) := \begin{cases} 1 & \text{if } f(x) > \text{REC}_f^\delta(f - \epsilon)(x) \\ 0 & \text{otherwise.} \end{cases} \qquad (3)$$

where we recall that ϵ is the minimum positive value in $\mathcal{V}$ and therefore the minimum absolute difference between distinct function values. By construction, every connected component of the binary image $\text{RMAX}(f)$ is a regional maximum.

Extended h-Maxima by Reconstruction. The *dynamic* of a regional maximum M (of level t) is the smallest height one needs to come down to reach a higher regional maximum M' (of level $t' > t$) (refer to Section 6.3.5 in [8], and see Fig. 2). By convention, the dynamic of the global maximum of f can be set to the difference between the maximum and the minimum values of f. The dynamic is usually used to distinguish between irrelevant maxima caused by noise and significant ones corresponding to underlying bright objects.

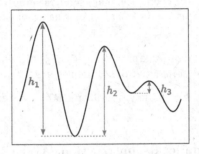

Fig. 2. The dynamics of three maxima in a one-dimensional function.

The *h-maxima transformation* suppresses all maxima with dynamic lower or equal to the given parameter value h, called also *h-domes* in [14]. This can be achieved by performing the reconstruction by dilation of f from $f - h$, i.e.:

$$\text{HMAX}_h(f) = \text{REC}_f^\delta(f - h) \qquad (4)$$

where $h \in \mathcal{V}$ is a parameter. To each regional maximum of f with dynamic strictly larger than h, corresponds exactly one regional maximum in $\text{HMAX}_h(f)$. This is illustrated in Fig. 3(a).

Therefore, we define the operator *extended maxima* as the regional maxima of the h-maxima transformation:

$$\text{EMAX}_h(f) := \text{RMAX}(\text{HMAX}_h(f)). \qquad (5)$$

By construction, the number of connected components of the binary image $\text{EMAX}_h(f)$ is the same as the number of regional maxima with dynamic strictly larger than h in f. This is illustrated by Fig. 3(b). We can see that this number n_c is a decreasing function of h, ranging from the total number of local maxima (for $h = 0$) to zero (for h exceeding the dynamic of the deepest maximum). This is why we expect that for each image, there is an h value yielding the right number of cells.

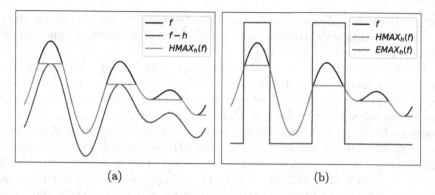

(a)   (b)

Fig. 3. (a) $\mathtt{HMAX}_h$ produces a regional maximum wherever there is a maximum with dynamic greater than h in f. (b) $\mathtt{EMAX}_h$ computes the regional maxima of $\mathtt{HMAX}_h(f)$.

3 Integration in a Deep Learning Architecture

The morphological pipeline described in the previous section depends on one crucial parameter, namely the minimum dynamic for a regional maximum to be considered a relevant object (in our case, a melanocyte). In this section we describe how this pipeline is implemented as a sequence of neural layers, and embedded in an architecture where it is preceded by a CNN which predicts an optimal dynamic for each image, in order to count the right number of cells at the output of the network.

Morphological Layers. This embedding relies on the library Morpholayers [11] which implements neural layers performing morphological operators [7,12]. This includes basic operators such as dilations, erosions, openings and closings, useful for the alternate filter in the first step of our pipeline. Furthermore, it also includes the geodesic reconstruction layers recently introduced in the deep learning framework [13], and on which relies most of the operations described in the previous section. It is important to note that the reconstruction by dilation (2), has no parameter because the value k depends on f and g. It can be included as a layer in a neural network since the subdifferential of this operation can be implemented in auto-differentiation software [6]. The Jacobian matrix of (2) has been explicitly calculated in [13]. Therefore, all the necessary layers are available to embed steps 1 and 2 of the morphological pipeline. The embedding of the third step is addressed in the next paragraph.

Counting Connected Components (CCC) Layer. Counting connected components is an important step in many of the classical methods in data processing [9], especially in images and graphs. In the context of deep learning, this problem has been used to evaluate the generalization capacity of neural networks [2]. However, to be used as a layer inside a network, it must be implemented so that gradient backpropagation can be computed in an auto-derivation software such as Pytorch or Tensorflow [6]. In this subsection, we present an algorithm to count the connected components of a binary image, relying mainly on the geodesic reconstruction defined in (2).

Let $f \in \mathcal{F}(\Omega, \{0,1\})$ be a binary image, and $\mathcal{U}_\Omega \in \mathcal{F}(\Omega, \mathcal{V})$ an injective and positive image, i.e. such that $\forall x, y \in \Omega, \quad x \neq y \Rightarrow \mathcal{U}_\Omega(x) \neq \mathcal{U}_\Omega(y)$ and $\mathcal{U}_\Omega(x) > 0$ (this is possible if and only if the number of values in $\mathcal{V}$ is strictly larger than the number of pixels in Ω, i.e. $\lfloor \frac{1}{\epsilon} \rfloor + 1 > MN$). Then it is clear that reconstructing $\mathcal{U}_\Omega \wedge f$ under f will label every connected component CC of f with the maximum value taken by $\mathcal{U}_\Omega$ in CC. Since $\mathcal{U}_\Omega$ is injective, this maximum is achieved only once in each connected component. Hence, setting to one each pixel $x \in \Omega$ such that $\text{REC}^\delta(\mathcal{U}_\Omega \wedge f, f)(x) = \mathcal{U}_\Omega(x)$ produces a binary image where exactly one pixel per connected component of f is lit. This is illustrated by Fig. 4.

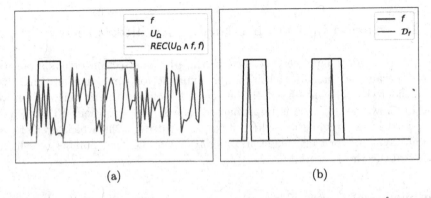

(a)　　　　　　　　　　　　　　　　　(b)

Fig. 4. Illustration of the CCC algorithm. (a) $\mathcal{U}_\Omega$ coincides only once with $\text{REC}^\delta(\mathcal{U}_\Omega \wedge f, f)$ inside each connected component, and never outside the connected components. (b) Hence $\mathcal{D}_f$ has only two pixels activated, one per connected component of f.

Therefore, we define

$$\mathcal{D}_f := \mathcal{B}(\mathcal{U}_\Omega, \text{REC}^\delta(\mathcal{U}_\Omega \wedge f, f)) \tag{6}$$

where the *binarization operator* $\mathcal{B}$ for two functions $f, g \in \mathcal{F}(\Omega, \mathbb{R})$ is:

$$\forall x \in \Omega, \quad \mathcal{B}(f, g)(x) := \begin{cases} 1 \text{ if } \quad f(x) = g(x) \\ 0 \text{ otherwise.} \end{cases} \tag{7}$$

$\mathcal{D}_f$ denotes the final detection result, represented by locations of the isolated connected components. Accordingly, one can count the number of connected components on f by simply summing (6):

$$\text{CCC}(f) := \sum_{x \in \Omega} \mathcal{D}_f(x). \tag{8}$$

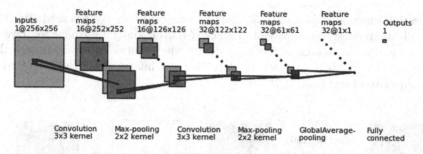

Fig. 5. CNN architecture used in the proposed pipeline

4   Experiments

Data. The TRP1 dataset [4] we used contains two sets, called *set*1 and *set*2, of 76 fluorescent melanocytes RGB images each, with resolution 1024×1024 pixels, which showed a high variability in density and shapes[1]. They come along with manual annotations of the cell coordinates. The images were acquired from *in vitro* pigmented reconstructed skin samples submitted to TRP1 fluorescent labeling [1]. Figures 1 and 8 show examples of images and manual annotations. In the present experiments we only worked with the green channel of images, rescaled at a resolution of 255×255 pixels. The image range was between 0 and $V_{\max} = 255$, quantized with a float precision ϵ. Hence the previously described method applies, taking $\mathcal{V} := [0, 255] \cap \epsilon \mathbb{N}$, and binary images with values in $\{0, 255\}$. Applying the morphological pipeline described in Sect. 2 for a wide range of dynamic values h, allowed to determine for each training image I the optimal dynamic h_{I}^{*}, i.e. the one minimizing the difference between true number of cells and the estimated one.

Overall Architecture. The proposed neural architecture[2] is illustrated in Fig. 6. Morphological layers apply an alternate filter (opening followed by closing) by a 3×3 square structuring element, to the resized green channel, which yields the first preprocessed image I. This image is passed as input to a CNN, which estimates the best dynamic $\hat{h}_{\mathrm{I}} = \mathrm{CNN}_{\theta}(\mathrm{I})$, where θ denotes the parameters of the CNN. This parameter is then used to compute the $\hat{h}_{\mathrm{I}}$-maxima of the filtered image, by first feeding it to the HMAX reconstruction layer, followed by the RMAX one to get the binary image $\mathrm{EMAX}_{\hat{h}_{\mathrm{I}}}$. Finally, the number of connected components of the latter is obtained at the output of the proposed connected components counting layer, $\mathrm{CCC}(\mathrm{EMAX}_{\hat{h}_{\mathrm{I}}})$.

CNN Architecture. The CNN architecture is summarized in Fig. 5. In particular, each convolutional layer[3] is followed by a BatchNormalization layer with Relu as

---

[1] The dataset is available at https://bit.ly/melanocytesTRP1.

[2] Code available at https://github.com/peter12398/DGMM2024-comptage-cellule.

[3] Convolutional layers are implemented as a double convolution with twice the same number of filters.

activation function. Following [13], global average pooling layer is used after the second Maxpooling layer instead of stacked flatten and fully connected layers, which has the advantage of further reducing the number of parameters of the network. What's more, extracting global feature information is coherent with the objective of our task.

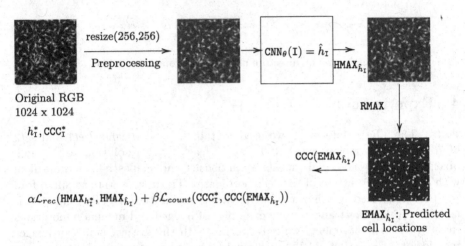

Fig. 6. Proposed end-to-end differentiable pipeline using a CNN and an HMAX layer to count melanocytes. The network is trained by minimizing a loss function that depends on 1) the difference between the geodesic reconstruction with the dynamic $\hat{h}$ estimated by the network, and the best possible one; and 2) the difference between the number of cells and the number of extended maxima of $\text{HMAX}_{\hat{h}}$.

4.1 Training Protocol

Training Data. To train the architecture presented earlier, hence fit the parameters of the CNN, we randomly split *set1*, containing 76 images, into 60 training images and 16 validation images. The parameters of the CNN were adjusted through gradient descent on the training set, and the best model weights were chosen through evaluation metrics on the validation set. The whole *set2*, which also contains 76 images, was used for the final test only.

Data Augmentation. Convolutional operations are translation invariant, but not rotation or flip invariant. Moreover, for the melanocyte image samples in the dataset, the flipped or rotated version still looks like a valid sample. Thus it is reasonable to use flip and rotation as data augmentation methods in the training stage. More precisely, we used both vertical and horizontal flip transformations with probability 0.5; a rotation with the same probability and the rotation angle was uniformly selected from $\{90°, 180°, 270°\}$.

Loss Function. The training phase aims at minimizing the joint loss function composed of a geodesic reconstruction loss and a cell counting loss. More precisely, the geodesic reconstruction loss is the mean squared error (MSE) between the predicted geodesic reconstruction image $\texttt{HMAX}_{\hat{h}_I}$ and the ground truth one $\texttt{HMAX}_{h_I^*}$:

$$\forall I \in \mathcal{F}(\Omega, \mathcal{V}), \quad \mathcal{L}_{rec}^I = \frac{1}{|\Omega|} \sum_{x \in \Omega} (\texttt{HMAX}_{\hat{h}_I}(x) - \texttt{HMAX}_{h_I^*}(x))^2. \tag{9}$$

The cell counting loss is the mean absolute error (MAE) between the estimated number of cells $\texttt{CCC}(\texttt{EMAX}_{\hat{h}_I})$ and the true number of cells $\texttt{CCC}_I^*$:

$$\mathcal{L}_{count}^I = \left| \texttt{CCC}(\texttt{EMAX}_{\hat{h}_I}) - \texttt{CCC}_I^* \right|. \tag{10}$$

The joint loss writes:

$$\mathcal{L}^I = \alpha \mathcal{L}_{rec}^I + \beta \mathcal{L}_{count}^I \tag{11}$$

where $\alpha > 0$ and $\beta > 0$ control the importance of each of the terms in the loss function. In our experiments we have used $\alpha = 1, \beta = 0.001$. These were set after five-folds cross validation experiments on $set1$, where 11 ratios β/α were tested, ranging from 10^{-4} to 10^2, including zero[4]. Additional loss terms were also tested, like the binary cross entropy between $\texttt{EMAX}_{\hat{h}_I}$ and $\texttt{EMAX}_{h_I^*}$, but no improvement was found.

Optimization. The joint loss function was minimized with stochastic gradient descent. More precisely, we used the RMSProp optimizer [10], with initial learning rate of 10^{-3}. Moreover, we used a batch size of 16 images and trained for 1600 training epochs. Only the model weights with lowest validation error were saved and used for final evaluation on the test set.

4.2   Evaluation Metrics

Following [13], we used two metrics for the test phase to evaluate our proposed pipeline. One intuitive metric is the *average relative error*, which averages the per-image relative counting error:

$$\mathcal{A}_{err}(\mathcal{S}) = \frac{1}{\mathcal{N}} \sum_{I \in \mathcal{S}} \frac{\left| \texttt{CCC}(\texttt{EMAX}_{\hat{h}_I}) - \texttt{CCC}_I^* \right|}{\texttt{CCC}_I^*} \tag{12}$$

where $\mathcal{S}$ denotes the test set with sample number $\mathcal{N}$. However, considering the variability in cell numbers of our dataset, we also monitor the *total relative error*.

---

[4] Tested values: $\alpha = 1$ and $\beta \in \{0\} \cup 10^{\{-4,-3,-2,-1,0,1,2\}} \cup 5 \times 10^{\{-4,-3,-2\}}$.

[3,4], which sees the whole dataset as one skin sample, but avoids compensations of errors by taking an absolute difference in each image:

$$\mathcal{T}_{err}(\mathcal{S}) = \frac{\sum_{I \in \mathcal{S}} \left| \text{CCC}(\text{EMAX}_{\hat{h}_I}) - \text{CCC}_I^* \right|}{\sum_{I \in \mathcal{S}} \text{CCC}_I^*}. \tag{13}$$

We report evaluation results on both metrics to prevent comparison bias. It is worth noting that finding the right number of cells does not imply correct detections, as false positives can compensate false negatives, like illustrated in Fig. 1. However, we have deliberately limited our analysis to counting errors, as counting is the primary objective. All counting methods benefit from this type of compensation, which is only possible if the method is not overly biased towards over or underestimation. Furthermore, future work could analyse reconstruction metrics, as a better reconstruction implies more accurate counting.

5   Results

We present our quantitative results on the test set, along with numbers of parameters and pipeline interpretability, in Table 1, comparing them with the current state-of-the-art method proposed by Lazard et al. [4]. Although our proposed pipeline does not surpass the latter in the two prediction metrics, the error rates we obtained are close and comparable to their methods. Furthermore, the absolute values of the error rates are around or below 10%, a threshold considered satisfactory by final users, as reported in [4]. It's noteworthy that the number of parameters in our pipeline is significantly lower than the state-of-the-art method using U-net (only less than 1/100 of theirs). This characteristic makes our approach suitable for scenarios with small training sets or with limited computational or storage resources, such as embedded systems or mobile devices. Additionally, our proposed method provides better interpretability. While the final output density map of the U-net methods [4] may be considered interpretable, actually density maps do not locate precisely cells, especially when cells overlap. Furthermore, the process to derive a density map from the input melanocyte sample (namely, the U-net architecture) is end-to-end non-interpretable. This makes the criteria used by the network less easy to exploit by the final users. In

Table 1. Results of our proposed pipeline compared to the current state-of-the-art method.

Method	$\mathcal{A}_{err}$	$\mathcal{T}_{err}$	Network parameter#	Interpretability
Lazard et al. [4]	9.28%	8.72%	1,760,000	end-to-end not interpretable
Ours	11.8%	9.35%	16,675	partially not interpretable

contrast, for our proposed pipeline, only the CNN part (best h prediction module in Fig. 6) is not interpretable; for the other parts, only morphological criteria based on size and contrast are used. Detailed predicted and ground truth results for the dynamics and numbers of cells are displayed in Fig. 7. We observe a stronger positive correlation between predicted and true cell numbers than between optimal and estimated h. This further demonstrates the robustness of the proposed algorithm with respect to the dynamic, as errors in the estimation of the optimal h often give an estimated count close to the true one.

Qualitative examples from the test set can be found in Fig. 8. These examples already illustrate the soundness of our proposed algorithm. Even in difficult cases (low contrast or non-convex shapes), the predicted cell number and cell locations appear to be in good agreement with the ground truth ones, and the gap between them are always in our expectations under the given difficulty.

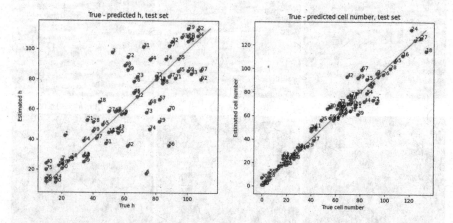

Fig. 7. The predicted $\hat{h}$ is compared to the ground truth h^* on the left, while the right hand plot compares the predicted and true cell numbers. Each dot represents an individual image, identified by its index in the test set. For a detailed analysis, refer to the corresponding discussion in the text of the paper.

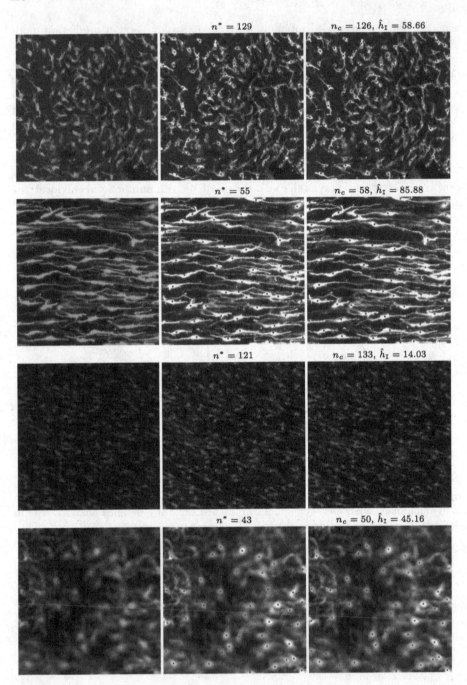

Fig. 8. Examples of results on the test set, with our proposed method. From left to right: Input resized RGB image, green channel image with ground truth cell locations and true number of cells n^*, green channel image with predicted cell locations and their estimated number n_c obtained for the estimated dynamic $\hat{h}_I$. (Color figure online)

6   Conclusion

In this paper, we have presented a new end-to-end trainable pipeline where a CNN predicts the dynamic threshold for h-maxima to be recognized as melanocytes, and a layer counts the connected components of the h-maxima images, yielding an estimated number of cells. Our pipeline is trained using a novel joint loss function, which comprises the geodesic reconstruction loss and a differentiable cell counting loss. Notably, our approach not only attains comparable results on the test set but also boasts a significantly reduced parameter count (1/100) and superior geometrical interpretability when compared to the state-of-the-art. Future endeavors could involve the prediction of *local h* values for smaller patches and the implementation of contrast enhancement for blurred regions. We posit that the synergy between neural networks and trainable morphological layers opens the door to a broader range of applications, such as counting astronomical objects in astronomical surveys.

References

1. Duval, C., Cohen, C., Chagnoleau, C., Flouret, V., Bourreau, E., Bernerd, F.: Key regulatory role of dermal fibroblasts in pigmentation as demonstrated using a reconstructed skin model: impact of photo-aging. PLoS ONE **9**(12), e114182 (2014)
2. Guan, S., Loew, M.: Understanding the ability of deep neural networks to count connected components in images. In: 2020 IEEE Applied Imagery Pattern Recognition Workshop (AIPR), pp. 1–7. IEEE (2020)
3. He, S., Minn, K.T., Solnica-Krezel, L., Anastasio, M.A., Li, H.: Deeply-supervised density regression for automatic cell counting in microscopy images. Med. Image Anal. **68**, 101892 (2021)
4. Lazard, T., et al.: Applying deep learning to melanocyte counting on fluorescent TRP1 labelled images of in vitro skin model. Image Anal. Stereol. (2022)
5. Najman, L., Talbot, H.: Mathematical Morphology: From Theory to Applications. Wiley, Hoboken (2013)
6. Rall, L.B.: Automatic Differentiation: Techniques and Applications. Springer, Heidelberg (1981)
7. Sangalli, M., Blusseau, S., Velasco-Forero, S., Angulo, J.: Scale equivariant neural networks with morphological scale-spaces. In: Lindblad, J., Malmberg, F., Sladoje, N. (eds.) DGMM 2021. LNCS, vol. 12708, pp. 483–495. Springer, Cham (2021). https://doi.org/10.1007/978-3-030-76657-3_35
8. Soille, P., et al.: Morphological Image Analysis: Principles and Applications, vol. 2. Springer, Cham (1999)
9. Tarjan, R.: Depth-first search and linear graph algorithms. SIAM J. Comput. **1**(2), 146–160 (1972)
10. Tieleman, T., Hinton, G., et al.: Lecture 6.5-RMSPROP: divide the gradient by a running average of its recent magnitude. COURSERA: Neural Netw. Mach. Learn. **4**(2), 26–31 (2012)
11. Velasco-Forero, S.: Morpholayers (2020). https://github.com/Jacobiano/morpholayers

12. Velasco-Forero, S., Angulo, J.: Learnable empirical mode decomposition based on mathematical morphology. SIAM J. Imaging Sci. **15**(1), 23–44 (2022)
13. Velasco-Forero, S., Rhim, A., Angulo, J.: Fixed point layers for geodesic morphological operations. In: British Machine Vision Conference (2022)
14. Vincent, L.: Morphological grayscale reconstruction in image analysis: applications and efficient algorithms. IEEE Trans. Image Process. **2**(2), 176–201 (1993)

Mathematical Morphology Applied to Feature Extraction in Music Spectrograms

Gonzalo Romero-García[1]([⊠]) [iD], Isabelle Bloch[2] [iD], and Carlos Agón[1] [iD]

[1] Sorbonne Université, CNRS, IRCAM, STMS, Paris, France
{romero,agonc}@ircam.fr
[2] Sorbonne Université, CNRS, LIP6, Paris, France
isabelle.bloch@sorbonne-universite.fr

Abstract. Mathematical Morphology has proven to be a powerful tool for extracting geometric information from greyscale images. In this paper, we demonstrate its application to spectrograms (two-dimensional greyscale images of sound) of music excerpts. The sounds of musical instruments exhibit particular shapes when represented as a spectrogram. These shapes are determined by the sound characteristics. In general, musical sounds contain three different components: the attack component, appearing as vertical lines; the sustain component, appearing as horizontal lines; and the stochastic component, appearing as a landscape of hills and holes. In this paper we propose a pipeline of morphological operators to separate these three components. This separation allows us to build a new sound similar to the input one.

Keywords: Mathematical Morphology · Spectrograms · Feature Extraction · Image Analysis

1 Introduction

Mathematical Morphology (MM) has been proven to be a useful tool for image analysis and segmentation. In this work, we show how to apply it to a particular type of images: music spectrograms. Spectrograms are the most popular time-frequency representations of audio [9, p. 23]. Since they present the audio content as a function with time and frequency as domain and intensity as codomain, they can be viewed as greyscale images.

Audio signals exhibit various shapes when represented through a spectrogram. However, in general, music instruments have a particular way of being laid down inside a spectrogram; we can, in most of the cases, find a combination of three main components: the sinusoidal component, the transient component and the noise component. These three components appear with characterizing

---

This work was partly supported by the chair of I. Bloch in Artificial Intelligence (Sorbonne Université and SCAI).

shapes: the sinusoidal component appears as horizontal lines, the transient component appears as vertical lines, and the noise component appears as a landscape of hills and holes (see Fig. 1).

In order to detect these components and to be able to characterize the sound, we need to extract some features that allow us to recompose the signal; in the case of the lines (*i.e.*, the sinusoidal and transient components) we want to obtain the times and frequencies (that is, the pixels) that characterize the lines, alongside with the amplitude (*i.e.*, the greyscale level). In the case of the noise, since it is a stochastic component, we do not want the precise values of the hills and the holes. We rather want a mask to be applied to a white noise through a pointwise multiplication. With these data, we are able to reconstruct the sound with the help of a synthesis model.

In this work, we propose a pipeline of MM operators to extract these components. MM operators are well suited to detect the lines in the spectrogram and also to remove them such that a mask for the white noise is obtained.

The paper is organized as follows: we first expose the related work in spectrogram feature extraction in Sect. 2; then, we expose the mathematical definition of spectrograms in Sect. 3. After this is set up, we propose our morphological pipeline in Sect. 4 and show the results of our experiments in Sect. 5. We expose our conclusions and future work in Sect. 6.

2   Related Work

The feature extraction from a spectrogram of audio was impulsed by the work of X. Serra [15,16], where a model of sound synthesis was proposed, based on a sum of a sinusoidal component plus a stochastic component, and peak detection methods were used to perform the analysis. Later, T. Verma and T. Meng proposed a sinusoids plus transients plus noise (STN) model in [18], that is the model we use in this work.

There has been a lot of work in the estimation of one or several of these three components in the case of music sounds [1,8,14,18,20]. The use of MM for feature extraction in spectrograms has also been done, but rather in speech spectrograms [3,17,22]. A similar approach is used for detecting lines in videos [6]. The first use of MM for analyzing music sounds is [12].

3   Spectrograms

In this section we expose the main formulas to generate spectrograms, following the book [5]. We expose them in a continuous framework such that they remain general and elegant. However, to apply morphological operators, we sample the result on a uniform grid with a specific time-frequency resolution[1].

---

[1] In the experiments exposed in this work, we chose a 10 ms step for time and a $\frac{44100}{4096} \approx$ 10.77 Hz step for frequency. These values are common values for music applications.

For our applications, we focus on the spectrogram of the Short-time Fourier transform (STFT). The STFT is defined as follows: let $f \in \mathcal{L}_\infty(\mathbb{R}; \mathbb{C})$ and let $g \in \mathcal{L}_1(\mathbb{R}; \mathbb{C})$. The STFT of f with respect to the window g is defined by

$$\begin{aligned}
\mathrm{STFT}_g[f] : \mathbb{R} \times \mathbb{R} &\to \mathbb{C} \\
(t, \xi) &\mapsto \int_{\mathbb{R}} f(x)\overline{g(x-t)}e^{-2\pi it\xi}\, dx
\end{aligned} \tag{1}$$

with i being the imaginary unit and $\overline{g(x-t)}$ the complex conjugate of $g(x-t)$.

The STFT is then a functional operator (it takes a function as input and returns a function as output). The input function f is associated to the input signal (an audio signal in our case) whose variable t represents time. The resulting function $\mathrm{STFT}_g[f]$ can be interpreted as the Fourier transform of a neighborhood of the function f around t (the neighborhood being determined by the window function g). The time will be measured in seconds (abbreviated s) and the frequency will be measured in Hertz (abbreviated Hz).

The function $\mathrm{STFT}_g[f]$ has as domain the time-frequency plane $\mathbb{R} \times \mathbb{R}$ and as codomain $\mathbb{C}$. Complex values are difficult to handle and to represent. In practical applications, we usually drop the phase information of the complex value and keep only the amplitude. This is what happens when using a spectrogram, that is defined as

$$\mathrm{SPEC}_g = |\mathrm{STFT}_g|^2. \tag{2}$$

This leads to an interpretation of the spectrogram as a power value: its value at a time-frequency point (t, ξ) is the power of frequency ξ at time t. We commonly use the logarithmic scale to represent it and thus measure it in decibels, *i.e.*

$$|\mathrm{SPEC}_g|^2 = 10\log_{10}|\mathrm{SPEC}_g|^2\, \mathrm{dB} = 20\log_{10}|\mathrm{SPEC}_g|\, \mathrm{dB}. \tag{3}$$

Since the input signal is bounded by 1 and we normalize the window function, we have that $0 \leq |\mathrm{SPEC}_g|^2 \leq 1$, which turns out to

$$-\infty \leq 20\log_{10}|\mathrm{SPEC}_g|\, \mathrm{dB} \leq 0 \tag{4}$$

in logarithmic scale, obtaining as codomain $\overline{\mathbb{R}}$.

Figure 1 shows the spectrogram of an audio excerpt that will be used as example; we can see the horizontal lines corresponding to sinusoids and the vertical lines corresponding to transients. We can also see the landscape of hills and holes for the noise component.

4   Morphological Pipeline

Based on a pipeline of simple MM operators, we propose a method for exploiting the structure of the spectrogram and the geometry of its components to achieve the detection of the three mentioned components. Let us expose the pipeline and illustrate it with the spectrogram from Fig. 1.

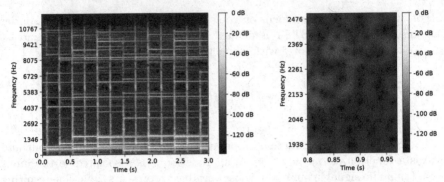

(a) The sinusoidal and transient component appear as horizontal and vertical lines

(b) The noise component appears as a landscape of hills and holes

Fig. 1. Spectrogram of an audio excerpt.

The pipeline is decomposed into three steps: pre-processing, processing and post-processing. All three steps use MM operators. The pre-processing is performed in order to adapt the image for the processing operations, and the main idea is to *fill the holes* of the spectrogram to have an image that is well adapted for the processing step. In the processing step, we use three different chains, one per component. Finally, for the components associated with lines (the sinusoidal and transient component) we use a post-processing step that allows removing small lines that might appear as residuals. The full morphological pipeline is exposed in Fig. 2.

4.1 Pre-processing

The main goal of the pre-processing is to *fill the holes* of the spectrogram. These holes correspond to the zeros of the spectrogram ($-\infty$ in logarithmic scale) and are present when there is noise.

To do that, we apply first a closing and then a reconstruction by erosion. The closing writes, as usual [2,11,13], $\varphi_B = \varepsilon_B \circ \delta_B$ where the dilation δ_B and the erosion ε_B are defined for a function $S \in \overline{\mathbb{R}}^{\mathbb{R} \times \mathbb{R}}$ and a structuring element $B \subseteq \mathbb{R} \times \mathbb{R}$ as $\delta_B[S] = S \oplus B$ and $\varepsilon_B[S] = S \ominus B$ with

$$
\begin{aligned}
S \oplus B : \mathbb{R} \times \mathbb{R} &\to \overline{\mathbb{R}} & S \ominus B : \mathbb{R} \times \mathbb{R} &\to \overline{\mathbb{R}} \\
p &\mapsto \bigvee_{x \in B} S(p - x) & p &\mapsto \bigwedge_{x \in B} S(p + x)
\end{aligned}
\tag{5}
$$

We use a closing with a structuring element that is a square of width 25 ms and height 75 Hz[2]. These values have been determined experimentally such that they are sufficiently large for filling the holes.

---

[2] These continuous values are sampled according to the grid, and become 7×3 in our case.

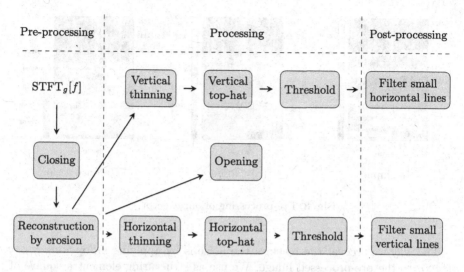

Fig. 2. Pipeline for the morphological processing.

While the closing manages to fill the holes (if B is sufficiently big), it deforms the shapes. To avoid this effect, we apply as next step a reconstruction by erosion. The reconstruction by erosion is defined as follows: we choose an image S to be the marker and we choose an image R to be the reference. Then, the geodesic erosion is defined as

$$\varepsilon_{R,B}[S] = \varepsilon_B[S] \vee R \tag{6}$$

where B is the structuring element of radius 1. The reconstruction by erosion is thus defined as the iteration of this operator until stability, *i.e.*

$$\varepsilon_{R,B}^{\infty}[S] = \varepsilon_{R,B}^{(n)}[S] \tag{7}$$

where $\varepsilon_{R,B}^{(n)}[S] = \underbrace{(\varepsilon_{R,B} \circ \cdots \circ \varepsilon_{R,B})}_{n \text{ times}}[S]$ and we have $\varepsilon_{R,B}^{(n)}[S] = \varepsilon_{R,B}^{(n+1)}[S]$.

For our purpose, we use as marker the result of the closing and as reference the input image. We use a structuring element B corresponding to the 8-connectivity, *i.e.* a square of size 3×3.

The results of our pre-processing on our example are shown in Fig. 3; we see that the holes disappear in the closing step but with a deformation of the shapes, and the reconstruction by erosion recovers the shapes of the lines.

4.2 Extracting the Noise Mask

Now that we have an input image without holes, we can apply our processing pipeline to recover the mask of the noise and the lines. Let us start by exposing how to recover the mask of the noise since it is the simpler part as it only involves a single operation.

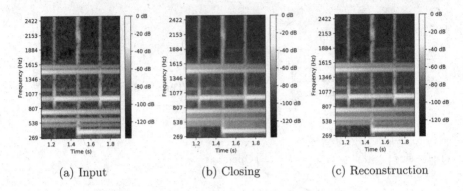

(a) Input   (b) Closing   (c) Reconstruction

Fig. 3. Pre-processing of our excerpt.

The way we recover the mask of the noise is by applying an opening $\gamma_B = \delta_B \circ \varepsilon_B$ to the pre-processed image. We use as structuring element a square of width 44 ms and height 193 Hz which are the values that ensure a -60 dB drop both in time and frequency in the shape of the window function.

After recovering the mask of the noise, we can filter a white noise with it by pixel-wise multiplication and produce a filtered noise that is similar to the one present in the input spectrogram. This can be remarked in Fig. 4.

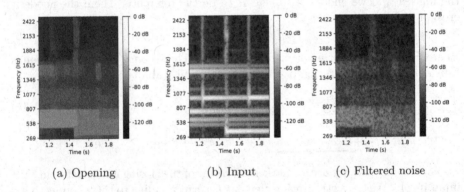

(a) Opening   (b) Input   (c) Filtered noise

Fig. 4. Processing of the noise component.

4.3 Recovering the Lines

The process of recovering horizontal and vertical lines are dual; we use the same operation but with different structuring elements. For the horizontal lines, we apply a vertical thinning, followed by a vertical top-hat and finally a threshold. For the vertical lines, we apply a horizontal thinning, followed by a horizontal top-hat and finally also a threshold. Let us describe these three operations independently of the kind of lines we search for.

Thinning. The first step is to thin the pre-processed image; this way we obtain lines that are one-pixel thin. However, since our image is greyscale, we shall use a greyscale approach. We apply the one presented in [4] based on the notion of destructible point. From an implementation point of view, this is equivalent to apply a hit-or-miss transform with some particular structuring elements.

The hit-or-miss transform that we are using is one of the ones presented in [10] and is defined as follows: for an input image S and a pair of structuring elements C and D, the hit-or-miss transform $S \circledast (C, D)$ is defined for all $p \in \mathbb{R} \times \mathbb{R}$ as

$$(S \circledast (C, D))(p) = ((S \ominus C)(p) - (S \oplus \check{D})(p)) \vee 0 \qquad (8)$$

where $\check{D} = \{-d \in \mathbb{R} \times \mathbb{R} : d \in D\}$.

Now, we can define the elementary thinning of S by C and D as

$$S \bigcirc (C, D) = S - (S \circledast (C, D)). \qquad (9)$$

We call (C, D) a template.

Applying successive elementary thinnings

$$((((S \bigcirc (C_1, D_1)) \bigcirc (C_2, D_2)) \bigcirc \cdots) \bigcirc (C_n, D_n) \qquad (10)$$

is called a thinning and is denoted as

$$S \bigcirc ((C_1, D_1), (C_2, D_2), \cdots, (C_n, D_n)). \qquad (11)$$

If we apply the thinning operation until stability, we have an ultimate thinning.

The type of thinning we apply depends on the templates we choose. In our case, we want to perform two different types of thinning: one that thins the lines vertically (and then produces horizontal lines of one pixel length) and another that thins the lines horizontally (and then produces vertical lines of one pixel length).

For the vertical thinning we use the following templates:

$$(C, D)_N, (C, D)_S, (C, D)_{NE}, (C, D)_{SW}, (C, D)_{NW}, (C, D)_{SE} \qquad (12)$$

in this (arbitrary) order, and for the horizontal thinning we use the templates

$$(C, D)_E, (C, D)_W, (C, D)_{NW}, (C, D)_{SE}, (C, D)_{NE}, (C, D)_{SW} \qquad (13)$$

in this (arbitrary) order, where the templates corresponding to each cardinal point are

$$N : \begin{bmatrix} 0 & 0 & 0 \\ - & 1 & - \\ - & 1 & - \end{bmatrix} \quad E : \begin{bmatrix} - & - & 0 \\ 1 & 1 & 0 \\ - & - & 0 \end{bmatrix} \quad S : \begin{bmatrix} - & 1 & - \\ - & 1 & - \\ 0 & 0 & 0 \end{bmatrix} \quad W : \begin{bmatrix} 0 & - & - \\ 0 & 1 & 1 \\ 0 & - & - \end{bmatrix}$$

$$NE : \begin{bmatrix} - & 0 & 0 \\ 1 & 1 & 0 \\ - & 1 & - \end{bmatrix} \quad SE : \begin{bmatrix} - & 1 & - \\ 1 & 1 & 0 \\ - & 0 & 0 \end{bmatrix} \quad SW : \begin{bmatrix} - & 1 & - \\ 0 & 1 & 1 \\ 0 & 0 & - \end{bmatrix} \quad NW : \begin{bmatrix} 0 & 0 & - \\ 0 & 1 & 1 \\ - & 1 & - \end{bmatrix}$$

These patterns should be interpreted in the following manner: ones correspond to the elements of set C, zeroes correspond to the elements of set D, and $-$'means that the point is not considered in the structuring element. The origin is located at the center pixel.

The result of the two types of thinnings are shown in Fig. 5; we see that the lines have been reduced to one pixel width but only in one direction.

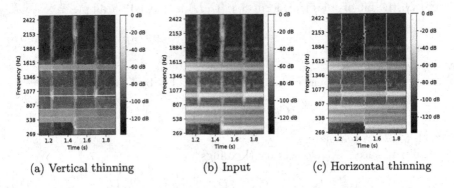

(a) Vertical thinning   (b) Input   (c) Horizontal thinning

Fig. 5. Thinnings for reduce the lines to one pixel width.

Top-Hat. The top-hat operation allows us now to isolate these lines of one pixel width. Similarly as in the case of thinning, we use a type of top hat for each type of thinning (vertical and horizontal). The top hat operation is defined as: $\forall p \in \mathbb{R} \times \mathbb{R}$,

$$S(p) - \gamma_B[S](p) \tag{14}$$

where γ_B is an opening with structuring element $B \subseteq \mathbb{R} \times \mathbb{R}$.

The structuring elements we choose are 3×1 and 1×3 for vertical and horizontal top-hats, respectively.

Threshold. After applying the top-hat, we want to recover the values that are above a certain threshold (that we set to 5 dB in our applications). The amplitude values we recover are not the ones of the top-hat but the ones of the result of the pre-processing, to be able to have the correct amplitudes. The operation is described as follows: if we denote the result of the pre-processing as S_0 and the result of the top-hat as $S_{\mathbf{Id}-\gamma}$, the threshold $S_>$ is defined for all $p \in \mathbb{R} \times \mathbb{R}$ as

$$S_>(p) = \begin{cases} S_0(p) & \text{if } S_{\mathbf{Id}-\gamma}(p) > \tau \\ -\infty & \text{if } S_{\mathbf{Id}-\gamma}(p) \leq \tau \end{cases} \tag{15}$$

4.4 Post-processing

With this processing we manage to isolate the horizontal and vertical lines. However, we get a lot of residuals that are too small to represent actual sinusoids and transients. This is why we apply a post-processing operation to remove these small lines.

The operation consists of two steps: first, we apply a thinning that shrinks the lines and eventually make them disappear (if they are below a threshold in length). Then, we apply a reconstruction by dilation with the thinned image as marker and the processed image as reference. This way, the lines that are sufficiently long will reappear in their full length.

Thinning. The thinning is done now in the same direction as the lines: we apply an horizontal thinning for the horizontal lines and a vertical thinning for the vertical ones. We actually only use the templates $(C, D)_E$ and $(C, D)_W$ for the horizontal thinning and $(C, D)_N$ and $(C, D)_S$ for the vertical one. The number of iterations is fixed to eliminate lines of a certain width or height; if τ_s (resp. τ_{Hz}) is the threshold for horizontal lines and Δt (resp. $\Delta \xi$) is the time (resp. frequency) resolution of the spectrogram, the number of iterations N_s (resp. N_{Hz}) is given by $N_s = \left\lceil \frac{\tau_s}{2\Delta t} \right\rceil$ and $N_{Hz} = \left\lceil \frac{\tau_{Hz}}{2\Delta \xi} \right\rceil$.

Reconstruction by Dilation. The reconstruction by dilation is the dual of the reconstruction by erosion. The geodesic dilation is defined as

$$\delta_{R,B}[S] = \delta_B[S] \wedge R \tag{16}$$

where S is the marker and R the reference. The reconstruction by dilation is the iteration of the geodesic dilation until stability and is denoted by $\delta_{R,B}^\infty[S]$. We use a structuring element B corresponding to the 8-connectivity, i.e. a square of 3×3.

At the end of the process, we recover a number of lines with which we can re-synthesize the input signal. Figure 6 shows the result on our excerpt, that yields very good results.

5 Experimental Results

In this section, we present the outcomes of our experiments, which highlight the effectiveness of our proposed pipeline. We acknowledge that while our chosen example features a glockenspiel (a musical instrument known for its distinct transient and sinusoidal components) the applicability of our pipeline may vary across different instruments.

Our pipeline was applied to a range of musical instruments, resulting in diverse outcomes (Fig. 7). In most cases, our pipeline performed considerably good without the need for parameter adjustments. However, there were instances

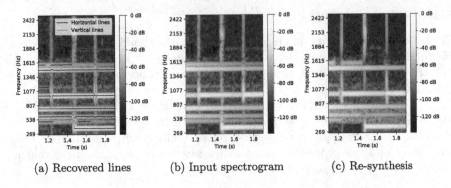

(a) Recovered lines   (b) Input spectrogram   (c) Re-synthesis

Fig. 6. Result of the processing in the case of our example.

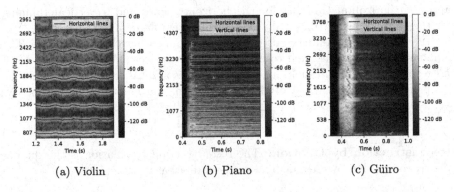

(a) Violin   (b) Piano   (c) Güiro

Fig. 7. Lines detection for other instruments.

where fine-tuning was necessary to achieve optimal results. A deeper analysis of these parameters and quantitative evaluation are left for future work.

In scenarios involving certain instruments or complex musical compositions with multiple instruments, the separation of the components (sinusoids, transients, and noise) proved challenging. This often resulted in sparse recovered lines, with the noise component dominating the information.

We performed the operations using the Python libraries NumPy [7], SciPy [19] and scikit-image [21]. However, greyscale thinning was not implemented in these libraries. Consequently, we implemented it through iterative processing, which became the bottleneck of our pipeline.

Furthermore, initially, we opted for a time resolution of 1 ms, resulting in very large images that noticeably slowed down the process. In an effort to improve efficiency, we experimented with increasing the time resolution to 10 ms. Remarkably, this change did not negatively impact the results while significantly enhancing the processing speed.

Audio examples and all the code used for the processing the spectrograms and generating the images are available at https://github.com/Manza12/DGMM-2024.

6   Conclusions and Future Work

In this work, we have proposed a morphological pipeline for extracting components from spectrograms of musical audio signals. We have shown how to detect the lines that correspond to sinusoids and transients through the use of a morphological pipeline, as well as how to extract a mask for the noise component.

While in general the method works well for simple excerpts, we noticed that when we apply it to very dense spectrograms it performs badly. This is probably caused by a difficulty in deciding if something is a line or if it belongs to the noise component in such situations.

While working on this topic, we found that the border between noise and signal (understanding signal as sinusoids or transients) is fuzzy. In future work we will go more in depth about the nature of spectrograms, proposing another model different from the STN model and more adapted to the detection ridges and holes, and making a more clear distinction between noise and signal.

References

1. Amatriain, X., Bonada, J., Loscos, A., Serra, X.: Spectral processing. In: DAFX, chap. 10, pp. 373–438. Wiley (2002)
2. Bloch, I., Heijmans, H., Ronse, C.: Mathematical morphology. In: Aiello, M., Pratt-Hartmann, I., Van Benthem, J. (eds.) Handbook of Spatial Logics, pp. 857–944. Springer, Dordrecht (2007). https://doi.org/10.1007/978-1-4020-5587-4_14
3. Cadore, J., Gallardo-Antolín, A., Peláez-Moreno, C.: Morphological processing of spectrograms for speech enhancement. In: Travieso-González, C.M., Alonso-Hernández, J.B. (eds.) NOLISP 2011. LNCS, vol. 7015, pp. 224–231. Springer, Heidelberg (2011). https://doi.org/10.1007/978-3-642-25020-0_29
4. Couprie, M., Bezerra, F.N., Bertrand, G.: Topological operators for grayscale image processing. J. Electron. Imaging $10(4)$, 1003–1015 (2001)
5. Gröchenig, K.: Foundations of Time-Frequency Analysis. Birkhäuser, Boston (2001)
6. Guimarães, S.J.F., Couprie, M., de Albuquerque Araújo, A., Jerônimo Leite, N.: Video segmentation based on 2D image analysis. Pattern Recogn. Lett. **24**, 947–957 (2003)
7. Harris, C.R., et al.: Array programming with NumPy. Nature **585**(7825), 357–362 (2020)
8. Keiler, F., Marchand, S.: Survey on extraction of sinusoids in stationary sounds. In: Digital Audio Effects (DAFx) Conference, Germany, pp. 51–58 (2002)
9. Klapuri, A., Davy, M.: Signal Processing Methods for Music Transcription. Springer, Cham (2007)
10. Naegel, B., Passat, N., Ronse, C.: Grey-level hit-or-miss transforms—part I: unified theory. Pattern Recogn. **40**(2), 635–647 (2007)
11. Najman, L., Talbot, H.: Mathematical Morphology: From Theory to Applications. Wiley-ISTE, London (2010)
12. Romero-García, G., Agón, C., Bloch, I.: Estimation de paramètres de resynthèse de sons d'instruments de musique avec des outils de morphologie mathématique. In: 19th Sound and Music Computing Conference, Zenodo, Saint-Etienne, France, pp. 653–662 (2022)

13. Ronse, C., Heijmans, H.J.A.M.: The algebraic basis of mathematical morphology: II. Openings and closings. CVGIP: Image Underst. **54**(1), 74–97 (1991)
14. Salamon, J., Gomez, E.: Melody extraction from polyphonic music signals using pitch contour characteristics. IEEE Trans. Audio Speech Lang. Process. **20**(6), 1759–1770 (2012)
15. Serra, X.: Musical sound modeling with sinusoids plus noise. In: Musical Signal Processing, pp. 91–122. Routledge, New York (1997)
16. Serra, X., Smith, J.: Spectral modeling synthesis: a sound analysis/synthesis system based on a deterministic plus stochastic decomposition. Comput. Music J. **14**(4), 12–24 (1990)
17. Steinberg, R., O'Shaughnessy, D.: Segmentation of a speech spectrogram using mathematical morphology. In: IEEE International Conference on Acoustics, Speech and Signal Processing, pp. 1637–1640 (2008)
18. Verma, T.S., Meng, T.H.Y.: Extending spectral modeling synthesis with transient modeling synthesis. Comput. Music J. **24**(2), 47–59 (2000)
19. Virtanen, P., et al.: SciPy 1.0: fundamental algorithms for scientific computing in Python. Nat. Methods **17**, 261–272 (2020). https://doi.org/10.1038/s41592-019-0686-2
20. Virtanen, T., Klapuri, A.: Separation of harmonic sound sources using sinusoidal modeling. In: IEEE International Conference on Acoustics, Speech, and Signal Processing, pp. II765–II768 (2000)
21. Van der Walt, S., et al.: scikit-image: image processing in python. PeerJ **2**, e453 (2014)
22. Xu, S., et al.: A mathematical morphological processing of spectrograms for the tone of Chinese vowels recognition. Appl. Mech. Mater. **571–572**, 665–671 (2014)

A Discrete Geometry Method for Atom Depth Computation in Complex Molecular Systems

Sara Marziali[1]([✉])[ID], Giacomo Nunziati[1,2][ID], Alessia Lucia Prete[1,3][ID], Neri Niccolai[4][ID], Sara Brunetti[1][ID], and Monica Bianchini[1][ID]

[1] Department of Information Engineering and Mathematics (DIISM), University of Siena, Via Roma 56, 53100 Siena, Italy
`{sara.marziali,giacomo.nunziati,alessialucia.prete}@student.unisi.it`,
`{sara.brunetti,monica.bianchini}@unisi.it`
[2] Department of Information Engineering (DINFO), University of Florence, Via S. Marta 3, 50139 Florence, Italy
`giacomo.nunziati@unifi.it`
[3] Toscana Life Sciences Foundation, Mass Spectrometry Unit (MSU), Via Fiorentina 1, 53100 Siena, Italy
[4] Department of Biotechnology, Chemistry and Pharmacy, University of Siena, Via Aldo Moro 2, 53100 Siena, Italy
`neri.niccolai@unisi.it`

Abstract. The field of structural biology is rapidly advancing thanks to significant improvements in X-ray crystallography, nuclear magnetic resonance (NMR), cryo-electron microscopy, and bioinformatics. The identification of structural descriptors allows for the correlation of functional properties with characteristics such as accessible molecular surfaces, volumes, and binding sites. Atom depth has been recognized as an additional structural feature that links protein structures to their folding and functional properties. In the case of proteins, the atom depth is typically defined as the distance between the atom and the nearest surface point or nearby water molecule.

In this paper, we propose a discrete geometry method to calculate the depth index with an alternative approach that takes into account the local molecular shape of the protein. To compute atom depth indices, we measure the volume of the intersection between the molecule and a sphere with an appropriate reference radius, centered on the atom for which we want to quantify the depth.

We validate our method on proteins of diverse shapes and sizes and compare it with metrics based on the distance to the nearest water molecule from bulk solvent to demonstrate its effectiveness.

Keywords: Morphological Algorithms · Voxel Model · Structural biology · Atom depth · Molecular shape · Depth index · Discrete Geometry

S. Marziali, G. Nunziati and A. L. Prete—Equal contribution.

S. Brunetti et al. (Eds.): DGMM 2024, LNCS 14605, pp. 443–455, 2024.
https://doi.org/10.1007/978-3-031-57793-2_34

1   Introduction

In this paper, we propose an application of discrete geometry and mathematical morphology techniques to address a problem in the field of structural biology. In particular, we exploit voxel-based modelling to represent protein structure and to design related algorithms.

The importance of atomic depth lies in its capacity to provide detailed insights into protein structures. It has been discovered to correlate with a variety of molecular and atomic features, such as average protein domain size, stability, the free energy of protein complex formation, amino acid hydrophobicity, residue conservation, and hydrogen/deuterium amide proton exchange rates [14]. Residue depth, obtained by combining the depths of atoms of the same residue, correlates significantly better than accessibility with the effects of mutations on protein stability and protein-protein interactions [2].

While surface features are important for understanding protein function, the capacity of a protein to fold correctly and maintain stability is mostly determined by its core. Atom depth extends existing descriptors such as solvent-accessible area and buried surface area by providing information about atoms and residues deep within the protein core. For example, as proteins get deeper into the nucleus, parameters such as solvent accessibility and buried surface area lose sensitivity and correlation. On the other hand, atom depth fails to differentiate among different exposition and accessibility levels of surface atoms.

An alternative approach to compute atom depth indices has been proposed by Varrazzo et al. [18], taking into account the molecular shape of the proteins by measuring the volume of the intersections between the molecule and a sphere of an "appropriate" radius centred on the atom under investigation. This method, called Simple Atom Depth Index Calculation (SADIC), provides a more accurate assessment of atom depth by considering the three-dimensional (3D) insertion of the atom inside the molecular structure.

In this paper, we investigate the problem of determining the 3D atom depth analysis by revising the one that has been previously proposed in [18]. The novelty of the method we propose here, SADIC v2, lies in the use of a discrete geometry approach that exploits voxel-based modelling to represent the protein's structure and related algorithms.

Voxelised representations of protein surfaces have attracted interest for various analyses. Kihara et al., for example, have developed protein docking, shape comparison, and interface identification algorithms based on 3D Zernike descriptors (3DZD) computed across circular surface patches of voxelised macromolecular surfaces [9]. Similarly, dot surfaces have been used to generate invariant descriptors for defining protein surfaces, making tasks like protein functional categorisation and retrieval of protein surfaces from large datasets easier [5]. Furthermore, using a dot-based model of molecular surfaces, invariant surface fingerprints have been established for the quick and efficient comparison of local protein surface similarities [19]. Grid representations of protein surfaces have also been used to detect cavities and identify binding sites [6].

In our method, we employ a volumetric representation of the protein to compute the depth index of its atoms proposed in [18]. Differently, in [18] the assumed model is an assembly of spheres (see Sect. 3) and a suitable sampling of points is performed. A comparison of the programs is provided in Sect. 5.

The paper is organised as follows: Sect. 2 briefly reports some existing methods for calculating the atom depth while Sect. 3 describes the model employed for protein representation and the definition of depth index that we assume. Section 4 outlines the discretisation of the molecule and the different steps of the algorithm, and Sect. 5 shows the results obtained for proteins of different sizes and shapes, comparing them with the outputs of other methods. In the final Section, the conclusions of our study are clarified.

2   Related Works

Atom depth, a key geometric descriptor of a protein's interior, has been widely investigated and many methods for its calculation have been proposed [2,12]. One commonly used approach involves the "nearest hypothetical water molecule" method, which consists of placing the protein inside a 3D lattice containing water molecules [12]. This method calculates the depth of the atom based on its distance to the nearest water molecule, excluding those within cavities or surface grooves. Another strategy, similar to the previous one, also considers the protein dynamic, which is emulated by consecutive rotations and translations, through Monte Carlo simulations [2]. In this case, the atom depth is the average distance between the atom and its closest water molecule, over the different configurations. Atom depth can also be defined as the distance between an atom and a molecular or solvent-accessible surface, which requires the computation of a dot surface [4], where dot points serve as virtual markers to outline the protein surface. Another method, which is thought to be the most precise but also the most computationally expensive, determines atom depth during molecular dynamics simulations of a solvated protein by calculating the distance between each atom and its nearest surface dot [2]. Depth is defined in a more modern way as the distance between a buried atom and its nearest solvent-accessible protein neighbour, with solvent-accessible atoms determined using the rolling sphere technique [10,13]. This method, however, introduces arbitrary parameters such as the radius of the probe and a cut-off to define an atom as "accessible".

All these methods offer valuable insights into the fundamental structural properties of proteins, but it is crucial to bear in mind that, regardless of differences in computational methodologies, they generally fail to capture the contributions of the local 3D molecular shape to actual atom depth. This limitation holds significant importance because the complex 3D structure of a protein is often intricately linked to its specific functions and interactions. Interactions involving proteins and other biomolecules, such as substrates, cofactors, or other proteins, frequently occur at active sites or specific regions of the protein, where atoms may reside at varying depths within the structure. The absence of detailed information regarding variations in atomic depth can, therefore, impede a comprehensive understanding of how these phenomena take place within the protein itself. This observation highlights the need for more precise methodologies that can effectively capture the influence of molecular structure on atom depth, a gap that our work tries to fill by adopting a definition of atom depth index that takes into account the position occupied by atoms in space.

3   The Molecular Model

The model we consider assumes each protein to be represented by a solid composed of the union of spheres centred on each atom [10]. This represents an approximation in the continuous space of the protein 3D structure. We exploit data obtained from the Protein Data Bank (PDB), the largest publicly accessible and free collection of polypeptide chains, to identify the structure of the protein, including the positions of the atoms in 3D space. According to the numbering introduced in the PDB file, each atom is indexed with an integer $i = 1, \ldots, N$, where N is the number of atoms in the protein. For each atom i, we construct the sphere $B(c_i, r_i)$ of centre c_i and radius r_i, where $c_i = (x_i, y_i, z_i)$ corresponds to the coordinates of the i-th atom and the radius r_i, expressed in ångström (Å), is calculated by increasing the Van der Waals radius of atom i ($r_{VdW}(i)$) by a constant C [18], i.e.

$$r_i = r_{VdW}(i) + C. \tag{1}$$

The most fundamental elements that can be found in proteins are hydrogen (H), carbon (C), nitrogen (N), oxygen (O), sulphur (S), and phosphorus (P), and their Van der Waals radii used by default by the algorithm [1] are reported in Table 1.

Table 1. Van der Waals radii as determined by the Rowland and Taylor system [1].

Atom	Van der Waals radius (Å)
H	1.10
C	1.77
N	1.64
O	1.58
S	1.81
P	1.87

In our model, we set the constant C equal to the Van der Waal radius of the water molecule, which can be approximated with the radius of the oxygen atom. Increasing each atomic radius by such a value ensures that convex regions that do not allow water or other molecules to access it are filled, as shown in Fig. 1, and contribute to the internal volume of the protein. Therefore, non-accessible cavities are considered internal, whereas accessible cavities are external to the protein.

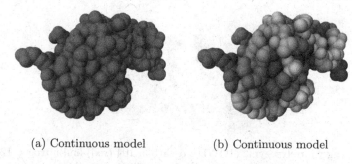

(a) Continuous model   (b) Continuous model

Fig. 1. Different representations of the hen egg white lysozyme (1GWD): (a) represents the protein according to the model defined in Sect. 3; (b) qualitatively shows the level of exposition of each residue, quantified as the exposed area of the residue. The proposed depth index aims to better characterise the exposition at the atomic granularity and consider the local shape of the molecule.

3.1   The Atom Depth Index

The atom depth index is a quantitative measure employed to characterise the 3D positioning of atoms within the protein structure. This index is determined by assessing each atom's distance from the centre to the solvent-exposed protein surface in all directions simultaneously. The goal is to distinguish atoms located on surface protrusions from those positioned equally close to the surface but within accessible cavities. To this aim, we adopt a definition of atom depth index based on volumes rather than linear distances [18].

Given a solid representation S of a protein according to the model described in Sect. 3, let r_{max} denote the radius of the largest sphere inscribed within S and centred on any atom, say j: then, let $B(c_j, r_{max})$ be such a reference sphere and $V_{r_{max}}$ be its volume. Let $B(c_i, r_{max})$ be the reference sphere centred in the i-th atom. The atom depth index for atom i is defined as follows:

$$D_{i,r} = 2 * \left(1 - \frac{V(B(c_i, r_{max}) \cap S)}{V_{r_{max}}} \right) \tag{2}$$

where $V(B(c_i, r_{max}) \cap S)$ denotes the volume of the intersection between the reference sphere centred in the i-th atom and the solid S. This index provides a relative measure of the atom's depth within the protein, emphasising variations in proximity to the protein surface. Note that it actually assigns the value zero to the deepest atom (atom j) and increasing values to atoms closer to the surface, in order to provide a measure of how accessible the atom is. This distinction ensures that atoms on bulges have a greater depth index than those inside the cavities (refer to Fig. 2).

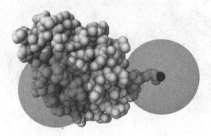

Fig. 2. Hen egg white lysozyme (1GWD): α-carbon 104 (red) and nitrogen 128 (blue) are both on the molecule surface, but the accessibility of the latter is much greater. (Color figure online)

4   The Proposed Method

Once the molecular model has been introduced in Sect. 3 and the depth index defined in Eq. 2, we proceed to the effective description of the algorithm for calculating the depth index of the atoms.

The protein is first discretised to create a voxel-based representation describing the structure of the molecule. Further on, the internal cavities that are not accessible to solvents from the outside are detected. In our model, these pockets can be considered as part of the protein since they are not relevant for interactions with other molecules. Finally, we calculate the reference radius and the depth indices of the atoms. All these steps, described in the following subsections can be computed in linear time in the size of the defined regular grid.

4.1   Volumetric Representation of the Protein

Given the multi-sphere solid described in Sect. 3, the first step is to create a voxel-based representation of the protein. First, the resolution parameter, which regulates the granularity of the voxel grid, is initialised. The resolution determines the size of each volume unit within the grid, effectively controlling the level of detail of the digital representation of the protein. To create the voxel grid, the protein, represented as the union of all the spheres $B(c_i, r_i)$, is placed in a reference coordinate system and a minimal bounding box, aligned with the axes, is constructed. The bounding box represents the smallest rectangular cuboid that contains the entire protein structure efficiently. Once the grid resolution has been established, we proceed to discretise each sphere $B(c_i, r_i)$ of the protein. Then, each voxel (unit cube) of the bounding box is indexed by x, y, z integer coordinates and is marked as foreground (1) if it is inside any sphere or as background (0) otherwise. Specifically, if the Euclidean distance between the centre of the voxel and the centre c_i of the sphere under consideration is less than the radius r_i of the sphere, the voxel falls within the sphere and is marked as inside the sphere. As a result, exemplified in Fig. 3, we get a 3D binary object, a finite set of voxels, representing the protein. This procedure is repeated for each sphere and the assignment of foreground and background to each voxel of the bounding box is obtained by merging the results.

4.2 Extraction of Connected Components and Holes Filling

To further analyse the structural features of the protein, we proceed with the identification of connected regions, also referred to as connected components, in its interior [8]. To extract such regions, an automatic morphological algorithm is applied with a symmetric structural element [15] using 6-connectivity. This process allows the identification of internal voids represented by groups of 6 adjacent voxels initially labelled as background and disconnected from the exterior of the protein. Hence, the output of the algorithm is a voxel labelling that assigns the same label to voxels in the same connected component. The identified internal voids correspond to regions within the protein that are not accessible to solvent molecules due to their enclosed nature and therefore, in our model, they should be considered as part of the foreground.

Once the internally connected components have been identified, a holes-filling algorithm [7] is applied: the voids are effectively filled by reassigning the voxel labels to be part of the 3D binary object rather than categorising them as background.

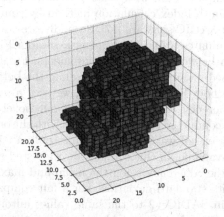

Fig. 3. Discrete representation of the hen egg white lysozyme (1GWD) using a voxel grid with a resolution of 2.0 Å. In this case, a coarse-grained grid has been chosen, for clarity of visualization.

4.3 Depth Index Calculation

In the following phase of the algorithm, we compute the depth atom index as defined in Eq. 2.

The first step of the procedure consists of determining the largest sphere, centred on any atom, that is contained in the protein. To compute its radius, the exact Euclidean Distance Transform (EDT) is exploited within the voxel representation to calculate the distance of each object voxel from the nearest one in the background by the algorithm [11]. The output of the EDT computation is a labelling of each voxel with the distance to the nearest background voxel and the maximum among the labels of the voxels corresponding to the centres of the atoms is selected.

Let $j \in \{1, \ldots, N\}$ be the index of the atom with the centre c_j labelled with the maximum EDT value: we take its label as the reference radius $r_{max} = EDT(j)$. As defined in Eq. 2, to compute the depth index of atom i the volume of the intersection between the reference sphere centred in c_i and the protein is calculated. This evaluation can be done by counting the object voxels whose centre has a distance from c_i less than r_{max}. In other words, for each voxel in $B(c_i, r_{max})$, the intersection is achieved through logical AND operations. Note that, since the depth index involves the ratio of the volumes, the computation is performed directly from the voxel model by summing up the voxels.

5 Results

The described algorithm is implemented in Python to ensure usability and platform compatibility. In this section, we present the results of the experiments we conducted and we compare them with other programs. Firstly, we validate the results of our atom depth index calculation method by comparing them to the results of the original SADIC software [17]. Secondly, we compared our method with another that evaluates atom depth using a standard approach, revealing an inverse correlation between the two measurements. Finally, to complete the comparison, we focused on surface atoms, revealing the better characterisation performed by our method. For all experiments, we used a resolution of 0.3 Å.

To assess the coherency and effectiveness of our novel approach to atom depth calculation, we conducted comparative tests with the original Simple Atom Depth Index Calculation software (SADIC). Since only a few results of the old implementation are available, in this comparison we tested a limited set of proteins. We evaluated the coherency using the mean and maximum absolute discrepancy, as well as its standard deviation. For a fair comparison, we fixed the reference radii used in SADIC v2 to the same values automatically computed by the old implementation. For the considered test proteins we obtained a mean absolute discrepancy of 0.010, with a standard deviation of 0.010. The maximum reported absolute discrepancy is 0.061. Figure 4 shows the absolute discrepancy distribution for the ATP-dependent DNA ligase from bacteriophage T7 (PDB code 1A0I).

To further validate the reliability of our atom depth calculations, we conducted a comparison of the results with ResidueDepth, a widely used method for measuring the depth of residues in protein structures. In ResidueDepth, the deepness is defined as the distance of any atom or residue to the closest bulk water molecule, and it serves as a parameterisation of the local protein environment [16]. We selected proteins with different shapes and structures, as reported

Absolute difference distribution

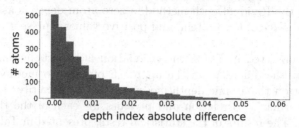

Fig. 4. Histogram representing the distribution of absolute discrepancy between the results of the old implementation and SADIC v2 for ATP-dependent DNA ligase from bacteriophage T7 (PDB code 1A0I), composed of 2903 atoms.

(a) PDB code 1UBQ   (b) PDB code 3ONS   (c) PDB code 3NSM

Fig. 5. Continuous 3D model for three representative proteins of the test set: erythrocytic ubiquitin (a), human ubiquitin (b) and beta-N-acetyl-D-hexosaminidase (Fig. 5c).

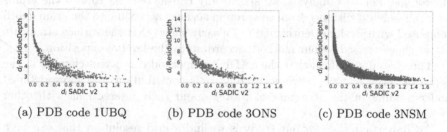

(a) PDB code 1UBQ   (b) PDB code 3ONS   (c) PDB code 3NSM

Fig. 6. Comparison between SADIC v2 and ResidueDepth: analysis of the correlation of the atom depth indices for proteins erythrocytic ubiquitin (a), human ubiquitin (b) and beta-N-acetyl-D-hexosaminidase (c).

in Fig. 5, and, for each molecule, we calculated the atom depth using both our approach and ResidueDepth.

We observed a clearly defined inverse correlation between the two methods, coherent with the fact that they exploit different approaches to calculate the depth of atoms (refer to Fig. 6). In fact, in the SADIC v2 method, the depth index is defined so that a zero value corresponds to the deepest atoms, with positive values indicating an expansion toward the surface of the protein. On the other hand, in the ResidueDepth method, a zero atom depth is assigned to the atoms on the surface of the protein, and positive values represent a penetration toward its core.

As can be observed in Fig. 6, the correlation is hyperbolic and non-linear, suggesting that the difference in the approach of the two methods has a significant impact on the atomic depth measurement. To quantify this aspect, we calculated Spearman's correlation and p-values for each of the three proteins under analysis. The results of the statistical tests, presented in Table 2, confirm the significance of the inverse correlation between the methods and indicate that this relationship is not random.

Table 2. Spearman's correlation of SADIC v2 and ResidueDepth atom depth indices for the three proteins under analysis.

Protein	Number of Atoms	Spearman's correlation	p_value
1UBQ	601	−0.94	<0.05
3ONS	574	−0.92	<0.05
3NSM	4615	−0.94	<0.05

Furthermore, we deepened the comparison between the two methods, SADIC v2 and ResidueDepth, by focusing on surface atoms, which are of particular interest for the characterisation of molecular interactions and structural properties of proteins. For this analysis, we specifically considered the 10% of the atoms with the smallest distance from the protein surface, according to the atom depth computed with ResidueDepth. Figure 7 clearly shows that the surface atoms are better characterised by our method, according to the level of exposition.

These results confirm that the SADIC v2 method can detect changes in the depth of surface atoms more accurately, thus contributing to a more detailed understanding of the structure of proteins and their interactions with other molecules.

An important task for our study is to find a grid resolution that can be a good trade-off between execution time and accuracy of results. Although the problem is complex to tackle, the experimental calculations we performed led us to suggest a resolution of 0.8 Å. As can be seen from the examples proposed in Fig. 8, when the resolution value reaches 0.8 Å, the execution time for calculating the depth index of a protein is drastically reduced (refer to Fig. 8a), and the average absolute difference with respect to the reference results remains low (refer to Fig. 8b).

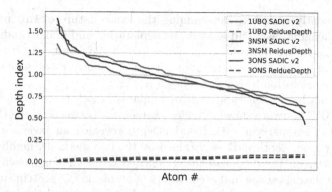

Fig. 7. Atom depth from ResidueDepth and SADIC v2 for the 10% more superficial atoms of the three proteins are compared. ResidueDepth atom depth is normalised between 0 and 2 for a fair comparison.

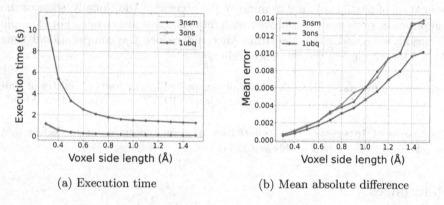

(a) Execution time   (b) Mean absolute difference

Fig. 8. Execution time (a) and mean absolute difference (b) on the depth indexes computed with different values of the voxel size, with respect to the values obtained with a reference voxel size of 0.2 Å.

6   Conclusions

In this paper, we proposed a new approach for calculating volume-based atom depth indices in protein structures, referred to as SADIC v2. The analysis of such indices is encouraged by the fact that atom depth is an important structural descriptor with a potential impact on the stability, interactions, and functional properties of proteins.

SADIC v2 is based on a discrete geometry approach, which uses a voxel-based model to accurately represent protein structures and allows us to employ efficient algorithms for the design of the phases composing our solution method. The resulting program runs in linear time in the size of the voxel grid and hence it can be executed efficiently on big molecular complexes. The relationship between the number of atoms in a molecular complex and the corresponding size of the voxel grid is inherently complex. Determining the exact nature of this relationship poses a challenging task, and as such, its exploration and understanding will be the subject of future work. Of course accuracy in the results can increase with the resolution but increasing the running time. A good trade-off seems to be the choice of the resolution to 0.8 Å as shown in Fig. 8.

The results have been validated in comparison with those obtained by the original SADIC software showing a clear agreement between the methods. Furthermore, the comparison with ResidueDepth revealed an inverse correlation between the atom depth indices obtained by the two methods, highlighting the difference between their approaches. Finally, SADIC v2 was shown to better characterise the exposition and accessibility of atoms on the protein surface than the methods based on linear distances, providing insight into the interaction of the protein with other molecules.

Further studies could be conducted to investigate the employment of multi-resolution to reach the trade-off accuracy-efficiency, and the involvement of the estimation of partial volumes to improve the accuracy. Additionally, some of the computation steps can be improved with more efficient algorithms. For example Coeurjolly's revised EDT algorithm [3] could reduce the computational time needed for computing the reference radius.

Acknowledgments. This work is partially supported by the Italian Ministry of University and Research as part of the PNRR project "THE - Tuscany Health Ecosystem".

Disclosure of Interests. The authors have no competing interests to declare that are relevant to the content of this article.

References

1. Batsanov, S.S.: Van der Waals radii of elements. Inorg. Mater. **37**(9), 871–885 (2001)
2. Chakravarty, S., Varadarajan, R.: Residue depth: a novel parameter for the analysis of protein structure and stability. Structure **7**(7), 723–732 (1999)
3. Coeurjolly, D., Montanvert, A.: Optimal separable algorithms to compute the reverse Euclidean distance transformation and discrete medial axis in arbitrary dimension. IEEE Trans. Pattern Anal. Mach. Intell. **29**(3), 437–448 (2007)
4. Connolly, M.L.: The molecular surface package. J. Mol. Graph. **11**(2), 139–141 (1993)
5. Deeb, Z.A., Adjeroh, D.A., Jiang, B.H.: Protein surface characterization using an invariant descriptor. J. Biomed. Imaging **2011**, 2 (2011)
6. Fernández-Recio, J., Totrov, M., Abagyan, R.: Soft protein-protein docking in internal coordinates. Protein Sci. **11**(2), 280–291 (2002)
7. Gonzalez, R.C.: Digital Image Processing. Pearson Education India (2009)
8. He, L., Ren, X., Gao, Q., Zhao, X., Yao, B., Chao, Y.: The connected-component labeling problem: a review of state-of-the-art algorithms. Pattern Recogn. **70**, 25–43 (2017)

9. Kihara, D., Sael, L., Chikhi, R., Esquivel-Rodriguez, J.: Molecular surface representation using 3D Zernike descriptors for protein shape comparison and docking. Curr. Protein Pept. Sci. **12**, 520–530 (2011). https://doi.org/10.2174/138920311796957612

10. Lee, B., Richards, F.M.: The interpretation of protein structures: estimation of static accessibility. J. Mol. Biol. **55**(3), 379–IN4 (1971)

11. Maurer, C.R., Qi, R., Raghavan, V.: A linear time algorithm for computing exact Euclidean distance transforms of binary images in arbitrary dimensions. IEEE Trans. Pattern Anal. Mach. Intell. **25**, 265–270 (2003). https://api.semanticscholar.org/CorpusID:1708973

12. Pedersen, T.G., et al.: A nuclear magnetic resonance study of the hydrogen-exchange behaviour of lysozyme in crystals and solution. J. Mol. Biol. **218**(2), 413–426 (1991)

13. Pintar, A., Carugo, O., Pongor, S.: DPX: for the analysis of the protein core. Bioinformatics **19**(2), 313–314 (2003)

14. Pintar, A., Carugo, O., Pongor, S.: Atom depth in protein structure and function. Trends Biochem. Sci. **28**(11), 593–597 (2003). https://doi.org/10.1016/j.tibs.2003.09.004. https://www.sciencedirect.com/science/article/pii/S0968000403002287

15. Soille, P., et al.: Morphological Image Analysis: Principles and Applications, vol. 2. Springer, Cham (1999)

16. Tan, K.P., Nguyen, T.B., Patel, S., Varadarajan, R., Madhusudhan, M.S.: Depth: a web server to compute depth, cavity sizes, detect potential small-molecule ligand-binding cavities and predict the pKa of ionizable residues in proteins. Nucleic Acids Res. **41**(W1), W314–W321 (2013). https://doi.org/10.1093/nar/gkt503

17. Varrazzo, D.: Simple atom depth index calculator (2006)

18. Varrazzo, D., et al.: Three-dimensional computation of atom depth in complex molecular structures. Bioinformatics **21**(12), 2856–2860 (2005)

19. Yin, S., Proctor, E.A., Lugovskoy, A.A., Dokholyan, N.V.: Fast screening of protein surfaces using geometric invariant fingerprints. Proc. Natl. Acad. Sci. **106**(39), 16622–16626 (2009)

Author Index

S. Brunetti et al. (Eds.): DGMM 2024, LNCS 14605, pp. 457–458, 2024.
https://doi.org/10.1007/978-3-031-57793-2

Printed in the United States
by Baker & Taylor Publisher Services